Faraday Symposia of the Chemical Society

No. 9

Physical Chemistry of Oscillatory Phenomena

1974

SYMPOSIA OF THE FARADAY SOCIETY
NO. 9   1974

Physical Chemistry of Oscillatory Phenomena

THE FARADAY DIVISION
CHEMICAL SOCIETY
LONDON

...OSIA OF THE FARADAY SOCIETY
1974

Physical Chemistry of
Oscillatory Phenomena

THE FARADAY DIVISION
CHEMICAL SOCIETY
LONDON

A SYMPOSIUM ON

Physical Chemistry of Oscillatory Phenomena

11th and 12th December, 1974

A SYMPOSIUM on the Physical Chemistry of Oscillatory Phenomena was held at The Royal Institution on 11th and 12th December, 1974. The President of the Faraday Division, Professor T. M. Sugden, F.R.S., was in the chair; about 120 Fellows of the Faraday Division and visitors from overseas attended the meeting. Visitors included:

Dr. W. Adloch, *West Germany*
Dr. P. M. Allen, *Belgium*
Dr. G. Auchmuty, *Belgium*
Miss A. Babloyantz, *Belgium*
Prof. I. Balslev, *Denmark*
Dr. E. Beretta, *Italy*
Dr. A. Boiteaux, *West Germany*
Dr. M. Branimir, *Switzerland*
Dr. H. Busse, *West Germany*
Dr. V. Caprio, *Italy*
Prof. B. L. Clarke, *Canada*
Mr. M. Collins, *Australia*
Dr. H. C. Degn, *Denmark*
Dr. P. E. Depoy, *U.S.A.*
Dr. J. S. Dunnett, *Switzerland*
Dr. M. Feinberg, *U.S.A.*
Dr. R. Foon, *West Germany*
Prof. U. F. Franck, *West Germany*
Dr. A. Goldbeter, *Israel*
Dr. W. J. W. Hanna, *Australia*
Dr. J. D. Harvey, *New Zealand*
Mme. A. Herschkowitz, *Belgium*
Prof. B. Hess, *West Germany*
Dr. A. Insola, *Italy*
Prof. D. Janjic, *Switzerland*
Prof. W. Jost, *West Germany*
Dr. F. Kaiser, *West Germany*
Dr. B. Karvaly, *Hungary*
Prof. T. Keleti, *Hungary*
Dr. J. G. Kloosterboer, *The Netherlands*

Prof. E. Körös, *Hungary*
Dr. R. Lefever, *Belgium*
Dr. P. G. Lignola, *Italy*
Prof. D. M. Mason, *U.S.A.*
Prof. V. Mathot, *Belgium*
Prof. R. M. Noyes, *U.S.A.*
Prof. A. F. Peerdeman, *The Netherlands*
Dr. A. Perche, *France*
Dr. T. Plesser, *West Germany*
Prof. J. Ross, *U.S.A.*
Dr. O. E. Rössler, *West Germany*
Dr. W. Schneider, *West Germany*
Dr. A. Schuyff, *The Netherlands*
Prof. H. Schwegler, *West Germany*
Dr. M. S. Seshadri, *West Germany*
Dr. V. E. Shashoua, *U.S.A.*
Dr. M.-L. Smoes, *West Germany*
Dr. R. L. Somorjai, *Canada*
Dr. P. G. Sørensen, *Denmark*
Dr. P. Stroot, *Switzerland*
Dr. G. Szasz, *Switzerland*
Prof. J. Texter, *U.S.A.*
Dr. H. Tributsch, *Germany*
Dr. J. S. Turner, *U.S.A.*
Dr. Van Suchtelen, *The Netherlands*
Dr. M. G. Velarde, *Spain*
Dr. G. Weisbuch, *France*
Prof. A. T. Winfree, *U.S.A.*
Dr. A. Wunderlin, *West Germany*
Prof. C. H. Yang, *U.S.A.*

ISBN: 0 85186 878 9

ISSN: 0301–5696

CONTENTS

A. Introductory and Inorganic Oscillations

Thermodynamic Aspects and Bifurcation Analysis of Spatio-temporal Dissipative Structures

By G. Nicolis* and I. Prigogine†

Faculté des Sciences, Université Libre de Bruxelles, 1050 Brussels, Belgium

Received 6th December, 1974

The thermodynamic prerequisites for the emergence of spatio-temporal patterns of organization in nonlinear chemical systems are reviewed. Analytic expressions for steady-state, time-periodic and travelling wave like solution of reaction-diffusion equations are reported. A comparison with the results based on computer simulations is outlined.

1. INTRODUCTION—THE THERMODYNAMIC BACKGROUND

It is well-established that certain types of chemical reactions, subject to appropriate conditions, organize themselves *spontaneously* in space and time to give rise to regular steady state spatial patterns, or to time-periodic flashes of chemical activity.

Obviously, self-organization phenomena of this kind are *cooperative* in the sense that they require a strong coupling between the different subunits constituting the system. However, in contrast to phase transitions and other familiar examples of cooperative processes which come from equilibrium thermodynamics, self-organization in the context of chemical reactions seems always to involve large scale, *macroscopic* elements. Thus, the patterns which are eventually established in a reacting mixture reflect in many respects, the global properties of the system such as the size, the symmetry of the spatial domain and, most importantly, the nature of the *constraints* acting from the external world.

Because of this, thermodynamics of irreversible processes constitutes the natural framework for studying these phenomena. The principal goal of this discipline is to describe the properties of macroscopic systems in terms of the properties of certain *state functionals* like entropy or entropy production. Now, in the neighbourhood of the state of thermodynamic equilibrium, the behaviour of these functionals is determined by the *theorem of minimum entropy production* [1] : In a system subject to time-independent constraints, evolution leads to a steady state where the entropy production per unit time, P, takes a minimum value, compatible with the constraints acting on the system :

$$\frac{dP}{dt} \leqslant 0. \tag{1.1}$$

At the same time, by the second law,

$$P \geqslant 0. \tag{1.2}$$

Inequalities (1.1) and (1.2) imply the *asymptotic stability* of the branch of states

† Also Center for Statistical Mechanics and Thermodynamics, The University of Texas at Austin, Austin, Texas, 78712, U.S.A.

constituting the continuation of the equilibrium states, hereafter referred to as the
thermodynamic branch. Thus for a single phase system, where the equilibrium state
itself is stable, the emergence of new types of cooperative behaviour is ruled out.

The extension of thermodynamics away from equilibrium [2] to the so-called
nonlinear range of irreversible processes, has proved that inequality (1.1) cannot be
extended to this range. Instead, we have derived a stability condition for the thermodynamic branch of states. It is found that stability will be ensured whenever [2]

$$\tfrac{1}{2}\delta^2 P = \int dV \sum_\rho \delta J_\rho \delta X_\rho \geqslant 0 \qquad \text{for all } t \geqslant t_0. \tag{1.3}$$

Here $\delta^2 P$ is the second order excess entropy production, and δJ_ρ and δX_ρ are respectively the excess flows (e.g., reaction rates) and forces (e.g., chemical affinities) due
to the deviation of the system from the reference state, i.e., from the thermodynamic
branch. A close analysis shows that inequality (1.3) can be violated in *open systems*,
obeying *nonlinear kinetics* and driven *beyond a critical distance from equilibrium*.
In such systems, an instability of the thermodynamics branch will develop and evolve
subsequently to a new regime. The threshold for this instability will be determined
by the equation:

$$\delta^2 P(\{\lambda_c\}) = 0 \tag{1.4}$$

where $\{\lambda_c\}$ is a set of critical values of the parameters—such as the intensities of the
external constraints, the chemical rate constants or the diffusion coefficients—which
influence the evolution of the system.

In this case we can see that cooperative behaviour is allowed, as the system can
conceivably leave the thermodynamic branch and thus exhibit markedly non-
equilibrium behaviour. We arrive, therefore, at the conclusion that spatio-temporal
organization in chemical systems is a *supercritical phenomenon* accompanied by an
instability of the thermodynamic branch. This phenomenon is created and sustained
by the constraints, that is by the dissipative processes inside the system. Dissipation
becomes here an organizing factor, contrary to what common intuition would
suggest. To stress this point we have called these supercritical organized states
dissipative structures.[3]

Ironically, the first chemical model showing some form of ordering—the Volterra-
Lotka oscillator—turns out to be a non-typical example of cooperative behaviour.[2, 4]
It belongs to the class of conservative systems, which do *not* exhibit an instability of
the thermodynamic branch. As a result, the oscillatory behaviour is *not* asymptoti-
cally stable and can, therefore, be destroyed by the least external perturbation.
As we shall see in detail in the subsequent sections, the crossing of a critical point of
instability (see eqn (1.4)) is a sufficient prerequisite to guarantee the stability of the
subsequently emerging pattern.

We close this section by a short remark concerning the origin of the deviations
from the reference state, causing the appearance of an excess entropy production in
relations (1.3) and (1.4). A macroscopic system involving many degrees of freedom is
always endowed with an internal mechanism permitting such deviations, namely the
fluctuations. Thus, a system near the state determined by relation (1.4) will always
have a non-vanishing probability of reaching the unstable region through fluctuations.

An interesting and most unexpected result of our recent work on fluctuations in
nonlinear systems [5] shows that only those fluctuations whose range exceeds some
critical value will be able to amplify and induce cooperative behaviour. This critical
size is the result of a competition between the natural growth rate of a local fluctuation
and the influence of the surroundings tending to damp this fluctuation. We expect,
therefore, to have an interesting phenomenon of *nucleation* of a new kind, which

suggests surprising analogies between chemical instabilities and phase transitions.[6] Molecular dynamics calculations developed by Portnow [7] and aiming to substantiate these ideas are presently in progress. These methods have been applied recently to questions such as metastable and unstable states, as described in a discussion remark by R. Lefever, I. Prigogine and J. Turner in this Symposium.

2. MATHEMATICAL FORMULATION

Chemical reactions are the most important elements responsible for evolution beyond the thermodynamic branch. Because of their nonlinear character, they are capable of generating instabilities by amplifying small effects like small inhomogeneities induced by external disturbances or by internal fluctuations. The latter can then propagate into the medium through the long range coupling between spatial regions provided by *transport processes*. Thus, we expect that *diffusion* and *heat conduction* as well as *convection* should be important factors in the understanding of cooperative behaviour in chemical systems.

In this paper we focus attention on the phenomena originated by the chemical reactions themselves and by the spatial inhomogeneities. Hence, we shall assume that the system is isothermal and at mechanical equilibrium. Moreover, we shall ignore the influence of electric or of any other types of fields. Let $N_1, \ldots, N_n$ be the concentrations of the n chemical substances involved. These variables, which by our assumptions provide a complete macroscopic description of the system, evolve according to the reaction-diffusion equations:

$$\frac{\partial N_i}{\partial t} = v_i(\{N_j\}) + D_i \nabla^2 N_i \qquad (i = 1, \ldots, n) \tag{2.1}$$

where D_i are the diffusion coefficients of species i (assumed to be constant), and v_i the rates of production of i from the chemical reactions. In general, the presence of feedback processes of various kinds will cause v_i to be nonlinear functions of the $\{N_j\}$'s. Thus, eqn (2.1) constitute a nonlinear partial differential system of the parabolic type. In order to have a well-posed problem, we will have to supplement these equations with appropriate boundary conditions on the surface Σ of the spatial domain of volume V. Two types of conditions will be considered:

(i) Dirichlet conditions:
$$\{N_i^\Sigma\} = \{\text{const.}\} \tag{2.2a}$$

(ii) Neumann conditions:
$$\{n \cdot \nabla N_i^\Sigma\} = \{\text{const.}\}. \tag{2.2b}$$

The latter type of condition with $\{\text{const.}\} = \{0\}$ applies in most of the experimental investigations on the Belousov–Zhabotinski reaction although, quite recently, Marek [8] reported experiments on this reaction under open system conditions.

Conditions (2.2a) and (2.2b) represent the *constraints* acting on the system which, in the most general case, will maintain it away from the state of thermodynamic equilibrium. Note that the entire formulation adopted in this section assumes *bounded media*. One reason for this is that in chemistry and, to an even greater extent, in biology, the size and the surface of the system play a very important role owing to the long range nature of diffusion which establishes a means for communication inside the system as well as between system and environment. The other reason for dealing with bounded media is that the mathematics becomes much more transparent. In particular, one can construct *explicit expressions* for the various types of solution of system (2.1) and study their *stability*. The requirement of stability is of course extremely important, as it will determine, among the solutions available, those which will actually be realized.

Owing to the nonlinearity, the mathematical theory of systems of equations of the type (2.1) is still at a rather primitive stage.* At best, one has some qualitative information about the behaviour of the solutions based on *bifurcation theory* [10] supplemented with stability considerations. The purpose of this theory is to study the possible *branchings of solutions* that may arise under certain conditions. This is linked, of course, to the point of view of thermodynamics developed in the previous section, according to which the emergence of dissipative structures implies the instability of the thermodynamic branch.

Under appropriate boundary conditions, eqn (2.1) admit a *uniform steady state solution* on the thermodynamic branch. In this paper we shall analyze the transitions from this state to new types of solutions such as: (*a*) space-dependent steady states; (*b*) time-periodic (and space-dependent) solutions, and (*c*) travelling waves and more general types of time-dependent solutions.

Part of our analysis will be performed on the general form of the eqn (2.1). However, quite often we will find it useful to illustrate the ideas on the simple trimolecular model [2, 4]

$$A \to X$$
$$2X + Y \to 3X$$
$$B + X \to Y + D \tag{2.3}$$
$$X \to E$$

which was analyzed recently in detail from the standpoint of bifurcation theory.[9]

The astonishing element of most of the transitions leading to patterns (*a*) to (*c*) is their *symmetry-breaking* character. Beyond a certain critical set of values of the parameters, defining a *bifurcation point* of the differential system (2.1), the most symmetrical solution of this system ceases to be stable. One then obtains a *macro-scopic quantization*† of the various new solutions, whose properties are determined by a set of a few " quantum numbers " expressing the influence of the rate constants and the diffusion coefficients, the symmetry of the spatial domain and the boundary conditions. In the subsequent three sections we shall review briefly the most characteristic phenomena arising in self-organizing chemical systems.

3. STEADY-STATE DISSIPATIVE STRUCTURES

3(i). DELOCALIZED STRUCTURES

Consider first the case for which the solutions on the thermodynamic branch are spatially uniform. Let $-k_m$, ϕ_m be respectively the eigenvalues and eigenfunctions of the Laplace operator within the spatial domain of interest. Their specific properties will of course depend on the details of the geometry. We shall come back to this point later in this subsection. The stability properties of the thermodynamic branch will be determined by the characteristic equation :

$$\det|(\partial v_i/\partial N_j)_0 - \delta_{ij}^{kr}(\omega + D_i k_m)| = 0 \tag{3.1}$$

---

* In this respect we may note that Thom's theory of catastrophes refers to systems described by *ordinary* differential equations. A brief comparison between dissipative structures and catastrophe theory has been attempted in ref. (9).

† This term is due to M. P. Hanson, *J. Chem. Phys.*, 1975.

where $(i, j) = 1, \ldots, n$ and the spectrum of k_m is determined by the Laplace operator and the spatial domain. Suppose eqn (3.1) predicts a critical set of bifurcation parameter values such that

$$\text{Re } \omega_\alpha = 0, \quad \text{Im } \omega_\alpha = 0, \quad \text{Re } \omega_{n \neq \alpha} < 0.$$

Then one will have bifurcation of a new steady state solution which will depend on space. The condition for this is:

$$\Delta = \det |(\partial v_i / \partial N_j)_0 - \delta_{ij}^{kr} D_i k_m| = 0. \tag{3.2}$$

The important point is that, for $n \geqslant 2$, eqn (3.2) may be fulfilled for *non-trivial* values of k_m, provided the bifurcation parameters $\{\lambda\}$ are within certain limits. Let $\{\lambda_c\}$ be the first values of $\{\lambda\}$ compatible with these conditions, k_m^c the corresponding values of k_m. Then, in the neighbourhood of $\{\lambda_c\}$, the emerging steady state structure will have a non-trivial space-dependence determined to first approximation by this k_m^c and by the corresponding eigenfunction ϕ_m^c. For any given model one can compute this, as well as higher order approximations, using bifurcation theory. The method is described in detail in ref. (9) and (10) and will not be reproduced here.

As an illustration we consider a one-dimensional system of length L. Then:

$$k_m = \left(\frac{m\pi}{L}\right)^2 \quad (m = 0, 1, \ldots) \tag{3.3a}$$

and

$$\phi_m \propto \sin \frac{m\pi r}{L} \quad 0 \leqslant r \leqslant L \tag{3.3b}$$

for the boundary condition (2.2a), whereas

$$\phi_m \propto \cos \frac{m\pi r}{L} \quad 0 \leqslant r \leqslant L \tag{3.3c}$$

for zero flux boundary condition. Thus, k_m describes the *wavelength* of the emerging dissipative structure. The critical value of this wavelength is determined by eqn (3.2).

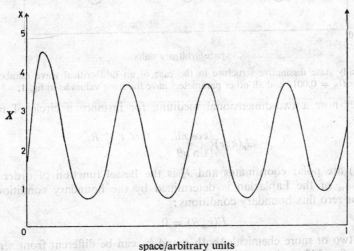

space/arbitrary units

FIG. 1.—One-dimensional steady state dissipative structure for X in the model reaction (2.3). $N_A = 2$, $N_B = 4.6$, $D_X = 0.0016$, $D_Y = 0.0080$ and $L = 1$. The boundary values for N_X, N_Y are: $\bar{N}_X = N_A = 2$, $\bar{N}_Y = N_B/N_A = 2.3$. The critical wave number $m_c = 8$.

The important point is first, that for two or more chemical substances m_c will be finite; and second, that the critical wavelength will depend on the *size* of the system. We find here a striking manifestation of the global character of dissipative structures. An additional important element is that the symmetry properties of m_c influence profoundly the form of the solutions. When m_c is even, the system exhibits a symmetry-breaking transition to *two* possible new states, both of which are stable. Fig. 1 represents the corresponding spatial pattern for the model reaction (2.3). For the same model the analytic calculation based on bifurcation theory yields [9]:

$$X(r) = A \pm a(B-B_c)^{\frac{1}{2}} \sin \frac{m_c \pi r}{L} + b(B-B_c) \sum_{m \text{ odd}} \frac{m}{(m^2 - m_c^2)^2} \frac{1}{m^2 - 4m_c^2} \sin \frac{m \pi r}{L} \quad (3.4)$$

where a, b are well-defined functions of the chemical parameters and the diffusion coefficients and the concentration B of the initial substance B has been used as bifurcation parameter.

When m_c is odd, one observes more complex phenomena like hysteresis. The corresponding spatial patterns for model (2.3) are shown in fig. 2. In both cases the emerging structures are shown to be stable.

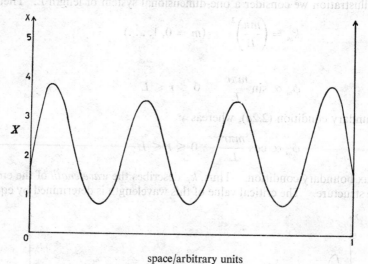

FIG. 2.—Steady state dissipative structure in the case of an odd critical wave number $m_c = 7$. $D_X = 0.0018$ and all other parameters have the same values as in fig. 1.

Consider now a two-dimensional medium, for instance a circle of radius R. Then:

$$\phi_{mn} \sim J_n(k_m r) \begin{cases} \cos n\theta \\ \sin n\theta \end{cases} \qquad 0 \leqslant r \leqslant R \quad (3.5a)$$

where (r, θ) are polar coordinates and J_n is the Bessel function of order n. The eigenvalue k_m of the Laplacian is determined by the boundary conditions. For instance, for zero flux boundary conditions:

$$J_n'(k_m R) = 0. \quad (3.5b)$$

Again, for two or more chemicals, both n and k_m can be different from zero at the point of the first bifurcation. Fig. 3 represents the spatial patterns resulting beyond bifurcation for the model system (2.3) in the case where $n_c = 0$, $k_m^c \neq 0$. This pattern is cylindrically symmetric. If, on the other hand, both k_m^c and n_c are different from zero, the resulting pattern exhibits *a polarity*, in the sense that a macroscopic gradient sets up *spontaneously* across the system.[11] Fig. 4 describes the spatial form of the solutions for the model reaction (2.3).

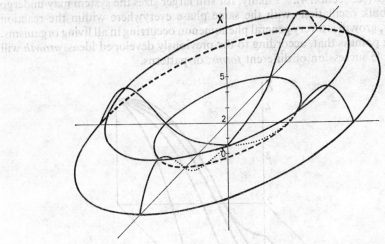

FIG. 3.—Cylindrically symmetric steady state dissipative structure on a circle for the trimolecular model (2.3). The radius of the circle is $R = 0.20$ and all other values are the same as in fig. 1.

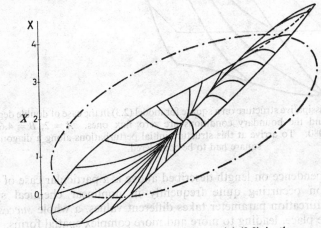

FIG. 4.—Polar dissipative structure on a circle for model (2.3) in the case of zero flux boundary conditions. $R = 0.10$, $D_X = 0.00325$, $D_Y = 0.0162$, $A = 2$ and $B = 4.6$.

In the simplest cases of bifurcation, the resulting pattern reflects to a good approximation the properties of the *critical modes*, i.e., of the eigenfunctions of the Laplacian *at* the bifurcation point. In situations involving high symmetries like, for example, a square domain, it is not uncommon to have, as first bifurcation, a state corresponding to a *degenerate* eigenvalue of the Laplacian. The spatial structures are then a complex mixture of the degenerate eigenfunctions, and the multiplicity of solutions can become very high. The situation is described in fig. 5.

A closer study reveals that, as the size of the system increases, different dissipative structures can be realized successively. Thus, below a critical size, R_{c_1}, the only possible solution is the uniform one on the thermodynamic branch. Between R_{c_1} and R_{c_2} the system may exhibit a non-polar pattern like the one in fig. 3. Then, between R_{c_2} and R_{c_3} the latter could be succeeded by a polar pattern such as that in fig. 4. Beyond R_{c_3} all steady state solutions may become unstable and a propagating wave can emerge (see section 4). Finally, for still larger sizes the system may undergo homogeneous bulk oscillations with the same phase everywhere within the reaction space.[12] Now, growth is a very general phenomenon occurring in all living organisms. The fascinating point is that, according to the previously developed ideas, *growth* will engender a whole succession of different *forms*, or patterns.

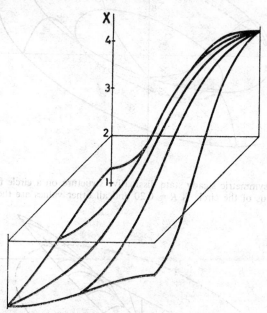

FIG. 5.—Steady state dissipative structure on a square for model (2.3) in the case of double degeneracy. The side $L = 0.132$ and the boundary conditions are zero flux ones. $A = 2$, $B = 4.6$, $D_X = 0.0016$, and $D_Y = 0.0080$. To arrive at this structure initial perturbations along a diagonal of the square had to be considered.

The intrinsic dependence on length described above is a particular case of a more general phenomenon occurring quite frequently in nonlinear chemical systems. Namely, when a bifurcation parameter takes different values, a whole *succession of instabilities* can take place, leading to more and more complex spatial forms. These can be either bifurcations from the thermodynamic branch, leading to a multiplicity of dissipative structures each one having its own domain of attraction of a set of initial conditions; or they can be the result of a *secondary bifurcation* from a previously established dissipative structure. Unfortunately, in both cases, bifurcation theory very rarely permits a rigorous study of the stability of the various states. An interesting thermodynamic property of these multiple transitions is that they may lead to an increase in dissipation as the wavelength of the structure becomes shorter.[13]

3(ii). LOCALIZED STRUCTURES

In many multicomponent systems the diffusion of some of the chemicals may create an inhomogeneous environment for other chemical intermediates. In this case the solutions on the thermodynamic branch are no longer space-independent. Computer simulations [14] have revealed the existence of spatial patterns which remain *localized* within a part of the reaction space. Recent theoretical analysis of these states [13, 15] has shown that these patterns are again of the dissipative structure type, as they arise beyond the instability of the thermodynamic branch. Fig. 6 describes the spatial distribution of concentrations in a one-dimensional system for the model reaction (2.3).

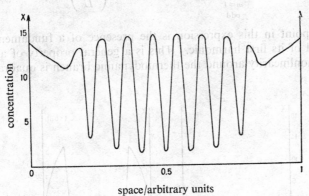

FIG. 6.—One-dimensional localized steady state dissipative structure for model (2.3). $D_A = 0.197$, $D_X = 0.001\,05$, $D_Y = 0.005\,26$, $B = 30$, $\bar{X} = \bar{A} = 14$, $\bar{Y} = B/\bar{A} = 2.14$.

It is important to realize that this localization phenomenon may be *spontaneous*. No spatial gradient needs to be preimposed. On the other hand, localization may result in a weaker dependence of the properties of the dissipative structure on the system as a whole, e.g., on the size. Still, complete independence of the size is difficult to imagine or to justify mathematically. It appears tempting to suggest that localized structures are good candidates to explain the formation of *leading centres* in some systems like the Belousov–Zhabotinski reaction. In this respect, however, it is necessary to point out that the complete independence of those leading centres of the size of the system postulated by Zaikin and Zhabotinski [16] is never fulfilled rigorously in localized structures.

4. TIME-PERIODIC SOLUTIONS

According to linear stability analysis, a time-periodic solution will bifurcate in the neighbourhood of the thermodynamic branch, provided the characteristic eqn (3.1) admits a pair of purely imaginary roots, $\pm i\omega_\alpha$, at the bifurcation point. In chemical systems involving only two variables, the first bifurcation leads necessarily to a state where $k_m^c = 0$.[17] If this state is allowed by the boundary conditions, then the system will presumably exhibit this most symmetrical solution, whose space dependence will be trivial. This actually happens in the case of zero flux boundary conditions, where the system exhibits limit cycle oscillations throughout the reaction space. In contrast, constant boundary conditions will in general rule out the state with $k_m^c = 0$. The first allowed bifurcating solution will then exhibit both space structure as well as temporal periodicity. Fig. 7 describes this solution for the model reaction (2.3).

Using bifurcation theory one can calculate explicitly these solutions in the neighbourhood of the bifurcation point. For contrast boundary conditions and for a one-dimensional system involving two chemical intermediates one obtains the following form for the deviation $x(r, t)$ from the thermodynamic branch:

$$x(r) = a(B-B_c)^{\frac{1}{2}} \sin\left(\frac{\pi r}{L}\right) \cos \mu t +$$

$$b(B-B_c) \sum_{\substack{m=1 \\ \text{odd}}}^{\infty} (a_m + \tilde{a}_m \cos 2\mu t) \sin\left(\frac{m\pi r}{L}\right) + \mathcal{O}((B-B_c)). \qquad (4.1)$$

The important point in this expression is the presence of a fundamental mode of period $2\pi/\mu$ and of its first harmonic. This is a general property of all systems in which the first nonlinearity around the thermodynamic branch is quadratic.[17]

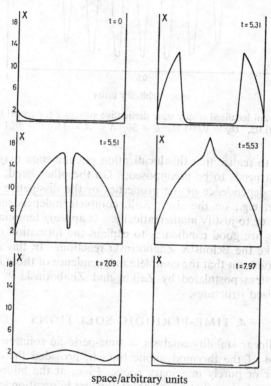

space/arbitrary units

FIG. 7.—Characteristic stages of evolution of the spatial distribution of X during one period of the wave for model (2.3) for a length $L = 0.80$ and for constant boundary conditions. $A = 2, B = 12.6$, $D_X = 0.0080, D_Y = 0.0040$.

Relation (4.1) describes a superposition of *standing waves*. However, the numerical evaluation of this formula for model (2.3) reveals a pattern which is strikingly similar to fig. 7. It appears, therefore, that in a bounded medium, propagating concentration fronts can appear during certain time intervals, but *not* during the entire period of the phenomenon. Waves of this kind are capable of transporting matter during macroscopic time intervals, as the (∇x) at a certain point retains a given sign during these intervals. However, they differ markedly from the waves encountered commonly in mechanics or in electromagnetic theory in the sense that it is meaningless to define for them a propagation velocity depending on x through the usual relationship:

$$u \propto \left(\frac{\partial x}{\partial t}\right)\bigg/\left(\frac{\partial x}{\partial r}\right).$$ (4.2)

In chemical systems involving more than two coupled variables an additional important element appears in that the first bifurcation can lead to states with $k_m^c \neq 0$.[17] Thus, even for zero flux boundary conditions, these systems will exhibit symmetry-breaking transitions leading to states with non-trivial space dependence. For instance, in a circular domain, the first bifurcating solution will in some cases be approximated by the function:

$$\phi_{m1} \propto J_1(k_m r) \cos(\theta + \omega t)$$ (4.3a)

with

$$J_1'(k_m R) = 0.$$ (4.3b)

At each instant of time, the equal concentration curves ϕ_{m1} = const. will be symmetrical with respect to a diameter of the circle. As time proceeds, this diameter will rotate at constant speed. Thus, the time-periodic solutions of this form represent rotating waves, although their form is not spiral in the approximation provided by relations (4.3). It is quite possible, however, that a bifurcation theoretical calculation leading to an expression such as eqn (4.1) will produce spiral waveforms, arising from the influence of higher order terms. The latter contain a series having $J_n(k_m r)$ as coefficients, which could break the symmetry of the equal concentration curves with respect to the diameter of the circle. The numerical evaluation of the bifurcation theoretical expression for a circle for the model reaction (2.3) is in progress.

Let us recall that the appearance of angle-dependent solutions beyond the first bifurcation from the thermodynamic branch occurs typically in systems with three or more variables. In this respect we may note that the Noyes–Field model for the Belousov–Zhabotinski reaction involves three intermediates and, therefore, fulfils these conditions. Additional variables which could play an important role in the Belousov–Zhabotinski reaction, as well as in other chemical systems, are the temperature and the electrical field. The latter could be especially crucial in the generation of electrical waves capable of interfering with or even triggering the chemical waves.

Other possibilities of wave forms having a symmetry less than the symmetry of the spatial domain include secondary bifurcations or multiple stable solutions as described at the end of subsection 3(i). At present it appears to be very difficult to investigate the stability of these solutions, although one can construct them explicitly using bifurcation theory.

5. TRAVELLING WAVES

The requirement of temporal periodicity imposed on the solutions calculated in the previous section turned out to be incompatible with the existence of a well-defined speed of wave propagation, although, as we saw, expressions such as eqn (4.1) could approximate propagating fronts as closely as may be desired. In this section we review briefly the problem of existence of travelling waves by requiring solutions of eqn (2.1) and (2.2) of the form $f(r - vt)$. For details we refer the reader to a recent paper by Auchmuty and Nicolis.[17] Work in this direction in the case of *unbounded media* has been reported recently by Kopell and Howard [18] and by Ortoleva and Ross.[19]

5(i). WAVES IN RINGS

Let 2π be the length of the ring. The Laplace operator reduces to a single derivative along the ring. We want to find solutions of the form :

$$N_i(r, t) = f_i(r - vt) \equiv f_i(\xi) \tag{5.1a}$$

subject to :

$$N_i(0, t) = N_i(2\pi, t). \tag{5.1b}$$

The reaction-diffusion equations take the form :

$$-vf_i'(\xi) = D_i f_i''(\xi) + v_i(\{f_j\}) \qquad (i = 1, \ldots, n). \tag{5.2}$$

We wish to determine those values of v for which there are non-trivial periodic solutions of (5.2) :

$$N_i = N_i^\circ + a_i \exp(im\xi)$$

$$m = 0, 1, \ldots \tag{5.3}$$

Note that both (5.1b) and (5.2) are compatible with the existence of a *constant* speed of propagation. A linear analysis of eqn (5.2) confirms this point and provides expressions for the critical values of the bifurcation parameter as well as for the speed of propagation. The solutions can then be constructed using bifurcation theory.[17] As in section 4, for systems with two variables the first bifurcation occurs at $m = 0$. Thereafter, the system undergoes a uniform limit cycle type oscillation. For three or more intermediates the first bifurcation can lead to a non-trivial spatial dependence in the form of a propagating pulse.

5(ii). MORE GENERAL ONE-DIMENSIONAL SYSTEMS

We now consider a system with open ends, subject to the boundary conditions :

$$N_i(0, t) = N_i(L, t) = N_i^\circ \tag{5.4a}$$

or

$$\frac{\partial N_i}{\partial r}(0, t) = \frac{\partial N_i}{\partial r}(L, t) = 0 \tag{5.4b}$$

where N_i° are the values of the (uniform) chemical concentrations on the thermodynamic branch. It can easily be seen [17] that such boundary conditions rule out automatically the existence of solutions of the form (5.1a) with v constant. We relax these conditions and seek solutions of the form :

$$N_i(r - v(r)t) = f_i(\xi). \tag{5.5}$$

A detailed analysis shows [17] that a solution of the form (5.5) can emerge beyond a *secondary bifurcation* of a steady-state dissipative structure. The solvability condition of the bifurcation equations will provide us with a set of partial differential equations which determine the propagation velocity v.

The ability of these waves to arise in a *non-uniform* background provides a mechanism of localization of these waves as well as a possible formation of *leading centres*.

5(iii). TRAVELLING WAVES IN SPACES OF HIGHER DIMENSION

By an argument similar to that developed in the previous subsection one can show that *plane waves* of the form $f(k \cdot r - \omega t)$ are ruled out in two or higher spatial dimensions. The possibility remains, however, of having superpositions of such waves resulting in a wave packet of the form:

$$N_i(r, t) = \int d\xi\, d\eta\, d\zeta\, f(\xi, \eta, \zeta)\, \exp[i(\mathbf{k}.\,\mathbf{r} - v(k)t)] \qquad (5.6)$$

where $\mathbf{k} = (\xi, \eta, \zeta)$. The group velocity $d\mathbf{r}/d\mathbf{k}$ of this wave packet will play the role of the propagation velocity. Again, when the geometry of the problem is specified, one can look for specific wave forms like cylindrically or spherically symmetric waves, spirals and so on.

6. CONCLUDING REMARKS

From this analysis of nonlinear chemical systems, we have seen that a wealth of structures, presenting a wide variety of sizes and shapes, becomes possible beyond an instability of the thermodynamic branch.

It is remarkable that, in spite of the extreme diversity of these far-from-equilibrium phenomena, it has been possible to reach general conclusions on two fundamental aspects of dissipative structures. On the one side, nonlinear thermodynamics and bifurcation theory have enabled us to provide a rough *classification* as well as to construct explicit expressions for the various possible types of organized states. On the other side nonlinear thermodynamics, supplemented with fluctuation theory, has enabled us to get a better understanding of the mechanisms prevailing in the *transitions* between states.

Finally, it is important to stress the analogies between these and similar phenomena encountered in various branches of natural sciences such as: fluid dynamics, shock waves and detonations, ecology and, even, social behaviour. One can hope that the progress accomplished during the last few years in the field of nonlinear chemical kinetics will provide the tools necessary to tackle complex problems in these fields, which until now have eluded quantitative analysis.

It is a pleasure to acknowledge that a substantial part of the mathematical results of this paper have been obtained in collaboration with Prof. J. F. G. Auchmuty. We are also indebted to Dr. Herschkowitz-Kaufman, Dr. Lefever, Mr. Malek-Mansour and Mr. Erneux for their important contributions in the research reported in this paper and for many stimulating discussions. The financial support of the " Ministère de l'Education Nationale et de la Culture française et le Ministère de l'Education Nationale et de la Culture Néerlandaise " from Belgium and of the Welch Foundation, Houston, Texas, is gratefully acknowledged.

[1] I. Prigogine, *Bull. Acad. Roy. Belg.*, 1945, **31**, 600.
[2] P. Glansdorff and I. Prigogine, *Thermodynamics of Structure, Stability and Fluctuations* (Wiley-Interscience, New York, 1971).
[3] For a historical survey of the concept of dissipative structure and of its precursors we refer to a recent review paper by R. Glansdorff and I. Prigogine, *Bull. Acad. Roy. Belg.*, 1973, **59**, 672.
[4] G. Nicolis and J. Portnow, *Chem. Rev.*, 1973, **73**, 365.
[5] G. Nicolis, M. Malek-Mansour, K. Kitahara and A. Van Nypelseer, *Phys. Letters*, 1974, **48A**, 217.
[6] G. Nicolis, *J. Stat. Phys.*, 1972, **6**, 195; G. Nicolis and I. Prigogine, *Proc. Natl. Acad. Sci.*, 1971, **68**, 2102; A. Nitzan, P. Ortoleva and J. Ross, this Symposium.
[7] J. Portnow and Svobodova, *Phys. Letters*, 1975, **51A**, 370.
[8] M. Marek, *Biophys. Chem.*, 1975, **3**, 263.
[9] G. Nicolis and J. F. G. Auchmuty, *Proc. Natl. Acad. Sci.*, 1974, **71**, 2748.
[10] D. Sattinger, *Lecture Notes in Mathematics*, vol. 309 (Springer Verlag, Berlin, 1973).
[11] A. Babloyantz and J. Hiernaux, *Bull. Math. Biol.*, 1975.

[12] A. Goldbeter, *Proc. Natl. Acad. Sci.*, 1973, **70**, 3255, M. P. Hanson, *J. Chem. Phys.*, 1974, **60**, 3210.

[13] J. F. G. Auchmuty and G. Nicolis, *Bull. Math. Biol.*, 1975, **37**,323.

[14] M. Herschkowitz-Kaufman, *Bull. Math. Biol.*, 1975.

[15] J. Boa, *Ph. D. Dissertation* (Calif. Inst. of Technology, 1974).

[16] A. M. Zhabotinski and A. N. Zaikin, *J. Theor. Biol.*, 1973, **40**, 45.

[17] J. F. G. Auchmuty and G. Nicolis, *Bull. Math. Biol.*, 1975.

[18] N. Kopell and L. Howard, *Studies in Appl. Math.*, 1973, **52**, 291.

[19] P. Ortoleva and J. Ross, *J. Chem. Phys.*, 1974, **60**, 5090.

A Model Illustrating Amplification of Perturbations in an Excitable Medium

By Richard J. Field and Richard M. Noyes

Department of Chemistry, University of Oregon,
Eugene, Oregon 97403 U.S.A.

Received 6th July, 1974

The oscillatory Belousov–Zhabotinskii reaction can be modelled approximately by five irreversible steps:

$$A+Y \rightarrow X \qquad \text{(M1)}$$
$$X+Y \rightarrow P \qquad \text{(M2)}$$
$$B+X \rightarrow 2X+Z \qquad \text{(M3)}$$
$$2X \rightarrow Q \qquad \text{(M4)}$$
$$Z \rightarrow fY. \qquad \text{(M5)}$$

These equations are based on the chemical equalities $X = HBrO_2$, $Y = Br^-$, $Z = 2Ce(IV)$, and $A = B = BrO_3^-$. If the rate constants k_{M1} to k_{M4} are assigned by experimental estimates from oxybromine chemistry, the kinetic behaviour of the model depends critically upon the remaining parameters k_{M5} and f. When f does not differ too greatly from unity, and when k_{M5} is not too large, the steady state is unstable to perturbation and the system oscillates by describing a limit cycle trajectory.

When f and k_{M5} lie outside the range of instability, the steady state is stable to very small perturbations. However, the steady state may still be excitable so that perturbation of the control intermediate Y by a few percent will instigate a single excursion during which concentrations of X, Y, and Z change by factors of about 10^5 before the system returns to the original steady state. This ability of a small perturbation of the steady state to trigger a major response by the system is just the type of behaviour necessary to explain the initiation of a trigger-wave by a heterogeneous " pacemaker " as has been observed by Winfree. The same type of excitability of a steady state has important implications for the understanding of biochemical control mechanisms.

The so-called Belousov–Zhabotinskii reaction occurs in one of the most versatile of chemical systems. If a sulphuric acid solution of bromate ion and malonic acid contains a catalytic amount of a one-equivalent redox couple (such as Ce(III)-Ce(IV) or Mn(II)-Mn(III)) with a reduction potential between about 1.0 and 1.5 V, several remarkable types of behaviour are possible.

In a stirred homogeneous solution, the degree of oxidation of the redox couple may oscillate repeatedly with time. In an unstirred solution containing a gradient in temperature or in the concentration of some reactant, variations in the phase of local oscillation can cause alternating bands of oxidation and reduction to propagate through the system; these phase-related oscillations have been called pseudo-waves by Winfree.[1] In an unstirred but initially homogeneous solution, local disturbances such as dust particles may generate regions of oxidation that propagate more rapidly than individual molecules could diffuse and leave refractory reduced regions behind them; these disturbances are called trigger-waves by Winfree.[1] Finally, if developing trigger-waves are disturbed in various ways, they may develop very complicated spiral structures called scroll-waves by Winfree.[2] References to these various types of behaviour may be found in a recent review.[3]

The detailed chemical mechanism of the temporal oscillations has now been elucidated,[4, 5] and a simplified mathematical model developed to exhibit the same type of limit cycle behaviour.[6] The same mechanism has been used both qualitatively [7] and quantitatively [8] to describe the chemical processes taking place in the sharp leading edge of a trigger-wave.

Although the previous discussion [8] demonstrated how a region of oxidation could propagate through the medium as a trigger-wave, it did not concern itself with the initiation of such oxidation. In the present paper, we show that the same model suggests how a modest perturbation in the concentration of bromide ion could initiate a transient region of oxidation that would then propagate as a trigger-wave.

CHEMICAL MECHANISM

The significant features of the chemical mechanism can be summarized by the following processes :

$$BrO_3^- + Br^- + 2H^+ \rightarrow HBrO_2 + HOBr \tag{C1}$$

$$HBrO_2 + Br^- + H^+ \rightarrow 2HOBr \tag{C2}$$

$$BrO_3^- + HBrO_2 + H^+ \rightarrow 2BrO_2 + H_2O \tag{C3a}$$

$$Ce^{3+} + BrO_2 + H^+ \rightarrow Ce^{4+} + HBrO_2 \tag{C3b}$$

$$2HBrO_2 \rightarrow BrO_3^- + HOBr + H^+ \tag{C4}$$

$$nCe^{4+} + BrCH(COOH)_2 \rightarrow nCe^{3+} + Br^- + \text{oxidized products.} \tag{C5}$$

Steps (C1), (C2), and (C4) are assumed to be bimolecular elementary processes involving oxygen atom transfer and accompanied by rapid proton transfers ; the HOBr so produced is rapidly consumed directly or indirectly with bromination of malonic acid. Step (C3a) is rate-determining for the overall process of (C3a) + 2(C3b). The Ce^{4+} produced in step (C3b) is consumed in step (C5) by oxidation of bromomalonic acid and other organic species with production of bromide ion.

The complete chemical mechanism is considerably more complicated,[5] but the simplified version presented here is sufficient to explain the oscillatory behaviour of the system.

COMPUTATIONAL MODEL

The significant kinetic features of the chemical mechanism can be simulated by the model we have called the Oregonator.[6]

$$A + Y \rightarrow X \tag{M1}$$

$$X + Y \rightarrow P \tag{M2}$$

$$B + X \rightarrow 2X + Z \tag{M3}$$

$$2X \rightarrow Q \tag{M4}$$

$$Z \rightarrow fY. \tag{M5}$$

This computational model can be related to the chemical mechanism by the identities $A \equiv B \equiv BrO_3^-$, $X \equiv HBrO_2$, $Y \equiv Br^-$, and $Z \equiv 2Ce(IV)$.

The Oregonator functions because the *switched intermediate*, X, is generated by steps (M1) and (M3) that are zero and first order in X and is destroyed by steps (M2) and (M4) that are first and second order in X. When the concentration of the *control intermediate*, Y, is sufficiently large, the concentration of X attains a steady state approximated by X_{small}.

$$X_{small} = \frac{k_{M1}AY}{k_{M2}Y - k_{M3}B}. \tag{1}$$

When the concentration of Y is sufficiently small, X attains a different steady state approximated by X_{large}.

$$X_{large} = \frac{k_{M3}B - k_{M2}Y}{2k_{M4}}. \tag{2}$$

The concentration of X is switched between the two steady states whenever the concentration of Y attains the critical value Y_{crit}.

$$Y_{crit} = k_{M3}B/k_{M2}. \tag{3}$$

The *regeneration intermediate* is produced in significant amount only when X is in the X_{large} steady state, and it regenerates Y until the critical concentration is attained and X is switched to X_{small}. The concentration of Z then decreases, and Y is destroyed until it again reaches Y_{crit} and permits the concentration of X to be switched again.

The differential equations describing temporal behaviour can be cast in terms of the dimensionless variables α, η, ρ, and τ and the dimensionless parameters q, s and w.

$$\alpha = \frac{k_{M2}}{k_{M1}A}X \tag{4}$$

$$\eta = \frac{k_{M2}}{k_{M3}B}Y \tag{5}$$

$$\rho = \frac{k_{M2}k_{M5}}{k_{M1}k_{M3}AB}Z \tag{6}$$

$$\tau = \sqrt{k_{M1}k_{M3}AB}\, t \tag{7}$$

$$q = \frac{2k_{M1}k_{M4}A}{k_{M2}k_{M3}B} \tag{8}$$

$$s = \sqrt{k_{M3}B/k_{M1}A} \tag{9}$$

$$w = \frac{k_{M5}}{\sqrt{k_{M1}k_{M3}AB}}. \tag{10}$$

These differential equations then become

$$d\alpha/d\tau = s(\eta - \eta\alpha + \alpha - q\alpha^2) \tag{11}$$

$$d\eta/d\tau = (1/s)(-\eta - \eta\alpha + f\rho) \tag{12}$$

$$d\rho/d\tau = w(\alpha - \rho). \tag{13}$$

We have examined the Oregonator under conditions such that the rate constants for the first four steps correspond to our best estimates [5] for the situation in a solution with 0.06 M $KBrO_3$ and 0.8 M H_2SO_4. Then $q = 8.375 \times 10^{-6}$, $s = 77.27$, and $w = 0.1610\, k_{M5}$. When these values are selected, the kinetic behaviour depends upon the two parameters f and k_{M5}.

APPLICATIONS TO UNSTABLE AND EXCITABLE SYSTEMS

It is possible to determine steady state values for the concentrations of the three intermediates such that $d\alpha/d\tau = d\eta/d\tau = d\rho/d\tau = 0$. Then a secular equation analysis of the linearized kinetic equations will reveal whether small perturbations from the steady state will decay or whether they will grow and lead to a limit cycle trajectory around the steady state.

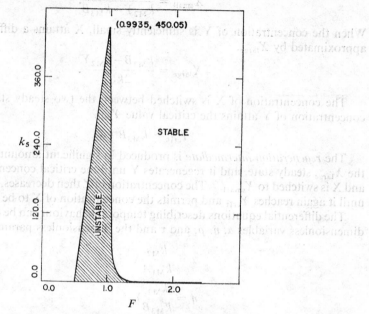

FIG. 1.—Regions of stable and unstable Oregonator steady states for $q = 8.375 \times 10^{-6}$, $s = 77.27$, and $w = 0.1610\,k_{M5}$. The steady state is stable to oscillation except for sufficiently small values of k_{M5} with f in the neighbourhood of unity.

The results of such an analysis are shown in fig. 1. This figure differs slightly from one presented previously [6] which was based only on a sufficient condition for instability of the steady state. It is impossible to have a steady state if $k_{M5} = 0$, and fig. 1 demonstrates that the steady state is always stable for sufficiently large values of k_{M5}. Unstable steady states are observed for some finite positive values of k_{M5} provided the stoichiometric parameter f does not differ too much from unity. Stable steady states with f appreciably less than unity correspond to $Y < Y_{crit}$ and $X = X_{large}$. Stable steady states with f appreciably greater than unity correspond to $Y > Y_{crit}$ and $X = X_{small}$.

Trigger-waves of the type studied by Winfree [9] involve regions of oxidation that advance into regions of excess bromide ion ($Y > Y_{crit}$) that are marginally stable to oscillation. Such a region can be modelled by a stable point near to the right boundary of the region of instability defined by fig. 1.

As Winfree [9] has pointed out, trigger-waves initiate at a " pacemaker " such as a speck of dust that locally perturbs the marginally stable solution and generates a region of oxidation that then propagates as a trigger-wave. If our model is appropriate, a small perturbation of the steady state should generate a single pulse of oxidation followed by return to the steady state. Moreover, the system should be particularly sensitive to perturbation of the control intermediate Y.

Exactly these predictions are demonstrated by fig. 2 to 5 illustrating the behaviour of a system initially at the stable but excitable steady state for $f = 1.5$ and $k_{M5} = 2\,s^{-1}$. Fig. 2 to 4 are logarithmic plots showing the behaviour of the various intermediates following a discontinuous reduction of Y (or η) by 6.5 % at time $\tau = 153$.

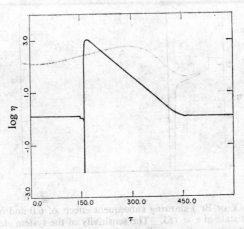

FIG. 2.—Logarithmic plot of $\eta(Y \text{ or } Br^-)$ following a discontinuous perturbation of -6.5% from the steady state. Perturbation occurred at $\tau = 153$.

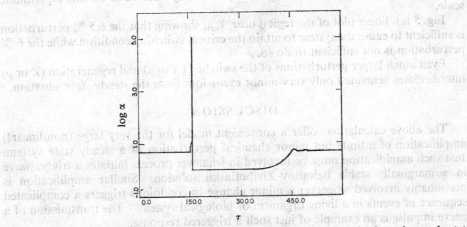

FIG. 3.—Logarithmic plot of $\alpha(X \text{ or } HBrO_2)$ following a 6.5% decrease of η from the steady state situation. Perturbation occurred at $\tau = 153$.

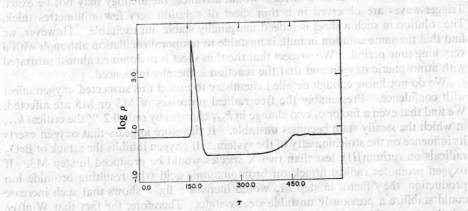

FIG. 4.—Logarithmic plot of $\rho(Z \text{ or } Ce(IV))$ following a 6.5% decrease of η from the steady state situation. Perturbation occurred at $\tau = 153$.

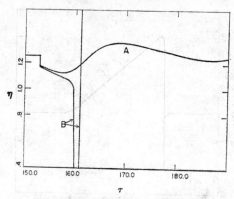

FIG. 5.—Linear plot of η(Y or Br$^-$) showing subsequent effects of 6.0 and of 6.5 % discontinuous decreases from the steady state at $\tau = 153$. The sensitivity of the system clearly is very dependent upon the magnitude of the perturbation.

A similar reduction of 6 % generated a ripple barely observable on this logarithmic scale.

Fig. 5 is a linear plot of the region near Y_{crit} showing that the 6.5 % perturbation is sufficient to cause the system to attain the critical switching condition while the 6 % perturbation is not sufficient to do so.

Even much larger perturbations of the switched (X or α) and regeneration (Z or ρ) intermediates generated only very minor excursions from the steady state situation.

DISCUSSION

The above calculations offer a convenient model for the very large (non-linear!) amplification of a finite but minor chemical perturbation of a steady state system. Just such amplification must be involved in whatever process initiates a trigger-wave in a marginally stable Belousov–Zhabotinskii solution. Similar amplification is presumably involved whenever a minor change in conditions triggers a complicated sequence of events in a living organism or biological system. The transmission of a nerve impulse is an example of just such a triggered response.

Although these calculations were initiated to model the excitable medium into which the trigger-waves of Winfree [9] would advance, the analogy may not be exact. Trigger-waves are observed in a thin sheet of solution very few millimetres thick. The solution in such a sheet is indeed marginally stable but excitable. However, we find that the same solution in bulk is unstable to temporal oscillation although with a very long time period. We suspect that the thin sheet is maintained almost saturated with atmospheric oxygen and that the reaction is thereby influenced.

We do not know enough detailed chemistry to model this suspected oxygen effect with confidence. Presumably the free radical processes of M3 or M5 are affected. We find that even a factor of two change in k_{M3} changes by only 0.2 % the critical k_{M5} at which the steady state becomes unstable. It therefore appears that oxygen exerts its influence on the stoichiometry of the system. If oxygen inhibits the attack of BrO$_2$ radicals on cerium(III), less than two X species would be produced in step M3. If oxygen promotes radical attack on bromomalonic acid with resulting bromide ion production, the f factor in step M5 would increase; fig. 1 shows that such increase could stabilize a previously unstable steady state. Therefore, the fact that Winfree solution is unstable to bulk oscillations does not affect the basic validity of the model we develop here for chemical amplification of a marginally stable system.

[1] A. T. Winfree, *Lecture Notes on Biomathematics*, ed. P. van den Driessche (Springer-Verlag, Berlin), in press.

[2] A. T. Winfree, *Science*, 1973, **181**, 937.

[3] R. M. Noyes and R. J. Field, *Ann. Rev. Phys. Chem.*, 1974, **25**, 95.

[4] R. M. Noyes, R. J. Field and E. Körös, *J. Amer. Chem. Soc.*, 1972, **94**, 1394.

[5] R. J. Field, E. Körös and R. M. Noyes, *J. Amer. Chem. Soc.*, 1972, **94**, 8649.

[6] R. J. Field and R. M. Noyes, *J. Chem. Phys.*, 1974, **60**, 1877.

[7] R. J. Field and R. M. Noyes, *Nature*, 1972, **237**, 390.

[8] R. J. Field and R. M. Noyes, *J. Amer. Chem. Soc.*, 1974, **96**, 2001.

[9] A. T. Winfree, *Science*, 1972, **175**, 634.

Chemistry of Belousov-type Oscillating Reactions

By E. Kőrös, M. Burger, V. Friedrich, L. Ladányi,
Zs. Nagy and M. Orbán

Institute of Inorganic and Analytical Chemistry,
L. Eötvös University, H-1443 Budapest, P.O.B. 123

Received 30th July, 1974

The heat output and the accumulation of monobromomalonic acid during chemical oscillation in the bromate, malonic acid, sulphuric acid and catalyst [Ce(III), Mn(II), Fe(phen)$_3^{2+}$, Ru(dipy)$_3^{2+}$] systems were measured and the results interpreted. The behaviour of the reacting oscillating systems depends to a certain extent on the catalyst applied. Both the rate of heat evolution and the rate of accumulation of monobromomalonic acid are periodic. The heat of the cerium-catalysed Belousov oscillating reaction is 130 ± 5 kJ mol^{-1} bromate.

Many teams are involved in studying either experimentally or theoretically the Belousov oscillating systems. Most of the experimental investigations have been performed on the cerium(III), bromate, malonic acid and sulphuric acid system and now much is known both about its chemistry and mechanism.

Recently Field and Noyes [1] have generalized the chemical mechanism of Field, Kőrös and Noyes [2] for the Belousov reaction by a model composed of five steps involving three independent intermediates (Br$^-$, Ce(IV) and HBrO$_2$). The model exhibits limit cycle behaviour. Kőrös [3] has treated chemical oscillation as a " monomolecular " reaction, regarded the oscillatory Belousov reaction a series of autocatalytic reaction bursts (bromate bursts) occurring with a certain frequency, and calculated the " activation parameters " for reacting oscillatory systems with different catalysts. Also the preliminary studies on the Ru(dipy)$_3^{2+}$-catalysed Belousov system have been reported.[4, 5] Kőrös *et al.*[6] have shown periodicity in the rate of heat evolution, and Degn [7] in the rate of carbon dioxide evolution during temporal chemical oscillation.

Despite the large body of information available on the Belousov systems there are still points to be cleared and revealed. Our researches have been directed (*a*) to perform calorimetric measurements on reacting Belousov systems, (*b*) to reveal the mechanism of the important composite reactions, and (*c*) to determine the products of the reaction early during oscillation.

Here we report on our calorimetric investigations on the Ce(III)-, Mn(II)-, Fe(phen)$_3^{2+}$- and Ru(dipy)$_3^{2+}$-catalysed bromate, malonic acid and sulphuric acid systems, and on our polarographic measurements aimed to follow the accumulation of monobromomalonic acid (BrMA).

EXPERIMENTAL

MATERIALS

All reagents used were of analytical grade.

POTENTIOMETRY

Chemical oscillation was followed either by monitoring the $M^{(n+1)+}/M^{n+}$ ratio [4] or by following the bromide ion concentration.[2] (M stands for the catalyst.)

POLAROGRAPHY

The concentrations of BrMA and bromate in the oscillation mixture were measured polarographically. The polarographic determination of BrMA has been described by us.[4] Bromate ion gives a well-defined wave on dropping mercury electrode, its $E_{\frac{1}{2}}$ value varies widely with the hydrogen-ion activity of the solution.[8] In order to separate the two waves an appropriate pH value was chosen. BrMA concentration was measured by withdrawing aliquots from the reaction mixture and adding to an acetate buffer of about pH 3.5 of the same volume. Then the BrMA wave was recorded. As a maximum suppressor 0.002 % Triton X-100 was used. In order to determine bromate concentrations the buffer samples were diluted 50 times with 0.5 M sulphuric acid and the bromate wave recorded. The actual concentrations were calculated from calibration curves.

CALORIMETRY

The calorimetric experiments were performed in a calorimeter of a constant temperature environment type. The measuring vessel was doubly jacketted and thermostatted by an ultrathermostat. A 6 kΩ thermistor forming one arm of a d.c. Wheatstone bridge was used as a sensor; the change in its resistance was continuously recorded during the reaction. This system allowed 0.2 J to be detected, equivalent to about 1×10^{-4} °C temperature change in the calorimeter vessel. The initial temperature of reactants and environment were kept at 25°C. The oscillating reactions were started by adding the catalyst to the reaction mixture. The total volume of the reaction mixtures was 200 ml. The solutions were stirred. The calorimeter was calibrated both chemically and electrically.

PROCEDURE

The chemical oscillation was initiated by adding an appropriate amount of a catalyst solution to a mixture of malonic acid, potassium bromate and sulphuric acid, thermostatted to 15.0, 25.0 and 35.0 ± 0.1°C, respectively. From the reacting oscillating reaction mixture aliquots were withdrawn at certain intervals and analysed for BrMA and bromate contents.

With $Fe(phen)_3^{2+}$ as a catalyst nitrogen was bubbled through the reaction mixture to remove dissolved atmospheric oxygen which is an inhibitor for the reaction.[9]

RESULTS

CALORIMETRIC MEASUREMENTS ON REACTING OSCILLATING SYSTEMS

The heat outputs of the different reacting Belousov systems are compiled in tables 1-4.

TABLE 1.—Ce(III) + MALONIC ACID + $KBrO_3$ + H_2SO_4 SYSTEM
MA = 0.40 M ; H_2SO_4 = 0.5 M

| | | | heat output/J | | | | |
$KbrO_3$/M	Ce(III) $\times 10^4$/M	induction period	1	2	3	5	10
0.100	5.74	602	21.3	20.1	19.7	18.9	17.2
0.100	11.5	582	39.0	38.1	37.2	36.1	31.5
0.100	46.0	569	89.5	83.3	79.1	72.4	64.0
0.050	11.5	259	22.4	21.6	20.9	18.9	14.2
0.163	11.5	837	53.1	51.0	49.4	47.3	41.8

ACCUMULATION OF BrMA IN REACTING OSCILLATING SYSTEMS

One of the main products of the oscillating Belousov reactions is BrMA. We regarded as of importance to follow the accumulation of BrMA in different reacting

TABLE 2.—Mn(II)+malonic acid+$KBrO_3$+H_2SO_4 system
MA = 0.40 M; H_2SO_4 = 0.5 M

$KBrO_3$/M	Mn(II)×10^4/M	induction period	heat output/J 1	2	3	5	10
0.100	1.15	305	3.4	3.3	3.2	3.1	2.7
0.100	4.60	310	23.5	22.6	21.5	29.8	17.7
0.100	11.5	305	49.8	46.4	43.5	41.4	35.1
0.100	46.0	308	111	105	97.9	89.5	78.2
0.025	11.5	17.6	17.2	14.9	13.6	6.4	1.9
0.050	11.5	97.1	34.5	30.9	29.5	26.2	20.5
0.163	11.5	458	62.3	57.7	54.8	49.0	45.6

TABLE 3.—Fe(phen)$_3^{2+}$+MALONIC ACID+$KBrO_3$+H_2SO_4 SYSTEM
MA = 0.4 M; H_2SO_4 = 0.25 M

$KBrO_3$/M	Fe(phen)$_3^{2+}$×10^4 /M	introductory period/min	induction period	heat output/J 1	2	3	5	10
0.100	1.44	1.5	11.9	8.4	8.5	8.6		
0.100	5.74	2.2	37.1	27.8	28.2	30.0	30.4	34.9
0.100	11.5	4.2	90.8	64.0	64.5	65.0	69.0	
0.050	5.74	>60	34.3	no oscillation even after 1 h				
0.060	5.74	18	37.2	34.8	35.6			
0.070	5.74	6.2	36.8	30.0	32.4	34.1	38.1	
0.130	5.74	1.6	38.1	28.8	29.7	30.8	32.0	37.7
0.175	5.74	1.1	38.2	28.0	28.5	29.0	29.3	31.8

TABLE 4.—Ru(dipy)$_3^{2+}$+MALONIC ACID+$KBrO_3$+H_2SO_4 SYSTEM
MA = 0.40 M; H_2SO_4 = 1.0 M

$KBrO_3$/M	Ru(dipy)$_3^{2+}$×10^4 /M	induction period	heat output/J 1	2	25
0.100	0.92	14.9	2.9	3.1	5.4
0.100	2.76	23.9	7.8	8.0	21.2
0.100	4.60	31.8	13.0	13.6	37.4
0.125	4.60	30.6	13.7	14.4	30.6
0.150	4.60	30.6	14.6	15.6	37.2

Belousov systems and at different temperatures early during oscillation when other brominated products (dibromomalonic acid, dibromoacetic acid) are not likely to be formed.

To meet this goal a polarographic method has been developed to determine even 10^{-4} M BrMA with a few percent accuracy in the presence of 100 to 500 times bromate. Also the concentration of bromate has been measured polarographically.

The results on the different Belousov systems are compiled in fig. 1-9.

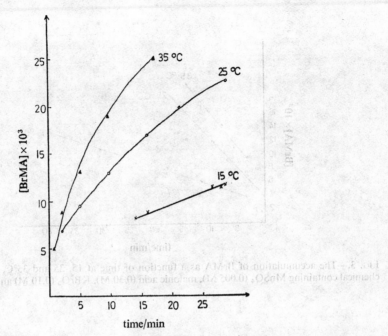

FIG. 1.—The accumulation of BrMA as a function of time at 15, 25 and 35°C, respectively. The chemical system containing $Ce(NO_3)_3$ (0.005 M), malonic acid (0.30 M), $KBrO_3$ (0.10 M) and H_2SO_4 (0.50 M).

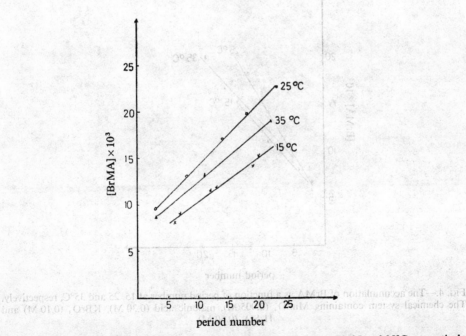

FIG. 2.—The accumulation of BrMA as a function of period number at 15, 25 and 35°C, respectively. The chemical system containing $Ce(NO_3)_3$ (0.005 M), malonic acid (0.30 M), $KBrO_3$ (0.10 M) and H_2SO_4 (0.50 M).

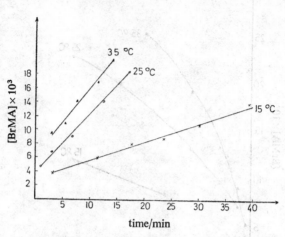

FIG. 3.—The accumulation of BrMA as a function of time at 15, 25 and 35°C, respectively. The chemical containing MnSO$_4$ (0.005 M), malonic acid (0.30 M), KBrO$_3$ (0.10 M) and H$_2$SO$_4$ (0.50 M).

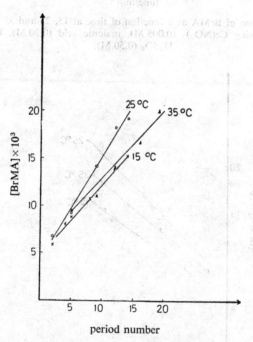

FIG. 4.—The accumulation of BrMA as a function of period number at 15, 25 and 35°C, respectively. The chemical system containing MnSO$_4$ (0.005 M), malonic acid (0.30 M), KBrO$_3$ (0.10 M) and H$_2$SO$_4$ (0.50 M).

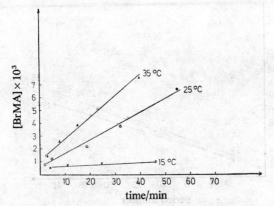

FIG. 5.—The accumulation of BrMA as a function of time at 15, 25 and 35°C, respectively. The chemical system containing $Fe(phen)_3^{2+}$ (0.0005 M), malonic acid (0.40 M), $KBrO_3$ (0.10 M) and H_2SO_4 (0.25 M).

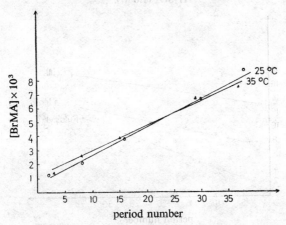

FIG. 6.—The accumulation of BrMA as a function of period number at 25 and 35°C, respectively. The chemical system containing $Fe(phen)_3^{2+}$ (0.0005 M), malonic acid (0.40 M), $KBrO_3$ (0.10 M) and H_2SO_4 (0.25 M).

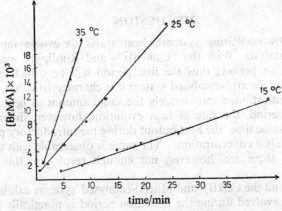

FIG. 7.—The accumulation of BrMA as a function of time at 15, 25 and 35°C, respectively. The chemical system containing $Ru(dipy)_3^{2+}$ (0.0005 M), malonic acid (0.40 M), $KBrO_3$ (0.10 M) and H_2SO_4 (1.0 M).

S 9—2

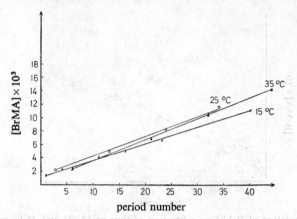

FIG. 8.—The accumulation of BrMA as a function of period number at 15, 25 and 35°C, respectively. The chemical system containing Ru(dipy)$_3^{2+}$ (0.0005 M), malonic acid (0.40 M), KBrO$_3$ (0.10 M) and H$_2$SO$_4$ (1.0 M).

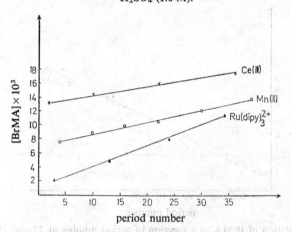

FIG. 9.—The accumulation of BrMA as a function of period number at 25°C. The chemical systems containing malonic acid (0.40 M), KBrO$_3$ (0.10 M), H$_2$SO$_4$ (1.0 M) and Ce(NO$_3$)$_3$ (0.0005 M) or MnSO$_4$ (0.0005 M) or Ru(dipy)$_3^{2+}$ (0.0005 M).

DISCUSSION

In the Belousov oscillating systems, heat starts to evolve immediately after addition of the catalyst. With the Fe(phen)$_3^{2+}$- and Ru(dipy)$_3^{2+}$-catalysed systems there is no induction period; thus the first period will be called the introductory period. For the Fe(phen)$_3^{2+}$-catalysed system it is characteristic that, independent of the bromate concentration, approximately the same amount of heat evolves during the introductory period, the rate of heat evolution, however, differs considerably (table 3). At the same time, the heat output during the introductory period increases with increasing catalyst concentration. The same is observable with the Ru(dipy)$_3^{2+}$-catalysed system; there are, however, not enough results on this system so far (table 4).

On the other hand the Ce(III)- and Mn(II)-catalysed systems exhibit an induction period. The heat evolved during the induction period is practically independent of the catalyst concentration, but increases, however, closely linearly with the bromate concentration (tables 1 and 2). This indicates that irrespective of the catalyst concentration in a particular system a certain amount of BrMA should accumulate (*vide infra*) the system to be converted to an oscillatory one. The BrMA concentration required for this depends on the initial bromate concentration. In other words, only after the necessary reactant concentrations are adjusted are conditions set for the initiation of chemical oscillation. With the same initial conditions higher BrMA concentration is required for the onset of oscillation in the Ce(III)-catalysed system as in the Mn(II)-catalysed one.

After the induction (or introductory) period, periodicity in the rate of heat evolution is observable. This is expected from the nature of oscillatory reactions; thus reports on temperature fluctuations [10, 11] or the claim that there is no overall heat production in the reaction are erroneous.[11]

During the oscillatory phase of the reaction the $Fe(phen)_3^{2+}$- and $Ru(dipy)_3^{2+}$-catalysed systems behave differently from the Ce(III)- and Mn(II)-catalysed ones. With the former systems early during oscillation the heat output in one period increases as the reaction proceeds, then it reaches a maximum and starts to decrease. (Similar features can be observed with the potential oscillations.) This phenomenon is more pronounced with the $Ru(dipy)_3^{2+}$-catalysed system and a detailed investigation of it is in progress. When $Fe(phen)_3^{2+}$ is used as a catalyst the decomposition of the complex should be considered and thus the concentration of the catalyst decreases in time. Namely, iron(II) formed by decomposition is not a catalyst for the Belousov system.[12] In spite of a decrease in catalyst concentration potential oscillation could be recorded throughout a period of 36 h. With the Ce(III)- and Mn(II)-catalysed systems, the heat evolved in one period decreases as the reaction proceeds, the damping factor being higher in the case of the Mn(II)-catalysed reaction.

With the $Fe(phen)_3^{2+}$- and $Ru(dipy)_3^{2+}$-catalysed systems, the heat output of one period increases closely linearly with the concentration of the catalyst—which means that during the autocatalytic reaction the total amount of the catalyst is oxidized—and only slightly or not at all depends on the bromate concentration. With the Ce(III)- and Mn(II)-catalysed systems the above linearity does not hold, i.e., only a certain proportion of the catalysts is oxidized in the autocatalytic reaction; there is, however, a considerable dependence on the bromate concentration. All these consequences follow from the redox potential data of the catalysts.

The results on the accumulation of BrMA provide us with valuable information concerning the reacting Belousov oscillating systems. It has been established that during the induction (or introductory) period a considerable amount of BrMA accumulates and this increases during the oscillatory phase. The ([BrMA], time) plots with the Ce(III)- and Mn(II)-catalysed systems show an excessive increase in the rate of accumulation of BrMA with increase of temperature (fig. 1 and 3); on the ([BrMA], period number) plots, however, the curves of 35°C lie below those of 25°C (fig. 2 and 4).

The same temperature dependences regarding the rate of BrMA accumulation are observable with the $Fe(phen)_3^{2+}$- and $Ru(dipy)_3^{2+}$-catalysed systems (fig. 5 and 7). The ([BrMA], period number) curves, however, have nearly the same slopes (see fig. 6 and 8). This indicates that although the period time decreases with increase of temperature, the amount of BrMA formed during one period is nearly the same. With the $Fe(phen)_3^{2+}$-catalysed system at 15°C the oscillatory conditions have not been met even after 100 min. The BrMA concentration reached a value of 6×10^{-4} M after 5 min and 1.6×10^{-3} M after 100 min.

The results of the experiments performed on systems of the same initial concentrations (fig. 9) show that the rate of accumulation of BrMA depends on the nature of the catalyst, however, the "activation energy" (E_ω) and "entropy of activation" ($\Delta S_\omega^\pm$) of chemical oscillation are the same in the case of the three systems.[3] (The $Fe(phen)_3^{2+}$-catalysed system does not oscillate in 1 M sulphuric acid.)

Polarographic measurements of BrMA and bromate in the same mixtures have revealed that early during the oscillating reaction the $BrO_3^- \rightarrow$ BrMA transformation holds. The changes in bromate concentration, however, could not have been followed accurately enough since very little differences were to be measured and thus only an accuracy of $\pm 10\%$ has been achieved.*

The method developed for the direct measurements of BrMA in low concentrations enabled us to gain more detailed information on the chemistry of the oscillatory Belousov reactions as obtained by Hess et al.[12-14] They identified the reaction products (BrMA and dibromoacetic acid in a ratio of about 6:1) extracted by ether from the reaction mixture after two hours reaction time by gas chromatography and mass spectrometry.[15] After a short reaction time (7 min) only BrMA was identifiable by thin-layer chromatography.[12]

BrMA is formed during the autocatalytic step in the reacting Belousov system and thus periodicity in its rate of formation is exceptable. To prove this, a run has been performed with Ce(III)-catalysed Belousov system thermostatted to $15.0 \pm 0.1°C$ where the period time is long enough—about 2 min. Samples were withdrawn from the reaction mixture after the autocatalytic step, before the autocatalytic step (at the end of the "restful" phase) and again after the autocatalytic step, and analysed for BrMA content. The results are given in fig. 10.

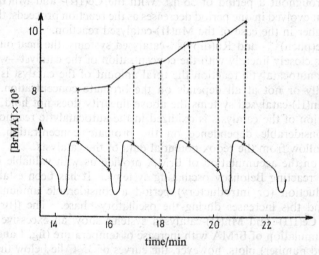

FIG. 10.—Bromide ion and BrMA concentration against time curves during the 6th and 7th periods. The initial chemical system containing $Ce(NO_3)_3$ (0.005 M), malonic acid (0.30 M), $KBrO_3$ (0.10 M) and H_2SO_4 (0.50 M). Temperature: 15°C.

Thus we were able to provide experimental evidence for the periodic rate of formation of BrMA.

By making use of the calorimetric and BrMA-concentration measurements we were able to calculate the heats of Belousov oscillating reactions referred to 1 mol bromate. For the cerium-catalysed Belousov system, the heat of the oscillating reaction in 130 ± 5 kJ mol^{-1} bromate.

* Recently an accuracy of $\pm 3\%$ has been achieved by iodometric titration.

With our aim in mind of a deeper insight into the heat effects, we have started to measure the heats of reaction of some of the composite reactions of the Belousov systems.

The results obtained so far are summarized briefly below: (a) The catalyst–bromate reactions have rather low heats of reaction. Considering the

$$BrO_3^- + 4M^{n+} + 5H^+ \rightarrow HOBr + 4M^{(n+1)+} + 2H_2O$$

reaction,[15] and this has been proved also for the $Fe(phen)_3^{2+}$-BrO_3^- reaction,[16] the heat of reaction in 0.25-1 M sulphuric and referred to 1 mol bromate is 10.1 kJ with Ce(III) and 6.70 kJ with $Fe(phen)_3^{2+}$ as reductant.

(b) The heat of reaction of the bromination of malonic acid ($MA + \frac{1}{2} Br_2 \rightarrow BrMA$) in 0.5 M sulphuric acid referred to 1 g atom of bromine is 138 kJ.

CONCLUSION

The driving force of the Belousov oscillatory processes is the high heat of reaction of the bromate–malonic acid reaction. This reaction, however, proceeds in steps following a rather sophisticated mechanism. The amount of bromate present in the system can deliver its stored energy only at that stage (phase) of the reaction (and only in small portions), when the concentration of bromide ion present in the system drops below a critical value. This provides a precondition for the autocatalytic reaction to be switched on. Thus bromide ion acts as a trigger, and its concentration changes periodically in time. This happens also with the oxidized form of the catalyst and with the bromous acid ($HBrO_2$).[2] However, this latter can not be followed analytically.

Besides these intermediates exhibiting temporal concentration oscillation, the end products of the reaction (e.g., CO_2, BrMA) are the non-recycling species. All non-recycling species show periodicity in their rates of formation. With carbon dioxide and BrMA this has been verified experimentally. It is, however, obvious that the rate of consumption of bromate and malonic acid should also be periodic in nature.

Finally the periodicity in the rate of reaction manifests itself in the periodicity of heat evolution.

[1] R. J. Field and R. M. Noyes, *J. Chem. Phys.*, 1974, **60**, 1877.
[2] R. J. Field, E. Kőrös and R. M. Noyes, *J. Amer. Chem. Soc.*, 1972, **94**, 8649.
[3] E. Kőrös, *Nature*, 1974.
[4] E. Kőrös, L. Ladányi, V. Friedrich, Zs. Nagy and Á. Kis, *Reaction Kin. Cat. Letters*, 1974, **1**.
[5] J. E. Demas and D. Diemente, *J. Chem. Ed.*, 1973, **50**, 357.
[6] E. Kőrös, M. Orbán and Zs. Nagy, *Nature* (*Phys. Sci.*), 1973, **242**, 30.
[7] H. Degn, *Nature*, 1967, **213**, 589.
[8] J. Heyrovsky and J. Kuta, *The Principles of Polarography* (Publishing House of the Czechoslovak Academy of Sciences, Prague, 1965).
[9] M. T. Beck and Z. Váradi, *Chem. Comm.*, 1973, 30.
[10] U. Franck and W. Geiseler, *Naturwiss.*, 1970, **58**, 52.
[11] H. G. Busse, *Nature* (*Phys. Sci.*), 1971, **233**, 137.
[12] L. Bornmann, H. Busse and B. Hess, *Z. Naturforsch.*, 1973, **28b**, 93.
[13] L. Bornmann, H. Busse, B. Hess, R. Riepe and C. Hesse, *Z. Naturforsch.*, 1973, **28b**, 824.
[14] L. Bornmann, H. Busse and B. Hess. *Z. Naturforsch.*, 1973, **28b**, 514.
[15] R. C. Thompson, *J. Amer. Chem. Soc.*, 1971, **93**, 7315.
[16] E. Kőrös, M. Burger and Á. Kis, *Reaction Kin. Cat. Letters*, 1974, **1**.

Two Kinds of Wave in an Oscillating Chemical Solution

BY A. T. WINFREE

Department of Biological Sciences, Purdue University, West Lafayette, Indiana 47907, U.S.A.

Received 6th September, 1974

Wavelike phenomena in a chemical solution oscillating at period T fall into two classes: (1) diffusion-independent but oscillation-dependent structures repeating at intervals T, and (2) diffusion-dependent but oscillation-independent structures derived from threadlike filaments of " scroll axis ", repeating at intervals T_0 much less than T.

This paper experimentally examines a typical example of each.

In unfiltered solution one usually observed a third class of wave source, forming " target patterns " at diverse periods intermediate between T and T_0.

KINEMATIC WAVES AND TRIGGER WAVES

Distributed throughout a large enough space, any unstirred chemical oscillation generally exhibits local parameter variations affecting the local period. Thus phase gradients develop and steepen, even from initial synchrony of the bulk oscillation. Unless reactants and products have identical optical properties, these changing phase gradients can be seen as moving bands of colour passing every point at regular intervals of time equal to the bulk oscillation period, T. Called " kinematic waves " by Kopell and Howard,[1] such waves have also been studied theoretically by Smoes and Dreitlein,[2] by Thoenes,[3] and by Ortoleva and Ross.[4, 5] Kopell and Howard,[1] and Thoenes[3] have additionally exhibited such waves in the oscillating reagent of Belousov[6] and of Zhabotinsky and Zaikin.[7]

Provided that the oscillating reagent is initially well stirred and left in a homogeneous environment, no chemical gradients remain. The oscillation period being therefore everywhere the same, any phase gradients subsequently introduced by a local disturbance will persist without change until molecular diffusion becomes significant on a scale embracing the affected area. In the diffusion-free uniform-period approximation, the corresponding kinematic waves have been called " pseudo-waves ".[8]

With or without spatial variations of period, such waves merely expose a shallow timing gradient in the spatially distributed but locally autonomous oscillation. Wave velocity is inversely proportional to the steepness of the phase gradient and so has no upper bound. Such waves do not involve diffusion, are not *conducted* through the medium, and are not impeded by impermeable barriers.

However, in sufficiently steep phase gradients, diffusion cannot be ignored. In the chemical reagent of Belousov[6] and of Zhabotinsky and Zaikin[7]—called Z reagent henceforth—a new phenomenon emerges from steep phase gradients which assigns a lower bound to pseudowave velocity: a pulse of chemical activity is triggered and propagates at a velocity, v, characteristic of the chemical medium. These have been called " trigger waves ".[9, 10] They are arrested at impermeable barriers and have spatially uniform velocity rather than spatially uniform repeat time. Both solitary trigger waves and regularly spaced wave trains appear in both oscillating and non-oscillating versions of Z reagent.

THE SLOWEST PSEUDOWAVES

Though pseudowaves slower than the conduction velocity are not observable (because they trigger the faster wave), pseudowaves at all velocities down to this limit are readily demonstrable in oscillating Z reagent. In fact, pseudowaves travelling at almost exactly the conduction velocity are the commonest sort. This is because every volume element turns blue spontaneously at intervals T after passage of a trigger wave.

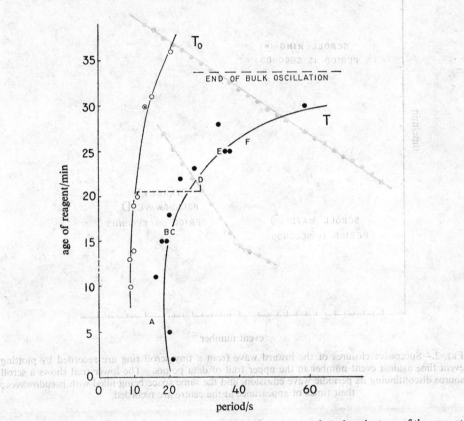

FIG. 1.—The periods of three kinds of wave in filtered Z reagent are plotted against age of the reagent. Open circles represent high-frequency waves deriving from scroll rotation : all have the same period T_0, which gradually increases. Solid circles represent bulk oscillation : its period T approaches infinity after half an hour. Letters represent concentric rings filling the holes left by extinct scroll sources : the period T identifies them as pseudowaves. Raw data underlying the dotted circle and the open circle connected to D by dashes are given in fig. 2.

For example, when a source of trigger waves abruptly ceases to radiate (see Appendix) the last wave emitted (at $t = 0$) sets up a radial phase gradient as it propagates away, leaving an initially wave-free hole. But within that hole, volume elements at distance r from the centre spontaneously turn blue at times $nT + r/v$. Thus behind the packet of trigger waves, all spaced apart by distance $T_0 v$, there appear concentric pseudowaves spaced apart by distance $Tv > T_0 v$, all following the last trigger wave outward at approximately* the conduction velocity, v.   This abrupt change of period is shown in fig. 1 and 2.

In fig. 1 are recorded the periods of various phenomena in a 2 mm deep layer of oscillating Z reagent at short intervals during a 40 min period after mixing of the ingredients (see Appendix for details).   The open circles record a slow increase in the

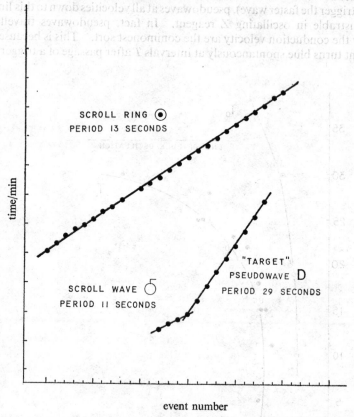

FIG. 2.—Successive closures of the inward wave from a tiny scroll ring are recorded by plotting event time against event number in the upper trail of data points.   The lower trail shows a scroll source discontinuing its periodic wave emission, and the same space being filled with pseudowaves; their times of appearance at the centre are recorded.

period of diverse trigger-wave sources: this is the period of scroll rotation.   The solid circles record an increase in the period of bulk oscillations in wave-free regions, and their eventual termination after half an hour (note that scroll sources continue). The letters A-F record the intervals between eruption of circular waves at the centre of the " hole " left by a receding packet of nearly circular trigger waves after its source vanishes: each letter represents the mean period of all the waves within one expanding hole.

* Since v does increase somewhat with wave spacing, it is possible that the conduction slightly outraces the pseudowave.   If so, then if an impermeable barrier could be implanted without deforming the liquid (I have not found a way!), the wave striking one side would continue from the other side after a slight delay proportional to the barrier's distance from the centre, or from the next wave approaching from the centre, whichever is less.

The observations underlying three of these data points are shown in fig. 2, where the time of each event (appearance of a wave, wave passing a marker, or collision of waves) is plotted in serial order.   The series of 12 observations plotted in the lower right depicts a scroll source emitting progressively less eccentric oval waves at an 11 s period, until the last one is nearly circular.   Thereafter circular waves appear at 29 s intervals, close to the simultaneously-measured period of bulk oscillations.   This transition is indicated by the dashed line in fig. 1.   The other 5 lettered cases show that the period of these hole-filling target patterns approximates to the bulk oscillation period.   After the end of bulk oscillations, no such waves appear in new holes.

The upper series of observations in fig. 2 shows the 13 s period of a horizontal scroll *ring*, approximately $\frac{1}{4}$ mm in diameter, shedding waves to the inside and outside, all of which appear circular in projection.   This observation is entered in fig. 1 as the dotted open circle.

In unfiltered reagent the space between curves T and T_0 in fig. 1 is typically filled with many trails of data points depicting the diverse and increasing periods of individual pacemaker centres, each generating a " target pattern " of concentric trigger-wave circles at its own period intermediate between T and T_0.

ROTATING PSEUDOWAVES?

The phase gradients constituting pseudowaves A-F were radial, spanning many cycles of bulk oscillation from centre to receding periphery.   But, unlike most thermodynamic variables, the phase of an oscillation is defined mathematically on the unit circle rather than on the real line.   Thus it seems possible, in principle, for phase to continually increase through any number of complete cycles around a closed path in space.   The pseudowave seen on such a circular gradient would rotate about a fixed centre.   Such pseudowaves have been invoked [3] to explain the rotating spiral waves seen in Z reagent, but these turn out to be trigger waves, by the criteria that their period T_0 is much less than T, that they are totally blocked by impermeable barriers, and that they appear in non-oscillating reagent as well as in oscillating reagent.   Rotating pseudowaves remain to be observed in chemically oscillating media, though there may be *biological* examples. [10]

ROTATING TRIGGER WAVE SOURCES

However, the principal mode of diffusion-dependent organization in Z reagent (at the ubiquitous period T_0) seems to be rotation of appropriately crossed concentration gradients.   A two-dimensional region (a thin film of liquid) organized in this way emits a rotating trigger wave shaped like an Archimedes' spiral.   Its colour pattern rotates about a slightly wobbly pivot.   In three dimensions the pivot becomes a 1-dimensional filament threading through the liquid like a vortex line in classical hydrodynamics.   However, here there is no fluid motion.   From this filament there emerges a scroll-shaped wave of excitation.   Its perpendicular cross-section resembles an Archimedes' spiral.   The geometry of this pivot-filament, the scroll axis, determines the spatial and temporal organization of the oscillating reaction.   The scroll axis tends to close in rings, except where a nearby interface prevents it.   Rings several centimetres in circumference are common.   The smallest rings yet seen are about one wavelength in circumference, leaving just room enough for organization of the scroll core (itself of circumference equal to one wavelength) around the circular scroll axis.

DISSECTION OF A SCROLL RING

In order to examine the anatomy of a scroll ring in more detail, it is convenient to use a version of the oscillating reagent which conducts waves well while absorbed in the pores of a Millipore filter (see Appendix for details). Impregnated filters can be stacked to make a horizontally-laminated three-dimensional medium with roughly isotropic conduction properties. By assembling such stacks from filters already bearing waves, it is possible to induce three-dimensional concentration patterns of peculiar geometry. The waveforms which eventually develop from such contrived initial conditions can be examined in detail by tossing the Millipore stack into a preservative bath, then reassembling an image of the wave from its fixed horizontal sections.

From computer simulations of two-dimensional media resembling a thin layer of Z reagent, one learns that crossed concentration gradients of the sort needed to create a scroll axis are formed when a solitary wave is suddenly brought into contact with inert medium, for example as in fig. 3. The pivot of a spiral wave appears near the

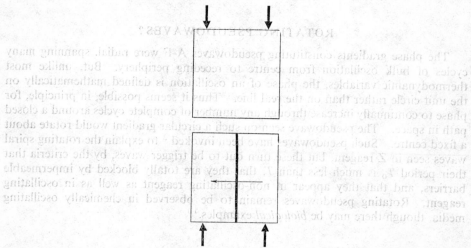

FIG. 3.—A spiral wave can be created by abutting a block of inert reagent against the endpoint of a plane wave in another block. The spiral's centre appears near the initial point of contact.

initial point of contact. Imagine fig. 3 spun about its right edge to form a pair of cylinders, the bottom one now containing an outward propagating cylindrical wave. This suggests a trick suitable for implementation in a Millipore stack. Two stacks, each about 0.7 mm deep, are prepared by stacking 5 filters permeated with Z reagent as described in the Appendix. A hemispherical wave is started in one stack by touching the centre of its upper surface with an electrically heated filament. After a minute the wave becomes a cylinder several mm in diameter, extending vertically through the stack and propagating outward. When the second stack, still inert, is set on top, the situation of fig. 3 is created in every vertical plane through the axis of the cylinder. The locus of pivots for the spiral wave emerging in each such plane is a horizontal circle roughly coincident with the upper edge of the cylindrical wave at the moment of contact. From this ring-shaped filament a scroll wave emerges, propagating in all directions throughout the Millipore stack (fig 4.)

Every horizontal cross-section of this wave consists of concentric circles. Every

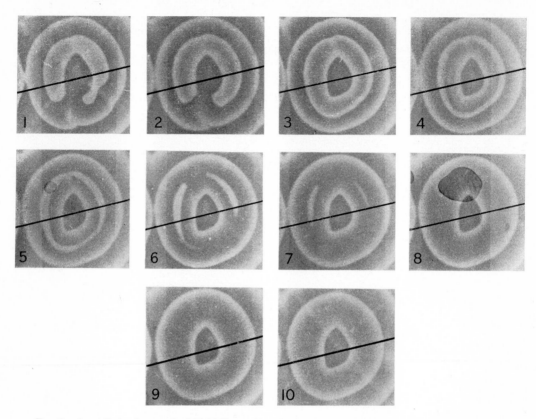

FIG. 5.—A scroll ring in a stack of 10 Millipore filters is shown in serial section parallel to the circular scroll axis, perpendicular to the axis of circular symmetry. The top side of each filter is shown. They are about 0.15 mm thick. Number 10 was on top. Each panel is a 1 cm square. The black slash across each panel shows the section line along which the view in fig. 6 was taken.

To face page 43]

vertical cross-section through the symmetry axis shows segments of two spirals, one on the left and a mirror image on the right . Fig. 4 attempts a schematic of 6 stages during one rotation of the spiral, showing only the left side spiral; the right edge of the box is the vertical symmetry axis of the Millipore stack. It will be noted that the wave erupts alternately at equal intervals of one half period through the top and bottom Millipores, above and below the initial ring of contact. The inside half of this wave continues inward to annihilation on the symmetry axis, while the outer half continues outward.

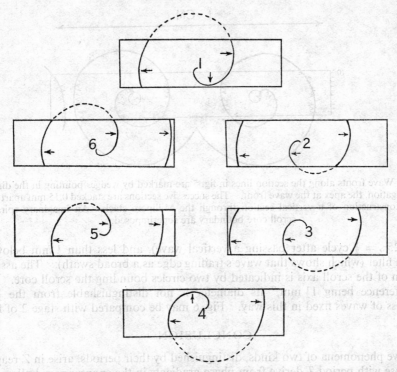

FIG. 4.—A spiral wave in a rectangular piece of medium is shown at 6 equally spaced times during the cycle. The successive panels were drawn by rotating an involute spiral 60° each time.

In the actual experiment, circular waves emerged through the opaque upper surface (filter No. 10), approximately above the initial cylinder, after 25 s, then after 50 s more, then after another 50 s, then after 40 s. The 40-50 s period of the 3-dimensional wave is close to the 45 s rotation period of a well-developed spiral in a parallel 2-dimensional control experiment in a single thickness of Millipore. (The bulk oscillation period is several minutes). 25 s later the stack was dispersed into cold fixative while the trailing edge of the fourth wave was still passing vertically through filter No. 1. Fixed in this way, wave thickness is about $\frac{1}{2}$ mm. Successive waves follow one another by $1\frac{1}{2}$ mm at a velocity of 2 mm/min.

Fig. 5 shows the upper surfaces of corresponding 1 cm squares cut out of the ten filters, in order from 1 (the bottom) to 10 (the top). Waves propagate with the sharp edge foremost : the inner circle is moving inward, the outer circles outward. In filters 1 and in 4-7 we also see vertically propagating waves : only the trailing edge of one in filter 1, but the full thickness of $\frac{1}{2}$ mm in filters 4-7. At the left edge a wave is entering from another source.   A gas bubble seems to have inhibited propagation between filters 3 and 2.   Filters 5 and 8 show air bubbles caught under the glass during photography.

Each of the 10 prints is marked with a section line : fig. 6 assembles a vertical cross-section through the stack along this line, shown as though seen from the top, the upper half of each print discarded.   The position of each wavefront in each layer is marked by a wedge pointing in the direction of propagation.   The imaginary spirals connecting these data include arcs about $\frac{1}{2}$ wave spacing above the top filter

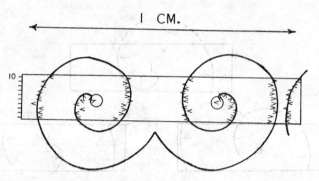

FIG. 6.—Wave fronts along the section lines in fig. 5 are marked by wedges pointing in the direction of propagation, the apex at the wave front.   The successive sections are stacked 0.15 mm apart in this scale reconstruction of a vertical section through the Millipore stack.   An imaginary spiral and scroll core boundary are superimposed.

(fixed 25 s = $\frac{1}{2}$ cycle after passing a vertical wave), and less than $\frac{1}{2}$ mm below the bottom filter (which shows that wave's trailing edge as a broad swath).   The assumed position of the scroll axis is indicated by two circles bounding the scroll core.   Core circumference being $1\frac{1}{2}$ mm,[9] its diameter is not distinguishable from the $\frac{1}{2}$ mm thickness of waves fixed in this way.   Fig. 6 may be compared with stage 2 of fig. 4.

CONCLUSION

Wave phenomena of two kinds, distinguished by their periods, arise in Z reagent :

Those with period T derive from phase gradients in the spontaneous bulk oscillation.   They have been analyzed extensively in the theoretical literature.[1-3]   They have been exhibited in Z reagent in the one-dimensional, long-wavelength case.[1, 3]   Their independence of diffusion has been shown experimentally by their passage through impermeable barriers.[1]   Fig. 1 and 2 describe such waves in the two-dimensional case in the short-wavelength limit.

Those with period $T_0 < T$ occur only in two- and three-dimensional situations, as spirals and scroll waves respectively.   Some theoretical discussion of such waves has appeared.[11, 12]   By obstructing their passage with impermeable barriers, they have been shown to be " trigger waves ", critically dependent on diffusion.   Fig. 5 and 6 show a ring-shaped scroll wave in Z reagent.

Wave phenomena at intermediate periods are much more commonly observed in unfiltered Z reagent.[7]   Since most of these trigger wave " target patterns " are eliminated by careful filtration and can be restored by deliberate contamination with dust, they are viewed as consequences of local shorter-period oscillation near a heterogeneous nucleus.

These experiments were made possible by NSF Grant 37947 and an NIH Research Career Development Award.

APPENDIX

The pseudowave experiment uses an oscillating cerium + malonate reagent similar to that of Kopell and Howard,[1] but with excess ferroin to enhance visibility :

2.15 g cerous nitrate hexahydrate in 250 ml water : 1 volume
36.0 g malonic acid in 250 ml water : 2 volumes
13.5 g potassium bromate in 250 ml water : 2 volumes
55.0 ml sulphuric acid plus 200 ml water : 4 volumes
1.0 ml Triton X-100 surfactant in 15 l. water : 1 volume
25 millimolar ferrous 1,10-phenanthroline sulphate : 2 volumes

It is important to avoid the E. Merck ferroin widely available in Europe. This is made with a chloride salt ; the chloride is not removed and poisons the reaction.

The solution is filtered through a Millipore GSWP 0.2 micron " Millex " filter into a fresh Falcolnware petri dish. Lining the dish with Sylgard silicone resin helps in eliminating pacemakers and the " target patterns " of trigger waves they emit. The dish is placed over a blue-green fluorescent light-box, baffled to avoid heating. The lid is coated with a film of 0.1 % Triton X-100 to prevent fogging. A magnifying glass is helpful.

The few remaining pacemakers produce circular waves (or a hot needle is used if there are none) which must be sheared by gently tilting the dish before the bulk oscillation annihilates them. From their wreckage a diversity of scroll sources emerge, generating diversely convoluted patterns (" intestines ") all characterized by the ubiquitous period and wavelength of the involute spiral wave. Some form tiny scroll *rings*, or fragments of rings interrupted by a single air/liquid interface. Some of these contract until they abruptly vanish, leaving a series of closely-packed waves to propagate away. The empty hole expanding around the annihilated source then fills, from centre outward, with concentric pseudowaves. The times of their appearances at the centre are noted to the nearest second with a stopwatch. Sometimes instead their passage by a fixed scratch or bubble on the floor of the dish is noted. The results are similar and are plotted in fig. 1 (A-F). The bulk oscillation period is measured by recording the times at which the reagent suddenly turns more transparent (in blue-green light) at a fixed, wave-free place in the dish, as well as in a separate wave-free dish of the same reagent. The interval between this event and the next appearance of a blue spot in the centre of one of the above holes remains the same within a few seconds, attesting the synchrony of both bulk-oscillations. Scroll rotation period is measured by recording the times when the inward circular wave in a scroll ring contracts to a point and vanishes ; or when waves from two or more separate scroll sources collide and annihilate each other : the intervals are the same within several percent.

Usually several of these diverse sources were watched simultaneously, noting event times in as many columns of a table. The times were later plotted and periods were read from the slopes of the resulting line segments as in fig. 2. The results from two independent experiments were indistinguishable ; fig. 1 combines them. Herman Gordon pointed out that these curves become much smoother when observations are restricted to one place in the dish. This is because the oscillation period varies several percent from place to place in the dish, and because intervals between wave collisions or wave passages reflect the (generally shorter) period of wave appearances at their source some time earlier.

The Millipore stack experiments use a published recipe [8] but with 1 ml instead of 2 ml of sulphuric acid. Consistent wave propagation depends on the purity of the sodium bromate ; even reagent grade samples sometimes need to be recrystallized. This reagent does not oscillate when poured in thin layers exposed to the air, presumably due to oxygen interfering with free-radical processes. However it does oscillate in bulk with a period of several minutes when confined in a sealed tube or in a thin layer under oil or in Millipores under a glass coverslip. (Sometimes the reagent poured in a petri dish will oscillate in the deeper meniscus though not in the shallower centre ; then a blue wave starts around the rim of the dish and propagates into the centre once every several minutes).   The more acid reagent [8] behaves similarly in air but must be spread in a much thinner layer or divided into fine droplets [8] to inhibit oscillation completely.

The filters are 23 mm diameter Millipore GSWP, numbered with tiny Magic Marker dots and filled with Z reagent by floating glossy-side-up on the liquid surface. With nylon forceps they are lifted free, stacked on plexiglas (Perspex), covered with a microscope cover-slip, and drained of excess liquid into the corner of a Kleenex tissue. The wave is started with a 3 V penlight, its bulb removed and the tungsten filament carefully stretched out to a point which becomes hot but does not glow. The two stacks are joined before a bulk oscillation erases this single cylindrical wave. Because carbon dioxide placques quickly accumulate between the filters, the whole process must be terminated within several minutes (at 20°C): the stack is impaled with a needle to assist in orienting the numbered filters later, then dispersed in cold 3 % perchloric acid. After several minutes each filter is sealed under a coverslip and photographed on high-contrast film in blue light.

About 20 such experiments were run. Scroll rings were caught at several of the stages of rotation depicted in fig. 4. Several stacks produced very complicated structures, possibly due to invisible gas bubbles separating the filters. In some of the very regular ones like fig. 5-6, the period of wave eruption on the visible surface of the stack was as long as 70 s. This sometimes happens in a single-Millipore spiral wave, too. The reason is not known but chemical contamination, perhaps from sodium chloride or perchloric acid fingerprints, is suspected.

[1] N. Kopell and L. Howard, *Science*, 1973, **180**, 1171.

[2] M. Smoes and J. Dreitlein, *J. Chem. Phys.*, 1973, **59**, 6277.

[3] D. Thoenes, *Nature (Phys. Sci.)*, 1973, **243**, 18.

[4] P. Ortoleva and J. Ross, *J. Chem. Phys.*, 1973, **58**, 5673.

[5] P. Ortoleva and J. Ross, *J. Chem. Phys.*, 1974, **60**, 5090.

[6] B. Belousov, *Sb. Ref. Rad. Med.*, 1958, 145.

[7] A. Zhabotinsky and A. Zaikin *Nature*, 1970, **225**, 535.

[8] A. Winfree, *Science*, 1972, **175**, 634.

[9] E. Zeeman, *Towards a Theoretical Biology*, ed. C. Waddington (Aldine, New York, 1972), vol. 4, pp. 8-67.

[10] A. Winfree, *Mathematical Problems in Biology, Lecture Notes in Biomathematics*, (Victoria Conference 1973), Vol. 2. ed. P. van den Driessche (Springer-Verlag, Berlin, 1974), p. 241.

[11] A. Winfree, *Sci. Amer.*, 1974, **230**, 82.

[12] A. Winfree, *Mathematical Aspects of Chemical and Biological Problems and Quantum Chemistry*, ed. D. Cohen (Amer. Math. Soc., Providence, 1974), vol. 8, in press.

Periodicity in Chemically Reacting Systems

A Model for the Kinetics of the Decomposition of Na$_2$S$_2$O$_4$

BY PHIL E. DEPOY AND DAVID M. MASON

Stanford University, Stanford, California, U.S.A.

Received 23rd July, 1974

Sodium dithionite undergoes thermal decomposition in aqueous solution to form sodium bisulphite and sodium thiosulphate. It has been observed previously that in the decomposition, under some conditions, the concentration of the reactant oscillates in time. Although various theoretical mechanisms, such as the Lotka system, have been proposed which would produce oscillations in the rate of reaction and in the concentrations of intermediate species in a reacting system, none of the theoretical mechanisms can explain the oscillations, including periodic increases, of the reactant concentration observed in the dithionite system.

In this paper, the observed behaviour of the dithionite system is described. Three mechanisms are discussed which have been proposed by others to explain various characteristics, other than the oscillations, of the decomposition. It is then shown how these mechanisms could, with certain modifications, produce oscillations such as are observed with the dithionite system.

Sodium dithionite (Na$_2$S$_2$O$_4$) is a powerful reducing agent used in the manufacture of various organic chemicals and in dyeing and bleaching processes. It undergoes thermal decomposition in aqueous solution to form sodium bisulphite (NaHSO$_3$) and sodium thiosulphate (Na$_2$S$_2$O$_3$) according to the stoichiometry:

$$2Na_2S_2O_4 + H_2O \rightarrow 2NaHSO_3 + Na_2S_2O_3. \tag{1.1}$$

Although relatively little has been published regarding the kinetics of the decomposition, those findings available are contradictory. The reaction has two distinct regimes, as shown by the data in fig. 1, taken from the work of Lem and Wayman.[1] There is an induction period in which the dithionite concentration decreases slowly with time, followed by a rapid autocatalytic reaction. Most of the published work

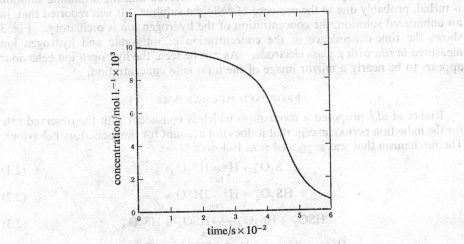

FIG. 1.—Sodium dithionite concentration against time.

deals with behaviour during the initial induction period. Different investigators have reported the order of the reaction in this initial period to be first, three-halves, and second with respect to the dithionite concentration, and one-half and first with respect to the hydrogen ion concentration.

Rinker, Lynn, Mason and Corcoran [2] first reported that, at some temperatures, marked oscillations in the dithionite concentration were observable, even in a system closed with respect to mass. Fig. 2 shows some of their results obtained at 60°C, where reproducible oscillations including periodic increases in the dithionite concentration as high as $\pm 15 \%$ are evident. At 70°C, the magnitudes of the oscillations were reported to be considerably less, and at 80°C, practically no oscillations were observed.

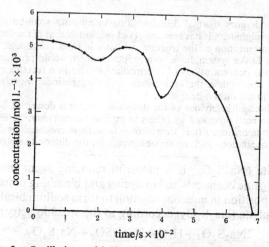

FIG. 2.—Oscillations with time in sodium dithionite concentration.[2]

Rinker and associates reported, as have other investigators, that the addition of the products of the reaction to fresh dithionite greatly increased the initial rate of decomposition. If products were added in sufficient concentration, the induction period was completely eliminated. It was also observed that the dithionite solution is turbid, probably due to the presence of colloidal sulphur. It was reported that, in an unbuffered solution, the concentration of the hydrogen ion is oscillatory. Fig. 3 shows the time-dependence of the concentration of dithionite and hydrogen ion measured *in situ* with a glass electrode. As can be seen, the hydrogen ion behaviour appears to be nearly a mirror image of the dithionite concentration.

PROPOSED MECHANISMS

Rinker *et al.*[2] proposed a mechanism which is consistent with the observed rate for the induction period, except that it does not account for the oscillatory behaviour. The mechanism that was suggested is as follows :

$$S_2O_4^= + H^+ \underset{\text{fast}}{\rightleftharpoons} HS_2O_4^- \qquad (2.1)$$

$$HS_2O_4^- + H^+ \underset{\text{fast}}{\rightleftharpoons} 2HSO_2 \cdot \qquad (2.2)$$

$$HSO_2 \cdot + HS_2O_4^- \underset{\text{controlling}}{\longrightarrow} HSO_3 \cdot + HS_2O_3^- \qquad (2.3)$$

$$HSO_3 \cdot + HSO_2 \cdot + H_2O \underset{\text{fast}}{\rightarrow} 2HSO_3^- + 2H^+ \qquad (2.4)$$

This reaction scheme leads to a rate expression of the form

$$r = k[H^+]^{\frac{1}{2}}[HS_2O_4^-]^{\frac{3}{2}}.$$ (2.5)

To be in accord with their observed kinetics, i.e., a rate first-order with respect to $[HSO_3^-]$, variable fractional order in $[H^+]$, and zero order in $[SO_3^{2-}]$, Spencer[3] and Burlamacchi, Guarini, and Tiezzi[4] have concluded that the rate-determining step must be

$$HSO_2\cdot + SO_2^- + HSO_3^- \rightarrow \text{intermediate products.}$$ (2.6)

Wayman and Lem[5] proposed the following sequence of reactions:

$$S_2O_4^= + H^+ \underset{\text{fast}}{\rightleftharpoons} HS_2O_4^-$$ (2.7)

$$S_2O_4^= + H^+ \rightarrow HSO_2^- + SO_2$$ (2.8)

$$HS_2O_4^- + H_2O \rightarrow HSO_2^- + HSO_3^- + H^+$$ (2.9)

$$HSO_2^- + HS_2O_4^- \rightarrow HSO_3^- + S_2O_3^= + H^+.$$ (2.10)

They assumed that the reaction step, eqn (2.10), is slow in the absence of a catalyst such as H_2S or colloidal sulphur. They have proposed three reactions of sulphoxylate (HSO_2^-) which might produce H_2S or S:

$$H^+ + 3HSO_2^- \rightarrow H_2S + 2HSO_3^-,$$ (2.11)

$$H^+ + 2HSO_2^- \rightarrow S + HSO_3^- + H_2O,$$ (2.12)

and

$$H^+ + HSO_2^- + H_2S \rightarrow 2S + 2H_2O.$$ (2.13)

They suggest that the catalytic action of hydrogen sulphide may be due to its ability to accept electrons and that its presence may improve the ability of the dithionite and sulphoxylate to react, by facilitating the formation of the sulphoxylate free radical to produce a more rapid reaction:

$$HSO_2\cdot + HS_2O_4^- \rightarrow S_2O_3^= + HSO_3\cdot + H^+.$$ (2.14)

The free radical product becomes the ion by reacquiring an electron from the H_2S:

$$HSO_3\cdot + e^- \rightarrow HSO_3^-.$$ (2.15)

Polysulphides, formed by the reaction of hydrogen sulphide and sulphur, would be even more effective radical stabilizers and explain the observed catalytic effect of colloidal sulphur.

Assuming that reaction (2.9) (the reaction of hydrogen dithionite with water) is rate-limiting during the induction period and that reaction (2.10) (the reaction of hydrogen dithionite ion with sulphoxylate) is limiting during the rapid decomposition, the overall rate equation is

$$r = k_1[H^+][S_2O_4^=] + k_2[H^+][S_2O_4^=][HSO_2^-].$$ (2.16)

Assuming that the catalytic effect of products of the reaction controls the rate during the fast decomposition, Wayman and Lem[5] obtain an overall rate expression of the form

$$-\frac{dC}{dt} = k_1[H^+]C + k_3[H^+]C(C_0 - C)$$ (2.17)

where C represents the concentration of dithionite.

The Wayman–Lem expression (eqn (2.17)) does appear to give reasonably good agreement with the observed behaviour, as shown in fig. 4 and 5. Fig. 4 compares their best fit of the rate equation with data obtained by them for a solution of 10 mM dithionite, buffered at a pH of 4 and at a temperature of 23°C. Fig. 5 is their comparison of the best fit of the rate equation to data obtained by Rinker *et al.*[2] for a solution of 11.5 mM dithionite, at a pH of 6 and a temperature of 60°C.

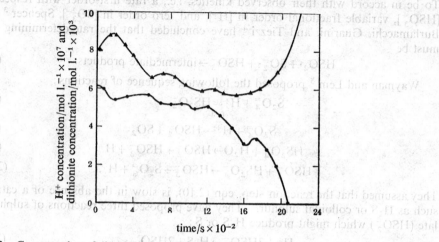

FIG. 3.—Concentration of dithionite and hydrogen ion against time [2]; ●, dithionite; ▲, hydrogen ion.

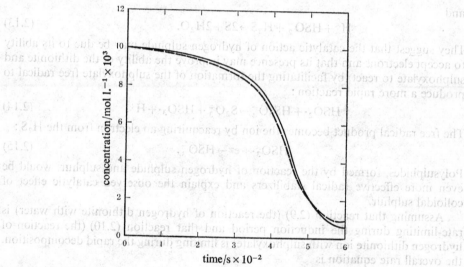

FIG. 4.—Comparison of Wayman–Lem data with mechanism of eqn (2.16); —— experimental, - - - best fit of eqn (2.16).

3. DISCUSSION

It can be easily shown that no oscillations can occur with the Wayman–Lem mechanism or the Rinker mechanism since the rate-controlling steps are not autocatalytic. It has been demonstrated theoretically by several investigators, e.g., Higgins,[6] that at least one autocatalytic step is required to produce damped oscillations of intermediates, and two or more autocatalytic steps are required to give undamped oscillations in *open systems*. It has also been shown that damped oscillations can be produced by autocatalytic reactions in systems which are *closed* with respect to mass.[7] The rate-controlling step proposed by Spencer [3] and by Burla-macchi, Guarini and Tiezzi [4] is autocatalytic, but with respect to one of the products, and this also would not produce oscillations.

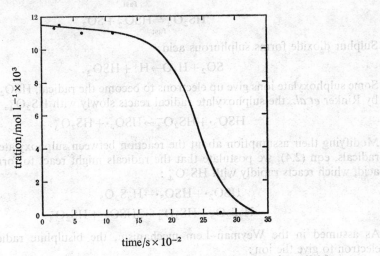

FIG. 5.—Comparison of the Rinker *et al.*[2] data with mechanism of eqn (2.16).

As mentioned earlier, the most unusual characteristic of the dithionite system is the periodic increases which are observed in the dithionite concentration itself (fig. 2). Although closed-loop " feedback " mechanisms, i.e., mechanisms in which one or more of the intermediates or products react to form the reactant, can be postulated which could produce such increases, it can be shown that these require that the concentrations oscillate through an equilibrium point, and this has been demonstrated in accordance with the Wegscheider constraint [8] to be thermodynamically unfeasible. Thus, no closed-loop system can produce such oscillations.

The most appealing explanation of the periodic increases in the dithionite concentration is that the dithionite does, in fact, behave as an intermediate during the decomposition process. This behaviour can be accomplished if the dithionite reacts to form some side-products (other than the products or intermediates of the principal decomposition reaction) and reaches equilibrium with them, possibly at a low temperature at which the concentrated dithionite solution is initially prepared. As the dithionite is consumed by the decomposition, the equilibrium with these side-products would be shifted, and they would react to form more dithionite, thus producing a singular or quasi-singular point in the dithionite concentration. In view of the observed appearance of what is believed to be colloidal sulphur in the solution, the formation of various complexes such as polythionic acids ($H_2S_xO_6$) observed in the decomposition of sodium thiosulphate in acid is a suspected side-reaction.

Several mechanisms for the decomposition can be postulated which produce oscillations. One, which appears to satisfy nearly all the observed characteristics of the reaction, is as follows. As postulated, the dithionite reaches equilibrium in a side reaction with various sulphur-compound complexes and possibly with colloidal sulphur :

$$S_2O_4^= \rightleftharpoons \text{complexes} + [S] + \ldots \ldots \qquad (3.1)$$

As proposed in the other mechanisms, the dithionite is in rapid equilibrium with

$HS_2O_4^-$, in a manner such as proposed by Rinker et al.[2] and the $HS_2O_4^-$ is in equilibrium with sulphoxylate ions and SO_2 similar to the stoichiometry of eqn (2.8):

$$H^+ + S_2O_4^= \underset{\text{fast}}{\rightleftharpoons} HS_2O_4^- \tag{3.2}$$

$$HS_2O_4^- \underset{\text{fast}}{\rightleftharpoons} HSO_2^- + SO_2 \tag{3.3}$$

Sulphur dioxide forms sulphurous acid:

$$SO_2 + H_2O \rightarrow H^+ + HSO_3^-. \tag{3.4}$$

Some sulphoxylate ions give up electrons to become the radical, $HSO_2\cdot$. As assumed by Rinker et al., the sulphoxylate radical reacts slowly with $HS_2O_4^-$, as follows:

$$HSO_2\cdot + HS_2O_4^- \rightarrow HSO_3\cdot + HS_2O_3^-. \tag{3.5}$$

Modifying their assumption about the reaction between sulphoxylate and bisulphite radicals, eqn (2.4), we postulate that the radicals might react to form disulphurous acid, which reacts rapidly with $HS_2O_4^-$:

$$HSO_2\cdot + HSO_3\cdot \rightarrow H_2S_2O_5 \tag{3.6}$$

$$H_2S_2O_5 + HS_2O_4^- \rightarrow 2HSO_3\cdot + HS_2O_3^-. \tag{3.7}$$

As assumed in the Weyman–Lem mechanism, the bisulphite radical absorbs an electron to give the ion:

$$HSO_3\cdot + e^- \rightarrow HSO_3^-. \tag{3.8}$$

Overall, this reaction is very similar to the theoretical oscillating mechanism first proposed by Lotka[9] in 1910, i.e.,

$$X_0 \rightarrow X_1 \tag{3.9}$$

$$X_1 + X_2 \rightarrow 2X_2 \tag{3.10}$$

$$X_2 \rightarrow \text{products}, \tag{3.11}$$

where the precursor, X_0, corresponds to various sulphur-compound complexes, the first intermediate, X_1, corresponds to the dithionite ion in equilibrium with the sulphoxylate radical, and the second intermediate, X_2, corresponds to the bisulphite radical.

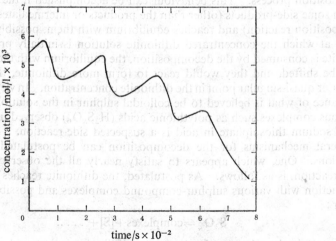

FIG. 6.—Behaviour of proposed mechanism.

Assuming a first-order decay of the precursor, the rate equations for our mechanism are:

$$\dot{x}_0 = -k_1 x_0,$$ (3.12)

$$\dot{x}_1 = k_1 x_0 - 2k_2 x_1/e - 2k_3 x_1 x_2/e,$$ (3.13)

$$\dot{x}_2 = k_2 x^2/e + k_3 x_1 x_2/e - k_4 e x_2,$$ (3.14)

and

$$\dot{x}_3 = k_2 x_1/e + k_3 x_1 x_2/e + k_4 e x_2$$ (3.15)

where x_0 is the concentration of the precursor, x_1 is the concentration of dithionite, x_2 is the concentration of the bisulphite radical, x_3 is the concentration of the bisulphite ion, and e is an electron availability factor. If the electron availability factor is constant, based on the results derived for the closed Lotka system,[7] it would be expected that this system would exhibit multiple damped oscillations when the factor $(k_3 x_0^0/k_1 e)^{\frac{1}{2}}$ is of the order of 5 or greater. If the electron availability factor is related to the precursor and product concentrations by

$$e = x_0/(x_0 + cx_3)$$ (3.16)

(which can be explained by assuming that the precursor, or a substance in equilibrium with it, is an electron donor, and that the bisulphite and another substance in equilibrium with the precursor are electron sinks), the autocatalytic behaviour of the dithionite is produced. This relationship was selected because it produces the observed autocatalytic-type behaviour (since the precursor concentration, x_0, falls off exponentially with time), and it does not dampen the oscillations, as is the case if the electron availability is a strong function of the dithionite concentration, x_1.

As an example of the behaviour of this system of reactions, fig. 6 shows the dithionite concentration when parameters are chosen to approximate the observed behaviour shown in fig. 2, i.e., a period of the oscillations of approximately 150-200 s and an induction period of about 600 s. Parameters on which the figure is based are $k_1 = 2 \times 10^{-3}$, $k_2 = 10^{-8}$, $k_3 = 290$, $k_4 = 1.6$, $c = 0.07$ and $x_0^0 = 4 \times 10^{-3}$ (all based on concentrations in mol/l^{-1} and time in seconds). As can be seen, the mechanism compares reasonably well with the observed behaviour.

This new mechanism is similar enought to the other three mechanisms presented by others that nearly all the observed characteristics of these mechanisms are preserved:

1. the reaction appears autocatalytic, i.e. the decrease in dithionite concentration is slow at first, then rapidly accelerates,

2. the reaction appears to be catalyzed by the addition of products (in the new mechanism, the addition of bisulphite ion would produce more of an electron sink, thereby more rapidly reducing the dithionite concentration), and

3. the reaction is catalyzed by the addition of sulphur (in the new mechanism, sulphur would also act as an electron sink).

In addition, other results observed by Rinker et al.,[2] e.g., the oscillation of the hydrogen ion concentration, are also consistent with our mechanism.

4. SUMMARY

In summary, we have briefly discussed the behaviour of an oscillating chemical system, the thermal decomposition of sodium dithionite. We have discussed three mechanisms which have been proposed by others to explain the general behaviour, not including oscillations, of the dithionite system. Finally, we have suggested modifications to those mechanisms which could produce the type of oscillations which have been observed.

Obviously, the mechanism of the dithionite decomposition is complex and a great deal more work would be required to gain a thorough understanding of it. Because of its uniqueness in the domain of oscillating chemical systems, further investigation is warranted.

[1] W. J. Lem and M. Wayman, *Canad. J. Chem.*, 1970, **48**, 776.

[2] R. G. Rinker, S. Lynn, D. M. Mason and W. H. Corcoran, *Ind. and Eng. Chem. Fund.*, 1965, **4**, 282; S. Lynn, *Ph.D. Thesis* (California Institute of Technology, 1952).

[3] M. S. Spencer, *Trans. Faraday Soc.*, 1967, **63**, 2510.

[4] L. Burlamacchi, G. Guarini and E. Tiezzi, *Trans. Faraday Soc.*, 1969, **65**, 496.

[5] M. Wayman and W. J. Lem, *Canad. J. Chem.*, 1970, **48**, 782.

[6] J. Higgins, *Ind. and Eng. Chem.*, 1967, **59**, 19.

[7] P. E. DePoy and D. M. Mason, *Combustion and Flame*, 1973, **20**, 127.

[8] R. Wegscheider, *Z. phys. Chem.*, 1902, **39**, 257.

[9] A. J. Lotka, *J. Phys. Chem.*, 1910, **14**, 271.

Mechanism of the Oscillatory Decomposition of Hydrogen Peroxide in the Presence of Iodate Ion, Iodine etc.

By Isao Matsuzaki and Tsuyoshi Nakajima

Dept. of Synthetic Chemistry, Faculty of Engineering, Shinshu
University, Nagano, Japan

AND

Herman A. Liebhafsky

Dept. of Chemistry, Texas A & M University, College Station, Texas, U.S.A.

Received 25th July, 1974

Various mechanisms are constructed with the aid of information on associated reactions and tested in the light of the theory of two-variable oscillating reactions as well as by means of computer simulations. Reactions were run mainly at $[HClO_4]_0 = 0.035$ to 0.080, $[KIO_3]_0 = 0.40$, and $[H_2O_2]_0 = 0.30$ mol l.$^{-1}$ and compared with computer results for a promising mechanism. Agreement is so good with respect to the induction period, the pulses of $[I^-]$ and the rate of O_2 evolution, the abrupt decrease in $[I_2]$ etc. for one to conclude the mechanism to be plausible. The plausible mechanism contains HIO_2, HIO, I^-, $H_2I_2O_3$, and $H_3I_3O_5$ as intermediates and the second-order back-activation step $2HIO_2 + HIO + H_2O_2 \rightarrow 3HIO_2 + H_2O$ as the key step for the oscillation, which results from the sequence of $HIO_2 + HIO \rightleftharpoons H_2I_2O_3$, $H_2I_2O_3 + HIO_2 \rightleftharpoons H_3I_3O_5$, and $H_3I_3O_5 + H_2O_2 \rightarrow 3HIO_2 + H_2O$. The oscillation source of the mechanism has been found to be in conformity with the Brusselator.

1. INTRODUCTION

Since Bray [1] discovered the oscillatory decomposition of H_2O_2 by the IO_3^-—I_2 couple, not a few investigations [2-9] have been carried out without giving any concrete mechanisms. Recently we have succeeded in finding a plausible mechanism which enables us to understand the capability of the system to oscillate.

Success was attained in the following way. First, information was collected on possible intermediates and elementary reactions. Second, mechanisms were constructed and examined for the capability of oscillation both theoretically and by means of a computer. Finally, experimental data were compared with computer results.

This paper is concerned with the process leading to the plausible mechanism, its nature and experimental evidence for it.

2. THE PROCESS LEADING TO THE PLAUSIBLE MECHANISM

A survey of the literature [2-9] has shown that aqueous solutions capable of oscillatory decomposition contain IO_3^-, H^+, I_2, I^-, and H_2O_2 as stable species. In such solutions the following reactions also can take place.

$$2IO_3^- + 2H^+ + 5H_2O_2 \rightarrow I_2 + 5O_2 + 6H_2O \quad [10] \tag{2.1}$$

$$I_2 + 5H_2O_2 \rightarrow 2IO_3^- + 2H^+ + 4H_2O \quad [11, 12] \tag{2.2}$$

$$IO_3^- + 5I^- + 6H^+ \rightleftharpoons 3I_2 + 3H_2O \quad [13-15] \tag{2.3}$$

$$2I^- + 2H^+ + H_2O_2 \rightarrow I_2 + 2H_2O \quad [16] \tag{2.4}$$

$$I_2 + H_2O \rightleftharpoons HIO + I^- + H^+ \quad [17] \tag{2.5}$$

$$H_2O_2 \rightarrow H_2O + 0.5\,O_2. \quad [18] \tag{2.6}$$

Here we adopt a view that if we find a mechanism which can explain all the above reactions mechanistically, the mechanism has a good chance of accounting for the oscillatory decomposition.

Previous investigations on the six reactions have suggested a number of intermediates such as HIO_2, HIO, H_2IO^+,[19, 20] and $H_2I_2O_3$.[12] Using such intermediates and supplementary intermediates, $H_2IO_2^+$*[ ] and $H_3I_3O_5$,† a comprehensive mechanism of an oscillatory nature was constructed by the trial and error method as shown in fig. 1, in which $\rightleftharpoons$ denotes so rapid a reversible step as to be usually in equilibrium

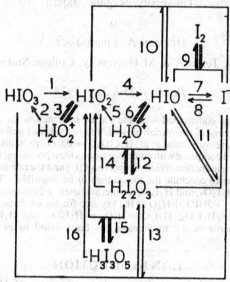

FIG. 1.—The comprehensive mechanism of oscillatory nature for the H_2O_2—IO_3^-—I_2 system. The numerals denote the step number.

and each step is specified by an Arabic figure; HIO_3, a medium acid, is used as a generic symbol for HIO_3 and IO_3^- also in what follows; species IO_2^-, IO^-, and I_3^- are neglected since HIO_2 and HIO are weak acids and no I_3^- is formed under the oscillation conditions. The detail of each step is listed below.

Step 1	$HIO_3 + H_2O_2 \rightarrow HIO_2 + H_2O + O_2$	(2.7)
Step 2	$H_2IO_2^+ + H_2O_2 \rightarrow HIO_3 + H^+ + H_2O$	(2.8)
Step 3	$H_2IO_2^+ \rightleftharpoons H^+ + HIO_2$	(2.9)
Step 4	$HIO_2 + H_2O_2 \rightarrow HIO + H_2O + O_2$	(2.10)
Step 5	$H_2IO^+ + H_2O_2 \rightarrow HIO_2 + H^+ + H_2O$	(2.11)
Step 6	$H_2IO^+ \rightleftharpoons HIO + H^+$	(2.12)
Step 7	$HIO + H_2O_2 \rightarrow H^+ + I^- + H_2O + O_2$	(2.13)
Step 8	$I^- + H^+ + H_2O_2 \rightarrow HIO + H_2O$	(2.14)
Step 9	$I_2 + H_2O \rightleftharpoons HIO + I^- + H^+$	(2.15)

* This species was devised by analogy to H_2IO^+ but its presence is probable from the occurrence of $XO_3^- + 2H^+ \rightleftharpoons H_2XO_3^+$ (X = Cl, Br, or I).[21–23]

† This species was devised without experimental support but its formation is probable since large molecules [24, 25] such as $H_2I_3O_{14}^{5-}$ and $H_4I_3O_{14}^{3-}$ and those [26] such as $[IO_3(HIO_3)_n]^-$ (n > 1) are suggested.

Step 10 $HIO_2 + H^+ + I^- \rightleftharpoons 2HIO$ (2.16)

Step 11 $H_2I_2O_3 + H^+ + I^- \rightleftharpoons 3HIO$ (2.17)

Step 12 $HIO_2 + HIO \rightleftharpoons H_2I_2O_3$ (2.18)

Step 13 $HIO_3 + H^+ + I^- \rightleftharpoons H_2I_2O_3$ (2.19)

Step 14 $H_2I_2O_3 + H_2O_2 \rightarrow 2HIO_2 + H_2O$ (2.20)

Step 15 $H_2I_2O_3 + HIO_2 \rightleftharpoons H_3I_3O_5$ (2.21)

Step 16 $H_3I_3O_5 + H_2O_2 \rightarrow 3HIO_2 + H_2O.$ (2.22)

According to the comprehensive mechanism, reactions (2.1) to (2.6) take place via the sequences of steps shown in fig. 2.

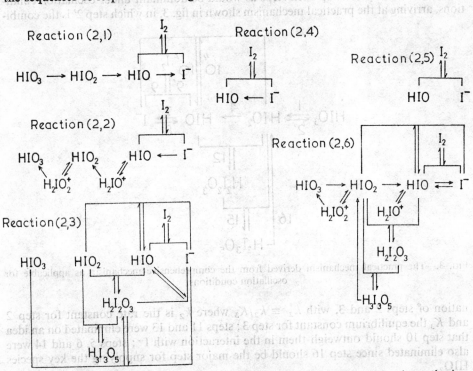

FIG. 2.—The sequences of steps for reactions (2.1) to (2.6) according to the comprehensive mechanism.

According to the theory [27] of oscillating reactions a feedback step is indispensable for oscillation, so in the trial-and-error search a mechanism equal to the comprehensive mechanism minus steps 15 and 16 was first subjected to computer simulations with 92 sets of values of rate constants for each step since it contains a feedback step $HIO_2 + HIO + H_2O_2 \rightarrow 2HIO_2 + H_2O$ (first-order back-activation) as a consequence of the sequence of steps 12 and 14, but without giving any oscillations. Here we examined the sequence theoretically (see section 3, (ii))

$$HIO_3 \rightleftharpoons HIO_2 \rightleftharpoons HIO \atop (act)^n$$ (2.23)

and found that the feedback step $HIO \rightarrow HIO_2$ should have an order n higher than 1. Based on this finding we chose 2, the smallest favourable integer for n, and added steps

15 and 16 so that the second-order back-activation step $2HIO_2 + HIO + H_2O_2 \rightarrow 3HIO_2 + H_2O$ might participate. Of course, it is possible to get a favourable sequence of steps using such intermediates as $IO\cdot$, $HO\cdot$, and $HO_2\cdot$, but at present we would like to retain the use of radical species as a possibility.

3. NATURE OF THE COMPREHENSIVE MECHANISM

(i) STEPS DOMINANT IN THE OSCILLATION

The failure of the associated mechanism to yield oscillation mentioned in section 2 suggests that any steps but steps 15 and 16 are not responsible for the oscillation. On this basis we picked up such steps as would be dominant under oscillation conditions, arriving at the practical mechanism shown in fig. 3, in which step 2′ is the combi-

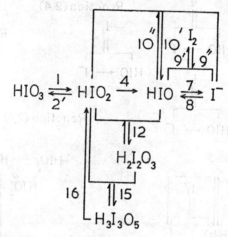

FIG. 3.—The practical mechanism derived from the comprehensive mechanism as applicable for oscillation conditions.

nation of steps 2 and 3, with $k_{2'} \equiv k_2/K_3$ where k_2 is the rate constant for step 2 and K_3 the equilibrium constant for step 3; steps 11 and 13 were eliminated on an idea that step 10 should outweigh them in the interaction with I^-; steps 5, 6 and 14 were also eliminated since step 16 should be the major step for supplying the key species HIO_2.

The practical mechanism was fed into computer simulation. The simulation is based on the Euler method.[28] Steps 12 and 15 were regarded as being in equilibrium, while steps 9 and 10 were each decomposed into two reversible steps, 9′ and 9″, and 10′ and 10″, respectively. The rate law for each step is determined according to the mass action law as exemplified by $v_8 = k_8[I^-][H_2O_2][H^+]$ for step 8 and $v_{16} = k_{16}K_{12}K_{15}[HIO_2]^2[HIO][H_2O_2]$ for step 16; that for step 1 is given by the experimental one [10] for reaction (2.1)

$$v_1/\text{mol l.}^{-1}\,\text{min}^{-1} = 2.6 \times 10^{-4}[HIO_3][H_2O_2] + 129 \times 10^{-4}[H^+][HIO_3][H_2O_2]. \quad (3.1)$$

Based on previous experimental results, the simplification was adopted that $[HIO_3]$, $[H_2O_2]$, and $[H^+]$ might be regarded as constant for a run as long as the switchover of the system to oscillation is concerned, leading to condensed rate constants $\bar{k}_i$ as exemplified by $v_1 = \text{eqn (3.1)} \equiv \bar{k}_1$, $v_8 = k_8[I^-][H_2O_2][H^+] \equiv \bar{k}_8[I^-]$, $v_{16} = k_{16}K_{12}K_{15}[HIO_2]^2[HIO][H_2O_2] \equiv \bar{k}_{16}[HIO_2]^2[HIO]$. The sets of k_i and $\bar{k}_i$ values employed for the simulation are summarized in table 1 and the time courses obtained are shown in fig. 4. In the sets of k_i and $\bar{k}_i$ values employed only $\bar{k}_i$ was changed systematically. In choosing the sets we did not adhere to experimentally determined k_i and K_i values because in this computer simulation we aimed at making the oscillatory nature of the mechanism visualizable.

It is clear from fig. 4 that the practical mechanism is so versatile as to reproduce almost all the aspects of the system such as the oscillatory decomposition, initial stage of reaction, catalytic decomposition and one-way I_2 formation.

TABLE 1.—SUMMARY OF THE COMPUTER RUNS CONDUCTED BASED ON THE PRACTICAL MECHANISM

run no.	$\bar{k}_1{}^a$	$\bar{k}_2{}'{}^b$	$\bar{k}_4{}^c$	$\bar{k}_{16}{}^d$	$\bar{k}_7{}^e$	$\bar{k}_8{}^f$	$\bar{k}_9{}^g$	$k_{9''}$	$k_{10'}$	$k_{10''}$	nit of conc.^{h} (mol l.$^{-1}$)
1	3	2	10	4	5	5	2	0.05	4	4	$v_1/\bar{k}_1$
2	4	2	10	4	5	5	2	0.05	4	4	$v_1/\bar{k}_1$
3	5	2	10	4	5	5	2	0.05	4	4	$v_1/\bar{k}_1$
4	6	2	10	4	5	5	2	0.05	4	4	$v_1/\bar{k}_1$

$^a k_1 = $ eqn (3.1). $^b \bar{k}_{2'} = (k_2/K_3)[H^+][H_2O_2]$. $^c \bar{k}_4 = k_4[H_2O_2]$. $^d \bar{k}_{16} = (k_{16}K_{12}K_{15})[H_2O_2]$. $^e \bar{k}_7 = k_7[H_2O_2]$. $^f \bar{k}_8 = k_8[H^+][H_2O_2]$. $^g \bar{k}_{9'} = k_{9'}[H^+]$. ^{h} For example, run 1 gives results for which $v_1/3$ corresponds to one unit of concentration where v_1 is the actual rate of step 1 given by eqn (3.1) (cf. the $[I_2]$ calculation in section 4, (ii)).

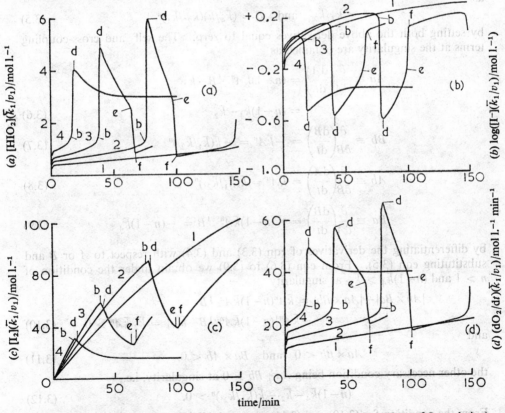

FIG. 4.—The computer-calculated time courses of $[HIO_2]$, $[I^-]$, $[I_2]$, and dO_2/dt (rate of O_2 evolution) for the runs of table 1. The numerals indicated denote the run number. See note (h), table 1 for $\bar{k}_1/v_1$.

(ii) OSCILLATION SOURCE

The failure of the associated mechanism mentioned in section 2, coupled with the success of the practical mechanism, indicates that the first-order back-activation step consisting of steps 12 and 14 is not effective, whereas the second-order one composed of steps 12, 15 and 16 is the oscillation source.

It is now necessary to make a theoretical check on the above indication. For this purpose the following sequence is the simplest including a nth order back-activation step. Since this sequence is a two-variable system (because $[HIO_3]$ = const. under consideration), its necessary condition for oscillation can be derived along the line by Higgins [27] as follows.

$$HIO_3 \underset{\bar{k}_{2'}}{\overset{\bar{k}_1}{\rightleftarrows}} HIO_2 \underset{\underset{(act)^n}{\bar{k}}}{\overset{\bar{k}_4}{\rightleftarrows}} HIO \tag{3.2}$$

The rates of increases in A and B are given by

$$dA/dt = \bar{k}_1 - (\bar{k}_{2'} + \bar{k}_4)A + \bar{k}A^nB \tag{3.3}$$
$$dB/dt = \bar{k}_4 A - \bar{k}A^n B \tag{3.4}$$

where A and B stand for $[HIO_2]$ and $[HIO]$, respectively. The singularity is obtained as

$$A = \bar{k}_1/\bar{k}_{2'} \quad \text{and} \quad B = (\bar{k}_4/\bar{k})(\bar{k}_{2'}/\bar{k}_1)^{n-1} \tag{3.5}$$

by setting both the above derivatives equal to zero. The self- and cross-coupling terms at the singularity are obtained as

$$Aa \equiv \frac{\partial}{\partial A}\left(\frac{dA}{dt}\right) = (n-1)\bar{k}A^{n-1}B - \bar{k}_{2'}$$
$$= (n-1)\bar{k}_4 - \bar{k}_{2'} \tag{3.6}$$

$$Bb \equiv \frac{\partial}{\partial B}\left(\frac{dB}{dt}\right) = -\bar{k}A^n = -\bar{k}(\bar{k}_1/\bar{k}_{2'})^n \tag{3.7}$$

$$Ab \equiv \frac{\partial}{\partial B}\left(\frac{dA}{dt}\right) = \bar{k}A^n = \bar{k}(\bar{k}_1/\bar{k}_{2'})^n \tag{3.8}$$

$$Ba \equiv \frac{\partial}{\partial A}\left(\frac{dB}{dt}\right) = -(n-1)\bar{k}A^{n-1}B = -(n-1)\bar{k}_4 \tag{3.9}$$

by differentiating the derivatives of eqn (3.3) and (3.4) with respect to A or B and substituting eqn (3.5). From eqn (3.6) to (3.9) we obtain under the condition of $n > 1$ and $(n-1)\bar{k}_4 > \bar{k}_{2'}$ at singularity

$$|Ab \times Ba| - |Aa \times Bb| = \bar{k}A^n(n-1)\bar{k}A^{n-1}B$$
$$-\bar{k}A^n\{(n-1)\bar{k}A^{n-1}B - \bar{k}_{2'}\} = \bar{k}_{2'}\bar{k}A^n > 0 \tag{3.10}$$

and

$$Aa \times Bb < 0 \quad \text{and} \quad Ba \times Ab < 0, \tag{3.11}$$

the other necessary condition being $Aa + Bb > 0$ at singularity, i.e.,

$$(n-1)\bar{k}_4 - \bar{k}_{2'} - \bar{k}(\bar{k}_1/\bar{k}_{2'})^n > 0. \tag{3.12}$$

From the condition for (3.10) and (3.11) or eqn (3.12) it is evident that the first-order ($n = 1$) back-activation step is ineffective, whereas the second-order ($n = 2$) can confer possibility of oscillation, in favour of the above-mentioned indication.

It should be pointed out that the part of the comprehensive mechanism corresponding to sequence (3.2) with $n = 2$ is equivalent to the Brusselator,[29-34] which according to Tyson [35] is the only known two-variable chemical scheme which admits limit cycle oscillations. This equivalence can readily be understood if we apply a replacement of A, E = HIO_3, X = HIO_2, B = H_2O_2, Y = HIO, and D = O_2 on the Brusselator

$$A \rightarrow X$$
$$B+X \rightarrow Y+D$$
$$2X+Y \rightarrow 3X$$
$$X \rightarrow E.$$

On the other hand, the unfavourable sequence (3.2) with $n = 1$ corresponds to the Brusselator with its step $2X+Y \rightarrow 3X$ replaced by $X+Y \rightarrow 2X$; with respect to such a replacement Tyson [35] has stated, in support of our unsuccess in getting oscillations, that the steady state of such a replaced scheme is always stable for positive values of the parameters.

4. EXPERIMENTAL EVIDENCE FOR THE PRACTICAL MECHANISM

(i) EXPERIMENTAL

The present experiment was designed to get time courses comparable with those of fig. 4 on the basis that the change in $\bar{k}_1$ corresponds to a change in $[H^+]_0$ when $[KIO_3]_0$ and $[H_2O_2]_0$ are kept constant.

All the chemicals used were of guaranteed grade and used without further purification. The hydrogen peroxide contained no inhibitor.

Two kinds of reaction vessels, open-type and closed-type, both made of Pyrex, were used. Both the vessels were used in previous experiments [6-9] and are to be described in detail elsewhere. The open-type was designed especially for the spectrophotometric $[I_2]$ measurement. The $[I^-]$ measurement [7] was made by means of an Orion I^- selective electrode and the dO_2/dt (rate of O_2 evolution) measurement [8] by means of a Matheson Company LF-20 mass flowmeter.

A solution containing KIO_3 and $HClO_4$ was kept at 323 K and then an amount of 30 % H_2O_2 was added to initiate the reaction.

In table 2 are summarized the runs of reaction conducted and the time courses

TABLE 2.—SUMMARY OF THE RUNS OF REACTION CONDUCTED AT 323 K

run no.	initial conc./mol l.$^{-1}$			volume (V/ml)	type of vessel used	quantities followed
	$[HClO_4]_0$	$[KIO_3]_0$	$[H_2O_2]_0$			
1'	0.035	0.40	0.30	150	closed	$[I^-]$, dO_2/dt
2'	0.051 6	0.40	0.30	150	closed	$[I^-]$, dO_2/dt
3'	0.060	0.40	0.30	150	closed	$[I^-]$, dO_2/dt
4'	0.070	0.40	0.30	150	closed	$[I^-]$, dO_2/dt
5'	0.080	0.40	0.30	150	closed	$[I^-]$, dO_2/dt
6'	0.070	0.40	0.30	140	open	$[I^-]$, $[I_2]$
7'	0.057 6	0.282	0.494	140	open	$[I^-]$, $[I_2]$
8'	0.057 6	0.282	0.494	150	closed	dO_2/dt

obtained are shown in fig. 5 to 7. A number of preliminary runs have concluded that the series of runs 1' to 5' has shown all kinds of time courses for $[I^-]$ and dO_2/dt. Run 6' is a separate run of run 4' for the time course of $[I_2]$. In fig. 4 and 5 to 7 lower case letters are used to make clear the synchronization among $[I_2]$, $[I^-]$, and dO_2/dt. The initial period $a-b$ where $[I^-]$, $[I_2]$, and dO_2/dt increase will be called the

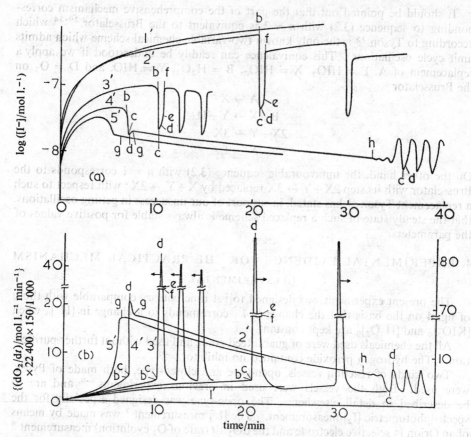

FIG. 5.—The experimentally obtained time courses of dO_2/dt and $[I^-]$ for runs 1' to 5' of table 2. The numerals indicated denote the run number

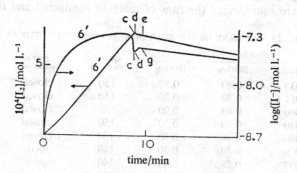

FIG. 6.—The experimentally obtained time courses of $[I_2]$ and $[I^-]$ for run 6' of table 2.

induction period and during the flat part after point g the smooth catalysis $H_2O_2 \rightarrow H_2O + 0.5O_2$ proceeds almost exclusively.

Fig. 5 to 7 coupled with observations in preliminary experiments characterize each time course as follows (all concentrations in mol l.$^{-1}$):

Run 1' ($[HClO_4]_0 = 0.035$)....$[I^-]$ and $[I_2]$ increase monotonously to reach a constant where I_2 begins to precipitate. dO_2/dt increases similarly but decreases as H_2O_2 is consumed.

Run 2' ($[HClO_4]_0 = 0.051\ 6$)..... The induction period is followed by pulse-like oscillations, where $[I^-]$ first decreases inducing $[I_2]$ to decrease and dO_2/dt to increase.

Run 3' ($[HClO_4]_0 = 0.060$) and runs 7' and 8' ($[HClO_4]_0 = 0.057\ 6$).... The same as in run 2', with oscillations appearing earlier with smaller amplitudes and shorter periods.

Run 4' ($[HClO_4]_0 = 0.070$).... The induction period is followed by smooth catalysis via a half-pulse and oscillations appear later via several wiggles.

Run 5' ($[HClO_4]_0 = 0.080$).... The induction period is followed by smooth catalysis.

Runs 1' to 5'.... The changes in the position and size of $[I^-]$ and dO_2/dt pulses in runs 2' and 3' and the descents of the line for the smooth catalysis in runs 4' and 5' result from the consumption of H_2O_2; the later oscillation in run 4' will be absent if H_2O_2 is supplied constantly.

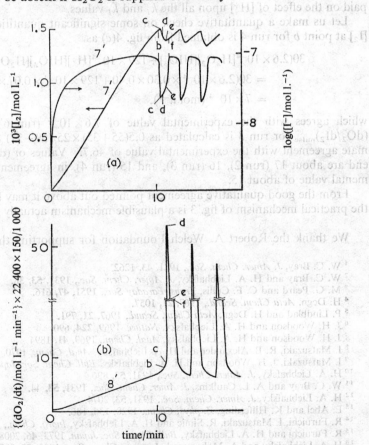

FIG. 7.—The experimentally obtained time courses of $[I^-]$ and $[I_2]$ for run 7' and that of dO_2/dt for run 8' of table 2.

The nearly linear increases in $[I_2]$ in the induction period are in accord with the rate law (3.1), which indicates that the formation of I_2 is due to reaction (2.1). One quantity of interest is the ratio $(dO_2/dI_2)_{ind.}$ of the amount of O_2 evolved to that of I_2 formed in the induction period. Values of this ratio estimated from the combined data of runs 4' and 6', and of runs 7' and 8', increase from 1 to about 15 near the end of the induction period, indicating that the catalytic decomposition resulting from elementary reactions involving intermediates becomes predominant before the start of a pulse or smooth catalysis.

Another quantity of interest is the ratio $(dO_2/dI_2)_{pulse}$ of the amount of O_2 evolved to that of I_2 disappearing during a pulse. For the second pulse in fig. 7 the amount of I_2 disappearing is 1.6×10^{-4} mol $l.^{-1}$ and that of O_2 evolved from 150 ml of reaction mixture is 25.08 ml (stp), hence the value of the ratio is as large as 46.7.

(ii) COMPARISON BETWEEN THE COMPUTER AND EXPERIMENTAL RESULTS

It is readily seen from a comparison between fig. 4 and fig. 5 to 7 that the change in the time courses of runs 1 to 4 with $\bar{k}_1$ is qualitatively similar to that of runs 1' to 5' with $[H^+]_0$. For quantitative comparison with the experimental results we have to carry out computer simulations with actual k_i and $\bar{k}_i$ values and with consideration paid on the effect of $[H^+]$ upon all the k_i and $\bar{k}_i$ values.

Let us make a quantitative check on some significant quantities. The value of $[I_2]$ at point b for run 4 is obtained from fig. 4(c) as

$$30(2.6 \times 10^{-4}[H_2O_2][HIO_3] + 129 \times 10^{-4}[H^+][HIO_3][H_2O_2])/6$$
$$= 30(2.6 \times 10^{-4} \times 0.30 \times 0.40 + 129 \times 10^{-4} \times 0.07 \times 0.30 \times 0.40)/6$$
$$= 7 \times 10^{-4} \text{ (mol } l.^{-1}\text{)},$$

which agrees with the experimental value of 7.6×10^{-4} (run 6'). The value of $(dO_2/dI_2)_{pulse}$ for run 3 is calculated as $0.5(55+35) \times 25/(59-27) = 37$, in approximate agreement with the experimental value of 46.7. Values of $(dO_2/dI_2)_{ind.}$ at the end are about 17 (run 2), 16 (run 3), and 15 (run 4), in agreement with the experimental value of about 15.

From the good qualitative agreement pointed out above it may be concluded that the practical mechanism of fig. 3 is a plausible mechanism actually in operation.

We thank the Robert A. Welch Foundation for supporting the experiment.

[1] W. C. Bray, J. Amer. Chem. Soc., 1921, **43**, 1262.
[2] W. C. Bray and H. A. Liebhafsky, J. Amer. Chem. Soc., 1931, **53**, 38.
[3] M. G. Peard and C. F. Cullis, Trans. Faraday Soc., 1951, **47**, 616.
[4] H. Degn, Acta Chem. Scand., 1967, **21**, 1057.
[5] P. Lindblad and H. Degn, Acta Chem. Scand., 1967, **21**, 791.
[6] J. H. Woodson and H. A. Liebhafsky, Nature, 1969, **224**, 690.
[7] J. H. Woodson and H. A. Liebhafsky, Anal. Chem., 1969, **41**, 1894.
[8] I. Matsuzaki, R. B. Alexander and H. A. Liebhafsky, Anal. Chem., 1970, **42**, 1690.
[9] I. Matsuzaki, J. H. Woodson and H. A. Liebhafsky, Bull. Chem. Soc. Japan, 1970, **43**, 3317.
[10] H. A. Liebhafsky, J. Amer. Chem. Soc., 1931, **53**, 896.
[11] W. C. Bray and A. L. Caulkins, J. Amer. Chem. Soc., 1931, **53**, 44.
[12] H. A. Liebhafsky, J. Amer. Chem. Soc., 1931, **53**, 2074.
[13] E. Abel and K. Hilferding, Z. phys. Chem., 1928, **136**, 186.
[14] R. Furuichi, I. Matsuzaki, R. Simic and H. A. Liebhafsky, Inorg. Chem., 1972, **11**, 952.
[15] R. Furuichi and H. A. Liebhafsky, Bull. Chem. Soc. Japan, 1973, **46**, 2008.
[16] H. A. Liebhafsky and A. Mohammad, J. Amer. Chem. Soc., 1933, **55**, 3977.
[17] E. Eigen and K. Kustin, J. Amer. Chem. Soc., 1962, **84**, 1355.
[18] H. A. Liebhafsky, J. Amer. Chem. Soc., 1932, **54**, 3504.

[19] R. P. Bell and E. Gelles, *J. Chem. Soc.*, 1951, 2734.
[20] I. Matsuzaki, R. Simic and H. A. Liebhafsky, *Bull. Chem. Soc. Japan*, 1972, **45**, 3367
[21] H. Taube and H. Dodgen, *J. Amer. Chem. Soc.*, 1949, **71**, 3330.
[22] A. F. M. Barton and G. A. Wright, *J. Chem. Soc. A*, 1968, 1747.
[23] M. Anbar and S. Guttmann, *J. Amer. Chem. Soc.*, 1961, **83**, 4741.
[24] H. Siebert, *Forts. Chem. Forsch.*, 1967, **8**, 470.
[25] M. Drátovský and L. Pačesová, *Russ. Chem. Rev.*, 1968, **37**, 243.
[26] J. G. Dawber, *J. Chem. Soc.*, 1965, 4111.
[27] J. Higgins, *Ind. Eng. Chem.*, 1967, **59**, (5), 19.
[28] H. Levy and E. A. Baggott, *Numerical Solutions of Differential Equations* (Dover, New York, 1950), p. 92.
[29] G. Nicolis, *Adv. Chem. Phys.*, 1971, **19**, 209.
[30] R. Lefever and G. Nicolis, *J. Theor. Biol.*, 1971, **30**, 267.
[31] I. Prigogine and R. Lefever, *J. Chem. Phys.*, 1968, **48**, 1695.
[32] R. Lefever, *J. Chem. Phys.*, 1968, **49**, 4977.
[33] M. Herschkowitz-Kaufman and G. Nicolis, *J. Chem. Phys.*, 1972, **56**, 1890.
[34] B. Levanda, G. Nicolis and M. Herschkowitz-Kaufman, *J. Theor. Biol.*, 1971, **32**, 283.
[35] J. J. Tyson, *J. Chem. Phys.*, 1973, **58**, 3919.

DISCUSSION REMARKS

Prof. J. Ross (*MIT*) said: The subject of chemical instabilities was first studied by means of macroscopic equations, then statistical theory, and is now being approached by means of the method of molecular dynamics. Some interesting results have been obtained by this method by Ortoleva and Yip at M.I.T. So far they have made calculations on a 32-particle system in which the reactions

$$X + Y \rightarrow 2X, \ T(X)$$
$$X \rightarrow Y, \ v$$
$$Y \rightarrow X, \ \mu$$

occur with the listed rate coefficients. Autocatalytic character is achieved in the first reaction by taking the transition probability for reactive collisions to be dependent on the number of X particles within a mean field radius, or just proportional to the concentration of X. Fig. 1 shows a comparison of the prediction of the calculation (points) with that of the macroscopic equation (solid line) for the steady state mole fraction of $X(\overline{F}_1)$ as a function of v for a value of $\mu = \mu_c$ such that the analogue of the critical isotherm is obtained (for further discussion on this point see *Fluctuations and Transitions at Chemical Instabilities: The Analogy to Phase Transitions* by A. Nitzan, P. Ortoleva, J. Deutch and J. Ross, *J. Chem. Phys.*, 1974, **61**, 1056). Fig. 2 shows a calculation of the mean square fluctuation in the number of X particles, $\Delta^2 = N^{-1} \langle N_x^2 - \langle N_x \rangle^2 \rangle$ as a function of the rate coefficient v. The calculation clearly shows an increase in the magnitude of fluctuations at the critical value of $v = v_c$; this analogue

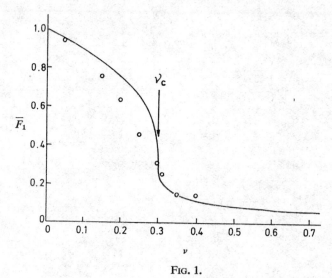

FIG. 1.

of critical opalescence was predicted in the quoted reference. Finally, in fig. 3 the calculations, now made for parameters such that multiple stable steady states are expected, show both fluctuations around one stable steady state, and transitions from one such state to another.

Prof. G. Ehrenfreund and Prof. J. C. Longuet-Higgins (Cambridge). The interest of a model based upon a three-dimensional study of dislocation packings became clear after the phase-space theory of fluctuations was developed by the theory... We believe that it is a relevant as well as in another of the abnormal increase of fluctuations at the critical point, namely, it was predicted for the first time. A molecular dynamics study of equilibrium of molecular dynamic systems, including liquid ones, has been developed during the last years by Fonnow. Some of his results are reported in a recent publication... Further results referring to the molecular model, the Vafer-Lennard-Jones and other static model involving fluctuations, have been obtained by... in a... We would also see that the increasing work of Duoton in the limits to an independent confirmation of our predictions.

Prof. A. Rice (Wisconsin). Referred to the nature of the relation of the crucial wave number of the emerging fluctuations relative to the size of the system, it may be useful to contrast a qualitative distinction between two types of instability breaking instabilities. As the system breaks to instability it is driven out of equilibrium,

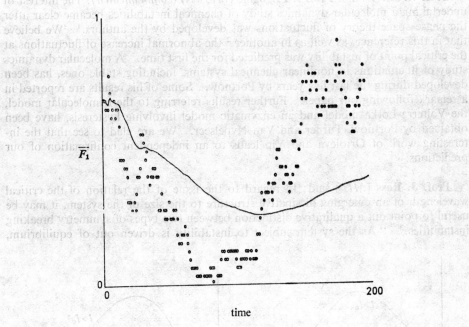

time

Fig. 2.

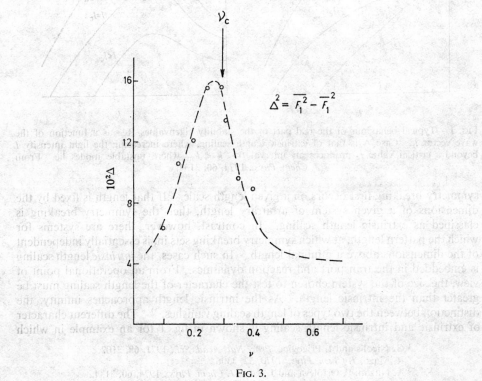

$$\Delta^2 = \overline{F_1{}^2} - \overline{F_1}^2$$

Fig. 3.

Prof. G. Nicolis and **Prof. I. Prigogine** (*Brussels*) (*communicated*): The interest of undertaking a molecular dynamics study of chemical instabilities became clear after the phase space theory of fluctuations was developed by the authors.[1]   We believe that in this reference, as well as in another [2] the abnormal increase of fluctuations at the critical point of instability was predicted for the first time.   A molecular dynamics study of fluctuations in nonlinear chemical systems, including stable ones, has been developed during the last few years by Portnow.   Some of his results are reported in a remark following our paper.   Further results referring to the trimolecular model, the Volterra-Lotka model and an enzymatic model involving hysteresis, have been obtained by Portnow, Turner and Van Nypelseer.   We are glad to see that the interesting work of Ortoleva and Yip leads to an independent confirmation of our predictions.

Prof. J. Ross (*MIT*) said; In regard to the issue of the relation of the critical wavelength of an emerging dissipative structure to the size of the system, it may be useful to point out a qualitative distinction between two types of symmetry breaking instabilities.[3]   " As the system subject to instability is driven out of equilibrium,

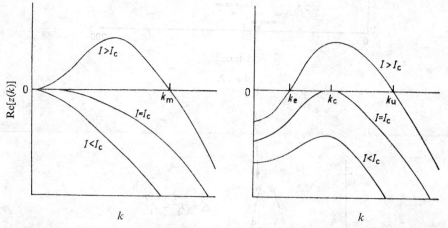

Fig. 1.—Typical behaviour of the real part of the stability eigenvalues Rez as a function of the wave vector k.   Case A is that of extrinsic length scaling, where increasing the light intensity I beyond a critical value I_c produces an interval $0 < k < k_m$ where unstable modes lie.   From *J. Chem. Phys.*, 1974, **60**, 3134.

symmetry breaking first occurs on a given length scale.   If that length is fixed by the dimensions of a given system of arbitrary length, then the symmetry breaking is classified as extrinsic length scaling.   In contrast, however, there are systems for which the pattern length, at which symmetry breaking sets in, is essentially independent of the dimensions above a minimal length.   In such cases, the *intrinsic* length scaling is embedded in the transport and reaction dynamics.   From an operational point of view, the size of the system chosen to test the character of the length scaling must be greater than the intrinsic length.   As the intrinsic length approaches infinity, the distinction between the two types of length scaling vanishes. "   The different character of extrinsic and intrinsic length scaling is shown in fig. 1 for an example in which

[1] G. Nicolis and I. Prigogine, *Proc. Nat. Acad. Sci.*, 1971, **68**, 2102.
[2] R. Mazo, *J. Chem. Phys.*, 1970, **52**, 3306.
[3] A. Nitzan, P. Ortoleva and J. Ross, *J. Chem. Phys.*, 1974, **60**, 3134.

instability is brought about by increasing light intensity. In the quoted article examples of both types of behaviour are given.

Prof. G. Nicolis and **Prof. I. Prigogine** (*Brussels*) (*communicated*): We believe that a qualitative distinction between " extrinsic " and " intrinsic " length scales may be misleading. In the terminology of our paper, the question at issue is whether at the first bifurcation predicted by the characteristic eqn (3.2), the wave number of the bifurcating solution is zero or finite (see comments following eqn (3.2) of our paper). Obviously, any finite wavenumber k satisfying the characteristic equation will depend solely on the intrinsic parameters of the system, as the size of the latter does not appear in this equation. Nevertheless, if one requires that the unstable modes be

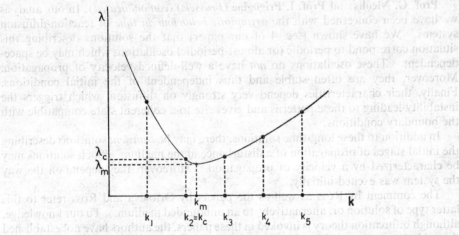

FIG. 1.—Marginal stability curve as a function of the wave number. k_m: a wave number k predicted by linear stability analysis. k_i: values of k compatible with the boundary conditions. $k_c = k_2$: critical wave number corresponding to the first bifurcating solution.

compatible with the boundary conditions, one will find a set of values $\{k_i\}$ of k which are *directly related* to the size of the system (see eqn (3.3a) to (3.3c) of our paper). Thus, near the critical point, the bifurcating solution will actually be dominated by a value k_c of k which agrees with the boundary conditions (and thus is related to the size of the system) and which is the closest one to the value k_m predicted by linear stability analysis. Computer simulations confirm this point entirely. The situation is represented schematically on fig. 1. For a more detailed description we refer to Lefever,[1] and to Nicolis and Auchmuty.[2]

In that sense, therefore, the length scale associated with symmetry breaking is always *extrinsic*, as long as the dissipative structure can extend throughout the system. The only *intrinsic* length scaling that appears to be possible is related to the possibility of localization of the dissipative structure, as discussed in section 3(ii) of our paper.

Dr. P. Ortoleva and **Prof. J. Ross** (*MIT*) said: The physical concept of propagation velocity in finite systems is familiar in atomic and nuclear scattering, shock waves and acoustic wave packet propagation, and waves in the Zaikin–Zhabotinsky reagent. As long as the characteristic structural dimensions of the propagating phenomena

[1] R. Lefever, *Thesis* (University of Brussels, 1970).
[2] G. Nicolis and J. F. G. Auchmuty, *Proc. Nat. Acad. Sci.*, 1974, **71**, 2748.

(i.e., the width of a front; (see fig. 4 in the article by Nitzan, Ortoleva and Ross, this Symposium) is much less than the dimensions of the system, then on the basis of causality one expects that boundaries of the system are not important in the concept of a velocity or of propagation (until the disturbance reaches a wall or another disturbance emitted from the wall).

Autonomous one-dimensional centres of wave emanation as well as circular and standing waves in two dimensions have been considered by us.[1] These phenomena are analysed with the aid of bifurcation theory and an alternative scheme, " phase diffusion theory ", based on the existence of a chemical oscillation in the rate mechanism.[2]

Prof. G. Nicolis and **Prof. I. Prigogine** (*Brussels*) (*communicated*): In our analysis we have been concerned with the *asymptotic behaviour in time* of reaction-diffusion systems. We have shown (sec. 4 of our paper) that the solutions describing this situation correspond to periodic (or almost-periodic) oscillations which may be space-dependent. These oscillations do *not* have a well-defined velocity of propagation. Moreover, they are often stable and thus independent of the initial conditions. Finally, their characteristics depend very strongly on diffusion, which triggers the instability leading to these patterns and gives rise to a coherent state compatible with the boundary conditions.

In addition to these long-time solutions, there may be *transient* solutions describing the initial stages of propagation of a disturbance in the medium. Such solutions may be characterized by a velocity of propagation. Moreover, they depend on the way the system was excited initially.

The comment by Ross as well as the papers by Ortoleva and Ross refer to this latter type of solution or, alternatively, to an unbounded medium. To our knowledge, although bifurcation theory is invoked in these papers, the authors have not established the existence or the stability of the various types of wave forms they list. Now, the requirement of stability is always a very stringent one : a solution whose existence is suggested by linear analysis can be rejected by the system on the basis of stability considerations. Thus, it is often necessary to ensure that a certain solution appears at the point of the *first* bifurcation from a reference state, in order to be able to guarantee its stability. This question is discussed in some detail in our paper as well as in the paper by Balslev and Degn at this Symposium.

Dr. M. Kaufman-Herschkowitz (*Brussels*) said: I would like to add some further remarks concerning the comparison between analytical calculations and computer simulations for the trimolecular reaction scheme.

From the linear stability analysis, performed for a bounded medium subjected to zero flux or fixed boundary conditions, one can infer the evolution of the homogeneous steady state, beyond a critical point, to new space dependent solutions characterized by a finite wavelength. Mathematically these transitions can be understood as a phenomenon of branching of the solutions of the non linear partial differential equations describing the system. Bifurcation theory enables one to construct analytically the form of the bifurcating solutions.[3]

However, the analytical expressions one can construct are limited to the neighbourhood of the marginal stability point. Computer simulations allow one to verify their predictions but also to investigate the behaviour of the system as one enters

[1] P. Ortoleva and J. Ross, *J. Chem. Phys.*, 1974, **60**, 5090.
[2] P. Ortoleva and J. Ross, *J. Chem. Phys.*, 1973, **58**, 5673.
[3] G. Nicolis and J. F. G. Auchmuty, *Proc. Nat. Acad. Soc.*, 1974, **71**, 2748.

more deeply into the unstable region. They show that new features can arise in these conditions, and I would like to discuss one of these features as an example.

Let us consider the case of fixed boundary conditions and use B as bifurcating parameters. Comparison between the analytical calculations (expression (3.4) in the paper by Nicolis and Prigogine) and computer simulations shows:

(i) that near the critical point ($B \simeq B_c$) the two approaches agree very well (fig. 1). The new steady state reflects to a good approximation the form of the critical mode. One observes in addition a certain distortion due to the contributions of the non-linear terms, which act to enhance the successive maxima (successive minima) from boundary at $r = 0$ to boundary at $r = L$.

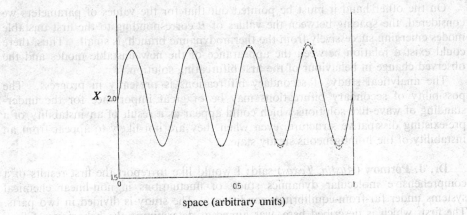

FIG. 1.—Steady state dissipative structure for fixed boundary conditions and $B \simeq B_c$. Dashed line: analytical curve; full line: result of the numerical integration on a digital computer. The following numerical values of the parameters have been chosen: $A = 2$, $L = 1$, $D_X = 1.6 \times 10^{-3}$, $D_Y = 6.0 \times 10^{-3}$, $B = 4.17$. The critical wavenumber is $n_c = 8$ and $B_c = 4.133$. The boundary values for X and Y are $X = A = 2$, $Y = B/A = 2.085$.

(ii) that the agreement between analytical and numerical results becomes poor when the calculations are performed for values of B which do not belong to the *direct* neighbourhood of the critical point (fig. 2).

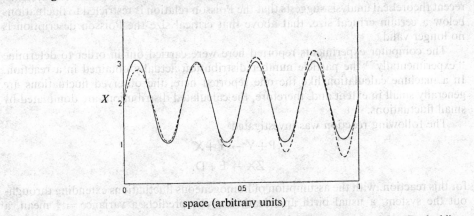

FIG. 2.—Steady state dissipative structure for fixed boundary conditions and $B > B_c$. Dashed lines analytical curve; full line: result of the numerical integration on a digital computer. $B = 4.5$; other parameters are as in fig. 1.

Moreover, one observes from the computer simulations that although the critical mode determines the number of extrema of the final steady state structure, the spatial asymmetry is now quite different as for $B \simeq B_c$: near the boundaries, the maximum is enhanced and minimum diminished. A great number of calculations performed for various initial conditions show that this new type of asymmetry does not depend on the initial perturbation.

These observations suggest the possibility of occurrence, beyond a certain distance from the critical point, of a *secondary bifurcation* which could be responsible for the observed change in behaviour of the solution: the first bifurcating space dependent solution becomes unstable and one observes the appearance of another space dependent dissipative structure.

On the other hand it must be pointed out that for the values of parameters we considered, the spacing between the values of B corresponding to the first unstable modes emerging successively from the thermodynamic branch, is small. Thus, there could exist a relation between the appearance of the new unstable modes and the observed change in behaviour of the first bifurcating solution.

The analytical study of secondary bifurcations is presently in progress. The possibility of secondary bifurcations may be of great importance for the understanding of wave-like solutions which could appear as a result of an instability of a pre-existing dissipative structure, even when they are not likely to appear from an instability of the homogeneous steady state.

Dr. J. Portnow (*Austin, Texas*) said: I would like to report the first results of a comprehensive molecular dynamics study of fluctuations in non-linear chemical systems under far-from-equilibrium conditions. The study is divided in two parts. The first, which is described here, was aimed at determining the behaviour of fluctuations under far-from-equilibrium conditions but far from instability points. The second part of the study, which is still in progress, is aimed at determining the growth of fluctuations at instability points, the mechanism of instability, and the critical size of fluctuations necessary to initiate instability.

Fluctuations in equilibrium systems, linear and nonlinear alike, are described by the Einstein relation. Thus, for a single intermediate, small fluctuations in the particle number density follow a Poisson distribution. For linear systems far from equilibrium, the Poisson relation is still maintained. But, for nonlinear systems, recent theoretical analysis suggests that the Poisson relation is restricted to fluctuations below a certain critical size, that above that critical size the Poisson description is no longer valid.

The computer experiments reported here were carried out in order to determine " experimentally " the particle number distribution actually obtained in a reaction. In a machine calculation like the one reported here, the observed fluctuations are generally small in extent and, therefore, the calculated distributions are dominated by small fluctuations.

The following reaction was investigated*

$$P + Y \rightarrow Y + X$$
$$ZX \rightarrow E + D.$$

for this reaction, with the assumption of homogeneous fluctuations extending throughout the system, a usual birth and death analysis predicts a variance $= \frac{3}{4}$ mean, a

---

* Several additional reactions were investigated as well and will be reported in a forthcoming paper. See also a report in *Phys. Letters*, in press.

significant deviation from Poisson behaviour. On the other hand, a local phase space description of the reaction (in which the local character of collisions is made explicit) predicts Poisson behaviour for small fluctuations.

Initially 75 hard spheres with Ar mass and radius are randomly placed in a cubic box, with $\rho = 0.001\ 783\ 7\ cm^{-3}$. They are given identical speeds, $\mu = [3kT/m]^{\frac{1}{2}}$ with $T = 273$ K, and random velocity directions. They interact with each other through hard sphere collisions, and undergo specular reflections off the walls of the box. Once thermal equilibrium, as measured by the H function, as attained, 25 spheres are given P identity, 25 X identity and 25 Y identity. Then, whenever two particles collide, their identities are changed in accordance with the reaction mechanism. There are neither activation energy nor configuration requirements for the reaction so that all collisions between reacting particles are reactive and all rate constants are equal. By neglecting all reactive criteria, and since the reacting particles have no internal degrees of freedom, the Maxwell–Boltzmann velocity distributions are not upset by the reactive collisions.

The number of P particles is kept constant at 25, and each time E and D particles are produced they are removed immediately from the reaction box. Whenever a P particle reacts, a new one is introduced into the box at a random position and given a speed $u = (3kT/m)^{\frac{1}{2}}$ with random velocity directions.*

The reaction was followed for 3.146×10^{-8} s during which time there were 1 941 reactive and 4 067 unreactive collisions. After 30 observations the mean and variance were

calculated mean	expected mean
25.2	24.5
calculated variance	Poisson variance
23.3	24.5

and, in terms of the χ-square test for goodness of fit the observed distribution showed no significant deviation from a Poisson distribution.

These results are strong evidence that a Poisson distribution is the correct description of the fluctuation behaviour in the present reaction, at least. Since the homogeneous birth and death analyses have predicted non-Poisson behaviour, the "experiments" dramatize the need for a local description of fluctuations in non linear far from equilibrium systems.

Dr. M.-L. Smoes (*Dortmund*) (*communicated*): 1. The claim is again made in the paper of Nicolis and Prigogine that steady spatial patterns have been observed in initially homogeneous chemical reactions. Although the authors do not mention any particular reaction, I presume that they might have in mind the case of the Zhabotinskii reaction. In this oscillatory system, nonpropagating waves have been reported by Busse [4] and a spatial dissipative structure is described by Kaufman-Herschkowitz.[5]

My own experience with the Zhabotinskii system leads me to think that such steady spatial structures have not really been found. The two observations reported are not supported by quantitative measurements of the position of the wavefronts as

* Boundary conditions requiring diffusion in from the walls have been used as well.
[1] G. Nicolis and A. Babloyantz, *J. Chem. Phys.*, 1969, **51**, 2632.
[2] G. Nicolis and I. Prigogine, *Proc. Nat. Acad. Sci.*, 1971, **68**, 2102; G. Nicolis, M. Malek-Mansour, K. Kitahara and A. Van Nypelseer, *Phys. Letters*, 1974, **48A**, 217.
[3] G. Nicolis, *J. Stat. Phys.*, 1972, **6**, 195; G. Nicolis, P. Allen and A. Van Nypelseer, personal communication.
[4] H. Busse, *J. Phys. Chem.*, 1969, **73**, 750.
[5] M. Kaufman–Herschkowitz, *Compt. Rend.*, 1970, **270C**, 1049.

a function of time. Is it not possible that those apparently immobile bands of oxidation are in fact slowly moving? Indeed, the speed of propagaion of the Zhabotinskii waves is known to depend on the period of bulk oscillations.[1] One can thus expect quasisteady structures in the Zhabotinskii system for an appropriate choice of the initial concentrations of the reactants.

I would like to suggest that steady structures in chemical systems have not yet been demonstrated, and that further claims should be based on careful quantitative measurements.

2. The mathematical models for oscillatory chemical reactions may lead to high expectations in terms of biological problems. However, the experimental facts do not seem to confirm these hopes. The chemical reactions by themselves do not have all the controlling properties postulated and required for biological applications. This is at least what must be concluded in the case of the spatio-temporal phenomena observed in the Zhabotinskii system. The leading centres, their period, and the speed of propagation of the wavefronts are all independent of the size of the system but depend on the initial concentrations of the reactants and on temperature. The spatial and temporal distribution of the centres appears to be random.

These experimental observations have been taken into account in our model of phase waves.[2] The model applies not only to the Zhabotinskii system but also to a large class of chemical oscillatory reactions in which the period depends on concentrations and/or tempreature. Although no structures have yet been observed in glycolysis, the dependence of the period of glycolytic oscillation on temperature and concentrations suggests the possibility of formation of phase waves in this case too.

Prof. G. Nicolis (*Brussels*) (*communicated*): 1. The claim attributed to us by Smoes is nowhere made in our paper. We are well aware of the controversial status of the horizontal bands in the Belousov-Zhabotinski reaction. The main purpose of our paper was to provide a mathematical classification of the various structures that become *possible* beyond an instability of the thermodynamic branch. In this respect, the existence of stable steady state solutions of reaction-diffusion equations is established rigorously for the first time. It is not unreasonable to expect that such solutions will describe the behaviour of real chemical systems under certain conditions.

2. It is important to distinguish between long-time behaviour and transient behaviour. In our analysis we have been concerned with the *asymptotic behaviour in time* of reaction-diffusion systems. The qualitative properties of dissipative structures like the dependence on size, the independence of the initial conditions, etc... hold only for the long time solutions. Smoes' remark refers on the other hand to the *transient* behaviour of an initial disturbance acting locally on a chemical system. Obviously, this behaviour will depend strongly on the initial conditions. Moreover, it will not be related to the size of the system. Eventually, however, the disturbance will reach the boundaries and it will evolve to one of the long-time solutions treated in our paper. Thus, there is no contradiction between Smoes' remark and our results.

Dr. M-L. Smoes (*Dortmund*) (*communicated*): (1) I refer to the claim made in the first paragraph of the Introduction. Since only theoretical steady-state patterns are well established, the wording of that paragraph is ambiguous.

The main problem concerning the physical realization of stable structures in chemical systems is the following. According to the theory of Nicolis and Prigogine,

[1] M-L. Smoes, unpublished work.
[2] M-L. Smoes and J. Dreitlein, *J. Chem. Phys.*, 1973, **59**, 6277.

Fig. 1.—System 1, 825 s after start.　　　Fig. 2.—System 2, 825 s after start.

Fig. 3.—After 22 min.　　　　　Fig. 4.—After 34 min.

Fig. 5.—After 52 min.　　　　Fig. 6.—After 63 min.

Fig. 3-6.—Progressive development of the waves.

To face page 75]

both constant parameters and appropriate nonlinearities in the rate equations are required. The two conditions are in practice mutually exclusive. Indeed, the nonlinearities are met only in very complicated chemical systems for which the constancy of the parameters in the rate equations becomes impossible. Even under mixing conditions, irregularities can be observed in the successive periods of the Zhabotinskii oscillations. Sudden variations in the period, which may be as large as 50 %, are observed [1] in homogeneous oscillations with long period. The irregularities are not observed in systems with short periods. Such anomalies are expected as a result of the dependence of the period on concentrations and temperature in the Zhabotinskii system.[1, 2] They are also expected in the glycolytic system.[3] This is the important fact which is not taken into account in the paper by Nicolis and Prigogine and which is accounted for in the work of Smoes and Dreitlein.[4]

(2) Transient behaviour is the only behaviour ever observed with the Zhabotinskii structures, even when the disturbances have reached the boundaries of the system. In our opinion, this is due to the instability of the period of oscillations with respect to the fluctuations in concentrations and/or temperature.[1, 4] Such an instability precludes the system from evolving toward one of the solutions treated by Nicolis and Prigogine.

To demonstrate the point, I introduce two sets of pictures. In the first set, is shown the influence of the homogeneous bulk period on the general aspect of the waves. System 1 (fig. 1) and System 2 (fig. 2) have a 2 : 3 ratio for the sulphuric acid concentration resulting in a much shorter period of bulk oscillations for System 2. The size of the Petri dishes, the thickness of the layer and the time elapsed since the start of the reaction (825 s) are identical.

In the second set, fig. 3-6, is shown an example of the renewal of the transients in the Zhabotinskii waves. Between fig. 3 and fig. 6, the waves from one centre progressively fill the whole space available, but new centres appear later and new transients develop. No long-time stable structures are observed. Our model of phase waves [4] accounts reasonably for the absence of a long-time stable solution in this chemical system.

Dr. J. S. Turner (*Austin, Texas*) said: As Nicolis and Prigogine point out, a natural mechanism for testing the stability of a macroscopic state exists in every many-body system in the form of fluctuations. Near a transition point, therefore, a macroscopic instability may be *nucleated* spontaneously by a small volume element which suddenly becomes unstable through a fluctuation. The response of the surrounding medium to such an evolving subvolume will be to damp the fluctuation by diffusion. If the unstable region is sufficiently large, however, diffusion may serve instead to propagate the disturbance. The fluctuation will then grow, leading ultimately to a new regime of macroscopic behaviour.

In order to make this notion quantitative, Nicolis and coworkers have proposed a stochastic model in which the effect of diffusion on a fluctuating volume element is treated in an average way.[5] If X is the number of molecules in a small volume (for

[1] M-L. Smoes, *Toward a Mathematical Description of the Phase Waves*, submitted for publication.
[2] *Oscillatory Processes in Biological and Chemical Systems* (Nauka, Moscow, 1967), p. 181 ff.
[3] B. Hess and A. Boiteux, *Ann. Rev. Biochem.* 1971, **40**, 237; A. Betz and B. Chance, *Arch. Biochem. Biophys.* 1965, **109**, 579.
[4] M-L. Smoes and J. Dreitlein, *J. Chem. Phys.*, 1973, **59**, 6277.
[5] G. Nicolis, M. Malek–Mansour, K. Kitahara and A. Van Nypelseer, *Phys. Letters*, 1974, **48A**, 217.

convenience consider a single reactive degree of freedom), then the appropriate stochastic master equation has the form, in number-of-particles space,

$$dP(X;t)/dt = \mathcal{R}\{X,P(X;t)\}+\mathcal{D}\langle X\rangle[P(X-1;t)-P(X;t)]+$$
$$\mathcal{D}[(X+1)P(X+1;t)-XP(X;t)]. \qquad (1)$$

Here the first term on the right-hand side denotes the reactive contribution, while the second and third account for diffusion into and out of the subvolume, respectively. The presence of the mean value $\langle X\rangle = \sum_x XP(X;t)$ in the second term expresses the fact that initially the system is globally homogeneous. The diffusion parameter is defined by $\mathcal{D} \equiv D/L^2$, with D a Fick-type diffusion coefficient and L a characteristic length over which the fluctuation maintains an approximately coherent character. [For more details see ref. (1).]

For chemical systems in which a single homogenoeus steady state exists, becoming unstable beyond a critical value of a system parameter (usually a measure of affinity), eqn (1), together with the definition of $\mathcal{D}$, yields a critical coherence length L_c giving the minimum size necessary to form an unstable nucleus in an initially homogeneous system.[1]   If more than one homogeneous steady state is accessible to the system, then several qualitatively new features are possible.   Typically, the macroscopic kinetic equations yield a region of a system parameter (e.g., λ) in which two such states are simultaneously stable and, therefore, predict a hysteresis in the transition between the two stable branches of the steady state solution (e.g., arrows, fig. 1).   In analogy to first-order phase transitions, however, one may ask whether for any value λ only one state is actually stable, the other, a *metastable* state, being unstable with respect to finite fluctuations occurring spontaneously in the medium.[2, 3]   To investigate this possibility, Lefever, Prigogine and I have applied eqn (1) to a simple chemical model which exhibits multiple steady states.

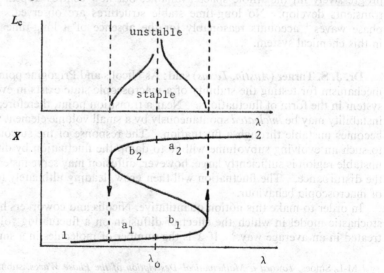

FIG. 1.—Critical coherence length L_c characterizing the stability of states on the metastable portions b_1, b_2 of the macroscopic coexistence region.

[1] G. Nicolis, M. Malek–Mansour, K. Kitahara and A. Van Nypelseer, *Phys. Letters*, 1974, **48A**, 217.
[2] J. S. Turner, *Phys. Letters*, 1973, **44A**, 395; *Bull. Math. Biol.*, 1974, **36**, 205.
[3] J. S. Turner, *Adv. Chem. Phys.*, 1975, **22**, 63.

Consider a Langmuir type of exchange process in which X atoms are adsorbed on a surface region of N binding sites. The cooperative nature of the adsorption-desorption process is expressed in the equilibrium constant for this reaction, which depends explicitly on the surface density, X/N, of adsorbed atoms, and on a parameter, λ. The macroscopic kinetic equations give an S-shaped steady-state curve as a function of λ, the middle branch of which is unstable (fig. 1). Applying now eqn (1) to a small surface element (finite N) in this system (with $\mathscr{D} \sim L^{-1}$ for surface diffusion), we find, for $\mathscr{D} = 0$, a unique stationary distribution which is bimodal in the multiple steady state region. Hence, the macroscopic predictions are recovered as far as small homogeneous fluctuations are concerned. The ratio of the peak heights is large except very near λ_0, the point at which the heights are equal. Away from λ_0, therefore, mean values $\langle X \rangle$ equal roughly $X_a(\lambda)$, the most probable value. For $\mathscr{D} \neq 0$, the master equation becomes nonlinear due to the presence of $\langle X \rangle$, implying the possibility of more than one stationary distribution. For $\mathscr{D} > 0$ small (L large), there remains a unique steady solution, but for $\mathscr{D}$ greater than a critical value $\mathscr{D}_c(\lambda)$ [or $L_c(\lambda)$], two such solutions are found, depending on the choice of initial distribution. For all $\mathscr{D} \geqslant 0$, an initial $\langle X \rangle$ near $X_a(\lambda)$ yields a final distribution with $\langle X \rangle \sim X_a(\lambda)$. For $\mathscr{D} < \mathscr{D}_c(\lambda)$, the same final distribution results from initial mean values near the smaller peak [at $X_b(\lambda)$] as well, implying that fluctuations are too large [$L > L_c(\lambda)$] for that branch to be stabilized. If $\mathscr{D} > \mathscr{D}_c(\lambda)$, however, the latter initial condition produces a stable stationary distribution having $\langle X \rangle \sim X_b(\lambda)$. This means that initial states on the branch " b " will not be de-stabilized by fluctuations of size $L < L_c(\lambda)$. These results are displayed schematically in fig. 1, where the stability properties of the *metastable* states " b " are indicated in a plot of the critical coherence length L_c. If fluctuations of size $L \geqslant L_c^{max} \equiv \max_\lambda L_c(\lambda)$ occur in the medium, then these metastable states will not be observed, any transition between branches 1 and 2 occurring at the point λ_0. If the largest fluctuations appearing frequently are smaller than L_c^{max}, then a corresponding portion of each metastable branch " b " will be stabilized, and an apparent hysteresis in the transition point recovered. In this case, the transition may still be *induced* by *external* perturbations of an appropriate type.

In summary, by treating explicitly the occurrence of spontaneous localized fluctuations we have verified the existence of metastable states and hence of a kind of first-order phase transition for chemical schemes exhibiting multistationary states. Moreover, we have obtained a quantitative characterization of metastability in terms of a critical coherence length of fluctuations necessary for the *spontaneous* formation of a critical nucleus of one " phase " within another. The evolution of an *existing* nucleus of one *pure* phase within another has been treated by Schlögl,[1] and is discussed in the paper by Nitzan, Ortoleva, and Ross in this Symposium. The idea of a non-equilibrium analogue to the Maxwell construction of equilibrium first-order phase transitions is also considered by these authors, and has been examined from the point of view of nonlinear irreversible thermodynamics by Kobatake[2] and by Turner.[3]

Mr. M. Collins and **Dr. R. G. Gilbert** (*University of Sydney*) said: " Trigger waves " have been described by Field and Noyes[4] as a concentration gradient moving under diffusion in a system containing effectively only a single autocatalytic reaction. On

[1] F. Schögl, *Z. Phys.*, 1972, **253**, 147.

[2] Y. Kobatake, *Physica*, 1970, **48**, 301.

[3] J. S. Turner, *Adv. Chem. Phys.*, 1975, **29**, 63.

[4] R. J. Field and R. M. Noyes, *J. Amer. Chem. Soc.*, 1974, **96**, 2001.

the other hand, in their paper at this Symposium, Nitzan, Ortoleva and Ross describe this as the moving boundary between two steady states of a multiple steady state system.   As Field and Noyes themselves point out, the former description is obviously too simple, as it allows the product concentration behind the propagating front to grow unbounded, resulting in excessively large propagation velocities and a quantitatively inaccurate picture.   Consider the single autocatalytic reaction $A + B \xrightarrow{k} 2B$, far from equilibrium in a closed isothermal system.   In one dimension, the equations of motion are

$$\partial A/\partial t = -kAB + D_A \partial^2 A/\partial x^2,$$
$$\partial B/\partial t = kAB \;\;\;\; + D_B \partial^2 B/\partial x^2.$$

In this system, the product concentration must remain finite and we avoid any unrealistic effects arising from infinite concentrations.   We have shown [1] that approximate analytic solutions in space and time to these equations may be obtained for $D_A \geqslant D_B$.   This condition ensures that the pulse or wave in B is significantly sustained by the system.   The analytical solutions then show explicitly that the velocity of the pulse maximum is never greater than its value in the absence of reaction, and if $D_A > B_B$, depletion of the reactants, by back-diffusion in front of the pulse, results in the velocity falling *below* that of simple diffusion.   Hence it appears unlikely that a single autocatalytic reaction can account for rapid pulse transmission in an unstable excitable system.   Thus two questions arise : (i) would complete numerical solution of the complete coupled partial differential equations for, e.g., the Oregonator mechanism account for such effects, or (ii) does the Oregonator mechanism allow multiple steady states, and if so is the pulse the boundary between these?

Prof. R. M. Noyes (*Oregon*) said: Gilbert has examined two equations in A and B associated with the reaction $A + B \rightarrow 2B$.   He asserts they represent autocatalysis and that the velocity *of the pulse maximum* is never greater than that for simple diffusion in the absence of reaction.   I see nothing surprising about this result.   The reaction that produces B at a rate proportional to its concentration simultaneously destroys a species needed to produce B; there is both autocatalysis and autoinhibition.   The effects are equal at the pulse maximum where A and B have very nearly equal concentrations.

I predict that Gilbert will find a very different result if he looks *at the leading edge* of B advancing into pure A.   I am sure he will find that edge (defined perhaps as $B = 0.01\,A$) is advancing faster than would be expected by simple diffusion without reaction.   The equations Gilbert is examining can hardly generate a wave front moving with uniform velocity along its profile.

Gilbert's equations could be modified to something very like ours if his A-equation were left unchanged and his B-equation were modified to read

$$\partial B/\partial t = -kAB + k'B - k''B^2 + D_B \partial^2 B/\partial x^2.$$

Calculations by Dr. J. D. Murray at the Mathematical Institute in Oxford show that these equations do develop a band profile that moves with uniform velocity.   The term in $k'B$ provides the autocatalysis, the term in kAB prevents the leading edge from running ahead of the main front the way it does in Gilbert's equation, and the term in $k''B^2$ ensures a finite concentration of B behind the front.

I am afraid that Gilbert confused our full and simplified equations in the paper to which he refers.   The full equations ((3) and (4) in that paper) *do* describe a moving

[1] M. A. Collins and R. G. Gilbert, *Chem. Phys.*, 1974, **5**, 49.

boundary between two steady states just as is claimed by Nitzan, Ortoleva, and Ross. The concentration of one variable rises by a factor of 10^5 during passage between those two states. In the same paper we made a very crude effort to obtain an approximate analytical solution for the full equations *just behind the leading edge* of the advancing front. Our simplified eqn (7) is essentially equivalent to omitting the kAB and $k''B^2$ terms in the equation above. We pointed out that this simplified equation would let B increase indefinitely behind the wave front, and we restricted any application to concentrations at least two powers of ten less than that attained in the upper steady state. We still believe that approximation was applicable *to the concentration range for which it was intended*.

Our full equations (with only two variables) generate a migration of a boundary between two steady states that are found in different regions of space. The Oregonator (with three variables) in a uniform space has a single true steady state which is unstable with respect to a limit cycle trajectory involving repeated switching between two pseudostationary states. Each of these pseudostationary states evolves so as to switch to the other one. I have complete confidence that a coupling of the full Oregonator to diffusion will model repeating trigger waves advancing in space. I am not attempting such computations myself, but I understand they are in progress elsewhere.

Dr. B. L. Clarke (*Alberta*) said: Whether or not the Belousov-Zhabotinski reaction can be modelled by the Oregonator is an experimental question. On the other hand, whether or not the Oregonator models the detailed kinetic mechanism given by Field, Körös, and Noyes in ref. (5) is a purely mathematical question. My paper, *Stability of Topologically Similar Chemical Networks, J. Chem. Phys.*, 1975, **62**, 3726, contains theorems proven to answer the latter question. I have developed a model of the Belousov-Zhabotinski system which is related to the detailed kinetics such that these theorems connect the stability of the model with the stability of the detailed mechanism. Surprisingly, this model has stable steady states for all rate constants and all concentrations of the pool chemicals! The stability of this model almost proves that the mechanism in ref. (5) is never unstable.

Noyes realized that the Oregonator and his detailed mechanism were only consistent when $f = \frac{1}{2}$. However, then the Oregonator is always stable. He has proposed that the reaction

$$HOBr + HCOOH \rightarrow Br^- + CO_2 + H_2O + H^+ \qquad (A)$$

be added to the original mechanism to increase f.

The expanded detailed mechanism can also be represented by the model I used and calculations on this model have mapped the stability domains analogous to fig. 1 in many dimensions. From the nature of these domains it is plausible that the model has limit cycle oscillations even in the case when reaction (A) is not strong enough to make the steady state unstable. Therefore, one should be cautious of using steady state stability as an argument for the validity of a model or mechanism— especially when there are nearby unstable pseudosteady states.

Prof. R. M. Noyes (*Oregon*) said: Clarke is correct that the original FKN mechanism [1] of the Belousov-Zhabotinskii reaction does not generate an unstable steady state. That mechanism was developed from qualitative arguments by experimental chemists who did not then know how to do a stability analysis of a steady

---

[1] R. J. Field, E. Körös and R. M. Noyes, *J. Amer. Chem. Soc.*, 1972, **94**, 8649.

state. We subsequently developed the Oregonator [1] model to incorporate the essential features of our mechanism. Examination of that model showed us that the stoichiometry of our original mechanism would have generated a stable steady state; Clarke has independently realized the same fact.

At the same time that the theoretical analysis was demonstrating the inadequacy of our original mechanism, experimental evidence was requiring a modification to add oxybromine oxidation of the formic acid that was inert to oxidation by cerium (IV). The revisions generate almost precisely the stoichiometry corresponding to maximum sensitivity of the Oregonator model; a paper has been accepted by *J. Amer. Chem. Soc.*

I am not sure of the significance of Clarke's examination of the revised model when his reaction (A) is "not strong enough" to make the steady state unstable. Formic acid will continue to accumulate until its average rates of formation and destruction during any sufficient period are equal. I believe the argument based on Oregonator stoichiometry will then indicate an unstable steady state. I am very interested in his suggestion that limit cycle oscillations might commence even before formic acid had accumulated enough to render the steady state unstable to a conventional linear analysis.

I am indebted to Clarke for pointing out it is possible to have a locally stable limit cycle trajectory around a locally stable steady state. He assures me this is possible even if all processes are unimolecular or bimolecular provided the concentrations of enough species are varying simultaneously. Much of the previous theoretical work has been concerned with only two simultaneous variables; we badly need the sorts of theorems Clarke is trying to develop for multi-component systems.

Dr. B. L. Clarke (*Alberta*) (*communicated*): The mechanism of the Zhabotinski system allows a number of *independent* overall reactions to be constructed. Those constructed from the mechanism of ref. (5) have formic acid as a product. Reaction A is part of another overall reaction in which formic acid is an intermediate and CO_2 is a product instead. When both types of overall reactions are occurring together the status of formic acid is ambiguous. The rate of Br^- production compared to $HBrO_2$ production is then not determined by stoichiometry alone but by the relative rates of the various possible overall reactions as well.

The stability of pseudosteady states plays an important role in the trajectory calculations shown in fig. 2-5 of the paper by Field and Noyes. By pseudosteady state stability I mean the stability of the matrix M, which appears in the linearization of eqn (11)-(13) about *arbitrary* values of α, η, ρ.

$$\frac{d}{dt} \begin{pmatrix} \Delta\alpha \\ \Delta\eta \\ \Delta\rho \end{pmatrix} = \begin{pmatrix} -(s\eta+s-2qs\alpha) & (s-s\alpha) & 0 \\ -\eta/s & -(1+\alpha)/s & \rho/s \\ w & 0 & -w \end{pmatrix} \begin{pmatrix} \Delta\alpha \\ \Delta\eta \\ \Delta\rho \end{pmatrix}$$

When M is stable, it is often possible to calculate the trajectory from simplified equations of motion. Otherwise the unstable normal mode of M plays a role in the dynamics.

First, we calculate the equation of the long slow decline in η shown in fig. 2. During this motion α and ρ are at a pseudosteady state for a given value of η. Solving eqn (11) and (13) when $d\alpha/d\tau = d\rho/d\tau = 0$ yields

$$\alpha = \rho = \eta/(\eta-1) - q\eta^2/(\eta-1)^3 + \ldots$$

for the pseudosteady state values of α and ρ. These two equations specify a curve

[1] R. J. Field and R. M. Noyes, *J. Chem. Phys.*, 1974, **60**, 1877.

(parametrized by η) which has been termed the slow manifold by Zeeman.[1] From eqn (12) the motion along the slow manifold is determined by

$$\frac{d\eta}{d\tau} = \frac{-2\eta}{s}$$

or $\log \eta = -2t/s + C$.

This prediction of the motion agrees with the calculations (fig. 2) because α and ρ approach pseudosteady state rapidly compared to the motion of η and, in addition, the pseudosteady state (i.e., M) is stable for all α, η, ρ on the trajectory.

We can obtain equations for curve A in fig. 5 by this method. α is a fast variable near steady state while η and ρ are both slow. The slow manifold for this situation is the *two* dimensional surface $\alpha = g(\eta, \rho)$ obtained by solving $d\alpha/d\tau = 0$. The motion on this manifold may be linearized about steady state to give the damped sinusoidal curve A. This procedure requires that the trajectory remains within (or almost within) the region of α, η, ρ space where M is stable.

Curve B shows what happens when the trajectory on the manifold enters a region where M is unstable. The fast variable α departs from its pseudosteady state rapidly and exponentially. It forces the slow variables to change rapidly also.

There are two additional complications involved in the spike of curve B. First, the one dimensional manifold $d\alpha/d\tau = d\alpha/d\tau = 0$ has a separatrix at $\eta = 1$ and as $\eta \to 1$ from below $d\eta/d\tau \to \infty$. Second, the rate of departure of α is dependent upon the depth of penetration of the trajectory into the unstable region. The faster α is relative to η and ρ, the less penetration is needed to leave the manifold. I have duplicated Field and Noyes' calculations and find that the trajectories of both curves A and B enter the unstable region near $\eta = 1.18$. Curve A grazes the unstable region but does not penetrate deeply. Curve B penetrates far enough for the instability in α to decrease η below $\eta \doteq 1$ where $d\eta/d\tau$ suddenly becomes large and negative.

The instability which causes the spike in fig. 2 to 5 appears mathematically as a negative term in the second Hurwitz determinant. When η is depressed about 30% below steady state this term is an order of magnitude larger than the next largest term. Thus the Hurwitz determinant is quite accurately represented by a single term. This term has a physical interpretation as the product of the feedback loops which cooperate to produce the instability. When the stability problem is set up properly, approximate equations of the surfaces which divide the manifold of pseudosteady states into stable and unstable regions may be obtained by equating the dominant terms of two adjacent stability domains.

The extension of the stability analysis of networks to pseudosteady states adds one additional parameter to the problem for each dynamical variable but it does not change the number of Hurwitz determinants which need to be examined. I have developed techniques which use a computer to obtain the equations of the boundary of the stability domains in cases where the algebra would be very tedious. It takes only a few seconds to do a complete pseudosteady state stability analysis on a very simple model like the Oregonator. These techniques are currently being used to study models which are closer to the detailed mechanism of the Belousov-Zhabotinski reaction.

Prof. R. M. Noyes (*Oregon*) (*communicated*): Clarke is not talking about the same situation that we were. The calculations reported in our paper concerned a system initially in a stable steady state defined approximately by $\alpha = \rho = 4.999$, $\eta = 1.250$. We then arbitrarily depressed η by a small percentage and let the system evolve

[1] E. C. Zeeman in *Towards a Theoretical Biology*, 4: *Essays* (Edinburgh University Press, 1972)

under the dynamic equations that had established the previous steady state. Clarke is discussing the evolution very near a point where $\alpha = \rho$ and where both are determined by the value of η.

As Clarke recognizes, almost immediately after the perturbation in our calculations α attained a pseudosteady state determined by the value of η. However, ρ responded more slowly, and Clarke is not justified in his assumption that he can set $\alpha = \rho$ in his pseudosteady state. A comparison of fig. 3 and 4 of our paper shows that the assumption $\alpha = \rho$ is even less justified for the slow decline in fig. 2 which Clarke discusses in his comment. If that assumption is indeed required, I have doubts about the utility of Clarke's method.

Clarke does not seem to have precise criteria for distinguishing curves A and B in our fig. 5. He talks of unstable pseudosteady states when $\eta \leqslant 1.18$ and admits that curve A penetrates the unstable region but says it does not do so " far enough " for exponential growth of α to become " significant ". We *do* have a precise criterion not discussed in our paper. If the trajectory is integrated along curve B in fig. 5, the quantity $d^2(d\alpha/dt)/d\alpha^2$ changes sign at a point approximated by $\alpha = 7.25$, $\rho = 5.3$, $\eta = 1.158$, and α then increases rapidly. The same quantity never changes sign during the decrease of η along the trajectory of curve A.

We shall prepare a paper in the near future defining our criteria for the conditions under which one pseudosteady state will switch to a different one. Perhaps our criteria will turn out to be mathematically equivalent to those for leaving the slow manifold in Clarke's treatment. However, neither treatment is yet defined with sufficient precision for a meaningful comparison.

Both of our treatments are based on a computation of the stability of a specific point in α, η, ρ space, and neither predicts whether a system in a stable pseudosteady state will evolve along a trajectory such that it will subsequently become unstable. Such prediction would be needed to determine whether the trajectory attained after a finite perturbation like that in fig. 5 will later grow to criticality as in curve B or will subside after a limited growth as in curve A.

Prof. J. Ross (*MIT*) said: A variety of models have been proposed which exhibit the property of threshold excitation. Here a system, far from equilibrium, is in one stable stationary state; upon excitation (change of concentration, temperature, etc) of the right magnitude and sign there occur much larger changes in concentrations (temperature) prior to the return to the stationary state. We [1] have studied the behaviour of such systems upon imposition of noise (fluctuations) in concentrations. At low noise levels (compared with the threshold excitation) random excitation and return to the steady state takes place, and at high noise levels the expected random variations in concentrations occur. At noise levels of the order of the threshold excitations, however, we find quasi-periodic concentration oscillations. Thus, with imposed noise or inherent fluctuations it may be possible to attain quasi-periodic behaviour under less stringent conditions that those necessary for a limit cycle.

Prof. R. M. Noyes (*Oregon*) said: The model we used for our calculations was strictly deterministic. However, if the control intermediate Y were subjected to random fluctuations of the order of the threshold magnitude for the deterministic model, we should also observe quasi-periodic excursions similar to the interesting ones reported by Ross. Because the concentration of Y during an excursion falls to a very small fraction of its steady state value, the random fluctuations imposed during such calculations should be by a percentage rather than by an absolute amount.

[1] H. Hahn, A. Nitzan, P. Ortoleva and J. Ross, *Proc. Nat. Acad. Sci.*, 1974, **71**, 4067.

Dr. P. Ortoleva and **Prof. J. Ross** (*MIT*) said: Consider a system subject to threshold excitation with subsequent fast and slow changes of concentration with time, as shown in fig. 2–4 in the article by Field and Noyes. We have studied wave propagation in systems with multiple time-scale kinetics; the propagation occurs upon local (heterogeneous) threshold excitation. In the formulation of the theory we take explicit account of the different time scales and related length scales in a consistent perturbation method. The theory, in lowest order, yields good estimates of the velocity and concentration profile of propagating pulses and transitions between stable stationary states (in systems with such multiple states).

Dr. A. Winfree (*Indiana*) (*partly communicated*): Carbon dioxide bubbles tragically limit experimental enquiry into the stability of various modes of spatial and temporal organization in Belousov-Zhabotinsky reagent. The problem is particularly crippling in studies of 3-dimensional wave propagation: the liquid must be at least 1 mm deep, so little CO_2 escapes through the surface directly, and when a bubble rises, it tears through a great volume of otherwise motionless liquid.

Is there any way to modify the organic acid's carboxyl groups (e.g., by forming a diester) or otherwise alter it to prevent decarboxylation without upsetting the β-keto group's reactivity or forming insoluble byproducts?

As a postscript, I am glad to pass on Trahanovsky's suggestion of replacing malonic acid by ethyl aceto-acetate, using acetic acid cosolvent to prevent precipitation of bromo derivatives. It works marvellously in a recipe similar to that given in A. Winfree, *Science*, 1972, **175**, 634.

Dr. A. Winfree (*Indiana*) said: Depending on pH, the Belousov-Zhabotinsky reagent can either oscillate spontaneously, or remain inert until triggered to execute a single oxidative pulse. Such parameter-sensitivity of behaviour is graphically portrayed by Field and Noyes' 2-component approximation to their own complete kinetic scheme: behaviour is oscillatory or excitable depending on the exact manner of intersection of the nullclines of the two components' rate equations. If those nullclines were bent a little, they would intersect *three* times, giving rise to reaction with two alternative stable steady-states and an intermediate unstable (threshold-like) steady state.

Such a reaction, diffusion-coupled in space, would provide *remarkable* opportunities for experimental study not only in connection with physical chemistry (see Ross and Ortoleva, this Symposium), but also as a model for the differentiation of living embryos into discrete tissues, each corresponding to a different stable steady-state of the biochemical-genetic machinery residing identically in each cell.

What parameters or reaction rates must be altered slightly to bend the reaction rate equations as required?

Dr. B. L. Clarke (*Alberta*) said: Winfree is asking how the mechanism of the Belousov-Zhabotinski reagent might be modified to give a multistable steady state situation which is called a Riemann–Hugoniot catastrophe by mathematicians. I think there is no simple modification of the FKN mechanism which would do this. Reaction (R7) of Field and Noyes' ref. (5) may be omitted because it is the back reaction of (R5) and (R6) and, therefore, it cannot affect the multiplicity of the steady states. If one omits " flow through reactants " from the remaining FKN mechanism the multiplicity is also not changed. Next, consider all the remaining reactions except (R3) and (R9). This reaction network can be proven to have a *unique* positive steady state which is always stable. However, if one adds Br^- from an external source at constant rate there are *two* steady states and they have a dynamical significance. The steady states of the network with Br^- being added at a constant rate correspond to dynamical situations of the actual network when Br^- is *decreasing at that instantaneous rate* with the other variables at a pseudosteady state for the instantaneous value of the Br^- concentration. These two steady states have high and low concentrations of $HBrO_2$ and correspond to the steady states involved in the switching of the Oregonator. One cannot have a third steady state for this network because the equation to be solved for the two steady states is quadratic. Next include reactions (R3) and (R9) in the network once again. Since they both consume Br^-, the second steady state is *removed* from the network. However, the new reaction Noyes has proposed (reaction (A) in my previous comment) adds Br^- in such a way that the modified FKN mechanism at steady state can now behave like the original FKN mechanism with a fixed input rate of Br^-. Whether or not it has two steady states depends on the relative importance of reactions (R3), (R9) and (A). In order to get three positive steady states the polynomial to be solved must have coefficients which alternate in sign $(+ - + -)$. I do not see any realistic way to modify the mechanism to achieve this.

Prof. R. M. Noyes (*Oregon*) said: Winfree has commented about three different matters that should be discussed separately. These are (*a*) the possibility of finding a system that does not produce carbon dioxide, (*b*) the transition of a system between oscillatory and inert-excitable conditions, and (*c*) the possibility that nullclines might be made to intersect three times.

(*a*) With regard to carbon dioxide evolution, I am sure Winfree is aware that Bowers, Caldwell, and Prendergast [1] observed oscillations with 2,4-pentanedione, $CH_3COCH_2COCH_3$. I gather this material is objectionable in other ways. The cyclopentanedione and cyclohexanedione ring systems might be more satisfactory. Kasperek and Bruice [2] report a number of other compounds that do or do not oscillate under their conditions, but they do not mention extents of carbon dioxide evolution. There is much merit to Winfree's suggestion to use an ester such as diethyl malonate. Our calculations with the Oregonator [3] indicate that stoichiometry is a very important factor. The desirable compound must brominate readily, and the resulting bromo derivative will be most effective if it liberates one bromide ion for each two $Ce(IV)$ or $Fe(phen)_3(III)$ ions reduced during the time scale of interest. Stoichiometric tests will probably provide the quickest way to screen the bromine derivatives of various organic compounds. The best organic substrate for a particular study of this reaction must ultimately be identified empirically.

(*b*) With regard to the transition between oscillatory and inert-excitable conditions, oxygen appears to be very important. As Winfree points out in his manuscript, the reagent remains for long periods in inert reducing condition when it is in a film a millimetre or two thick, but the same composition oscillates with slow frequency in bulk. Agitation of the bulk solution with a stream of air can also prevent oscillations.

Oxygen is most likely to attack radical species, and the $\cdot BrO_2$ intermediates are likely targets. A plausible mechanism is

$$\cdot BrO_2 + O_2 \rightleftarrows \cdot OOBrO_2$$
$$\cdot OOBrO_2 + \cdot BrO_2 \rightarrow 2 \cdot BrO_3$$
$$\cdot BrO_3 + Ce^{3+} \rightarrow BrO_3^- + Ce^{4+}.$$

[1] P. G. Bowers, K. E. Caldwell and D. F. Prendergast, *J. Phys. Chem.*, 1972, **76**, 2185.
[2] G. J. Kasperek and T. C. Bruice, *Inorg. Chem.*, 1971, **10**, 382.
[3] R. J. Field and R. M. Noyes, *J. Chem. Phys.*, 1974, **60**, 1877.

This mechanism is purely conjectural at present, but it is consistent with what is known about thermodynamic and kinetic behaviour of related species.   The net effect is that some $\cdot BrO_2$ is oxidised to BrO_3^- instead of being reduced to $HBrO_2$; reduction is essential for the autocatalytic behaviour that generates oscillations.

This revised mechanism can be modelled if the third step of the Oregonator [1] is altered to (M3′) with $0 \leqslant n \leqslant 2$.

$$B+X \rightarrow nX+Z. \qquad \text{(M3′)}$$

If $n \leqslant 1$, this revised model will have a stable steady state for all possible combinations of rate constants.   If the system is oscillatory for $n = 2$, accessibility of oxygen could reduce n sufficiently to make the system inert to oscillation.   However, a steady state that was barely stable should be excitable just like the state we obtained by manipulating the stoichiometric factor f.

(c) Winfree asks about the possibility that nullclines might intersect three times. The stiffly-coupled Oregonator [1] can model the Belousov-Zhabotinskii reaction in terms of the two intermediates Y and Z.   The curve $\dot{Z} = 0$ is monotonic in the Y-Z plane, but the curve $\dot{Y} = 0$ displays a pair of very sharp relative maxima and minima in Z.   However, even though there are certain values of Z such that $\dot{Y} = 0$ is satisfied for three values of Y, there is no combination of rate constants for which the coupled Oregonator model permits more than one solution such that $\dot{Y} = 0$ and $\dot{Z} = 0$ simultaneously.

The kind of situation Winfree is looking for would be obtained if the stoichiometric factor f could vary so that $f < 1$ when Y is a little less than $k_{M3}B/k_{M2}$ and $f > 2$ when Y is a little more than $k_{M3}B/k_{M2}$.   Such an effect introduces autocatalysis to step (M5) as well as that existing in step (M3).   Although such a model would permit interesting hysteresis effects of the kind Winfree is looking for, I do not see any way to realize it experimentally.

Dr. M-L. Smoes (*Dortmund*) (*communicated*): 1. I agree with Field and Noyes that the so-called Winfree solution is an oscillatory system with long period.   Indeed, Winfree has been trying since 1972 to eliminate the bulk oscillations in the distributed Zhabotinskii system.[2]   To do so, he increases the bromide ion concentration and decreases the sulphuric acid content of the reacting mixture.   I have shown recently that the period of homogeneous (bulk) oscillations in the ferroin-catalyzed Zhabotinskii system increases with an increase in bromide ions and a decrease in sulphuric acid.[3]   Moreover, when the homogeneous period of oscillations is sufficiently long, large variations in period are observed.   This may explain the fact that oscillations in the Winfree solution are observed only irregularly and that the system appears unstable but non-oscillatory.

2. The period of the Zhabotinskii oscillations depends on the initial reactant concentrations in a highly nonlinear manner.[3, 4]   This fact suggests another possibility for the amplification of small perturbations in the concentrations.   Indeed, any period of the form:

$$T = 2\pi/(a-b)$$

with a and b related to some parametric concentrations can be drastically reduced by very small perturbations in a or b as long as the unperturbed period is sufficiently large.

[1] R. J. Field and R. M. Noyes *J. Chem. Phys.*, 1974, **60**, 1877.
[2] A. T. Winfree, *Science*, 1972, **175**, 634.
[3] M-L. Smoes, *J. Chem. Phys.*, 1975.
[4] *Oscillatory Processes in Biological and Chemical Systems*, (Nauka, Moscow, 1967) p. 181ff.

It is this transformation of the small concentration perturbations into large period perturbations which justifies the phase waves interpretation of the spatio-temporal structures observed in the Zhabotinskii system.[1]

The amplification model of Field and Noyes is limited to nonoscillatory media while our model can be used for the waves observed in the oscillatory Zhabotinskii system.

Prof. R. M. Noyes (*Oregon*) (*communicated*): Smoes evidently does not appreciate the effect of oxygen on Winfree solution. When she talks about bulk oscillations with long period, she is considering compositions that are not maintained fully saturated with oxygen. If saturation is maintained, the medium can indeed become non-oscillatory. As I point out in my response to the comment by Winfree, we now believe the system could be modelled better by changing the stoichiometry of the third step of the Oregonator rather than the fifth step as we did in our paper. However, the effect should be very similar with either model.

When Smoes talks about the effect of bromide ion on the period of oscillation, she must be referring to *initial* concentration. Bromide ion is one of the species that undergoes limit cycle oscillations and is not a proper parameter by which to characterize a reactant composition.

We remain convinced our chemical mechanism can account for virtually all of the essential features of the Belousov-Zhabotinskii reaction. Smoes may attempt an alternative explanation based on phase waves if she wishes, but she should then explain why the period of bulk oscillations depends in a very non-linear way on reactant concentrations yet the rate of propagation of waves is closely proportional to $[H^+]^{\frac{1}{2}}$ $[BrO_3^-]^{\frac{1}{2}}$ and almost independent of other concentrations.[2]

Dr. M-L. Smoes (*Dortmund*) (*communicated*): (1) I do not exclude the possibility that oxygen too increases the period of bulk oscillation in the Zhabotinskii system. But the changes in oxygen concentration are probably negligible compared to the changes in concentrations of bromide ions and of sulphuric acid which have led to the progressive elongation of the bulk period culminating in the " Winfree solution ".

Although I was actually referring to initial concentrations of bromide ions, the fact that a concentration oscillates does not prevent it from having an average value. Moreover, a distinction between oscillatory compounds and constant parameters is a helpful approximation to the very complex kinetics involved in chemical oscillations. In any case, it is an experimental fact that an addition of bromide ions increases the homogeneous period of the Zhabotinskii oscillations.

(2) Our phase wave model [3] can account qualitatively for all experimentally well established phenomena in the Zhabotinskii system. Unlike Field and Noyes,[4] we do not have to consider as interferences all waves except one. Moreover, we have predicted the larger external wavelength which is observed in the waves at the centre of fig. 1 in the paper of Field and Noyes. Finally, we have found in a detailed study of the phase waves [5] that the speed of wave propagation is dependent on the homogeneous period of oscillations. This explains an increase in the speed with an increase in the concentrations of bromate and of malonic and sulphuric acid. It explains also the fact that the speed is insensitive to changes in ferroin concentration.

[1] M-L. Smoes and J. Dreitlein, *J. Chem. Phys.*, 1973, **59**, 6277.
[2] R. J. Field and R. M. Noyes, *J. Amer. Chem. Soc.*, 1974, **96**, 2001.
[3] M-L. Smoes and J. Dreitlein, *J. Chem. Phys.*, 1973, **59**, 6277.
[4] R. J. Field and R. M. Noyes, *J. Amer. Chem. Soc.*, 1974, **96**, 2001.
[5] M-L. Smoes, *Characteristic Properties of the Phase Waves*, in preparation.

The nonlinear dependence of the period of oscillations on concentration is matched in our model by a similarly nonlinear dependence of the speed of propagation on the homogeneous period. As a result, the experimental data of Field and Noyes are not in disagreement with the phase wave model. But the interest of our model comes from its generality. We are not limiting ourselves to the interpretation of the Zhabotinskii waves. Indeed, our results are expected to apply to a large class of chemical systems with concentration or temperature dependent period of oscillations. Since the glycolytic system belongs to this class, glycolytic phase waves can be expected. We are interested in the features that should be shared by the waves in glycolysis and in the Zhabotinskii system.

Dr. O. E. Rössler (*Tübingen*) said: The Oregonator (eqn (11)–(13)) is, as a 3-variable non-linear oscillator, not very easy to analyze mathematically.[1] It also involves the mathematically crucial, but chemically somewhat *ad hoc* assumption of $f > 1$, in order to account for monostability. The question therefore arises whether a 2-variable prototype does not also exist, especially since the verbal descriptions given (like " X being switched by Y's varying between 2 critical values ") all refer to such a model. Evidence [2] on the inorganic subsystem (sulphuric acid + bromate + cerium III) suggests that the autocatalytic reaction can be shifted through a hysteresis cycle depending on the value of an exogenous influx of bromide. The complete Noyes-Field-Körös scheme [3] indeed provides for such a mechanism : the fact that bromide is being regenerated (at least partially) from its own products (namely via the reverse reaction of R2 from hypobromous acid, and via R8 from bromine) [3] allows a weak influx from x to shift this " catalyst " up and down, without requiring the " accumulating " power of an intercalated variable Z. Stipulating that something like a Michaelis-Menten type approximation is valid, the following 2-variable model applies :

$$dx/dt = k_1 y - k_2 y \frac{x}{x+K} + k_3 x - k_4 x^2$$
$$dy/dt = k_5 x - k_1 y + k_6 \tag{1}$$

where x again refers to the autocatalytic cycle (represented by bromous acid, for example) and y is the total concentration of the catalyst (involving bromide, hypobromous acid and bromine). This equation, which is readily analyzed by phase-plane techniques, is of Bonhoeffer-van der Pol (BVP[4]) type : the second variable acts as a slowly changing " parameter variable " which drives the bistable subsystem (first variable) through a hysteresis cycle, and this is an either astable (self-oscillating) or monostable (repeatedly triggerable) way.[5] One adapted set of parameters : $k_1 = 1 = k_5 = 1$, K/k_2 (corresponding to the former k_2) $= 4 \times 10^{-4}$ (or less), $k_2 = 1$, $k_3 = 4.8 \times 10^2$, $k_4 = 4 \times 10^7$, $k_5 = 1$, $k_6 = 0$ (astability) or 0.12 (monostability) ; analogue computer results.

Prof. R. M. Noyes (*Oregon*) said: There is nothing particularly *ad hoc* about the assumption that $f > 1$. The experimental fact is that the excitable Winfree reagent is in the red reduced condition, which is the steady state generated by the Oregonator model when k_{MS} is sufficiently large and $f > 1$. (When $f < 1$, the steady state at

[1] J. D. Murray, *J. Chem. Phys.*, 1974, **61**, 3610.
[2] V. A. Vavilin and A. M. Zhabotinski, *Kinetika i Kataliz*, 1969, **10**, 83. (**10**, 65 of cover-to-cover translation).
[3] See ref. (4) and (5) of preceding paper.
[4] R. FitzHugh, *Biophys. J.*, 1961, **1**, 445.
[5] O. E. Rössler, *Lecture Notes in Biomathematics* (Springer–Verlag), 1974, **4**, 399 and 546.

large k_{M5} corresponds to the blue oxidised condition.)   Oxygen is apparently involved in maintaining this steady state, and a possible mechanism is discussed in my response to the comments by Winfree.

Rössler is right about the mathematical advantage of describing the system in terms of only two variables.   However, Tyson and Light[1] have shown that limit cycle behaviour can not be generated by a model based on bimolecular reactions of only two variables.   Rössler has introduced a term $k_2 xy/(x+K)$ that resembles a Michaelis–Menten situation.   However, by doing so he has created a third chemical variable.   His equation models an intermediate species $H_2Br_2O_2$ that is present in significant concentration compared to $HBrO_2$ and that is more likely to decompose to $HBrO_2 + Br^- + H^+$ than it is to 2HOBr.   There is no experimental evidence for such a species and no theoretical justification from what we know about chemical bonding and reactivity.

Rössler's equations also omit any effect of metal ion catalyst, which is modelled by Z in the Oregonator.   It is an experimental fact that such a catalyst must be present if oscillations are to be observed.

Fortunately, the Oregonator can indeed be modelled with only two variables. As we have shown elsewhere,[2] if X at all times is approximated by solving $\dot{X} = 0$ for existing values of Y and Z, the Oregonator equations generate limit cycle behaviour in Y and Z differing little from the results of more complicated calculations in three variables.   I am confident our excitable medium calculations would be little affected by repeating them in two variables at the same level of approximation.   The key pair of variables for approximating the Oregonator is Y and Z rather than X and Y as suggested by Rössler.

Dr. P. G. Sørensen (*Denmark*) said: Experimental studies of the Belousov reaction have been carried out in an isothermal stirred tank reactor with constant volume and time-independent external flows.   The reactants in the input flows were $KBrO_3$, $CH_2(COOH)_2$ and $Ce_2(SO_4)_3$.   At low flow rate the system shows periodic oscillations

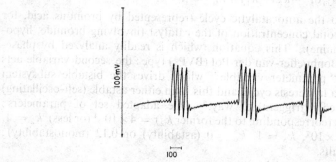

FIG. 1.—Potential variations of a platinum electrode relative to a calomel electrode.   Tank volume 8.1 ml.   Temperature 25.0°C.   Input flow 1: 0.09 M $KBrO_3$ in 3N H_2SO_4.   0.0037 ml/s.   Input flow 2: 0.5M $CH_2(COOH)_2$ in 3N H_2SO_4 0.0037 ml/s.   Input flow 3: 0.0008M $Ce_2(SO_4)_3$ in 3N H_2SO_4 0.0037 ml/s.

corresponding to an attracting limit cycle, but at higher flow rate the oscillations occur in bursts; fig. 1.   The proportion of time spent by the system in the oscillating phase decreases with increasing flow rate, and for a sufficiently high flow rate the

[1] J. J. Tyson and J. C. Light, *J. Chem. Phys.*, 1973, **59**, 4164.
[2] R. J. Field and R. M. Noyes, *J. Chem. Phys.*, 1974, **60**, 1877.

oscillations disappear, corresponding to an attracting steady state. Although the number of oscillations in each burst is almost constant for a certain flow rate, and the bursts occur with fairly constant time interval, there is no fixed proportion between the burst period and the oscillation period. This behaviour can not be explained by the existence of a limit cycle. I suggest that the attractor in this case is a surface in state space, and that the system shows almost periodic movements on this surface. The Belousov reaction is thus an example of a chemical system that may have attractors, which are neither fixed points nor limit cycles. The kinetic explanation for the pulsed oscillations is probably that the concentration of a compound which is essential for oscillatory behaviour, must be higher than a certain limit in order to allow transition from a non-oscillatory to an oscillatory phase, and that the compound is removed faster during the oscillatory phase than it is produced. When the concentration has decreased below a critical concentration, the oscillations stop, and regeneration of the compound occurs during the following non-oscillatory phase. I wish to ask if measurements have been made on the rate of production of $BrCH(COOH)_2$ in the closed system, immediately before the oscillations start, and immediately after. If the production rate is larger in the first case, $BrCH(COOH)_2$ is a probable candidate for the critical compound in the kinetic scheme described above.

Prof. R. M. Noyes (*Oregon*) said: Sørensen's observations are very interesting. Although his description of experimental conditions is not entirely clear, I gather that his platinum electrode has a more positive potential when the system is not oscillating than it does on average during the pulses of oscillation. If that is so, the non-oscillatory period is behaving just like the oxidising induction period that is observed when malonic acid is the only organic compound initially present. Sørensen's observations can then be explained rather easily by the mechanism we have already developed.[1-3] The explanation is a detailed amplification of the one he proposes in his comment:

Oscillations in a closed system involve rapid switching between an oxidising and a reducing condition depending upon the net direction of change in oxidation state of cerium. The reducing condition is characterized by significant but monotonically decreasing concentrations of both Br^- and $Ce(IV)$. The oxidising condition is characterised by a very much smaller concentration of Br^- and by a monotonically increasing concentration of $Ce(IV)$. When the system switches to the oxidising condition, $Ce(IV)$ is produced initially at a very rapid rate.

When the system is in the reducing condition, the important processes are those of (A-D).

$$BrO_3^- + 2Br^- + 3H^+ \rightarrow 3HOBr \tag{A}$$

$$HOBr + CH_2(COOH)_2 \rightarrow BrCH(COOH)_2 + H_2O \tag{B}$$

$$BrCH(COOH)_2 + 4Ce^{4+} + 2H_2O \rightarrow Br^- + 4Ce^{3+} + HCOOH + 2CO_2 + 5H^+ \tag{C}$$

$$HOBr + HCOOH \rightarrow Br^- + CO_2 + H^+ + H_2O. \tag{D}$$

The stoichiometry of (A)+2(B)+(C)+(D) yields (R) as the net reaction in the reducing condition.

$$BrO_3^- + 2CH_2(COOH)_2 + 4Ce^{4+} \rightarrow BrCH(COOH)_2 + 4Ce^{3+} + 3CO_2 + 3H^+ + H_2O. \tag{R}$$

[1] R. J. Field, E. Körös and R. M. Noyes, *J. Amer. Chem. Soc.*, 1972, **94**, 8649.
[2] J. J. Jwo and R. M. Noyes, *J. Amer. Chem. Soc.*, 1975.
[3] R. M. Noyes and J. J. Jwo, *J. Amer. Chem. Soc.*, 1975.

The kinetics in the reducing condition are somewhat complex. If there is insufficient bromomalonic acid (BrMA), process (C) is so slow that the depletion of bromide ion by process (A) reduces $[Br^-]$ below the critical value necessary to maintain the reducing condition. If there is somewhat more BrMA, $[Br^-]$ attains a steady state established by equal rates of processes (A) and (C). If there is still more BrMA, the cerium(IV) will first be consumed by (C) and the resulting bromide will then disappear by the effect of (A) + 2(B) + (D). If there is enough BrMA to generate a reducing condition, (R) describes the net stoichiometry for the overall period in that condition. To a good approximation,[1] the rate during a reducing period is given by eqn (1).

$$-\frac{d[BrO_3^-]}{dt} = \frac{d[BrMA]}{dt} = 2.1[H^+]^2[Br^-][BrO_3^-]. \tag{1}$$

The numerical value of the rate constant is based on concentrations in mol/l. and time in seconds. Approximate concentrations during such a period are about 1 M for hydrogen ion, a few times 0.01 M for bromate, and a few times 10^{-6} M for bromide ion.

When the system is in the oxidising condition, the net process (Ox) results from processes (E) and (B).

$$BrO_3^- + 4Ce^{3+} + 5H^+ \rightarrow HOBr + 4Ce^{4+} + 2H_2O \tag{E}$$

$$HOBr + CH_2(COOH)_2 \rightarrow BrCH(COOH)_2 + H_2O \tag{B}$$

$$BrO_3^- + CH_2(COOH)_2 + 4Ce^{3+} + 5H^+ \rightarrow BrCH(COOH)_2 + 4Ce^{4+} + 3H_2O. \tag{Ox}$$

The kinetics in the oxidising condition are also complex. If the intermediate $\cdot BrO_2$ radicals irreversibly oxidise cerium(III) as rapidly as they are formed, our previous analysis [1] indicates the rate of (Ox) is given by eqn (2).

$$-\frac{d[BrO_3^-]}{dt} = \frac{d[BrMA]}{dt} = 0.6[H^+]^2[BrO_3^-]^2. \tag{2}$$

If $\cdot BrO_2$ radicals are reversible oxidants, if they disproportionate with each other, or if they also reduce cerium(IV), the rate of (Ox) is somewhat less than that of eqn (2).

Reaction conditions are such that to a first approximation $[BrO_3^-] \approx 10^4[Br^-]$. Therefore, even if eqn (2) somewhat overestimates the rate, a comparison of eqn (1) and (2) shows that $d[BrMA]/dt$ during the oxidising condition is many times that during the reducing condition. Therefore, it is quite possible to design conditions for an oxidising condition in a stirred tank reactor such that [BrMA] is temporarily too small for process (C) to switch the system to reducing condition but such that $d[BrMA]/dt$ by eqn (2) is greater than the rate at which BrMA is being removed from the reactor by flow. As [BrMA] increases in the reactor, the system eventually switches to oscillation just as it does in a homogeneous closed system. However, most of the time during oscillation is spent in the reducing condition, and eqn (1) is so much slower than eqn (2) that net $d[BrMA]/dt$ during an oscillatory period is insufficient to compensate for loss of BrMA by flow from the reactor; the system will revert to a non-oscillating oxidising condition. This explanation is entirely consistent with the one proposed by Sørensen.

The above qualitative description can be modelled by the following revised Oregonator [2] and differential equations.

[1] R. J. Field, E. Körös and R. M. Noyes, J. Amer. Chem. Soc., 1972, 94, 8649.
[2] R. J. Field and R. M. Noyes, J. Chem. Phys., 1974, 60, 1877.

$$A+Y \rightarrow X+P \tag{N1}$$

$$X+Y \rightarrow 2P \tag{N2}$$

$$A+X \rightarrow 2X+Z \tag{N3}$$

$$2X \rightarrow P+A \tag{N4}$$

$$Z \rightarrow fY-fP \tag{N5}$$

$$f = \frac{P}{a+P} \tag{3}$$

$$dX/dt = k_1AY-k_2XY+k_3AX-2k_4X^2-vX \tag{4}$$

$$dY/dt = -k_1AY-k_2XY+k_5fZ-vY \tag{5}$$

$$dZ/dt = k_3AX-k_5Z-vZ \tag{6}$$

$$dP/dt = k_1AY+2k_2XY+k_4X^2-k_5fZ-vP. \tag{7}$$

In this model, the chemical significance of the letters is intended to be $A = BrO_3^-$, $X = HBrO_2$, $Y = Br^-$, $Z = 2Ce(IV)$, $P = BrMA$. The rate constants for steps (N1-N4) are determined by those of oxybromine chemistry.[1, 4] The stoichiometry of step (N5) is that of (C)+(D) if P represents either BrMA or its HOBr precursor and if process (C) is first order in Ce(IV).

The simplified model developed here assumes that cerium(IV) is consumed at an almost constant rate by an excess of malonic acid (MA) containing a smaller amount of bromomalonic acid (BrMA). Because the radicals from malonic acid may attack BrMA with liberation of bromide,[2] the stoichiometric factor f is calculated by eqn (3). As P increases, the resulting increase in f will cause the system to pass sharply from an oxidising steady state to an oscillating condition as happens in a closed homogeneous system.

It appears that a proper selection of v, k_5, and a could reproduce the main features of Sørensen's observations. At too great a v, the system would remain in an oxidising condition and would not oscillate, just as is observed. At too small a v, the system would go into continuous oscillation as is also observed.

However, it is not certain that this simplified model would generate packets of oscillation rather than single pulses of reducing condition followed by return to oxidising condition. Once [BrMA] has risen to the critical f necessary for oscillations to commence, there must be a delay before [BrMA] falls enough to shut them off again. It may be that the thermal effects associated with the different rates of (R) and of (Ox) will need to be coupled in order to reproduce the experimental observations. It may also be that the intermediate tartronic acid, $HOCH(COOH)_2$, changes at a sufficiently different rate to produce the necessary coupling.[2, 3]

The ideas suggested here can be tested by computations with the revised Oregonator proposed above, and additional experimental testing can be provided by adding bromomalonic or tartronic acid to the solutions entering the reactor and by changing the efficiency with which the reactor is thermostatted.

Dr. O. E. Rössler (*Tübingen*) said: The same result (periodic bursts of oscillation in a stirred open system version) has also been obtained by Junkers.[5] The stops

[1] R. J. Field, E. Körös and R. M. Noyes, *J. Amer. Chem. Soc.*, 1972, **94**, 8649.

[2] J. J. Jwo and R. M. Noyes, *J. Amer. Chem. Soc.*, 1975.

[3] R. M. Noyes and J. J. Jwo, *J. Amer. Chem. Soc.*, 1975.

[4] R. J. Field and R. M. Noyes, *J. Chem. Soc.*, 1974, **60**, 1877.

[5] G. Junkers, *Über die periodisch verlaufende Reaktion zwischen Malonsäure und Bromat in der Gegenwart von Cerionen* (Diploma Thesis, University of Aachen, 1969).

occurred in the " high Ce(IV) " state, and addition of Br⁻ immediately revived the oscillation.   In order to model this type of behaviour, at least three variables are required in any case.   For example, the following third equation may be added to eqn (1) (as indicated in my earlier discussion remark).

$$dz/dt = k_7 x - k_8 z,$$

with y being coupled to z by replacing the term k_6 on the right hand side of the second equation by $k_6 z$. Z may correspond to a bromide-releasing intermediate compound (even Br_2).   For example, $k_5 = 1.7$, $k_7 = k_8 = 3 \times 10^{-3}$, $k_6 = 0.1$; analogue computer results.

Prof. E. Körös (*Budapest*) said: A large number of measurements have been performed by us using the polarographic method [1] on the rate of production of $BrCH(COOH)_2$ in different Belousov systems during the non-oscillatory (induction) period of the reaction.   Especially suitable for this purpose is the bromate-malonic acid-cerium(III) + nitric acid system [2] where the non-oscillatory period is rather long and the consumption of the reagent during the non-oscillatory period is significant. (Approximately 40–50 % of the initial bromate content is consumed)—a few data are given in table 1.

TABLE 1.—$T = 15°C$.   MALONIC ACID 0.40 M, NITRIC ACID 5.0 M

[BrO₃⁻]/M	[Ce/III/]10³/M	(d[BrMA]/dt)ₜₙᵢₜᵢₐₗ/ M min⁻¹
0.10	4.00	3.32×10^{-3}
	1.00	1.25×10^{-3}
0.05	4.00	2.47×10^{-3}
	1.00	0.87×10^{-3}

When *catalyst was not present* the initial rate of production of $BrCH(COOH)_2$ was 0.16×10^{-3} M min⁻¹ in the following reaction mixture : $[BrO_3^-] = 0.10$, $[MA] = 0.40$ and $[HNO_3] = 5.0$ at 15°C.

It is my opinion that in a flow system continuous oscillation would be observable if $BrCH(COOH)_2$ were also added together with the other reagents.   Then the chemical system would be always in an excitable state.

Very recently we measured the rate of formation of bromomalonic acid both during the non-oscillatory and the oscillatory periods of the cerium(III)-, and manganese(II)-catalysed Belousov–Zhabotinsky reactions.   A typical curve is given in

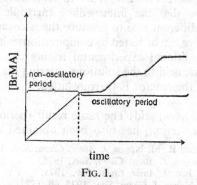

FIG. 1.

[1] See paper by Körös *et al.*, this Symposium.
[2] E. Körös and M. Burger, to be published.

fig. 1. From the curve it is obvious that at the transition from the non-oscillatory to the oscillatory stage a rather sharp decrease occurs in the rate of formation of bromomalonic acid. These results yield experimental support for Sørensen's explanation of the phenomena observed in his flow system.

Prof. R. M. Noyes (*Oregon*) said: The experiments by Körös and his co-workers contribute important new insights to the function of the metal-ion catalyst in the Belousov-Zhabotinskii reaction. By a somewhat unfortunate coincidence the strong oxidants Ce(IV) and Mn(III) with potentials of 1.4 V or over are liable to substitution of the oxygen in the inner sphere of coordination while the weaker oxidants Ru $(dipy)_3^{3+}$ and Fe(phen)$_3^{3+}$ with potentials of 1.2 V or less are inert to substitution of the organic species in the inner sphere. Species that were inert and labile to substitution would probably show very different relative kinetic behaviour with oxybromine species and with organic molecules like malonic acid even if they had very similar thermodynamic reduction potentials. It would be very helpful if a catalyst system could be found that would uncouple effects of changing reduction potential from effects of changing substitution lability.

In addition to the excellent experiments reported here, I can suggest some other types of measurement that would help to sort out the details of metal-ion reactions in this system:

(1) The oxidation of organic species is very complex. Jwo [1] at the University of Oregon has completed a thorough study of the cerium system. His observations show that malonyl radicals attack bromomalonic acid with liberation of bromide ion and that presence of the intermediate tartronic (hydroxymalonic) acid is important to the detailed behaviour of the Belousov-Zhabotinskii system. It would be useful to know the ways in which ferric phenanthroline differs from cerium(IV) as an oxidant of these organic species.

In addition to studies of direct metal ion oxidation of organic species, it would be useful to measure d ln [Br$^-$]/dt during slow bromide consumption periods with different ratios of malonic and bromomalonic acids. Such measurements (at known concentrations of bromate) would establish rates of bromide ion generation by metal-ion oxidation of the mixture of organic species in a Belousov–Zhabotinskii system.

(2) The reduction of bromate by metal ion also needs to be understood better for the catalysts discussed here. The studies by Thompson [2] were crucial to elucidating the mechanism [3-5] of the Belousov–Zhabotinskii reaction. However, they do not establish the relative reactivities of $BrO_2\cdot$ radicals with each other and with both the reduced and oxidized forms of the various metal ions. Neither do they establish whether metal-ion oxidation of $HBrO_2$ can compete with disproportionation of that species.

These effects will be difficult to sort out because the very reactive species $BrO_2\cdot$ and $HBrO_2$ exist in very low, kinetically established, steady state concentrations that can not yet be measured directly. It is suggested that mechanistically useful information can be obtained by measuring absolute values of [Br$^-$]$_{crit}$, which is the bromide concentration at which the system goes into rapid bromide consumption or production. In the cerium catalyzed system, the critical concentration for rapid bromide

[1] J. J. Jwo, *J. Amer. Chem. Soc.*, 1975.
[2] R. C. Thompson, *J. Amer. Chem. Soc.*, 1971, **93**, 7315.
[3] R. M. Noyes, R. J. Field and R. C. Thompson, *J. Amer. Chem. Soc.*, 1971, **93**, 7315.
[4] R. M. Noyes, R. J. Field and E. Körös, *J. Amer. Chem. Soc.*, 1972, **94**, 1394.
[5] R. J. Field, E. Körös and R. M. Noyes, *J. Amer. Chem. Soc.*, 1972, **94**, 8649.

consumption is considerably greater than that for rapid production. It is tentatively predicted that these two critical concentrations will be much closer to each other in the iron phenanthroline system.

The above suggestions merely indicate that the study of no chemical system is ever truly complete! Not all the suggested experiments are necessary or even desirable to carry out. However, the Belousov-Zhabotinskii reaction is so dramatic it has attracted considerable attention. Further effort is undoubtedly warranted in order to understand more about the detailed mechanisms in this remarkable system. It is to be hoped that people undertaking such studies will communicate enough among themselves so that unnecessary duplication is avoided.

Prof. E. Kőrös (*Budapest*) said : (1) Unfortunately tris(phenanthroline)iron(III) is unstable in dilute sulphuric acid; the complex partly decomposes, partly is reduced to tris(phenanthroline)iron(II), and for this reason it is not possible to look at its reactions with organic compounds.

(2) The reduction of bromate by tris(phenanthroline)iron(II) was investigated by us, and here I refer only to the original paper.[1]

Dr. H. G. Busse (*Kiel*) and **Prof. B. Hess** (*Dortmund*) said: The Belousov-type reactions are most suitable systems for investigating oscillatory chemical reaction mechanisms. Indeed the malonic acid + cerium sulphate + bromate reaction in aqueous sulphuric acid is currently being studied in several laboratories. Originally, only the range of initial concentrations in which oscillations could occur was determined apart from the shape of the oscillations within this range.[2] Later, intermediate and final products were also analysed. For some of them it was shown that the intermediates oscillate with the same frequency as the system itself. This behaviour would be expected from oscillations of the limit cycle type.

We analysed the final product of this reaction system and found, besides CO_2, monobromomalonic acid as well as bromoacetic acid.[3-5] Furthermore, we investigated the reactivity of the oscillating reaction system towards light perturbation.[6] From estimates of the quantities of products, the overall reactions can tentatively be formulated by the two following equations :

$$3H_2C(COOH)_2 + 2BrO_3^- + 2H^+ \rightarrow 2BrHC(COOH)_2 + 4H_2O + 3CO_2$$

$$2H_2C(COOH)_2 + 2BrO_3^- + 2H^+ \rightarrow Br_2HCCOOH + 4H_2O + 4CO_2.$$

[1] E. Kőrös, M. Burger and Á. Kis, *Reaction Kin. Cat. Letters*, 1974, **1**, 475.
[2] A. M. Zhabotinsky, in *Oscillatory Processes in Biological and Chemical Systems* (Puschino on Oka—1967).
[3] L. Bornmann, H. G. Busse and B. Hess, *Z. Naturforsch.*, 1973, **28b**, 93.
[4] L. Bornmann, H. G. Busse, B. Hess, R. Riepe and C. Hesse, *Z. Naturforsch.*, 1973, **28b**, 824.
[5] L. Bornmann, H. G. Busse and B. Hess, *Z. Naturforsch.*, 1973, **28c**, 514.
[6] H. Busse and B. Hess, *Nature*, 1973, **244**, 203.

FIG. 1.—Time course of the optical density changes as recorded in a double-beam spectrophotometer. Initial conditions are : malonic acid (0.1 M), $KBrO_3$ (0.1 M), $Ce(SO_4)_2$ (2×10^{-4} M) in 3 N H_2SO_4. (*a*) Simultaneous record of the optical density changes in time, analysed at 260 nm and 377 nm. (*b*) Record of the optical density changes at 377 and 260 nm plotted on a *X-Y*-recorder as given in fig. 1*b*. The experiment starts with the trace at the top side left. The time course is indicated by the stepwise shift of the record of one period towards the right. Each step corresponds to approximately 52 s. (*c*) Time course of the optical density difference analysed at the wavelengths of 270 and 356 nm. The oscillating contribution of the optical density changes of ceric ions is suppressed by recording the difference. The periodicity is indicated in a stepwise increase towards a higher optical density (downward deflection of the trace, see arrows) at 270 nm relative to 356 nm.

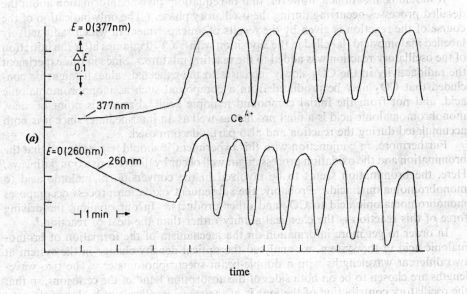

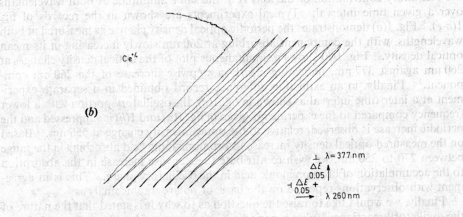

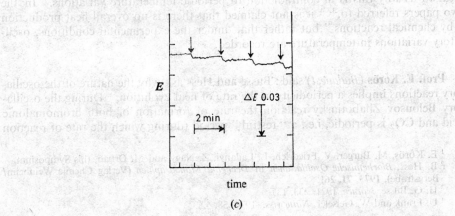

It should be mentioned, however, that the equations give no information about the detailed processes occurring during the oscillatory phase. The only indication of the course of the reaction is given by the results of an experiment in which radioactively labelled malonic acid (labelled in the 2 position with ^{14}C) 10 minutes after the initiation of the oscillatory reaction was added to the reacting mixture. Since in this experiment the radioactivity in the CO_2 slowly increases to the expected value, it might be concluded that CO_2 may be produced from a compound such as monobromomalonic acid, and not from the initial compound malonic acid. From this point of view, monobromomalonic acid is a final product as well as an intermediate, since it is both accumulated during the reaction and also partly decomposed.

Furthermore, in conjunction with this experiment it should be mentioned that the bromination and the oxidation process might well occur by different reaction pathways. Here, the bromination seems to be involved in the conversion of malonic acid to monobromomalonic acid. Probably, the subsequent oxidation process decomposes monobromomalonic acid to CO_2 and other products. In our opinion, the driving force of this reaction is the chemical affinity rather than the heat of reaction.[1]

In order to get more information on the mechanism of the formation of bromomalonic acid in the system, we analysed the optical density change in the system at two different wavelengths with a double-beam spectrophotometer. The two wavelengths are chosen to be on both sides of the absorption band of the ceric ions, so that the oscillatory contribution of the ions is of the same amplitude at both wavelengths over a given time interval. Typical experiments are shown in the records of fig. 1(a-c). Fig. 1(a) demonstrates the periodic optical density change as measured at both wavelengths, with the component absorbing at 260 nm slowly increasing in its mean optical density. Fig. 1(b) shows a simultaneous plot of the optical density changes at 260 nm against 377 nm, indicating clearly a stepwise increase of the 260 nm component.[2] Finally, in an extinction difference record obtained in a separate experiment at a later time interval as shown in fig. 1(c), the oscillatory portion with a lower frequency compared to the experiments given in fig. 1(a) and 1(b) is suppressed and the periodic increase is observed, relative to the optical density change at 356 nm. Based on the measured optical density increase of bromomalonic acid absorbing in the range between 270 to 250 nm, we wish to attribute the stepwise increase in the absorption to the accumulation of bromomalonic acid in the oscillatory phase. This is in agreement with observations reported on the basis of polarographic analysis.[1]

Finally, we would like to raise the question as to why it is stated that the nature of the oscillatory reactions implies a periodicity in the rate of heat evolution. It is not clear to us why this is in contradiction to periodic temperature variations. In the two papers referred to,[3,4] it is not claimed that there is no overall heat production " by chemical reactions " but rather that, under the experimental conditions, oscillatory variations in temperature are recorded.

Prof. E. Kőrös (*Budapest*) said : Busse and Hess ask why the nature of the oscillatory reactions implies a periodicity in the rate of heat evolution. During the oscillatory Belousov–Zhabotinsky reaction the rate of formation of both bromomalonic acid and CO_2 is periodic, i.e., a " restful " period (during which the rate of reaction

[1] E. Kőrös, M. Burger, V. Friedrich, L. Ladanyi, Z. Nagy and M. Orban, this Symposium.
[2] B. Hess, *Biochemische Oscillationen* in *Dechema Monographien* (Verlag Chemie Weinheim/ Bergstrabe), 1973, **71**, 261.
[3] H. G. Busse, *Nature*, 1971, **233**, 137.
[4] U. Frank and W. Geiseler, *Naturwiss.*, 1970, **58**, 52.

is low) is followed by a " burst " period (during which the rate of reaction is high).[1] Our calorimetric investigations on the different Belousov–Zhabotinsky systems unambiguously proved that the amount of heat evolved has a linear relationship with the amount of bromomalonic acid formed. Thus it can be expected that the rate of heat evolution should exhibit the same periodicity as the rate of formation of bromomalonic acid, provided that the heat transfer between the reaction mixture and its environment is slow. Our calorimetric measurements provided the experimental proof for that expectation.

Dr. J. R. Bond (*Leeds*) (*communicated*) Körös correctly disposes of previous suggestions (ref. (10 and (11) of his paper) (1) that temperatures fall below ambient as well as rising above it during oscillations in the Belousov reaction, and (2) that there is no overall heat production. However, temperature peaks are entirely possible ; temperatures only rise monotically if heat transfer is slow, and the fact that Körös recorded only monotonically increasing steps is simply due to this. We have measured temperatures in the Belousov reaction in rapidly stirred conditions, and have observed peaks and valleys in the temperature-time record. Like Körös, we find no evidence for cooling below ambient temperature ; overall, the reaction is strongly exothermic.

Dr. O. E. Rössler (*Tübingen*) said: Stirred-flow results [2, 3] suggest that an " upper state " excitable version of the Z-reagent (being blue in the resting state) may also be possible. Even more tricky, though perhaps still possible, would be the realization of a *doubly* excitable Z-reagent (switching readily not only from red to blue, but also from blue to red). In it, two actively propagated moving folds (trigger waves) chasing each other would be possible, thus allowing of new nontrivial spatial patterns. Such a reagent would in the simplest case again be described by my eqn (1) (as indicated in the discussion following the paper by Field and Noyes), with $k_5 = 2.5$ and $k_6 = 0.3$; analogue computer results. In biology, a doubly excitable system has been detected recently [4] (an optic nerve propagating impulses of variable length).

Prof. R. M. Noyes (*Oregon*) said: Rössler's suggestion of a blue resting state is stimulating. It is significant he observes such a blue state only in a stirred flow reactor from which bromomalonic acid is constantly being removed. The reasons are discussed in my response to Sørensen's comment.

A blue resting state will be generated only if $f < 1$ (and probably also $f < 0.5$) in the Oregonator model. It occurs to me that a free radical trap such as the polymerizable monomer acrylamide might lower f enough to create a blue excitable state. Excitation would require a mechanism that suddenly generated bromide ion instead of depleting it as in my paper. I do not see how a solution could be made to switch from blue excitable to red excitable unless a catalyst could be found for which the equivalent of k_{M5} is several hundred times that for cerium(IV).

Dr. P. Ortoleva and **Prof. J. Ross** (*MIT*) said: The concept that kinematic waves in oscillatory systems depend weakly on diffusion has been analysed by a perturbation series in the effects of diffusion and the weakness of imposed gradients and hetero-

[1] E. Körös, *Nature*, 1974, **251**, 703.
[2] G. Junkers, *Über die periodisch verlaufende Reaktion zwischen Malonsäure und Bromat in der Gegenwart von Cerionen* (Diploma Thesis, University of Aachen, 1969).
[3] P. G. Sørensen, this Symposium (discussion remark following the paper of Körös *et al.*).
[4] F. Zettler and M. Järvilechto, *J. comp. Physiol.*, 1973, **85**, 89.

geneities. The developments in ref. (1)-(3) in Winfree's paper appear as a natural consequence in lowest order of this phase diffusion theory.[1]

In addition, the presence of heterogeneities in an oscillatory reagent has been shown by means of this theory to lead to the emission of localized wave patterns. Expressions were derived for the wavelength of the emitted waves. It was shown that to first order in the strength of the heterogeneity, the period is just that of the bulk reagent, T. We note that if the heterogeneity tends to slow down the bulk dynamics, a pattern of incoming waves emerge, although to our knowledge this has not been observed in Z-reagent.

Dr. M-L. Smoes (*Dortmund*) (*communicated*): 1. It is not correct that the work of Smoes and Dreitlein [2] describes kinematic waves. What we have proposed is an interpretation of the spatio-temporal structures which appear spontaneously in the distributed Zhabotinskii oscillatory system. The model reproduces qualitatively all the well-established features of this system: bulk oscillations, leading centres with periods shorter than the bulk period from which waves propagate, non-uniformity of the wavelengths, annihilation of colliding wavefronts, formation of simple and double spirals in two dimensions.

The fundamental hypothesis is that the period of oscillation, depending on temperature and parametric concentrations, is subject to perturbations. The leading centre appears as the result of a fluctuation in parametric concentration, for instance. The waves originating from the centre are due to the regression of the initial perturbation through diffusion. Diffusion is thus important in this model. Although in the work cited above we neglected the diffusion of the oscillating intermediate concentrations, this was done only in order to save computer time. We have now evaluated the effect of the diffusion of the intermediates; the results show that the speed of propagation of the wavefronts becomes very constant except near the leading centre and just before annihilation by the bulk oscillations or colliding wavefronts.[3]

2. The distinction between " pseudo-waves " and " trigger waves " suggested by Winfree on the basis of the behaviour of these waves in the presence of impermeable barriers is unwarranted. As we have shown elsewhere,[2, 4] all chemical waves are blocked by truly impermeable barriers. However, if two points of the distributed system are in such a state that they will become oxidised in succession, due to an established phase difference, the introduction of a barrier between the two points will not prevent the successive oxidations to take place as expected.

3. I have verified that no special increase in the number of leading centres can be recorded when dust is added directly into the Zhabotinksii oscillatory system. (One must keep in mind that new centres occur spontaneously and randomly during the whole reacting time). Although no positive results are obtained with dust, a local small contamination of the solution by concentrated sulphuric acid has been shown to produce propagating waves that cannot be distinguished from the spontaneous waves which are simultaneously observed.

More experiments of this kind should be done in order to determine if a dust theory can really explain the most interesting waves in the Zhabotinksii system.

Dr. J. F. G. Auchmuty and **G. Nicolis** (*Brussels*) (*partly communicated*): The classification of waves into " pseudowaves " and " trigger waves " based on

[1] P. Ortoleva and J. Ross, *J. Chem. Phys.*, 1972, **58**, 5673 ; 1974, **60**, 5090.
[2] M-L. Smoes and J. Dreitlein, *J. Chem. Phys.*, 1973, **59**, 6277.
[3] M-L. Smoes, to be submitted.
[4] M-L. Smoes, *Ph.D. Dissertation* (University of Colorado, 1973).

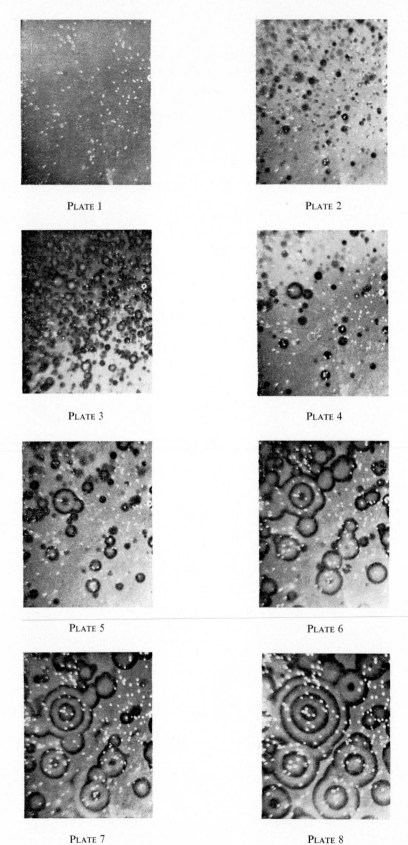

PLATE 1

PLATE 2

PLATE 3

PLATE 4

PLATE 5

PLATE 6

PLATE 7

PLATE 8

To face page 99]

the very interesting experiments by Winfree seems to be appropriate for describing the propagation of disturbances in chemical systems.

However, it seems that there can be other types of oscillations in such systems. In particular, chemical systems may undergo periodic oscillations, which may even be spatially dependent (cf. the paper of Nicolis and Prigogine). These oscillations are different from those appearing in classical electromagnetism and elsewhere, in that they are often independent of the initial conditions. Such oscillations appear as the asymptotic behaviour in time of these systems, and are very similar to limit cycle behaviour for ordinary differential equations. Their characteristics depend very strongly on diffusion, which is responsible for initiating the instability leading to these patterns and for synchronizing the local limit cycles.

We believe that it is important to distinguish, in chemical systems, between *transient* behaviour and *long-time* behaviour and to point out that chemical systems can evolve to stable time-periodic and space-dependent states which are not typical propagating waves.

Prof. E. Körös (*Budapest*) said: Here I should like to report on our observations with the unstirred tris(dipyridine)ruthenium(II)-[Ru(dipy)$_3^{2+}$]-catalysed Belousov system. It is known that Ru(dipy)$_3^{2+}$ is a catalyst in the Belousov oscillating system,[1, 2] and our investigation on the Ru(dipy)$_3^{2+}$-catalysed reacting Belousov system have revealed its close resemblance to the ferroin-catalysed one.[3] This fact encouraged us to look at the spatial behaviour of the former system.

The experiments were performed in a Petri dish of about 10 cm diameter, at ambient temperature with the following reagent concentrations in the final mixture: 0.3 M sodium bromate, 0.065 M monobromomalonic acid, 0.05 M malonic acid, 0.3 M sulphuric acid and 0.003 M Ru(dipy)$_3$Cl$_2$. To a reaction mixture of about 9 ml one drop of 0.1 % TritonX-100 was added, and swirled until the mixture was homogeneous. The temporal oscillation started immediately after mixing the reagents, simultaneously a large number of pacemaker (trigger) centres formed, and from some of them oxidation bands started to travel. This phenomenon is visible even in laboratory illumination; the colour contrast, however, is not marked enough. For this reason the photographs were taken in u.-v. light illuminating the solution with a 360-nm radiation. (Namely, Ru(dipy)$_3^{2+}$ exhibits luminescence when irradiated with a 360-nm radiation; however, Ru(dipy)$_3^{3+}$ can not be excited by u.-v. energy.) A series of 8 photographs demonstrates the generation and propagation of the chemical waves. Plate no. 1 was taken first, and the following photographs were taken at suitable intervals. Plate no. 8 was taken approximately after 10 minutes. (The small bubbles on the photographs are from carbon dioxide.)

With the Ru(dipy)$_3^{2+}$-catalysed system the development and propagation of trigger waves is accompanied by the occurrence of pseudo waves. During the early period, pseudo waves annihilate most of the trigger centres and only a few of them can develop further. Plate 1. shows the system without trigger centres and when the catalyst is in the oxidized form, [Ru(dipy$_3^{3+}$]; on plate 2 a few trigger centres (black spots) are already discernible, their development can be seen on plate 3, pseudo waves propagating through the medium, however, annihilate most of the trigger centres (plate 4). Plates 5–8 show the trigger waves in the progressively developed phases. (Especially easily

[1] J. N. Demas and D. Diemente, *J. Chem. Ed.*, 1973, **50**, 357
[2] E. Körös, L. Ladanyi, V. Friedrich, Zs. Nagy and A. Kis, *Reaction Kin. Cat. Letters*, 1974, **1**, 455.
[3] E. Körös, M. Burger, V. Friedrich, L. Ladányi, Zs. Nagy and M. Orbán, paper at this Symposium.

observable is the propagation of chemical waves from a trigger centre at the upper left part of the plates.)

Similar to the ferroin-catalyzed system, the leading edge of an oxidizing band is sharp, the trailing edge of the band, however, is diffuse, i.e., there is a continuous change from the totally oxidized form of the catalyst to the totally reduced one. $Ru(dipy)_3^{2+}$ as a catalyst has a great advantage over ferroin, the former being an inert complex and stable even in highly acid medium over a very long period of time. On the other hand ferroin is subjected to protolytic decomposition.[1]

The chemical mechanism controlling the band migration is the same as that proposed by Field and Noyes [2, 3] for the $Fe(phen)_3^{2+}$-catalysed system.

Dr. A. Winfree (*Indiana*) said: My answers to various questions which have been posed informally are as follows

(1) Could the observed scroll waves be transient rather than stable solutions to the reaction/diffusion equation? To which I would reply

(a) Numerical solutions to a reaction/diffusion equation not unlike Z reagent include spiral waves rotating without detectable change of shape or angular velocity in a square box for 10 cycles.

(b) Fig. 2 shows no change of period in 26 cycles of a scroll *ring* in Z reagent. Observations were terminated by a committee meeting, not by an instability of the reaction.

(c) Scroll waves seem to be attracted to interfaces and to counter-rotating scrolls less than 1/2 wavelength away. Tiny scroll *rings* may, therefore, be slowly contracting (without change of period, according to fig. 2). Twisted and knotted scroll rings may have escaped detection up to now because they are violently unstable; but possibly only because the required initial conditions have never been arranged.

(d) The whole reaction dies out after enough malonate has been decarboxylated, so in that sense all the waveforms are transients.

(2) Busse asked what are the fuzzy red ripples sometimes seen near the end of an experiment with Z reagent.

Well, I don't know what these are. They are not anything like scroll axes though. They seem to be less excitable regions: plane-wave trains lying obliquely to these ripples are interrupted at crossings. They resemble thermal convection cells, and may be regions where oxygen is transported into the medium from the air above.

(3) Is the Z reaction really homogeneous, or are there suspended particles, gas bubbles surface films, etc. of importance in determining wave geometry? I would state in answer to this question that

(for reasons stated in the manuscript) I believe that the circular-wave pacemakers with diverse periods between T and T_0 are heterogeneous nuclei. I have found no threadlike heterogeneity along the scroll wave's axis. In reagent filtered at 0.2 μm and covered against dust, scroll rings behave normally while microscopic observation reveals no turbidity, gas bubbles, surface films or suspended precipitate of brominated ferroin.

[1] B. Z. Shakhashiri and G. Gordon, *J. Amer. Chem. Soc.*, 1969, **91**, 1103.
[2] R. J. Field and R. M. Noyes, *Nature*, 1972, **237**, 390.
[3] R. J. Field and R. M. Noyes, *J. Amer. Chem. Soc.*, 1974, **96**, 2001.

Prof. R. M. Noyes (*Oregon*) said: It is unfortunate the dithionite measurements of De Poy and Mason required titration of samples removed by pipette from a solution very sensitive to oxygen-initiated autocatalytic reaction. For the observations reported, points from no more than two successive titrations deviated in the same direction from a smooth curve representing non-oscillatory autocatalytic disproportionation of dithionite. Hence it is not established with certainty that oscillations even exist, although the observations of pH and of turbidity are certainly very suggestive. Let us hope it will be possible to develop an analytical procedure that will continuously monitor dithionite concentration.

Even if the reported oscillations in titre are real, it is doubtful that they solely reflect changes in concentration of the dithionite starting material. If several percent of starting material is consumed and then regenerated, a still greater amount of some intermediate must build up and then break down by disproportionation. The radical intermediates proposed here could not attain the concentrations necessary to produce the observed changes in titre. It therefore seems probable the points in fig. 2 represent composite concentrations of dithionite and of some yet unidentified intermediate species.

Furthermore, with regard to Matsuzaki's paper, I am disturbed about the proposed mechanism because it employs unprecedented species like $H_2I_2O_3$ and $H_3I_3O_5$, because it regards elementary oxygen as an inert product, and because it produces oxygen only by reduction of iodine containing species. It is well established that oxygen is produced most rapidly at times when iodine species are undergoing net oxidation.

Sharma[1] has recently shown that visible light can shift the system from a presumably nonradical condition characterized by relatively high iodine ion concentration and slow evolution of oxygen to a radical condition characterized by much less iodide ion and by much faster oxygen evolution. He has also shown that the system becomes grossly supersaturated with oxygen during oxidation of iodine-containing species and that the pressure of oxygen has an important effect on reaction behaviour.

These observations indicate that the mechanism involves autocatalytic switching between a nonradical condition during which iodate is reduced and a radical condition during which iodine is oxidized. A detailed mechanism of such type will be documented in a longer manuscript.

Prof. D. M. Mason (*Stanford*) (*communicated*): We concur with Noyes that continuous observatons of the dithionite system are highly desirable, and plan to use e.s.r. to follow some of the free radicals with time to determine if their concentrations also oscillate. We feel that the reproducibility of Rinker's data, coupled with the continuous pH measurements, does provide strong evidence of the existence of oscillations of the dithionite concentration.

We also agree with Noyes that other intermediates may affect the titre but that they could not exist in sufficient concentrations necessary to make the observed oscillations. It is for this reason that we feel that the dithionite is, in effect, an intermediate, being consumed by the thermal decomposition and formed from side products, with which it has reached equilibrium at the low temperature at which it is initially prepared.

Prof. I. Matsuzaki (*Japan*) (*communicated*): In response to Noyes, I would like to make the following remarks

[1] K. R. Sharma and R. M. Noyes, *J. Amer. Chem. Soc.*, 1975, **97**.

1. On the existence of $H_2I_2O_3$.

This species was postulated by H. A. Liebhafsky.[1]

2. On the existence of $H_3I_3O_5$.

This species is used in the paper as the key species for the oscillation. However, its existence is left to be examined, that is, I do not at present intend to claim anything concrete about it more than is described in the text. From the standpoint of oxidation number it had better been considered to be a complex $H_2I_2O_3 \cdot HIO_2$. We merely devised this chemical formula as one possibility in order to embody the theoretical conclusion that the autocatalytic back-activation step to be contained should have an order higher than 1.

3. On the effect of light resulting in the decrease in $[I^-]$.

Let me consider the effect of light on the basis of our non-radical mechanism. According to our mechanism a decrease in $[I^-]$ results in an increase in $[HIO]$ which in turn causes $[HIO_2]$ to increase; these increases will favour the functioning of the autocatalytic back-activation step $2HIO_2 + HIO + H_2O_2 \rightarrow 3HIO_2 + H_2O$. In connection with the mechanism in which light causes $[I^-]$ to decrease, a few runs of experiments with a 150 W tungsten lamp have been made with results for which the well-known reaction

$$2I^- + \tfrac{1}{2}O_2 + H_2O \xrightarrow{\text{light}} I_2 + 2OH^-$$

might be considered to be responsible. With this finding I have come to think that addition of the above process to our current mechanism would extend the range of application of the mechanism to reaction conditions with irradiated light of varying intensities.

4. On the validity of a mechanism.

I have not established the mechanism for the complex oscillation but merely proposed a plausible mechanism. To establish a mechanism for complicated phenomena we need to get a number of successful checks on its capability of accounting for a variety of phenomena.

Prof. R. M. Noyes (*Oregon*) (*communicated*) : Because the hydrolysis of elementary iodine is a rapidly established equilibrium, I must disagree with Matsuzaki's claim that a decrease in $[I^-]$ will immediately cause a decrease in $[HOI]$; the effect will be in the opposite direction. However, I agree with Matsuzaki that any decision about alternative mechanisms will require a more detailed analysis of kinetic data than is possible here.

[1] H. A. Liebhafsky, *J. Amer. Chem. Soc.*, 1931, **53**, 2074.

B. Thermokinetic Oscillations

Thermokinetic Oscillations accompanying Propane Oxidation

By Peter Gray, J. F. Griffiths and R. J. Moule

Department of Physical Chemistry, The University, Leeds LS2 9JT

Received 7th August, 1974

Gaseous hydrocarbon oxidations are often accompanied by remarkable non-isothermal phenomena, such as oscillatory cool flames and complex ignitions. These owe their existence to an interplay between kinetics and self-heating via thermal feedback in a system involving chain branching. The present work illustrates these phenomena and provides a quantitative assessment of thermokinetic concepts applied to them.

Propane is oxidised in a stirred closed reactor. Characteristically, time-dependent oscillations occur in the pressure range 40-100 kN m^{-2} at vessel temperatures between 570 and 660 K. Substantial temperature changes accompany the oscillating reaction rate, and these changes are measured by a fine (25 μm) thermocouple and rapid recording equipment. Oscillating temperatures may vary in form from roughly sinusoidal with damping, to steep cusp-like maxima of hardly diminished amplitude, multiple events being terminated abruptly by the consumption of fuel.

Thermokinetic oscillations depend vitally on thermal feedback and hence on heat transfer: by the deliberate variation of heat transfer properties such as heat capacities and thermal conductivities (with diluting gases), and intensities of stirring, we are able to test this dependence.

The spontaneous oxidation of hydrocarbons in the gaseous phase not only shows " slow " and explosive modes of reaction but also may be accompanied by periodic pulses of light emitted during repetitive bursts of enhanced reactivity.[1-3] The light emission is very feeble and an associated pulse in gas temperature [4] is generally less than 200 K; these phenomena are called " cool flames " to distinguish them from normal explosive burning.[5]

Cool flames are now recognised as oscillatory reactions: in closed reactors up to 11 successive events are observed,[6] and in open systems they may be sustained indefinitely.[7] The temperature attained in a cool flame pulse varies between about 5 and 200 K, according to the initial reactant temperature and concentration,[8] though in closed reactors successive flames become progressively weaker since moderate proportions of reactants are consumed in each of them (up to 20 %).[9]

The earliest recognition of oscillatory reaction attracted contrasting interpretations. At first, isothermal explanations were offered [10-12] based on superficial analogy to the Lotka–Volterra autocatalytic scheme already applied in some electrochemical and biological contexts. However, this simple isothermal explanation fails to offer a satisfactory interpretation of hydrocarbon cool flames. Amongst the reasons why the Lotka–Volterra mechanism does not survive detailed scrutiny are (i) its failure to predict an alternate growth and decay in concentrations of intermediates that is *not* determined by the initial composition (it predicts conservative oscillations, not limit cycles),[13] and (ii) the impossibility of making a satisfactory chemical identification with the participants of the isothermal kinetic model.[2] Recent, more elaborate, isothermal models are hardly more acceptable since schemes that predict limit cycles have to invoke chemically unlikely elementary reactions.[13, 14] (The best success so far with an isothermal scheme seems to be that of Field and Noyes [15] who have fitted the rigorous criteria for isothermal oscillatory models to the chemical framework of the Belousov [16] reaction. Autocatalysis in their kinetic mechanism occurs via chain branching.)

In hydrocarbon oxidation there is an obvious alternative feedback mechanism to isothermal autocatalysis. This is thermal feedback, which takes account of the strongly exothermic properties of reaction. The earliest thermal feedback model was proposed by Salnikoff [11, 17] who showed that two consecutive first order reactions (viz. $A \rightarrow B \rightarrow C$) will generate sustained oscillations in the concentration of the intermediate (B) provided that certain (plausible) criteria apply to the exothermicities and Arrhenius parameters of each step.

Present day thermokinetic models, such as that of Gray and Yang [18] or of Halstead, Prothero and Quinn [19] remain simple, though now they feature chain branched autocatalysis. Through it, these schemes predict not only oscillatory regimes but also other striking and unusual features of hydrocarbon oxidation such as the negative temperature-dependent heat-release rate, the single and multiple-stage ignitions, and the "lobes" associated with the ignition limit in a pressure-temperature ignition diagram.

Implicit in each of these treatments is that simply to devise a plausible model chemical scheme is not enough : proper analytical methods must be applied to prove that stable oscillatory behaviour is possible. Moreover, extension beyond a qualitative test requires not only correct information on stoichiometries, kinetic constants and thermochemistry but also correct thermal parameters of the system. We need to supplement conventional kinetic investigations with measurements of heat release and loss rates and of how they depend on temperature. Where theoretical treatments have been simplified, such as by assuming spatially uniform temperatures and concentrations, we need to devise experimental techniques in which nature is made to imitate art.

Accordingly, the present paper describes new investigations of the thermal effects accompanying propane oxidation. A closed reactor is used in which the temperature excess of reactants is made uniform by mechanical stirring. Heat-loss rates are measured and, by deliberately changing them, the dependence of cool flame oscillations on thermal feedback is tested.

THEORETICAL FOUNDATION

Most order has been brought into the classification and analysis of oscillatory reactions by phase-plane analysis. Even in systems that are chemically as complex as hydrocarbon oxidations many of the features can be rationalised in terms of a single key intermediate participating in branching and termination processes.[18, 19] The starting point for analysis of a thermokinetic model is the pair of expressions for the generation and loss of heat and for the generation and loss of the chain branching species (X). These conservation equations may take the general forms :

$$\text{for energy}; \quad c\dot{T} = \mathcal{R}(T, P, X) - l(S/V)(\bar{T} - T_0) \tag{1}$$

$$\text{for mass}; \quad \dot{X} = -f(T, X) \tag{2}$$

where $\mathcal{R}(T, P, X)$ is the rate of heat release per unit volume and $f(T, X)$ is a function involving the rates of reactions from which X is formed or in which X is removed. Each of these parameters depends upon the concentration of X. l is the heat transfer coefficient between gas and reactor walls, $(\bar{T} - T_0)$ is the spatially averaged temperature excess of the reactants, c is their heat capacity per unit volume and S/V is the surface to volume ratio for the reactor.

Solutions of these equations for T and X begin by location of singular points in the (T, X) plane and identifying the behaviour close to them. Closed-curve trajectories in the (T, X) plane correspond to oscillatory behaviour in time, illustrated by either $T(t)$ or $X(t)$. This connection causes us often to describe the experimental, time-dependent observations of undiminished, sustained oscillations by phase-plane terminology, i.e., as " limit cycle behaviour ".

Experimentally $T(t)$ may be observed via direct measurements of temperature (as in our experiments [8, 20] and in those of others [4, 21]) or $X(t)$ by following the changing concentration of one or more appropriate intermediate species. So far, experimental evidence demonstrating the alternate growth and decay of chemical intermediates in cool flames is scant [21, 22]; this is because the important branching intermediates are extremely difficult to identify and measure continuously.

EXPERIMENTAL

MATERIALS

Propane (instrument grade B.O.C. Ltd.) and acetaldehyde (Analar, B.D.H. Chemicals Ltd.) were distilled *in vacuo* before use. Oxygen and diluent gases (B.O.C. Ltd.) were taken directly from cylinders.

APPARATUS AND PROCEDURE

The apparatus and general procedure have been described previously.[20, 23] Good mixing of reactants was achieved by a double-vaned rotor spinning round a vertical axis within the reactor (Pyrex, 0.5 dm³, spherical). The rotor, magnetically driven, was made of stainless steel previously coated with a ceramic layer to minimise surface reactions. The reactor was thermostatted in a re-circulating air furnace to ± 1 K over its surface in the range 550-650 K.

Propane, oxygen and reactive additives or inert diluents were pre-mixed and stored in a conventional Pyrex vacuum line. Reactants were admitted to the reactor via an electromagnetic valve opened reproducibly for 0.1 s. Initial pressures in the reactor (typically 30-100 kN m⁻²) were measured from the output e.m.f. of a pressure transducer (Ether Ltd. UP4).

To detect and follow accurately the temperature changes in gaseous reactions the measuring device must have a fast response, a very small thermal capacity, low thermal conductivity, adequate sensitivity, and it must be of robust construction. Our thermocouples, gas welded from very fine Pt–Pt/13 % Rh wire (25 μm diam.), come close to satisfying these criteria. Their response time is less than 20 ms in moving gas [24] and so they are able to give a faithful record of all behaviour except hot ignitions.

A very fine junction, coated with a thin layer of silica from a methanol+silicon oil flame, was situated within the reactor and the reference junction (100 μm diam. wire) was placed on the outside of the vessel wall. The e.m.f. generated from these junctions was amplified and displayed on a light sensitive chart by an ultra-violet recorder (Southern Instruments Ltd.). The interior probe was moveable across a horizontal diameter of the reactor so that temperature-time histories could be obtained at any position across the reactor. Temperature–time records for the same initial conditions, but for different positions of the probe, were combined to allow temperature–position profiles to be mapped at successive instants of reaction. Records for the same propane+oxygen+additive mixtures at different initial temperatures and pressures were combined to map the various types of non-isothermal behaviour on pressure–temperature ignition diagrams.

Heat transfer coefficients (l) for all of the reactant mixtures were measured at various temperatures and pressures in additional experiments. With knowledge of vessel dimensions and of reactant heat capacities, l may be evaluated from the measured temperature excess of reactants (eqn (1)). Two routes are open; either via a steady state ($\dot{T} = 0$), if energy is supplied at a known rate,[20, 23] so that

$$l(S/V)(T' - T_0) = \mathscr{R}'_{ss} \tag{3}$$

or by a dynamic method from the rate of exchange of heat between the gas and reactor walls. If $\mathscr{R} = 0$, as is the case when inert gas is heated or cooled adiabatically,[20]

$$\dot{T} = (lS/cV)(T - T_0) = (1/\tau)(T - T_0) \tag{4}$$

$$\Delta T = \Delta T_0 \exp(-t/\tau). \tag{5}$$

The heat transfer coefficient has a natural identity with the characteristic relaxation time (τ) for heat dissipation between the gas and reactor walls.[25] We have chosen this dynamic method to measure l. Gas was expanded adiabatically from the reaction vessel, and the temperature–time history was followed as the residual gas warmed back to ambient temperature. The characteristic relaxation time (τ) was derived from the gradient of the graph of $\log \Delta T$ against time.

RESULTS

TEMPERATURE UNIFORMITY, HEAT LOSS RATES AND HEAT TRANSFER COEFFICIENTS

We have measured [23] the spatial variation of temperature in the reactor. Except near the walls, the effect of stirring is to make temperature excesses very nearly uniform across the reactor. It is this which is the experimentalists' justification for expressing heat loss rates in a Newtonian form (eqn (1), (3) and (4)) and is consistent with some theoretical models for cool flame [17-19] and other [26, 27] non-isothermal gaseous reactions. Similar spatial temperature distributions prevail for stirring speeds in the range 1200-2400 r.p.m.

Heat loss coefficients are determined for different gas mixtures in a variety of conditions by a dynamic method. A typical adiabatic quenching and relaxation curve is shown in fig. 1, and with it a semi-logarithmic plot of $\log \Delta T$ against time.

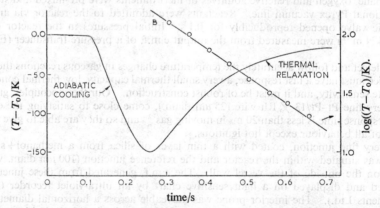

FIG. 1.—Curve (a) An adiabatic cooling followed by thermal relaxation back to ambient temperature (620 K) for nitrogen at 50 kN m^{-2} in a stirred reactor (0.5 dm^3). Temperature changes are measured directly by a Pt—Pt/Rh (13 %) thermocouple (25 μm diam. wire). Curve (b) A plot of $\log(T - T_0)$ against time for thermal relaxation.

The relaxation time, from which l is evaluated (eqn (4)), is determined from the gradient of this line. Values for τ and l at various conditions are given in table 1.

The actual values of l may be expected to depend on transport properties, especially λ (here altered by dilution), on density (here varied by varying pressure), and on stirring efficiency (here varied by the speed of the rotor). The lowest values of l prevail at the lowest pressures, and stirring rates (table 1). High concentrations of propane yield high values of l, thus the simple expedient of dilution with inert gases of low thermal conductivity and heat capacity gives scope to reduce the heat transfer coefficient. τ is hardly dependent on pressure: the variation of l arises mainly from the pressure dependence of c.

TABLE 1.—HEAT TRANSFER COEFFICIENTS FOR GASEOUS MIXTURES IN A SPHERICAL VESSEL ($0.5\ dm^3$) AT VARIOUS CONDITIONS

composition/mol %	pressure/kN m^{-2}	temp./K	rotor speed/r.p.m.	τ/s	l/W K^{-1} m^{-2}
$N_2 = 100$	40.0	623	2400	0.16	19.9
	53.3				26.2
	66.6				32.5
$C_3H_8 = 8$	40.0	623	2400	0.18	27.9
$N_2 = 92$	53.3				34.4
	66.6				45.2
$C_3H_8 = 33$	40.0	623	2400	0.26	28.8
$N_2 = 67$	53.3				38.6
	66.6				50.2
$C_3H_8 = 33$	40.0	623	1200	0.32	23.0
$N_2 = 67$	53.3				28.8
	66.6				36.0
$C_3H_8 = 50$	40.0	623	2400	0.27	36.4
$N_2 = 50$	53.3				44.8
	66.6				54.0
$C_3H_8 = 50$	40.0	623	1200	0.34	26.8
$N_2 = 50$	53.3				35.6
	66.6				44.4

TEMPERATURE-TIME RECORDS FOR EXOTHERMIC OXIDATION

The four main types of non-isothermal behaviour during propane oxidation are depicted in our (temperature, time) records. These are: (i) a damped oscillatory approach to a quasi-steady state (fig. 2a, b and c), (ii) nearly undamped oscillations ending abruptly (fig. 2d), (iii) a monotonic approach to a quasi-steady state (fig. 2e), and (iv) two-stage ignition (fig. 2f).

In damped oscillations, successive amplitudes diminished appreciably even after the first temperature peak. For pure propane+oxygen mixtures all multiple cool flames show damped characteristics, their damping factors depending upon the initial conditions of temperature and pressure. Fig. 2a shows cool flame oscillations typical of those observed at low temperatures (roughly in the range 580-620 K). Although amplitudes clearly diminish, the damping factor is low; oscillations occur up to the end of reaction. Initial amplitudes are large (>100 K) and the peaks have a cusp-like shape interspersed by periods longer than 1 s. At higher temperatures (beyond about 620 K), damping is sufficiently high for oscillations to have died away before the fuel is completely consumed (fig. 2c). These temperature histories are roughly sinusoidal, starting with a maximum amplitude that is usually less than 100 K and sometimes as low as 5 K. Periods are generally very short (<2 s).

Cool flame oscillations with barely diminished amplitudes occur at low temperatures in the oxidation of propane to which acetaldehyde is added (even small amounts, say <0.5 mol %). They have steep cusp-like maxima interspersed by shallow minima,

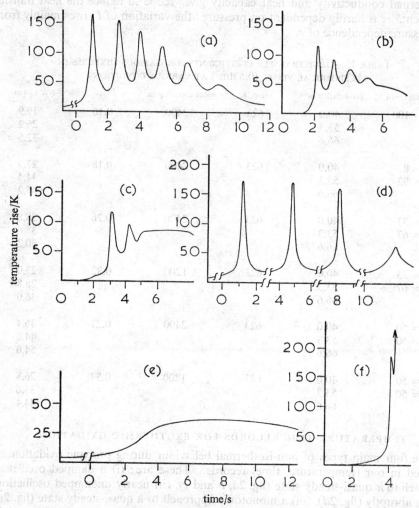

FIG. 2.—(Temperature, time) histories for propane oxidation at different conditions in a stirred reactor (0.5 dm³). (a) to (d) multiple oscillations; (a) C_3H_8 (50 mol %), $T_0 = 610$ K, $P = 75$ kN m⁻²; (b) C_3H_8 (47 mol %)+CH_3CHO (3 mol %), $T_0 = 635$ K, $P = 65$ kN m⁻²; (c) C_3H_8 (50 mol %), $T_0 = 640$ K, $P = 84$ kN m⁻²; (d) C_3H_8 (47 mol %)+CH_3CHO (3 mol %), $T_0 = 600$ K, $P = 60$ kN m⁻²; (e) slow reaction; C_3H_8 (50 mol %), $T_0 = 640$ K, $P = 60$ kN m⁻²; (f) two-stage ignition; C_3H_8 (47 mol %)+CH_3CHO (3 mol %), $T_0 = 590$ K, $P = 85$ kN m⁻².

and they give the impression of sustained oscillations. That is to say, if fuel were to be fed continuously to the system they would propagate indefinitely; in our closed reactor they terminate abruptly when no fuel remains (fig. 2d). Amplitudes vary between 50 and 150 K and periods from 3 to 15 s, the highest temperature excess and largest times being associated with the lowest reactor temperatures (<590 K). These "sustained" oscillations occur below about 630 K, above which there is a marked change to damped sinusoidal oscillations (fig. 2b).

Slow reaction may be sufficiently exothermic to increase the reactant temperature monotonically (by up to 35 K) to a quasi-steady state (fig. 2e). In two-stage ignition the temperature rises by about 200 K, characteristic of a cool flame. This is followed instantly, however, by a violent hot ignition, sometimes after the temperature has begun to fall from its first maximum (fig. 2f).

Induction times to the occurrence of these non-isothermal phenomena may extend from several seconds to several hours. For this reason, fig. 2a, d and e are drawn with an arbitrary time zero. Generally, the longest induction times are associated with the lowest initial temperatures and pressures, but they are reduced dramatically by the addition of acetaldehyde (or other reactive compounds [28]). Induction times in hydrocarbon oxidations are rarely exactly reproducible; they are susceptible to changes of surface, such as the deposition of carbon during a hot ignition.

PRESSURE-TEMPERATURE IGNITION DIAGRAMS

It is most convenient to display these varieties of non-isothermal behaviour according to the initial reactor temperature and the initial reactant pressure for each reactant composition, i.e., as a (P, T_0) ignition diagram (fig. 3 and 4). Because the

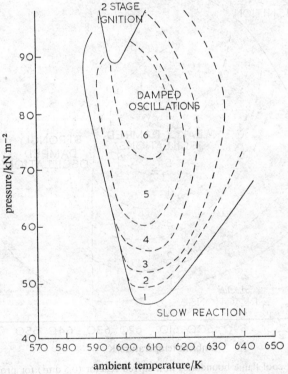

FIG. 3.—Ignition and cool flame boundaries in a stirred reactor (0.5 dm³) for propane (50 mol %) and oxygen.

precise location of boundaries is affected by the reactor dimensions and its surface treatment, each experimentalist has to map his own diagram: experimental ignition diagrams usually agree qualitatively but rarely quantitatively.

The ignition diagram is divided into the main regions of slow reaction, ignition and oscillatory reaction (cool flames); it is this last region that is of particular interest to us. Between 1 and 7 consecutive cool flames occur during propane oxidation, and fig. 3 and 4 indicate the approximate pressure and temperature locations of their boundaries. However, as the temperature histories show (fig. 2a and d), at least for low temperatures in closed conditions, the number of successive oscillations appears to be limited by complete consumption of fuel: the exact number is of less significance than their existence. At high temperatures, oscillations characteristically die away before reaction is complete (fig. 2b and c).

The ignition diagram for pure propane + oxygen mixtures (fig. 3) differs from that when acetaldehyde is added to the reactants (fig. 4). When CH_3CHO is present the cool flame zone extends to lower temperatures (570 K compared with 590 K for pure C_3H_8/O_2 mixtures), moving with it the two-stage ignition boundary. Moreover, there is sufficient distinction between weakly damped and strongly damped oscillations to justify a boundary between them in the (P, T_0) ignition diagram. For C_3H_8/O_2 mixtures, damping increases progressively through the oscillatory region.

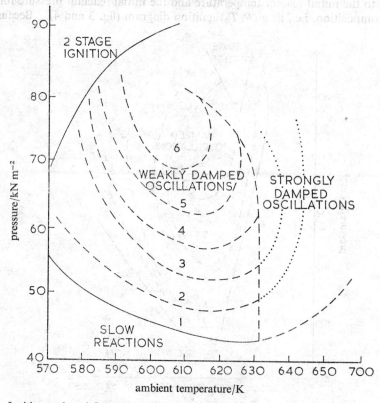

FIG. 4.—Ignition and cool flame boundaries in a stirred reactor (0.5 dm³) for propane (47 mol %) and oxygen with acetaldehyde (3 mol %) added.

When diluting inert gases (N_2 and Ar) are added the cool flame boundary moves to higher pressures. As far as we can tell, the new position is determined exactly by the partial pressure of inert diluent. This restricts our studies because the region in which more than 3 cool flames occurs is moved beyond the pressure limit of our system when sufficient diluent is added to affect l markedly.

THE EFFECT ON TEMPERATURE HISTORIES OF VARIATIONS IN HEAT TRANSFER RATES

The dependence of the non-isothermal behaviour on the magnitude of l is tested most effectively in the closed reactor by altering the stirring rate. As table 1 shows, a decrease of 50 % in the speed of the rotor causes an approximately 30 % decrease of l. Fig. 5 exemplifies how multiple oscillations in the combustion of a 1 : 1 propane + oxygen mixture are changed when l decreases from 45 W K^{-1} m^{-2} (at 2400 r.p.m.) to 35 W K^{-1} m^{-2} (at 1200 r.p.m.). In particular, as l decreases so the damping factor decreases and amplitudes of temperature oscillations are increased.

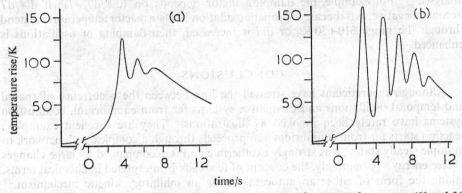

FIG. 5.—(Temperature, time) histories in a stirred reactor (0.5 dm^3) at 625 K for propane (50 mol %) and oxygen at 60 kN m^{-2}. Curve (a) stirring rate = 2400 r.p.m. Curve (b) stirring rate = 1200 r.p.m.

DISCUSSION

HEAT TRANSFER RATES

Over our ranges of temperature, pressure, reactant compositions and stirring rates, l varies from 20-55 W K^{-1} m^{-2} for the corresponding relaxation times 0.16-0.34 s. These values for l agree satisfactorily with those that we have determined in other ways using stationary [29] and quasi-stationary [23] methods. A value for $l = 25$ W K^{-1} m^{-2} implies that when a reaction is producing 1.5 W dm^{-3} ($\equiv 0.15$ kJ mol^{-1} s^{-1} at 0.5 atm and 620 K) a 1 K temperature excess will be maintained in a spherical reactor. Altering the stirring speed changes heat losses; but even in an unstirred system at appreciably less than 0.5 atm., convection is significant,[30] so that further forced convection enhances the heat loss coefficient less than linearly.

Stirring of the reactants not only produces temperature uniformity and enhanced heat loss rates but also destroys any flame structure. (Temperature, time) histories thus become a better mirror of a uniform system.[8]

PRESSURE-TEMPERATURE IGNITION DIAGRAMS

So far as (P, T_0) diagrams are concerned, the areas on them are identified by different temporal behaviour, the properties of which correspond to those of the singularities in the (T, X) phase-plane that are experimentally accessible from fixed initial temperatures in closed reactors, i.e., limit cycles ("sustained" oscillations), stable foci (damped oscillations), stable nodes (slow reaction), and unstable saddle points (two-stage ignition). The boundaries between each of these in the (P, T_0)

diagram which represent criticality in the experimentalists' sense (viz., marginal achievement of ignition, or marginal achievement of oscillation) in the phase-plane correspond to the merging of singularities.

THE DEPENDENCE OF THERMOKINETIC OSCILLATIONS ON THERMAL PARAMETERS

Multiple cool flames are sensitive to changes of reactor temperature and heat transfer coefficients : increases of each cause decreases in amplitudes and increases in frequency and damping factors (fig. 2 and 5). A logical connection between the cause and effect, which can be tested by our results, is provided by phase-plane analysis.[18] For example, the damping factor depends on $(d\mathscr{R}/dT - l)$; if $d\mathscr{R}/dT$ becomes negative, as is the case in propane oxidation[20] when reactor temperatures extend through the range 610-650 K, or if l is increased, then damping of oscillations is enhanced.

CONCLUSIONS

Although theoreticians have stressed the link between the occurrence of spatial and temporal oscillations and dissipative systems far from equilibrium, combustion systems have rarely been invoked as illustrations. They are excellent examples : reaction starts far from equilibrium and proceeds through a complex path network to multiple steady states ; it is strongly exothermic and is accompanied by large changes of free energy. Commonly, the concept of feedback is interpreted in chemical terms, taking the form of either an autocatalytic or an inhibitory kinetic mechanism.[31] In oscillatory combustion systems chemical autocatalysis, even with the unusually responsive changes of rate induced by chain branching, only partly describes what is happening. There is also a thermal contribution to a feedback mechanism which, because of the Arrhenius temperature dependence of elementary reaction rates, is very strongly non-linear.

Isothermal oscillatory systems are apt to depend on artificial reaction schemes with elementary steps of order greater than two in active intermediates ; the corresponding degree of non-linearity may be more easily and naturally attained via the temperature dependence of velocity constants.

[1] G. J. Minkoff and C. F. H. Tipper, *Chemistry of Combustion Reactions* (Butterworth, London, 1962), p. 200.

[2] R. Ben-Aïm and M. Lucquin, *Oxidation and Combustion Reviews*, Vol. 1, ed. C. F. H. Tipper (Elsevier, Amsterdam, 1966) p. 1.

[3] B. Lewis and G. von Elbe, *Combustion, Flames and Explosions of Gases* (Academic Press, New York, 1962).

[4] R. Hughes and R. F. Simmons, Twelfth Symposium (International) on Combustion (The Combustion Institute, 1969) p. 449.

[5] M. Prettre, *Bull. Soc. Chim. France*, 1932, **51**, 1132.

[6] R. E. Ferguson and C. R. Yokley, Seventh Symposium (International) on Combustion (Butterworth, London, 1959) p. 113.

[7] P. G. Felton and B. F. Gray, *Combustion and Flame*, 1974.

[8] J. F. Griffiths, B. F. Gray and P. Gray, Thirteenth Symposium (International) on Combustion (The Combustion Institute, 1971) p. 239.

[9] V. Ya Shtern, *The Gas Phase Oxidation of Hydrocarbons*, English Translation, B. P. Mullins (Pergamon, London, 1962).

[10] D. A. Frank-Kamenetskii, *Zhur. Fiz. Khim.*, 1940, **14**, 30.

[11] D. A. Frank-Kamenetskii, *Diffusion and Heat Transfer in Chemical Kinetics*, English Translation, J. P. Appleton (Plenum, New York, 1969) p. 508.

[12] A. D. Walsh, *Trans. Faraday Soc.*, 1947, **43**, 305.

[13] J. J. Tyson and J. C. Light, *J. Chem. Phys.*, 1973, **59**, 4164.

14 J. J. Tyson, *J. Chem. Phys.*, 1973, **58**, 3919.

15 R. J. Field and R. M. Noyes, *J. Chem. Phys.*, 1974, **60**, 1877.

16 B. P. Belousov, *Sb. Ref. Radiats. Med.*, 1959, **1958**, 145.

17 J. E. Salnikoff, *Zhur. Fiz. Khim.*, 1949, **23**, 258.

18 B. F. Gray and C. H. Yang, (*a*) *Trans. Faraday Soc.*, 1969, **65**, 1614. (*b*) *J. Phys. Chem.*, 1969, **73**, 3395.

19 M. P. Halstead, A. Prothero and C. P. Quinn, (*a*) *Proc. Roy. Soc. A*, 1971, **322**, 377 ; (*b*) *Combustion and Flame*, 1973, **20**, 211.

20 J. F. Griffiths, P. G. Felton and P. Gray, Fourteenth Symposium (International) on Combustion (The Combustion Institute, 1973) p. 453.

21 J. A. Barnard and A. Watts, Twelfth Symposium (International) on Combustion (The Combustion Institute, 1969) p. 365.

22 J. Bardwell and C. N. Hinshelwood, *Proc. Roy. Soc. A*, 1951, **205**, 375.

23 J. F. Griffiths, P. Gray and K. Kishore, *Combustion and Flame*, 1974, **22**, 197.

24 D. Thompson, unpublished results.

25 A. M. Grishin and O. M. Todes, *Doklady Akad. Nauk SSSR*, 1963, **151**, 365.

26 N. N. Semenov, *Some Problems in Chemical Kinetics and Reactivity*, Vol. 2 trans. M. Boudart (Princeton University Press, 1958), p. 87.

27 P. Gray and P. R. Lee, *Oxidation and Combustion Reviews*, Vol. 2, ed. C. F. H. Tipper (Elsevier , Amsterdam, 1967), p. 1.

28 M. D. Scheer and H. A. Taylor, *J. Chem. Phys.*, 1952, **20** 653.

29 D. Thompson and P. Gray, *Combustion and Flame*, 1974.

30 B. J. Tyler and A. F. Tuck, *Int. J. Heat Mass Trans.*, 1967, **10**, 251.

31 G. Nicolis and J. Portnow, *Chem. Rev.*, 1973, **73**, 365.

Oscillatory and Explosive Oxidation of Carbon Monoxide

By Ching H. Yang

Department of Mechanics, State University of New York at Stony Brook,
Stony Brook, New York 11794, U.S.A.

Received 30th July, 1974

The glow and explosion limits of the CO and O_2 system are calculated in wide ranges of composition, water content, surface efficiency, temperature and pressure. The boundary of the oscillatory region in the P-T plane is theoretically predicted. Results are compared with the existing experimental data. The well-known experimental observation that traces of water sensitively affect the oxidation process of carbon monoxide is quantitatively examined and verified.

The kinetic oscillation in the low temperature oxidation of carbon monoxide reported by Dickens, Dove, Harold and Linnett [1] is, perhaps, one of the most striking observations in gas kinetics. Obviously, the autocatalytic chain mechanism in its classical form does not explain this intriguing phenomenon.

In an attempt to construct the kinetic oscillation mechanism mathematically a scheme which focused on the interaction between an autocatalytic chain carrier (O atoms) and an inhibitive intermediate (CO_2 molecules) was postulated.[2] A binary non-linear model was derived from the scheme. Analysis of this simple model not only yielded solutions for sustained oscillation, it also predicted the glow and explosion phenomena that are associated with the oxidation of CO.

In order to examine the detailed kinetic mechanism with realistic rate constants, a general model [3] which conserves all reactants, catalytic impurities (H_2O) and ntermediate species was formulated to contain all proposed reactions without invoking the assumption of steady states. Machine-integrated solutions again predicted the various phenomena. Particular effort was directed to the study of the characteristics of kinetic oscillation.

The wealth of experimental data on CO oxidation, on the other hand, is concentrated on the measurements of glow and explosion limits. It is evident that an extensive calculation programme is needed to map out the limits and oscillatory regions in the P-T plane so that the theory can be quantitatively tested and compared with experimental measurements. This is the primary objective of our present work.

The kinetic scheme previously proposed [3] is slightly modified. The glow and explosion limits are calculated with composition of the mixture, and wall efficiencies of the vessel varied. The boundaries of the oscillatory regions in the P-T plane are delineated. Computed limits with water content, pressure and temperature varied in wide ranges ($[H_2O] \sim 0.001 - 10\%$, $P \sim 1$–400 Torr and $T \sim 400$–1100°C) are compared with the existing data. The well-known observation that explosion limits are sensitively affected by traces of water [1,4-6] is theoretically verified. Wide agreement between the data [7-9] and the theoretical predictions argues convincingly for the validity of the currently proposed kinetic mechanism for the low temperature oxidation of carbon monoxide.

KINETIC MECHANISM AND MATHEMATICAL MODEL

The kinetic scheme contains all reactions proposed previously [3] with the addition of only one new reaction (reaction (18)):

$$CO_2 + O_2 \xrightarrow{k_1} CO_2 + O \tag{1}$$

$$CO + O + M \xrightarrow{k_2} CO_2^*(CO_2) + M \tag{2}$$

$$CO_2^* + O \xrightarrow{k_3} CO + O_2 \tag{3}$$

$$CO_2^* + M \xrightarrow{k_4} CO_2 + M \tag{4}$$

$$CO + OH \xrightarrow{k_5} CO_2 + H \tag{5}$$

$$H + O_2 \xrightarrow{k_6} OH + O \tag{6}$$

$$O + H_2O \xrightarrow{k_7} 2OH \tag{7}$$

$$H \xrightarrow{k_8} \text{wall destruction} \tag{8}$$

$$O \xrightarrow{k_9} \text{wall destruction} \tag{9}$$

$$HO \xrightarrow{k_{10}} \text{wall destruction} \tag{10}$$

$$H + O_2 + M \xrightarrow{k_{11}} HO_2 + M \tag{11}$$

$$HO_2 + HO_2 \xrightarrow{k_{12}} H_2O_2 + O_2 \tag{12}$$

$$H_2O_2 + M \xrightarrow{k_{13}} 2OH + M \tag{13}$$

$$HO_2 \xrightarrow{k_{14}} \text{destruction on wall} \tag{14}$$

$$H_2O_2 \xrightarrow{k_{15}} \text{destruction on wall} \tag{15}$$

$$H + H_2O_2 \xrightarrow{k_{16}} H_2O + OH \tag{16}$$

$$O + OH \xrightarrow{k_{17}} O_2 + H \tag{17}$$

$$H + HO_2 \xrightarrow{k_{18}} 2OH \tag{18}$$

Let $y_1 = [CO_2]_t$; $y_2 = [OH]$; $y_3 = [H]$; $y_4 = [O]$; $y_5 = [HO_2]$; $y_6 = [H_2O_2]$; $y_7 = [CO]$; $y_8 = [H_2O]$; $y_9 = [O_2]$; $m = [M]$. The kinetic equations are then constructed as follows:

$$\frac{dy_1}{dt} = k_2 m y_4 y_7 - k_3 y_1 y_4 - k_4 m y_1 \tag{19}$$

$$\frac{dy_2}{dt} = k_6 y_3 y_9 - k_5 y_2 y_7 + 2k_7 y_4 y_8 - k_{10} y_2 + 2k_{13} m y_6 + k_{16} y_3 y_6 - $$
$$k_{17} y_2 y_4 + 2k_{18} y_3 y_5 \tag{20}$$

$$\frac{dy_3}{dt} = k_5 y_2 y_7 - k_6 y_3 y_9 - k_{11} m y_3 y_9 - k_{16} y_3 y_6 + k_{17} y_2 y_4 - $$
$$k_8 y_3 - k_{18} y_3 y_5 \tag{21}$$

$$\frac{dy_4}{dt} = k_1 y_5 y_9 - k_2 m y_4 y_7 - k_3 y_1 y_4 + k_6 y_3 y_9 - k_7 y_4 y_8 - k_9 y_4 - k_{17} y_2 y_4 \quad (22)$$

$$\frac{dy_5}{dt} = k_{11} m y_3 y_9 - 2k_{12} y_5 y_5 - k_{14} y_5 - k_{18} y_3 y_5 \quad (23)$$

$$\frac{dy_6}{dt} = k_{12} y_5 y_5 - k_{15} y_6 - k_{13} m y_6 - k_{16} y_3 y_6 \quad (24)$$

$$\frac{dy_7}{dt} = -k_1 y_5 y_9 - k_2 m y_4 y_7 + k_3 y_1 y_4 - k_5 y_2 y_7 \quad (25)$$

$$\frac{dy_8}{dt} = -k_7 y_4 y_8 + \tfrac{1}{2}(k_8 y_3 + k_{10} y_2 + k_{14} y_5) + k_{15} y_6 + k_{16} y_3 y_6 \quad (26)$$

$$y_9 = y_9^\circ + \tfrac{1}{2}(y_8^\circ - y_8 - y_7^\circ + y_7 - y_2 - y_4) - y_5 - y_6. \quad (27)$$

$[CO]_t$ represents the sum of the concentrations' of both ground state CO_2 and excited CO_2^* molecules produced in reaction (2) and y_2° represents the initial concentration of the species y_i. All chain termination reactions are assumed in the kinetic region. The third body efficiencies for all species are assumed to be unity with only one exception; a value of 2.3 is selected for CO_2 molecules. Other details of the calculation method are reported in ref. (3).

Selected values of the rate constants for all gas phase reactions are listed in table 1. No systematic attempts had been made to iterate the numerical values of rate constants to achieve the optimum fitting of the data.

TABLE 1

reaction	activation energy/kcal mol^{-1}	Arrhenius factor */ cm^3 mol^{-1} s^{-1}	remarks
(1) $CO + O_2$	60	2.25×10^{11}	ref. (3)
(2) $CO + O + M$	0	1.0×10^{14}	ref. (10)
(3) $CO_2^* + O$	0	5.0×10^{10}	ref. (3)
(4) $CO_2^* + M'$	0	1.25×10^5	ref. (3)
(5) $CO + OH$	1.08	4.2×10^{11}	ref. (10)
(6) $H + O_2$	16.8	2.24×10^{14}	ref. (10)
(7) $O + H_2O$	19.5	4.2×10^{14}	ref. (10)
(11) $H + O_2 + M$	-13	5.0×10^{15}	ref. (10)
(12) $HO_2 + HO_2$	0	1.8×10^{12}	ref. (10)
(13) $H_2O_2 + M$	45	1.17×10^{10}	ref. (10)
(16) $H_2O_2 + H$	0	3.18×10^{14}	ref. (10)
(17) $OH + O$	9	1.3×10^{13}	ref. (10)
(18) $H + HO_2$	0	3.0×10^{13}	(a)

* Arrhenius factor for termolecular reactions has the dimensions of cm^6 mol^{-2} s^{-1}.

(a) A value smaller by a factor of two than the one suggested in ref. (11) is selected.

COMPUTATION RESULTS

The definitions of the various phenomena such as glow, oscillation, etc. are generally vague. In the laboratory they are used to differentiate the different characteristics displayed by the bluish light emission from the reacting gases. In general, if the pulse of emission is visible for an extended period of time, a few seconds or more, it is taken to be a glow. The temperature and the pressure of the mixture is then called the "glow limit". When the emission is intense and brief, lasting no longer than a split second, it is identified as an explosion and the corresponding temperature and pressure is denoted as the "explosion limit". Oscillations are usually referred to a series of glows that appear periodically. The limits of oscillation, however, had not been systematically measured in previous work. Before the presentation of our results it is desirable to define these terms a little more precisely.

Glow—a single stroke of visible emission that lasts a few seconds. After the passage of a glow only a small portion of the reactants in the mixture is consumed. Invisible reactivities are supported for a long time after the glow. In the P–T plane the point where the glow occurs is marked with the sign "$\bar{x}$" as the glow limit in our presentation of data that follows.

Sustained glow—an extended stroke of emission which lasts from a second to several minutes. All reactants are consumed after the passage of a sustained glow. The sustained glow limit is marked with "x".

Oscillation—periodic glows with virtually undamped amplitudes. The pressure and temperature at which this occurs is marked with the sign "$\bigcirc$" in the P–T plane.

After glow—a sustained glow which is preceded by a glow or several cycles of oscillation.

It will be shown that there is no real difference between the solutions of sustained glow and explosion. The criterion to differentiate them by the length of their derivation is arbitrarily chosen. It will also be demonstrated that glows and oscillations are all represented by periodic solutions. For a glow the oscillatory solution is sharply damped with only the first one of its emission peaks visible.

In all cases calculated trajectories for all reactants and all intermediate species are obtained. As expected the concentration of reactants decays with time. The trajectory of either O_2 or CO can be conveniently used to indicate the degree of completion of the reaction at any point in time. O atom concentration is always the highest among all radicals. The implication is that O atoms are involved in the slowest reaction path in the oxidation process. Water concentration is always 180° out of phase with O concentration. Hydrogen containing radicals or intermediates are assumed to form water molecules in the gas phase as soon as they are terminated on the walls. According to a bimolecular association theory ($CO + O \rightarrow CO_2 + h\nu$) proposed by Broida and Gaydon [14] the emission intensity from the reacting gas is proportional to the concentrations of CO and O. A threshold intensity must exist below which the emissions become undetectable. An O atom concentration of 10^{-11} mol cm^{-3} was selected previously [2,3] to be the threshold value and it has been retained in the present work. Since the visible emission always coincides with the O atom concentration peaks, the presentation of the O atom concentration trajectory usually is sufficient to convey the complete kinetic behaviour of the system.

The wall efficiencies are assumed to have the values listed in table 2.

The first set of limits is calculated with $CO + 2O_2$ mixtures; results are plotted in fig. 1. Curves A, B, C, D and E correspond to mixtures with water contents of 10, 1, 0.1, 0.01 and 0.001 % by volume, respectively. The computing procedure follows the same sequence of steps used in the heating method for determining limits in the laboratory. Trajectories of eqn (19)–(27) are integrated for mixtures with fixed water content, composition and pressure. Bath temperatures are raised with increments of 1 or 2 K each time until the desired limits are reached. In fig. 1 solid curves join the points of the lowest temperature at which either a sustained glow or an explosion is calculated. On many occasions, these calculated limits are preceded by glows, oscillations or both at slightly lower bath temperatures. The boundary of the oscillation region in the P–T plane is drawn with a closed curve of broken lines.

No oscillation has been obtained for mixtures containing 10 % water, or above the pressure of 120 Torr. It is interesting to note that the highest pressure at which Dove [12] observed oscillation is 110 Torr.

TABLE 2.—WALL EFFICIENCIES

reaction	wall efficiency
(8) H	10^{-4}
(9) O	10^{-4}
(10) OH	10^{-4}
(14) HO_2	10^{-3}
(15) H_2O_2	10^{-3}

Fig. 1. shows that the largest pressure range (40-110 Torr) within which oscillations are calculated is for mixtures with 1 % water content. This does not mean, however, that oscillations are more likely to be found in wetter mixtures experimentally. It may be that the reverse is true. Thus for a wet mixture, with the pressure fixed, the bath temperature range within which oscillation can be calculated is extremely narrow, often no more than 1 or 2 K. Unless the heating rate is very slow, the heated mixture passes this range and reaches sustained glow or explosion

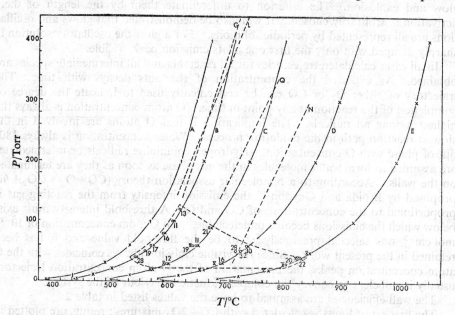

FIG. 1.—Explosion limits and oscillation region of $CO + 2O_2$ mixtures. Calculated curves: A Water content = 10 %, B Water content = 1 %, C Water content = 0.1 %, D Water content = 0.01 %, E Water content = 0.001 %. Experimental results: G Hadman, Thompson and Hinshelwood,[7] Q Gordon and Knipe,[6] L Lewis, von Elbe and Roth,[8] N Dickens.

temperatures before oscillations are fully developed. In contrast, with the drier mixtures that contain water from 0.1 to 0.01 % the oscillatory bath temperature range extends up to 20 K in span. To observe oscillations experimentally in these mixtures will probably not be difficult even with an average heating rate. Dove's [12] discovery of oscillations in the CO system is thus probably linked with his use of very dry mixtures at low pressures. Measured explosion limits by several authors [6-9] are replotted in dotted lines, though the exact water content for the experimental data is unknown. Dickens [9] apparently used the driest mixtures. His data fall in between the calculated limits with water contents of 0.01 and 0.1 %. This seems to be high for his supposedly well dried mixtures. Gordon and Knipe's [6] data match the limits calculated for mixtures containing 0.1 % water which is one or two orders of magnitude higher than their estimate; one is dubious about the meaning of these discrepancies when the water content in the mixture has not been determined accurately. It should be noted that the experimental limits are merely straight lines drawn through sets of scattered data points. There is no reason to expect the actual limits being straight lines in the P-T plane. The slopes of the limits are generally steeper for the wetter mixtures. Figures around the points where oscillations are calculated indicate the period of oscillation in seconds. Dove [12] observed no clear dependence of period of oscillation on composition, pressure and temperature in his experiments. Our results in fig. 1 show the same random character.

EXPLOSION AND SUSTAINED GLOW LIMITS

In experimental studies explosion and glow limits are obtained by heating the mixture to a critical temperature. Below this temperature, the reactivity of the system is usually unmeasurable. We found this to be the case in our calculations with very wet mixtures or at very high pressures. Two trajectories are plotted in fig. 2. They

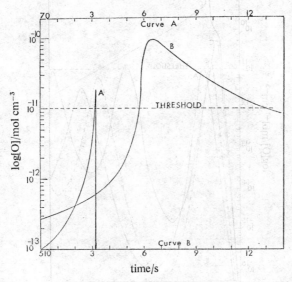

FIG. 2.—Trajectories of explosion and sustained glow. Water content = 10 %, composition = $CO + 2O_2$. Calculated trajectories: A temperature = 441°C and pressure = 10 Torr, B temperature = 400°C and pressure = 3 Torr.

are calculated with $CO + 2O_2$ mixtures containing 10 % water. The pressure for trajectory A is 10 Torr and the temperature is 441°C. The concentration of O atoms rises to exceed the threshold after an induction time of 73 s. Within a period of 0.3 s, CO concentration is reduced to 10 % of its initial value. The trajectory describes an explosion according to our definition. If the calculation is repeated with the bath temperature one degree lower than the preceding case, the calculated O atom concentration remains, indefinitely, many orders of magnitude below the threshold.

No reactivity will be detectable, in correspondence with this trajectory. The explosion and non-explosion regions in the P–T plane are thus sharply divided by the limit. Trajectory B is calculated at a pressure of 3 Torr and a temperature of 400°C. The O atom concentration exceeds the threshold after an induction period of 516 s. It is reduced to 5 % of its initial value in 10 s. The visible emission will probably last 5 to 6 s. It is a sustained glow limit. Again, results show the reactivity of the system to be unmeasurable at a temperature one degree lower than the limit.

GLOW AND SUSTAINED GLOW TRANSITION

The glow and sustained glow phenomena cannot be differentiated in experiments if the reactant consumption is not measured. Hoare and Walsh [4] reported that if stationary conditions of pressure and temperature were used a glow would appear and then slowly fade away in time; but on reducing the pressure slightly, the glow would brighten again. Obviously, this could not have happened at a sustained glow limit where the reactants would be exhausted in one stroke. In calculated trajectories glow and sustained glow are clearly different. The trajectory for a glow is an oscillatory solution and the trajectory for a sustained glow is not. A series of trajectories that maps the transition from glow to sustained glow is presented in fig. 3. Mixtures of

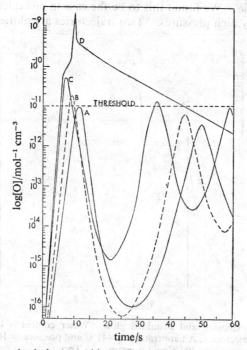

FIG. 3.—Glow and sustained glow transition. Composition = $CO + 2O_2$, water content = 0.1 %, pressure = 20 Torr. Calculated trajectories: A temperature = 575°C, B temperature = 577°C, C temperature = 580°C, D temperature = 581°C.

$CO + 2O_2$ with 0.1 % of water are used and the pressure is fixed at 20 Torr. Trajectories A, B, C and D correspond to bath temperatures of 575, 577, 580 and 581°C, respectively. The O atom concentration represented by trajectory A never exceeded the threshold. It is, therefore, undetectable. The first peak of trajectory B and the first and second peaks of trajectory C are above the threshold level. Visible glows will accompany these peaks. The results also show that after the passage of the first peak of trajectory B only 0.4 % of CO concentration is consumed and after the passage of the first and second peak of trajectory C the CO consumption is increased to 3 %.

When the bath temperature is raised only one degree to 581°C, a sustained glow, trajectory D, is attained. CO concentration is consumed to 8 % of its initial value in 20 s. The visible glow will last 15 to 20 s and then fade away as reactants are completely exhausted.

GLOW, OSCILLATION, SUSTAINED GLOW AND EXPLOSION TRANSITION

Calculations show that as the mixtures become drier the temperature interval in which the transitions take place (glow → oscillation → sustained glow → explosion) becomes wider. A series of trajectories is calculated and presented to show such transitions in fig. 4. $CO + 2O_2$ mixtures containing 0.1 % of water are used, and the

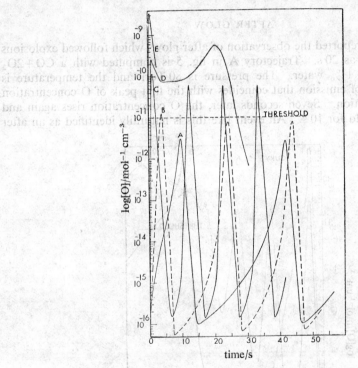

FIG. 4.—Glow, oscillation, sustained glow and explosion transitions. Composition = $CO + 2O_2$, water content = 0.1 %, pressure = 60 Torr. Calculated trajectories: A temperature = 646°C, B temperature = 648°C. C temperature = 657°C, D temperature = 662°C, E temperature = 682°C.

pressure is fixed at 60 Torr. Trajectories A, B, C, D and E correspond to bath temperatures of 646, 648, 657, 662 and 682°C, respectively. Trajectory A is presumably undetectable and a glow must accompany the first peak of trajectory B. Within the temperature range from 649 to 658°C all solutions are oscillatory and trajectory C ($T = 675°C$) represents a typical example. The period of oscillation decreases from 18 to 10 s as the bath temperature is raised from one end to the other. Oscillations of this type may continue for many cycles without appreciable change in their period and amplitude. In ref. (3) one hundred cycles were calculated for one case and McCaffrey and Berlad [13] observed over two hundred cycles of oscillation in their experiments with the CO system. In a binary system, this type of behaviour is generally associated with oscillations about a limit cycle. Trajectory C represents a sustained glow, the visible emission in this case will last about 20 s. The surging second peak of O atom concentration is reminiscent of the " pic d'arret " [15] phenomenon in hydrocarbon oxidation. The kinetic mechanism for " pic d'arret " is still unclear, but the surge of O atom concentration in the final phase of CO oxidation, such as shown here, may be attributed to the depletion of CO concentration in the mixture. Reactions (2) and (3) are the basic inhibitive steps in the kinetic scheme. The depletion of CO weakens both reactions and chain carriers are no longer prevented from a temporary divergence which is finally checked by complete fuel exhaustion. The duration of the sustained glow becomes shorter as the temperature is raised higher above 662°C. At $T = 682°C$ the sustained glow is compressed into a flash or an explosion.

AFTER GLOW

Linnett et al.[16] reported the observation of after glows which followed explosions and lasted as long as 20 s. Trajectory A in fig. 5 is computed with a $CO + 2O_2$ mixture containing 1 % water. The pressure is 80 Torr and the temperature is 605°C. The flash of emission that coincides with the first peak of O concentration is a glow, by definition. Seven seconds later, the O concentration rises again and exceeds the threshold for 10 s. An event like this is naturally identified as an after

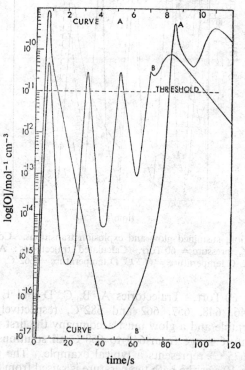

FIG. 5.—Trajectories of after glow: A water content = 1 %, pressure = 80 Torr, temperature = 605°C, composition = $CO + 2O_2$. B water content = 1 %, pressure = 30 Torr, temperature = 525°C, composition = $CO + 5O_2$.

glow in the laboratory. The first peak will probably be identified as an explosion due to its brief duration ($\frac{1}{2}$ s). After glows may also follow many cycles of oscillation. A typical example is calculated with a $CO+5O_2$ mixture containing 1 % water and plotted in fig. 5 (trajectory B). The bath temperature and pressure are 525°C and 30 Torr, respectively. The glow will last over 20 s.

Like the second peak of trajectory C in fig. 4, after glows are caused by the depletion of CO concentration in the final phase of oxidation.

EFFECTS OF PRESSURE

Egerton and Warren [17] reported that, using a withdrawal method, a blue glow appeared as the explosion limit of their moist CO and O_2 mixtures was approached, that on crossing the explosion limit at a fast rate the glow suddenly increased to a brilliant flash; but that with sufficiently slow evacuation no such transition from glow to flash occurred, the glow rising to a constant intensity throughout the explosion region. Their experiment evidently dealt with the transition between a sustained glow and an explosion. Three trajectories are calculated to show the transitions

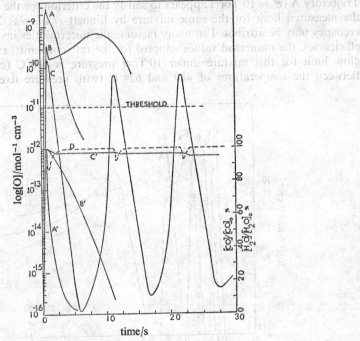

FIG. 6.—Transition among oscillation sustained glow and explosion with pressure varied. Composition = $CO+2O_2$, water content = 0.1 %, temperature = 652°C. Calculated trajectories: A(A′) pressure = 45 Torr, B(B′) pressure = 50 Torr, C(C′ and D) pressure = 45 Torr.

among oscillation, sustained glow and explosion by varying the pressure. The composition of the mixture is $CO+2O_2$ and its water content is 0.1 %. The temperature is fixed at 652°C. In fig. 6, trajectory A, B and C are calculated at pressures 45, 50 and 55 Torr, respectively. The fuel consumption rate in terms of CO concentration over its initial value, $[CO]/[CO]_0$, is also presented for all three cases (curves A′, B′ and C′).

Under 45 Torr pressure trajectory A represents a marginal explosion with a

duration of about one second.   The corresponding fuel consumption curve A' shows
that the fuel concentration falls rapidly one second after the reaction started.
Trajectory B ($P = 50$ Torr) represents a typical sustained glow which probably
would be visible for about 8 or 9 s.   Under the pressure of 55 Torr trajectory C is an
oscillatory solution.   The corresponding fuel concentration curve C' shows the
steps of falling off that coincide with the peaks of O atom trajectory.   Water concentration over its initial value for the case $P = 55$ Torr is represented by curve D in
fig. 6.   It shows that 8-12 % of the total water molecules are dissociated at the peaks
of O atom concentration.

Linnett et al.[16] defined a quasi limit according to the time required for the visible
emission to build up to the peak intensity.   If the required time is equal to or less
than 0.15 s, then the system is said to be on or to have surpassed the quasi explosion
limit.   A series of trajectories is calculated with pressures across the entire explosion
peninsula B in fig. 1.   The composition for the mixture is $CO+2O_2$ and the water
content is 1 %.   The bath temperature is fixed at 624°C which was also the value used
in the experiments of Linnett et al.   Trajectories A, B, C, D, E, F and G corresponding
to pressures of 10, 15, 20, 40, 80, 90 and 110 Torr, respectively, are presented in fig. 7.
Trajectory A ($P = 10$ Torr) appears to satisfy the criterion for the " quasi limit " but
the measured limit for the same mixture by Linnett et al. was 20 Torr.   The discrepancy may be attributed to many factors: the uncertainties involved with surface
efficiencies, the numerical values adopted for the rate constants, etc.   The sustained
glow limit for this mixture under 10 Torr pressure is 480°C (curve B in fig. 1).
Between the temperatures of 480 and 624°C (with pressure fixed at 10 Torr) the

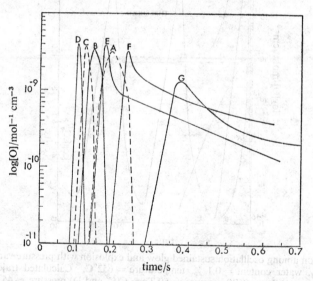

Fig. 7.—Trajectories across explosion peninsula.   Composition = $CO+2O_2$, water content = 1 %,
temperature = 624°C.   Calculated trajectories: A pressure = 10 Torr, B pressure = 15 Torr,
C pressure = 20 Torr, D pressure = 40 Torr, E pressure = 80 Torr, F pressure = 90 Torr, G
pressure = 110 Torr.

calculated O atom trajectory starts as a sustained glow and accelerates gradually to
become a " quasi explosion " at 624°C.   The transition is always smooth without
sudden disruption at any point when the temperature is raised.   This is consistent
with their conclusion that a lower explosion limit does not exist inside the glow limit.

The mixture is most reactive at the pressure of 40 Torr (trajectory D). The flash of emission in this case lasts no longer than 0.04 s.

Curve B in fig. 1 shows that the upper limit for the mixture at the temperature of 624°C is 120 Torr. The rate of reaction is gradually decreasing as the pressures are raised to approach the upper limit as shown in fig. 7 (trajectories E, F and G).

EFFECTS OF COMPOSITION

Experimental results [12] showed that the reaction rates of lean CO mixtures are usually greater than that of the stoichiometric mixture. Limits for stoichiometric mixtures are calculated and presented in fig. 8. No oscillations can be obtained for

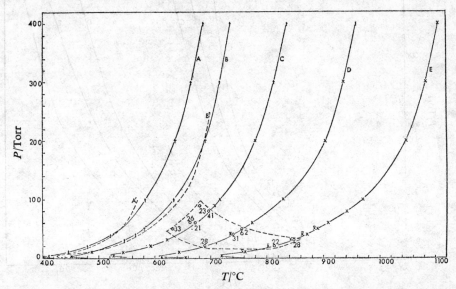

FIG. 8.—Explosion limits and oscillation region of stoichiometric mixtures. Calculated curves: A water content = 10 %, B water content = 1 %, C water content = 0.1 %, D water content = 0.01 %, E water content = 0.001 %. Experimental results from Hoare and Walsh : A' water content = 5 %, B' dry.

such mixtures with water content in excess of 0.5 %. Hoare and Walsh's [4] data for mixtures of the same composition are replotted as dotted lines (curves A' and B'). Their curve A' (5 % water content) appears to fall on the calculated curve A with a water content of 10 %. Results for their dry mixtures (curve B') fit the calculated curve B (1 % water) quite well. It is noteworthy that the peculiar curvature of the measured upper limit at low pressures is closely matched by the calculated one. The shape of the peninsula tip calculated with different water contents does not change appreciably.

Hoare and Walsh [4] noted that upper limits for the dry and wet mixtures merge together and become one at very low pressures. Our calculated results show that the effect of water on the temperature part of the limit is still distinct, only the pressure scale between the different limit curves is compressed at low pressures. It must be pointed out that the glow limits reported by Hoare and Walsh are probably a combination of glow and sustained glow limits, as the two are indistinguishable as long as fuel consumptions are not measured.

Limits for very lean and rich mixtures are plotted in fig. 9. Curves A, B and C are calculated with mixtures containing 1 % water while curves D, E and F are for drier mixtures with only 0.01 % of water. The experimental results of Dickens *et al.*[1] are reproduced in dotted lines A′, B′ and C′. Theoretical results are, at least, consistent with the experimental measurements in predicting higher reaction rate for the leaner mixtures.

The glow and oscillation limits for curves B and E are not shown in fig. 9 to avoid crowding. It can be concluded from the results that oscillations are less likely for rich mixtures and the period of oscillation is shorter for leaner mixtures.

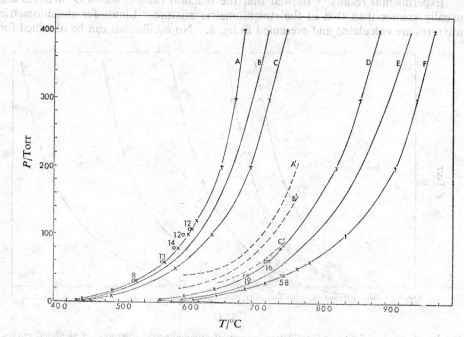

FIG. 9.—The effects of composition on explosion limits. Calculated curves: A, B, C water content = 1 %, D, E, F water content = 0.01 %, A, D composition = $CO+5O_2$, B, E composition = $CO+2O_2$, C, F composition = $4CO+O_2$. Experimental results from Dickens, *et al.*[1]: A′ composition = $CO+9O_2$, B′ composition = $CO+2O_2$, C′ composition = $2CO+O_2$.

EFFECTS OF WALL EFFICIENCIES

It is assumed that the kinetic role of the walls is limited to the destruction of chain carriers and intermediates in this work. Wall efficiencies (listed in table 2) which have been used for all calculations so far are intended for mildly reflective vessel surfaces. Limit curves B and D of fig. 1 are replotted in fig. 10, with the labelling retained for comparison with limits curves A and C. Curves A and C are calculated exactly as for curves B and D, respectively, except all the wall efficiencies are one order of magnitude greater. These higher efficiencies are selected to simulate effective vessel surfaces. Curves A and C show the limits shifted to higher temperatures, at low pressures where heterogeneous termination is most effective. One surprise result is that the difference between the limits for reflective and effective surfaces are greater at higher pressures than at intermediate pressures. Curves A′ and C′ represent data from Dickens *et al.*[1] for $CO+2O_2$ mixtures in two different vessels (vessel No. 1 and 3). Qualitative features of the experimental measurements are clearly followed by the calculated ones. One intriguing question remains, namely, why the

wall efficiencies for the two vessels which were made of same materials (quartz) and
had been treated the same way before tested, are so different.

The calculated period of oscillation is generally shorter for effective surfaces.
The pumping mechanisms due to wall site saturation proposed previously [2, 3] are not
invoked in all the calculations presented here. Our other calculations show that the
oscillatory region in the P–T plane is enlarged towards the higher and lower pressures
if the wall site saturation mechanism is imposed on the scheme. At the present time,
this region in the P–T plane has not been mapped out by measurements. It is
probably premature to either accept or reject this pumping mechanism.

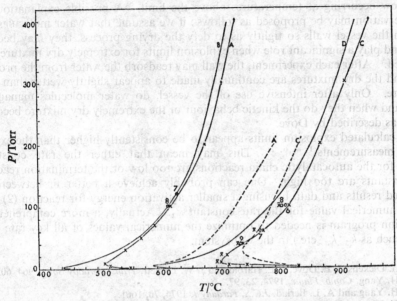

FIG. 10.—The effects of surface on explosion limits. Calculated curves: B, D reflective surface
(wall efficiencies listed in table 2), A, C effective surface (10 × wall efficiencies listed in table 2).
Experimental results from Dickens [9]: A′ vessel No. 1, C′ vessel No. 3.

DISCUSSION

It was suggested [18] that the explosive reaction of CO oxidation is retarded by some
product formed in the reaction. Earlier, Jono [19] found that the reaction rate of a
CO and O_2 mixture fell off sharply after a fast initial start. This was confirmed later
by Knipe and Gordon.[20] Minkoff and Tipper [18] proposed that, in the early phase of
oxidation, carbon suboxides such as C_2O and C_3O_2 may have been produced which
inhibit the reaction by removing the chain carrier, O atoms, from the system. Using
C_2O as an intermediate, Gray [21] constructed a theory quite similar in mathematical
structure to that of ref. (2) to explain the oscillations and glow limits in CO oxidation.
The calculated trajectories for the cases of explosion and sustained glow in fig. 2, 3,
4 and 7 clearly confirm this observation. The reaction rate, which is proportional to
O atom concentration, decreases by as much as several orders soon after the
reaction starts. Excited CO_2^* molecules are clearly the inhibitors. Reaction (3), the
rate retardation step, becomes effective once sufficient CO_2^* concentration is accumulated shortly after commencement of the reaction. The oscillation phenomenon
is, of course, another product of this autocatalytic and inhibitive reaction mechanism.

Dove [12] raised the question : is the supposed effect of water merely to depress an explosion which already exists at high temperature or is the limit entirely dependent on the presence of water? Our answer is affirmative to the second part. It was shown in ref. (2) that water plays a key role in the autocatalytic chain. In the absence of the autocatalytic reaction both the oscillation and explosion phenomena disappear. Our calculations for a mixture with 0.0003 % water have yielded only glow with long lasting slow reaction or extended sustained glow at very high temperature. Dove [12] made similar observations in his experiments with extremely dry mixtures.

Dove [12] also reported that intensive use of one vessel caused the limits to become very indistinct or to be replaced by a relatively slow reaction accompanied sometimes by a glow occurring at temperatures above the limit. A possible explanation for his observation may be proposed as follows : if we assume that water molecules can cling on the vessel walls so tightly as to defy the drying process, they may become active and play a significant role when explosion limits for extremely dry mixtures are measured. After each experiment, the wall may readsorb the water from the product gases and the dry mixtures are continually made to appear slightly wetter than they really are. Only after intensive use of the vessel, do water molecules manage to escape and when they do the kinetic behaviour of the extremely dry mixture becomes evident as described by Dove.

The calculated explosion limits appear to be consistently higher than the experimental measurements.[1, 4, 6, 16] This may mean that either the rate constants selected for the autocatalytic chain reactions are too low or the termination reaction rate constants are too high. One can probably achieve a better fit between the calculated results and data by using a smaller activation energy for reaction (2) or a smaller numerical value for the rate constant k_2. Actually, a more comprehensive calculation program is needed to optimize the numerical values of all key rate constants (such as k_3, k_4, etc.) in the CO system.

[1] P. G. Dickens, J. E. Dove, J. E. Harold and J. W. Linnett, *Trans. Faraday Soc.*, 1964, **60**, 539.
[2] C. H. Yang, *Comb. Flame*, 1974, **23**, 97.
[3] C. H. Yang and A. L. Berlad, *J.C.S. Faraday I*, 1974, **70**, 1661.
[4] D. E. Hoare and A. D. Walsh, *Trans. Faraday Soc.*, 1954, **50**, 37.
[5] A. S. Gordon, *J. Chem. Phys.*, 1952, **20**, 340.
[6] A. S. Gordon and R. H. Knipe, *J. Phys. Chem.*, 1955, **59**, 1160.
[7] G. Hadman, H. W. Thompson and C. N. Hinshelwood, *Proc. Roy. Soc. A*, 1932, **137**, 87; 1932, **138**, 297.
[8] B. Lewis, G. von Elbe and W. Roth, *5th Int. Symp. Combustion* (The Combustion Institute, Pittsburgh, Pennsylvania, 1955), p. 610.
[9] P. G. Dickens, *Dissertation* (Oxford, 1956).
[10] D. L. Baulch, D. D. Drysdale, D. G. Hoare and A. C. Lloyd, *High Temperature Reaction Rate Data* (Leeds University, 1968), No. 1 and 2.
[11] R. R. Baldwin, L. Mayer and P. Doran, *Trans. Faraday Soc.*, 1967, **63**, 1665.
[12] J. E. Dove, *Dissertation* (Oxford, 1956).
[13] B. McCaffrey, personal communication.
[14] H. P. Broida and A. G. Gaydon, *Trans. Faraday Soc.*, 1953, **49**, 1120.
[15] M. Lucquin, *J. Chim. Phys.*, 1968, **55**, 827.
[16] J. W. Linnett, B. G. Reuben and T. F. Wheatley, *Comb. Flame*, 1968, **12**, 325.
[17] A. C. Egerton and D. R. Warren, *Nature*, 1952, **170**, 420.
[18] G. J. Minkoff and C. F. H. Tipper, *Chemistry of Combustion Reactions* (Butterworth, London, 1962).
[19] W. Jono, *Rev. Phys. Chem. Japan*, 1941, **15**, 17.
[20] R. H. Knipe and A. S. Gordon, *J. Chem. Phys.*, 1957, **27**, 1418.
[21] B. F. Gray, *Trans. Faraday Soc.*, 1970, **63**, 1118.

Small Parasitic Parameters and Chemical Oscillations

By B. F. Gray and L. J. Aarons

School of Chemistry, University of Leeds,
Leeds LS2 9JT

Received 23rd July, 1974

In treating chemical oscillators, it is common practice to use no more than two concentrations as dependent variables due to the mathematical difficulties with more than two. Other variables, necessarily involved to make the system nontrivial—the " pool " chemicals—are treated as parameters which are independent of time. We investigate here under what conditions the " pool " chemicals can be treated as constants without *qualitatively* effecting the behaviour of the system. Mathematical methods developed by Tikhonov are used to study the effect of a small parameter on the roots of the characteristic equation. This small parameter may be chosen as the ratio of the initial concentrations of the reactants to the " pool " chemicals or for dilute solutions, the reciprocal of the heat capacity. A number of interesting results are obtained in which the slow variation of the " pool " chemicals can either produce a limit cycle where there was none previously or place severe restrictions on the rate constants so as to exclude regions where interesting instabilities have been found in the two variable case. Multistability in the two variable system lends itself very well to the production of limit cycles of the relaxation type. Finally it is shown that it is possible to devise thermokinetic oscillators with very small temperature oscillations provided the energy equation is highly nonlinear.

---

1. INTRODUCTION

It has become common practise in designing models of chemical oscillators to limit the description to two dependent variables. The reasons for this are obvious, being due to the formidable mathematical apparatus that exists for treating non-linear differential equations with only two variables.[1] Several two dimensional kinetic schemes that exhibit undamped oscillations have been developed including the so-called " Brusselator ".[2, 3] However, the latter model employs a physically unrealistic termolecular step. In fact, Hanusse [4] has shown that it is impossible for any two variable kinetic scheme which has only unimolecular and bimolecular steps, to exhibit undamped oscillations. The one exception to this, the Lotka–Volterra scheme,[5] is conservative and so is again physically unacceptable.

It is implied in all these models that certain substances, generally referred to as " pool " chemicals (such as fuel and end products), are held constant, either by being in large excess or by flowing them in (out) as soon as they are used (produced). It is the purpose of this paper to investigate under what conditions the " pool " chemicals can be treated as constant and what effect their variation may have.

We will be concerned with a system of equations typically of the form

$$\mu\dot{x}_i = F_i(x_1, \ldots, x_s; y) \qquad i = 1, \ldots, s \qquad (1)$$
$$\dot{y} = G(x_1, \ldots, x_s; y)$$

where μ is a small positive parameter. The solutions of these equations fall into two regions [1] : the region of slow motion and the region of rapid motion. In the first region a representative point moves comparatively slowly ($\dot{x}$ and $\dot{y}$ bounded) within a small neighbourhood (of order μ) of the curve $F(x; y) = 0$. Outside this neighbourhood the representative point moves in rapid jumps and in the limit as $\mu \to 0$ the equations of rapid motion can be written

$$y \equiv y^\circ = \text{constant}, \qquad \dot{x} = \frac{1}{\mu}F(x; y^\circ). \qquad (2)$$

Furthermore, for a point to stay within a small neighbourhood of $F(x; y) = 0$, then this region must be one of stable equilibrium for the equations of rapid motion, (2). This will be true if all s roots of the characteristic eqn (3) have negative real parts.

$$
\begin{vmatrix}
\dfrac{\partial F_1}{\partial x_1} - \lambda & \dfrac{\partial F_1}{\partial x_2} & \cdots & \dfrac{\partial F_1}{\partial x_s} \\[2ex]
\dfrac{\partial F_2}{\partial x_1} & \dfrac{\partial F_2}{\partial x_2} - \lambda & \cdots & \dfrac{\partial F_2}{\partial x_s} \\[1ex]
\cdot & \cdot & & \cdot \\
\cdot & \cdot & & \cdot \\
\dfrac{\partial F_s}{\partial x_1} & \dfrac{\partial F_s}{\partial x_2} & \cdots & \dfrac{\partial F_s}{\partial x_s} - \lambda
\end{vmatrix} = 0.
\tag{3}
$$

The possibility of discontinuous oscillations then occurs if a representative point alternates between regions of slow and rapid motion. The points of transition between the regions of slow and rapid motion—the " jump " points—are given by the intersection of the curves $F(x; y) = 0$ and $D(x; y) = 0$ where $D(x; y)$ is the Jacobian

$$
D(x; y) = \frac{\partial(F_1, F_2, \ldots, F_s)}{\partial(x_1, x_2, \ldots, x_s)}.
\tag{4}
$$

The remainder of this paper will be given over to four examples. First we consider the Lotka–Volterra scheme and show that, if the fuel is allowed to vary, the system can show non-conservative rather than conservative oscillations. In section 3 it is shown that when the " pool " chemicals are allowed to vary in the " Brusselator " severe restrictions are placed on the rate constants so as to exclude certain regions where it was thought to oscillate. In section 4 a kinetic scheme is devised in which slow variation of one of the " pool " chemicals gives rise to discontinuous oscillations, of the type described above. Finally, in section 5, a scheme originally proposed by Edelstein [6] is modified and discontinuous oscillations are demonstrated when the temperature is allowed to vary slowly.

2. THE LOTKA–VOLTERRA SCHEME

The set of differential equations originally proposed by Lotka [7] in an ecological context can be reset in a chemical context via the following kinetic scheme

$$
A + X \xrightarrow{k_1} 2X
$$

$$
X + Y \xrightarrow{k_2} 2Y
$$

$$
A + Y \xrightarrow{k_3} B.
\tag{5}
$$

The rate equations for this scheme are

$$
\frac{dA}{dt} = -k_1 AX - k_3 AY + k(A_0 - A)
$$

$$
\frac{dX}{dt} = k_1 AX - k_2 XY
\tag{6}
$$

$$
\frac{dY}{dt} = k_2 XY - k_3 AY
$$

where species A diffuses into the system from the outside where its value is assumed constant at A_0. It is convenient to change to new dimensionless variables $A' = A/A_0$, $X' = X/X_0$, $Y' = Y/Y_0$ where A_0, X_0, Y_0 are the initial values of A, X, Y respectively and let $X_0/A_0 \sim Y_0/A_0 = \xi \ll 1$. Eqn (6) become

$$\frac{\mathrm{d}A'}{\mathrm{d}t} = -k_1 X_0 A' X' - k_3 Y_0 A' Y' + k(1-A')$$

$$\xi \frac{\mathrm{d}X'}{\mathrm{d}t} = k_1 X_0 A' X' - k_2 \xi X_0 X' Y' \tag{7}$$

$$\xi \frac{\mathrm{d}Y'}{\mathrm{d}t} = k_2 \xi X_0 X' Y' - k_3 Y_0 A' Y'.$$

It can be seen that eqn (7) only reduce to the usual Lotka–Volterra scheme as $\xi \to 0$ provided $k_2 X_0$ is of order $1/\xi$ while $k_1 X_0$ and $k_3 Y_0$ are of order 1. Under these conditions, eqn (7) possess a singularity at the point $X' = Y' = A' = \frac{1}{2}[-k/2 + (2k+k^2/4)^{\frac{1}{2}}]$ which is easily shown to be an unstable focus.[8] These equations were programmed on a Solatron HS7 analogue computer and for a wide range of initial conditions the trajectories always wound onto a stable limit cycle. A typical run is shown in fig. 1. Thus, by including the slow variation of the " pool " chemical A, the system is made structurally stable. In the next section we will give an example of the opposite effect where the inclusion of a small parameter changes an oscillatory scheme to a nonoscillatory one.

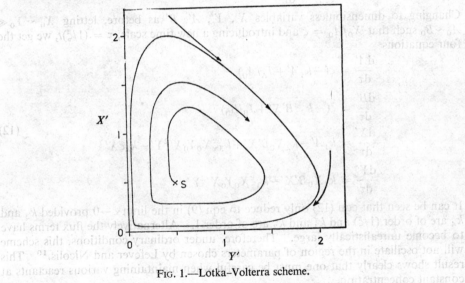

FIG. 1.—Lotka–Volterra scheme.

3. THE BRUSSELATOR

The following kinetic scheme was devised and has been exhaustively studied by the Brussels school [2, 3]

$$A \xrightarrow{k_1} X$$

$$B + X \xrightarrow{k_2} Y + D \tag{8}$$

$$2X + Y \xrightarrow{k_3} 3X$$

$$X \xrightarrow{k_4} E$$

where in the original model A, B, D, E were assumed to be constant in time. Tyson later removed the necessity of the termolecular step by introducing another variable. The rate equations for the two dimensional model are

$$\frac{dX}{dt} = k_1 A - k_2 BX + k_3 X^2 Y - k_4 X$$

$$\frac{dY}{dt} = k_2 BX - k_3 X^2 Y. \tag{9}$$

The equations have a singularity for $X_s = k_1 A/k_4$, $Y_s = k_2 k_4 B/k_1 k_3 A$. It is easily deduced that the necessary condition for this point to be unstable is

$$k_2 B > \frac{k_1^2 k_3}{k_4^2} A^2 + k_4. \tag{10}$$

Lefever and Nicolis [10] have shown that under certain conditions this stationary point is surrounded by a stable limit cycle.

If A and B are allowed to vary subject to external fluxes we get the two additional equations

$$\frac{dA}{dt} = -k_1 A + J_A$$

$$\frac{dB}{dt} = -k_2 BX + J_B. \tag{11}$$

Changing to dimensionless variables X', Y', A', B' as before, letting $X_0 \sim Y_0 \ll A_0 \sim B_0$ such that $X_0/A_0 = \xi$ and introducing a new time scale, $\tau = (1/\xi)t$, we get the four equations

$$\frac{dA'}{d\tau} = \xi(-k_1 A' + J_A/A_0)$$

$$\frac{dB'}{d\tau} = \xi(-k_2 \xi B' X' + J_B/B_0)$$

$$\frac{dX'}{d\tau} = k_1 A' - k_2 X_0 B' X' + k_3 \xi X_0 Y_0 X'^2 Y' - k_4 \xi X' \tag{12}$$

$$\frac{dY'}{d\tau} = k_2 X_0 B' X' - k_3 \xi X_0 Y_0 X'^2 Y'.$$

It can be seen that eqn (12) only reduce to eqn (9) in the limit $\xi \to 0$ provided k_3 and k_4 are of order $(1/\xi)$ and k_1 and k_2 are of order 1. Alternatively the flux terms have to become unrealistically large. Therefore, under ordinary conditions, this scheme will not oscillate in the region of parameters chosen by Lefever and Nicolis.[10] This result shows clearly that one must be careful when maintaining various reactants at constant concentration.

4. A CHEMICAL DISCONTINUOUS OSCILLATOR

Consider the following kinetic scheme

$$\begin{align}
A + X &\xrightarrow{k_1} 2X \\
X + X &\xrightarrow{k_2} X_2 \\
X + B &\xrightarrow{k_3} \text{products} \\
B &\xrightarrow{k_4} \text{products}.
\end{align} \tag{13}$$

The rate equations are

$$\frac{dX}{dt} = k_1 AX - k_2 X^2 - k_3 XB$$

$$\frac{dB}{dt} = -k_3 XB - k_4 B + \phi_B \tag{14}$$

$$\frac{dA}{dt} = -k_1 AX + \phi_A$$

where ϕ_A and ϕ_B are flow rates. Changing to dimensionless variables X', A', B' and letting $X_0/A_0 \sim B_0/A_0 = \xi \ll 1$ we obtain

$$\xi \frac{dX'}{dt} = k_1 X_0 A' X' - k_2 \xi X_0 X'^2 - k_3 \xi X_0 X' Y'$$

$$\xi \frac{dB'}{dt} = -k_3 \xi X_0 X' B' - k_4 \xi B' + \phi_B' \tag{15}$$

$$\frac{dA'}{dt} = -k_1 X_0 A' X' + \phi_A'$$

where $\phi_A' = \phi_A/A_0$ and $\phi_B' = \phi_B/A_0$. If $k_2 X_0$, k_4, ϕ_A and ϕ_B are all of order $1/\xi$ and $k_1 X_0$ is of order unity, eqn (15) become

$$\xi \frac{dX'}{dt} = A' X' - X'^2 - X' B'$$

$$\xi \frac{dB'}{dt} = -X' B' - B' + \phi_B' \tag{16}$$

$$\frac{dA'}{dt} = -A' X' + \phi_A'.$$

They are then in the form of eqn (1). The equation of slow motion is given by $F(x ; y) = 0$, which yields

$$X' = 0, \, B' = \phi_B' \tag{17}$$

and

$$X'^2 + (1 - A') X' + (\phi_B' - A') = 0, \, B' = \phi_B'/(X' + 1). \tag{18}$$

These solutions are sketched in fig. 2 in the X', A' plane. The branch $X' = 0$ is stable for A between 0 and P ($A' = \phi_B'$) at which point it becomes unstable. The upper branch is stable down to the " jump " point R ($X' = -1 + \sqrt{\phi_B'}$; $A' = -1 + \sqrt{\phi_B'}$) and the middle branch, between P and R, is unstable. Thus for $\phi_A' < 1 - 2\sqrt{\phi_B'} + \phi_B'$ a representative point will perform discontinuous oscillations as depicted in fig. 2. This limit cycle was verified for a wide range of initial conditions at the analogue computer.

Discontinuous oscillators of this kind have been proposed by Lavenda, Nicolis and Herschkowitz–Kaufman [11] in connection with the Brusselator and by Rössler [12] for an autocatalytic scheme with feed back inhibition. However this scheme is perhaps simpler and is well suited as a model of a branched chain reaction of the type occurring in many combustion reactions. [13]

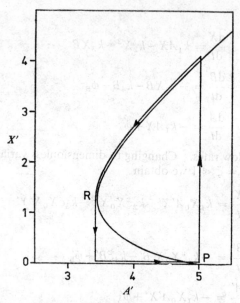

FIG. 2.—Equation of slow motion and limit cycle for eqn (16) with $\phi' = 5$.

5. EDELSTEIN'S EQUATIONS

A novel biochemical scheme that shows multiple steady states has been devised by Edelstein.[6] The scheme is characterised by the following reactions

$$A + X \underset{k_{-1}}{\overset{k_1}{\rightleftharpoons}} 2X$$

$$X + E \underset{k_{-2}}{\overset{k_2}{\rightleftharpoons}} C \qquad (19)$$

$$C \underset{k_{-3}}{\overset{k_3}{\rightleftharpoons}} E + B$$

where E is an enzyme and C is the enzyme-substrate complex. It is assumed that the total amount of enzyme is conserved, i.e., $E + C = $ constant (E_T). It is trivial to produce a discontinuous oscillator of the type described in the last section merely by letting A vary slowly subject to a flux into the system. We have in fact succeeded in producing oscillations at the analogue computer this way. However, a more interesting case arises if the first eqn in (19) is made exothermic and the reverse reaction has a strong temperature dependence. The rate equations, for this system are (assuming we choose the rate constants such that A and B can be treated as constant)

$$\frac{dX}{dt} = k_1 AX - k_{-1}X^2 - k_2 XE + k_{-2}E_T - k_{-2}E$$

$$\frac{dE}{dt} = -k_2 XE - k_{-3}EB + (k_3 + k_{-2})(E_T - E) \qquad (20)$$

$$C\frac{dT}{dt} = \sum_i \Delta h_i(R_i^+ - R_i^-) - L(T - T_0)$$

where T is the reactant temperature, T_0 is the ambient temperature, C the heat capacity of the mixture, Δh_i is the enthalpy of the ith reaction, R_i^+ and R_i^- are the rates of the forward and reverse ith reaction, respectively, and L is the heat loss proportionality constant.[14] Assume that k_{-1} is the only significant temperature dependent rate constant and that it can be written as $Z \exp(-E_a/T)$. If the reactants are in dilute solution such that C is large we can choose $1/C$ as our small parameter. The equation of slow motion is given by (with all rate constants other than k_{-1} put equal to unity)

$$Z \exp(-E_a/T)X^3 - [A - (2+B)Z \exp(-E_a/T)]X^2 +$$
$$[E_T - A(2+B)]X - BE_T = 0 \qquad (21)$$

$$E = \frac{2E_T}{X+B+2}.$$

For certain values of the parameters a hysteresis curve results, such as the one shown in fig. 3. The variation of T is given by

$$C\frac{dT}{dt} = \Delta h_1(AX - Z \exp(-E_a/T)X^2) - L(T-T_0). \qquad (22)$$

Again for particular choices of the parameters a discontinuous limit cycle results typified by the one shown in fig. 3.

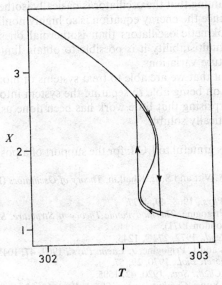

FIG. 3.—X, T phase plane and limit cycle for Edelstein's equations, (20). Parametric values chosen: $A = 10.2$, $B = 0.2$, $E_T = 35$, $E_a = 10^4$, $Z = 3 \times 10^{14}$, $\Delta h_1 = 1$, $1/C = 10^{-4}$, $T_0 = 300$, $L = 4$.

One may be misled into thinking that an oscillator is basically isothermal merely on the grounds that the temperature oscillations are of small amplitude. As has been shown here that may not be the case and one suspects that oscillators of this type may be more common than suspected, due to the highly nonlinear energy equation.

6. CONCLUSIONS

By way of conclusion we wish to stress several points made earlier. Our main point is that one should be careful when setting as constant the concentrations of various " pool " chemicals.   It may be possible that the inclusion of a small parameter may change a stable equilibrium point into an unstable one (the reverse certainly cannot occur).   More commonly, if the rate constants are not treated correctly the scheme including the small parameter will not reduce to the degenerate scheme as the small parameter is reduced to zero.

The results obtained are interesting and include the discovery of a number of counterexamples to Hanusse's conjecture (e.g., very small percentage variations in the pool chemicals are sufficient to allow sustained oscillations in the active chemicals). Multistability of the two variable system seems to lend itself very well to the production of limit cycles of the relaxation type and bistable systems are shown to oscillate in this manner when the necessary variation of " pool " chemicals is taken into account.   Biochemical models employing allosteric enzymes are particularly good candidates for discontinuous oscillators of this type.[15, 16]

Conditions are derived under which the " pool " chemicals can be treated as constants as a zero'th approximation and these conditions place severe restrictions on the rate constants, and may exclude regions where interesting instabilities have been discussed in the two variable case; e.g., the " Brusselator " appears to be unlikely to " Brusselate " in the region where it is permissible to treat the two variable system.

Finally it is shown that very small amplitude temperature oscillations in solution are not grounds for assuming that the oscillator is basically isothermal with secondary temperature effects.   Since the energy equation is so highly nonlinear it should be far easier to devise thermokinetic oscillators than isothermal ones.   Again if the two variable system shows multistability it is possible to obtain limit cycle behaviour by allowing small temperature variations.

It is highly significant that we are able to treat systems of more than two variables exactly.   This depends on being able to separate the system into regions of slow and rapid motion.   It is surprising that little work has been done using this theory when the equations are analytically soluble.

One of us (L. J. A.) is grateful to I.C.I. for the support of a postdoctoral fellowship.

[1] A. A. Andronov, A. A. Vitt and S. E. Khaikin, *Theory of Oscillators* (Pergamon Press, Oxford 1966).
[2] R. Lefever, *J. Chem. Phys.*, 1968, **49**, 4977.
[3] P. Glansdorff and I. Prigogine, *Thermodynamic Theory of Structure, Stability and Fluctuations* (Wiley-Interscience, London 1971).
[4] P. Hanusse, *Compt. Rend.*, 1972, **274C**, 1245.
[5] R. Lefever, G. Nicolis and I. Prigogine, *J. Chem. Phys.*, 1967, **47**, 1045.
[6] B. B. Edelstein, *J. Theoret. Biol.*, 1970, **29**, 57.
[7] A. J. Lotka, *J. Amer. Chem. Soc.*, 1920, **42**, 1595.
[8] B. F. Gray and C. H. Yang, *Combustion and Flame*, 1969, **13**, 20.
[9] J. J. Tyson, *J. Chem. Phys.*, 1973, **58**, 3919.
[10] R. Lefever and G. Nicolis, *J. Theoret. Biol.*, 1971, **30**, 267.
[11] B. Lavenda, G. Nicolis and M. Herschkowitz-Kaufman, *J. Theoret. Biol.*, 1971, **32**, 283.
[12] O. E. Rössler, *J. Theoret. Biol.*, 1972, **36**, 413.
[13] B. F. Gray, *Kinetics of Oscillatory Reactions* (Specialist Periodical, ed. P. G. Ashmore Reports, The Chemical Society, London, 1974).
[14] R. Aris, *Elementary Chemical Reactor Analysis* (Prentice-Hall, New Jersey, 1969).
[15] H. R. Karfunkel and F. F. Seelig, *J. Theoret. Biol.*, 1972, **36**, 237.
[16] L. J. Aarons and B. F. Gray, to be published.

Kinetic Feedback Processes in Physico-Chemical Oscillatory Systems

By U. F. Franck

Institute for Physical Chemistry, Technical University of Aachen, West Germany

Received 29th July, 1974

On the basis of a generalized feedback concept it is shown that the kinetic phenomena occurring in oscillatory systems can be understood as a result of two simultaneous counteracting feedback mechanisms. The positive feedback in particular brings forth excitability and propagation phenomena. The counteracting negative feedback brings forth recovery and overshoot phenomena. The cooperation of both feedback effects causes oscillation pulse formation and dissipative structures.

1. PHENOMENOLOGY

Non-linear oscillations occurring in biological, chemical, electrochemical and other physico-chemical systems (fig. 1) appear to be quite regular and uncomplicated. But on closer inspection the mechanisms bringing forth those rhythmic phenomena turn out to be unexpectedly intricate. In spite of the fact that they are of quite different physico-chemical nature the recordings of the oscillations look remarkably similar in their general shape, suggesting that there may exist a common kinetic principle for their occurrence. As is well known, the difficulties of elucidating oscillatory phenomena arise from the fact that the systems in question are essentially

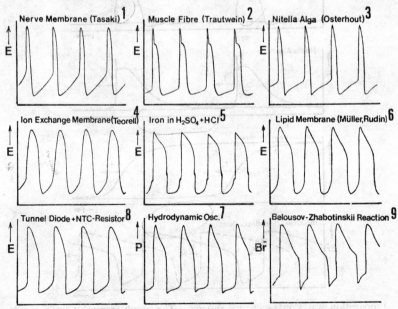

FIG. 1.—Oscillations of biological and non-biological systems.

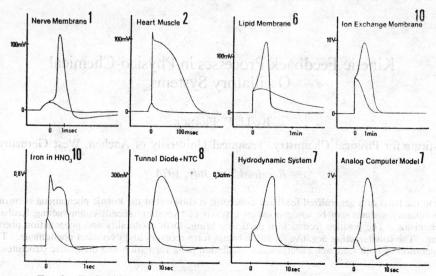

FIG. 2.—Excitability (pulse formation) of biological and non-biological systems.

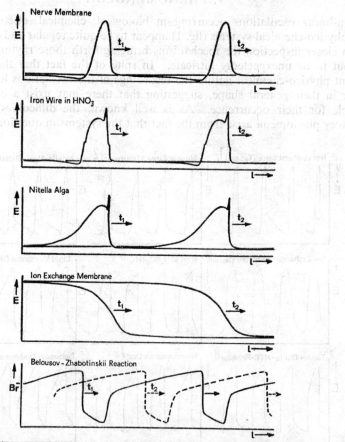

FIG. 3.—Propagation waves in oscillatory systems. The graphs represent wave profiles at two successive points of time (abscissa : length *l* in direction of propagation).

multi-variable systems with extremely non-linear kinetic relationships and complicated coupling mechanisms. By changing the environmental conditions slightly in such a way that oscillatory behaviour just ceases, all these systems can be made to exhibit excitation phenomena, and can be triggered by adequate stimuli obeying the all-or-none-law of excitation (fig. 2). A superthreshold stimulus, depending on its strength and duration, releases an excitation pulse having the characteristic shape of the so-called action potential of nerve membranes.

The third essential property of these systems is the phenomenon of decrement-free propagation, signifying that a locally triggered excitation state or pulse propagates respectively along or through the entire system by means of eddy currents or fluxes (fig. 3).

According to the phenomenology of nerve excitation, refractoriness, accommodation, adaptation etc. are also common features of all systems capable of oscillation.[10]

2. THE PROBLEM OF OSCILLATING VARIABLES

Systems exhibiting sustained oscillations are thermodynamically open systems in which for the most part forces and fluxes behave rhythmically. Between the forces and fluxes there exist the well-known conjugation relationships and cross-effects. The kinetic variables appear in the mathematical description in form of temporal differential quotients combined in sets of simultaneous differential equations.

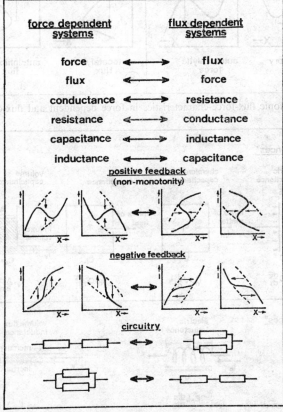

Fig. 4.—The duality relationships between variables, parameters, feedback and circuitry of force dependent and flux dependent systems.

From the kinetic point of view it must be borne in mind that there are two distinct classes of oscillatory systems; those in which the fluxes are the result of driving forces and those in which the forces are the result of force generating fluxes. Accordingly, classification of oscillatory systems with respect to force-dependent and flux-dependent interrelationships is essentially a matter of kinetic causality. Between the variables, parameters and the "circuitries" of the system of both classes there exist defined duality-relationships as shown in fig. 4.

Most of the known oscillations, in particular those of chemical, electrochemical and biological systems belong to the class of force-dependent oscillators. Here they

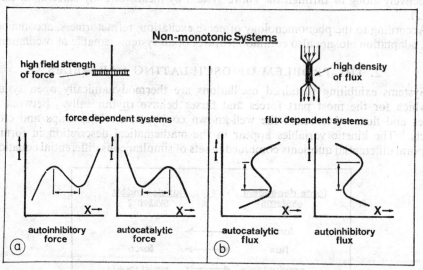

FIG. 5.—Non-monotonic flux-force-characteristics in force dependent and flux-dependent systems.

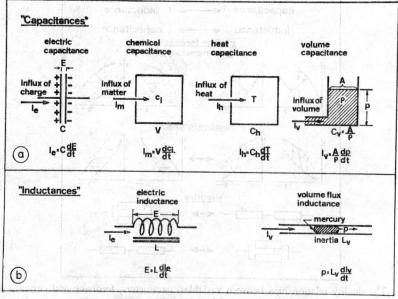

FIG. 6.—Generalized capacitances and inductances.

are brought forth by force-depending non-monotonic flux-force-relationships (fig. 5a). Such kinetic characteristics arise frequently in systems containing structures where high field strength of forces can occur such as in membranes and at interfaces.

On the other hand, flux-dependent characteristics (fig. 5b) appear in systems in which high densities of fluxes occur. In biological and chemical systems such kinetic situations are extremely unlikely, or even impossible. This, obviously, is the reason why flux-dependent oscillators are much less abundant than force-dependent oscillators.

The time dependence of the oscillating variables arises from two kinds of intrinsic properties of the systems:

(a) STORAGE PROPERTIES [11]

They may be of capacitance or inductance type whether the time dependence concerns the forces or fluxes: fig. 6 gives examples of " generalized " capacitances and inductances. They are defined by the well-known differential relationships:

<table>
<tr><td>capacitances</td><td>inductances</td></tr>
<tr><td>$$I_c = C(\mathrm{d}X/\mathrm{d}t)$$</td><td>$$X_l = L(\mathrm{d}I/\mathrm{d}t)$$</td></tr>
<tr><td>I_c: flux into the capacitor (e.g., electric current, molar flux, heat flux, volume flux)</td><td>X_l: force of the inductor (e.g., electromotive force, pressure)</td></tr>
<tr><td>C: generalized capacitance, (e.g., electric capacitance, volume of solvent, heat capacitance, volume capacitance)</td><td>L: generalized inductance (e.g., electric inductance, inertia of fluid)</td></tr>
<tr><td>X: force of the capacitor (e.g., electromotive force, concentration, temperature, pressure)</td><td>I: flux of the inductor (e.g., electric current, volume flux)</td></tr>
</table>

In the case of capacitances, as illustrated in fig. 6a, an influx of electric charges into a capacitor leads to a temporal increase of voltage, an influx of matter into a volume of solvent leads to an increase of concentration, an influx of heat into a heat capacitor leads to an increase of temperature and an influx of volume of fluid into a volume capacitor leads to an increase of pressure. In most cases the capacitances may be regarded as constants. They may, however, also depend on the forces, but they are essentially positive. That follows from the well-known laws of conservation of charges, matter, volume of incompressible fluid and energy.

(b) DISSIPATIVE PROPERTIES

Whilst time dependence of variables effected by capacitances and inductances respectively concerns the storage properties of free energy, dissipative properties, such as conductivities and chemical reactions may also give rise to time dependences. That is, in particular, the case if such dissipative properties depend on their own driving forces or fluxes respectively. Evidently such behaviour is always a result of a feedback mechanism consisting of consecutive reactions and/or transportation processes, which always cause a corresponding lag of time. This is the intrinsic reason why all phenomena effected by feedback mechanisms are time dependent and play a substantial role in all non-linear oscillatory systems.[12]

Strictly speaking, any physico-chemical oscillator has as many variables as fluxes and reactions occur in the entire oscillatory process. Chemical oscillators in particular have as many chemical variables as chemical species or reactions take part in the oscillation mechanism. Biological and chemical oscillators, therefore, are usually systems of relatively many kinetic variables, each of them giving rise to a differential equation of its own.

For mere qualitative understanding of oscillatory behaviour, the entirety of existing variables can mostly be reduced to only two " essential variables ". However, if the specific shape of a given oscillation has to be described and interpreted in detail, more than two variables must usually be taken into consideration.

3. THE FEEDBACK CONCEPT OF NON-LINEAR OSCILLATIONS

Coupling effects as well as autocatalytic and autoinhibitory mechanisms may lead, as is well-known, to positive or negative feedback in physico-chemical systems. In the case of oscillatory behaviour both kinds of feedback must obviously be present simultaneously.

A direct consequence of strong positive feedback is the occurrence of non-monotonic flux-force-characteristics. As will be discussed later more in detail, such characteristics cause the instability behaviour and excitation phenomena mentioned above. In contrast to positive feedback, negative feedback has a stabilizing effect and causes recovery and adaptation behaviour in oscillatory systems. Variables involved in negative feedback mechanisms are always free from instability situations.

Obviously, a general kinetic pattern [13] which is valid for all physico-chemical oscillatory systems can be designed according to fig. 7. Let X be a kinetic variable (e.g., voltage, concentration, pressure, temperature, etc.) which increases by a " formation process " and decreases by a simultaneous " consumption process ". X becomes constant when both processes have equal turnover. As already mentioned, oscillatory behaviour may arise if two antagonistic feedback mechanisms are effective inside the system simultaneously :

    (a) a fast positive feedback in a " favouring loop"

    (b) a slow negative feedback in a " counteracting loop ".

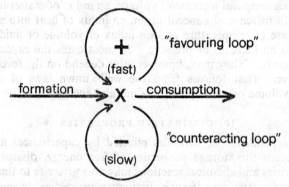

FIG. 7.—General pattern of oscillatory systems.

Each loop as a rule includes several variables, which may be arranged in direct stoichiometric reaction chains or in non-stoichiometric coupling mechanisms. Only the actual total effects and rates of both kinetic loops are essential here. It is obvious that for the occurrence of oscillations the lag of time in both feedback loops has to be sufficiently different, otherwise the system would gain a stable steady state.

Favouring and counteracting effects can arise by autocatalytic or autoinhibitory action. It depends on whether the loop in question concerns the formation or the consumption process. As a consequence the general pattern of fig. 7 can be realized in four different ways (fig. 8). Examples for all of these four types of oscillatory systems are known. For the elucidation of oscillatory systems it may be advantageous as a first step to find out to which of the four types the system in question belongs.

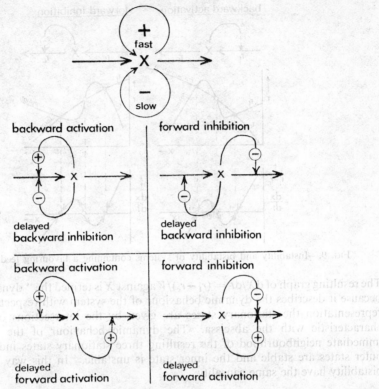

FIG. 8.—The four kinetic types of dissipative oscillatory systems ($+$ and $-$ stands for autocatalytic or autoinhibitory effects respectively).

4. EXCITABILITY PHENOMENA

Instability as a result of positive feedback by backward activation or forward inhibition [14] is a necessary but not sufficient presupposition of oscillatory behaviour. However, systems containing positive feedback only may already bring forth excitability and propagation phenomena. Fig. 9 illustrates how instability is brought forth by the action of a favouring loop. In both of the possible cases the result of autocatalysis or autoinhibition is, likewise, the occurrence of a force—(i.e., concentration) dependent non-monotonic kinetic characteristic. [10] The stationary states of $X(dX/dt = 0)$ are given by the intersections of the characteristics of formation and consumption of X. In the non-stationary states, the difference of the reaction rates of formation and consumption corresponds directly to the time variation dX/dt according to the generalized capacitance equation:

$$r_f - r_c = V(dX/dt)$$

(r_f, r_c: reaction rates of formation and consumption respectively
V: volume
X: concentration).

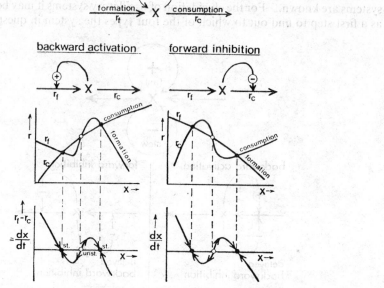

FIG. 9.—Instability and bistability of systems containing a favouring feedback loop.

The resulting graph of $dX/dt = (r_f - r_c)/V$ against X is termed the "dynamic diagram" because it describes the dynamic behaviour of the system with respect to X. In this representation the stationary states are given by the intersections of the dynamic characteristic with the abscissa. The dynamic behaviour of the system in the immediate neighbourhood of the resulting three stationary states indicates that the outer states are stable and the inner state is unstable. In this way instability and bistability have the same causality.

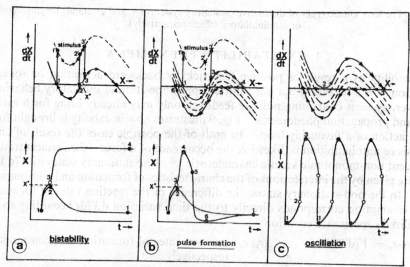

FIG. 10.—Bistability, pulse formation and oscillation as a result of feedback represented in the dynamic and the force-time diagrams.

Fig. 10a illustrates the effect of sub-threshold and super-threshold stimuli according to a concept which Bonhoeffer [16] introduced as early as 1943 into the theory of excitation kinetics. Some typical examples of systems in which non-monotonic force-dependent characteristics occur are shown in fig. 11.

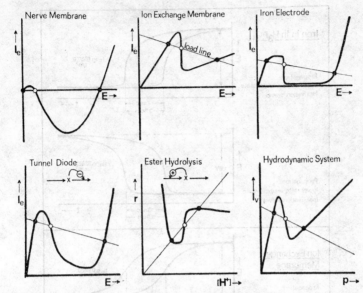

FIG. 11.—Examples of systems exhibiting non-monotonic flux-force-characteristics by positive feedback.

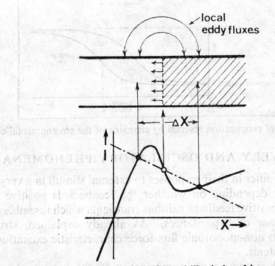

FIG. 12.—Propagation as a result of bistability induced by positive feedback.

Bistability exhibited by a favouring loop is the intrinsic cause of propagation phenomena (fig. 12). The two stable states of the bistable system correspond to two different values of driving force. As a consequence, at the boundary between ranges of different states, transportation processes take place driven by the difference of the forces of both ranges. The local fluxes across the boundary act as stimuli for both ranges. The propagation of state conversion then proceeds in that direction determined by the mutual stimulation process which succeeds first. By appropriate alteration of the environmental conditions it is possible to reverse the direction of propagation (fig. 13).

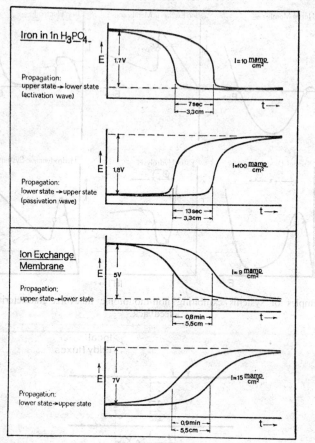

FIG. 13.—Examples of propagation reversal by changing of the environmental conditions.

5. RECOVERY AND OSCILLATORY PHENOMENA

Feedback systems differ in their responses to external stimuli in a very distinct and characteristic manner depending on whether the feedback is positive or negative. As shown in fig. 14a, positive feedback exhibits transients which resemble qualitatively the temporal behaviour of capacitances. As already explained, strong positive feedback giving rise to non-monotonic flux-force characteristic excitation transitions may occur in such systems.

On the other hand, systems containing negative feedback respond in the form of overshoot phenomena (fig. 14b). In this respect, the transients of negative feedback loops resemble inductances in their temporal behaviour. In other words negative feedback tends—with a certain temporal delay—to counteract or to compensate disturbances or alterations of the system.

By combining a positive feedback with an appropriate negative feedback of sufficient long time delay then oscillations and excitation pulses are possible. In this case, the non-monotonic dynamic characteristic of the favouring loop also depends simultaneously on the state of the counteracting loop. Fig. 10b, c illustrate schematically how pulses and oscillations are brought forth by cooperation of both antagonistically reacting feedback loops.[10] Here, for the sake of simplicity, only two kinetic

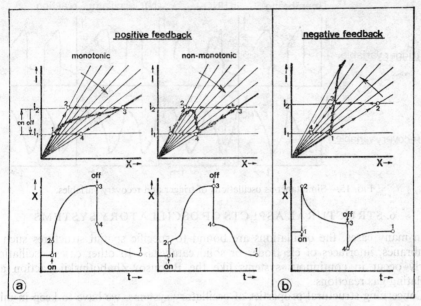

FIG. 14.—The time behaviour of dissipative systems exhibiting force-dependent conductance by positive and negative feedback. In the upper row of graphs the slope of the beams correspond to the conductance $G(X) = I/X$ depending on the force X. In case of positive feedback (forward inhibition) the conductance decreases and in case of negative feedback (forward activation) it increases with increasing force. As a result of the time-lag of the feedback sudden changes of flux bring about the characteristic transients of the lower row of graphs.

variables are taken into consideration, X being the "trigger variable", characterizes the action of the favouring loop and Y, being the "recovery variable", characterizes the action of the counteracting loop. The primary excitation of the triggerable favouring loop recovers spontaneously by the delayed action of the counteracting loop, in the course of which an unstable state may be attained too. In other words, excitation and recovery similarly imply instability events. In the case of pulse generation, excitation triggering is provoked by external disturbances (stimuli) and recovery is an internal, spontaneous process.

In case of oscillations, however, both trigger processes are internal events which occur spontaneously inside the system. Oscillatory behaviour, considered from that point of view, then consists of successive trigger events caused by the cooperation of both feedback loops. This idea, that physico-chemical oscillators might imply two intrinsic trigger processes was suggested by Bonhoeffer [15] in 1948.

In connection with the concept of antagonistic feedback loops in oscillatory systems, it might be noticed here that the oscillating variables can be divided clearly into two classes depending on whether they belong to the positive or to the negative feedback loop. In most cases, it is easy to decide experimentally to which loop a recorded oscillation variable belongs (fig. 15). Variables involved in the positive feedback as a rule exhibit characteristic breaks or steps in the course of their oscillation recordings. These breaks are the result of the unstable states occurring periodically in the kinetic mechanism of the positive feedback loop.   Variables belonging to the negative feedback loop never have such breaks in their oscillograms because of the absence of instabilities in their kinetics.

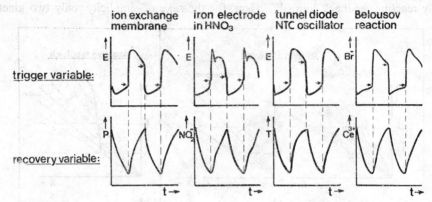

FIG. 15.—Simultaneous oscillations of trigger and recovery variables.

6. STRUCTURAL ASPECTS OF OSCILLATORY SYSTEMS

In many cases, the oscillations are bound to specific spatial structures such as membranes, interfaces of electrodes or solid catalysts.   In other cases, oscillations clearly occur in continuous systems like the Belousov-Zhabotinskii-reaction and oscillating gas reactions.

Concerning structural properties of oscillatory systems, we have to keep in mind that all systems which oscillate in macroscopic spatial ranges necessarily imply macroscopic propagation processes.   Otherwise, only microscopic fluctuations or dissipative structures can occur.   As mentioned before, propagation generally is brought about by local transportation processes caused by local gradients of driving forces.   These local transportation processes are decisive for the structural requirements of oscillatory systems.   For instance, all systems in which electric local currents are involved as propagation-inducing transportation processes, necessarily consist of interfacial structures, since local currents require closed circuits which only can occur at interfaces.   For this reason electrochemical and electrobiological oscillations are strictly bound to membranes or interfaces.

On the other hand, transportation processes inducing propagation in chemical systems (diffusion, heat conduction) are independent of the existence of closed circuits.   Here oscillatory behaviour may occur in continuous systems as well.

7. SUMMARY

It has been shown that :

(1) The causality concerning the relationship between fluxes and forces in nonlinear dissipative systems leads to two distinct classes of oscillators :

        (a) force-dependent oscillators
        (b) flux-dependent oscillators.

(2) Time dependence in those systems arises from :

        (a) storage properties of free energy
        (b) feedback properties of dissipative processes
           (chemical reactions and transportation processes).

(3) Feedback properties are brought about by autocatalytic and autoinhibitory processes concerning chemical reactions and/or transportation processes.

(4) Non-linear dissipative oscillators require the simultaneous cooperation of a strong (i.e., non-monotonic) positive feedback with a negative feedback being relatively delayed yet strong enough to recover excitation transitions of the positive feedback.

(5) Non-linear dissipative oscillators can, in principle, be distinguished as four different types of systems depending on whether the formation process and the consumption process are influenced by autocatalytic or autoinhibitory effects. From the eight possible combinations four of them may induce oscillatory behaviour:

> (a) forward inhibition—backward inhibition
> (b) forward inhibition—forward activation
> (c) backward activation—backward inhibition
> (d) backward activation—forward activation.

(6) Positive feedback induces pseudo-capacitance behaviour (fig. 14a) and in the case of non-monotonity:

> (a) excitability, instability, bistability, all-or-none-behaviour
> (b) propagation phenomena.

Negative feedback causes pseudo-inductance behaviour, in particular:

> (a) overshoot phenomena
> (b) adaptation behaviour, accommodation, refractoriness
> (c) recovery of excited states.

Simultaneous cooperation of both kinds of feedback may lead to:

> (a) oscillatory behaviour
> (b) pulse formation
> (c) dissipative structures.

(7) All of the non-linear dissipative oscillatory systems are genuine " nerve models " exhibiting under appropriate environmental conditions the entire phenomenology of excitable nerve and muscle membranes.

(8) Oscillations including electrical transportation processes are strictly bound to heterogeneous interfacial structures, whereas chemical and thermal oscillations may also occur in continuous systems.

[1] I. Tasaki, *Handbook of Physiol. Neurophysiol.*, 1959, **I**, 75.
[2] W. Trautwein, *Ergebn. Physiol.*, 1961, **51**, 131.
[3] W. J. V. Osterhout, *J. Gen. Physiol.*, 1943, **26**, 457.
[4] T. Teorell, *J. Gen. Physiol.*, 1959, **42**, 831.
[5] U. F. Franck, unpublished.
[6] P. Müller and D. Rudin, *Nature*, 1968, **217**, 713.
[7] U. F. Franck and F. Kettner, to be published.
[8] U. F. Franck and F. Kettner, *Ber. Bunsenges.*, 1964, **68**, 875.
[9] R. J. Field, E. Körös and R. M. Noyes, *J. Amer. Chem. Soc.*, 1972, **94**, 8649.
[10] U. F. Franck, *Biological and Biochemical Oscillators* (Academic Press, New York, 1972), p. 7.
[11] U. F. Franck, *Chem. Ing. Techn.*, 1972, **44**, 228.
[12] G. Viniegra-Gonzalez, *Biological and Biochemical Oscillators* (Academic Press, New York, 1972), p. 71.
[13] U. F. Franck, Abstracts of contributed papers Congr. International Biophysics IUPAB Moscow (1972).
[14] J. Higgins, *Ind. Eng. Chem.*, 1967, **59**, 19.
[15] K. F. Bonhoeffer, *Z. Elektrochem.*, 1948, **51**, 24.
[16] K. F. Bonhoeffer, *Naturwiss.*, 1943, **31**, 270.

GENERAL DISCUSSION

Dr. B. F. Gray (*Leeds*) said: We [1] have studied propane oxidation in a stirred flow reactor, an apparatus in which questions of fuel consumption etc., do not arise, because either perfectly steady states or limit cycles can be achieved exactly, and their behaviour with respect to perturbations studied experimentally. Results which can only be guessed at from closed vessel studies (such as the existence of a stable focus) can be studied exactly in this case, and in particular limit cycle oscillations can be made to persist indefinitely. Hysteresis and multistability are also shown in this system. In the closely related case of acetaldehyde oxidation the advantages of the flow system are relatively greater, since the reaction parameters are such that in a closed system the behaviour is dominated first by the arbitrary initial conditions selected and then almost immediately by the monotonic decay towards equilibrium. The system never gets close to the kinetic stationary state (or sustained oscillation) and thus little information is obtained from experiments in closed vessels. On the other hand, in the flow system the kinetic stationary state can be achieved exactly, as can conditions of sustained oscillation. The system also exhibits hysteresis and multistability.[2] Also as the kinetics of this system are reasonably well understood on a semi-quantitative basis, computations have been performed including the energy conservation equation,[3] excellent agreement with experiment being obtained considering the uncertainties in some of the experimental values of the rate constants used. This agreement leads one to feel that this thermokinetic oscillation is at least as well understood as the Belousov reaction, following the detailed work of Noyes *et al.*, but at the same time it displays a considerably richer phenomenology than the latter (which is not sufficiently nonlinear to show multistability or hysteresis). By analogy it would be interesting to pursue the possibility of obtaining spatial patterns in the acetaldehyde system.

Dr. L. J. Kirsch (*Thornton*) said: At Thornton we have been working for some time on the mathematical modelling of hydrocarbon oxidation phenomena. These models, starting with the acetaldehyde oxidation model of Halstead, Prothero and Quinn [4] have more recently been developed in a generalised form to describe alkane oxidation phenomena over a wide range of experimental conditions. They employ a degenerate branched chain mechanism, with competing reactions of differing activation energies controlling the supply and consumption of branching agent and removal of radicals from the system. This competition leads to oscillation under certain conditions, and the models are therefore examples of the thermokinetic type described by P. Gray *et al.* in their paper. In both analytical and numerical treatments of the model we have taken fuel consumption into account. The models give excellent simulations of all the patterns of behaviour illustrated in fig. 2 of the paper by Gray *et al.* I should like to mention some conclusions resulting from our experience of working with these models which are of relevance to the points raised by these authors.

[1] B. F. Gray and P. G. Felton, *Comb. Flame*, 1975.
[2] B. F. Gray and P. G. Felton, *Comb. Flame*, 1975.
[3] B. F. Gray, P. G. Felton and N. Shank, *Comb. Flame*, 1975.
[4] M. P. Halstead, A. Prothero and C. P. Quinn, *Proc. Roy. Soc. A*, 1971, **322**, 377.

First, with reference to the " sustained " oscillation illustrated in fig. 2(d), we have noted in our simulations that cool flame behaviour often terminates when a sizeable proportion of reactant remains unconsumed. There follows a lengthy period of slow combustion. Secondly, we have observed during the course of a single simulation, transitions from the damped oscillations of fig. 2(a) to the sustained oscillation of fig. 2(d).

Both of these observations illustrate a property of the thermokinetic model we have considered—namely that the nature of the stationary point solutions are properties not only of the initial conditions and heat transfer properties, but also of reaction co-ordinate, in as much as reactant consumption influences the appropriate rate coefficients.

It would, therefore, be of interest to learn whether the statement that cool flame behaviour terminates abruptly when no fuel remains is based upon direct experimental evidence. Also, whether Gray *et al.* have found any experimental evidence for a change in the nature of oscillatory behaviour during the course of a single experiment. This might particularly be anticipated in their experiments where a small quantity of acetaldehyde was added, and clearly had a pronounced influence on the pattern of behaviour. Variation of the relative quantities of propane and acetaldehyde present may occur during the course of the reaction due to unequal reaction rates.

Some rationalisation of the fact that our models predict changes in the nature of oscillation due to fuel consumption, whereas this was not observed in the experiments of Gray *et al.*, may lie in the orientation of the experimentally determined line separating weakly and strongly damped oscillations (fig. 4). If the chemical mechanism does not involve any pressure dependent (termolecular) steps, the effect of reactant consumption can to the first approximation be envisaged in terms of motion of the point describing the appropriate time-dependent reactant conditions parallel to the pressure axis as the reaction proceeds. The line separating weakly and strongly damped oscillations lies parallel to this direction of motion over much of its range. Thus transitions from one kind of oscillation to another will be unlikely in this system, although an eventual transition into the region of slow combustion seems probable.

Prof. P. Gray (*Leeds*) said: The calculations which Kirsch has made attempt the most complete modelling of thermokinetic effects in hydrocarbon oxidations. They are based on the assumption of spatially uniform temperature and concentrations, i.e., they represent a perfectly stirred system. The quantitative validation of the model, therefore, requires comparison with experiments done in well-stirred conditions. Unfortunately, nearly all classical investigations of cool flames have not been made under such conditions : they are all affected by natural convection and flame propagation usually from a hot region.[1] We believe that our experiments in a stirred reactor offer Kirsch experimental results with which he can compare his computer analysis ; we can also supply him with experimental values of heat transfer coefficients for this type of reactor.

In answer to Kirsch's point about cessation of reaction, we find that temperature excesses, and hence rates of heat release by reaction, fall to zero almost discontinuously as soon as the oscillations have died away. (The time axis of our fig. 2 is too brief to display this.) The " piq d'arrêt " is often exhibited ; Lucquin[2] identifies this with the complete consumption of oxygen and the end of combustion. We have not measured

[1] J. F. Griffiths, B. F. Gray and P. Gray, Thirteenth International Symposium on Combustion, (The Combustion Institute, 1971), p. 239.

[2] L. R. Sochet, J. P. Sawersyn and M. Lucquin, *Advances in Chemistry Series*, No. 76, *Oxidation of Organic Compounds—II* (The American Chemical Society, 1968), p. 111.

reactant concentrations by direct chemical analysis. Kirsch's calculations that a large proportion of fuel remains and that combustion continues for a lengthy period even after the last undamped oscillation seem to show a real difference between experiment and the model.

Alteration of the amount of acetaldehyde added has a pronounced effect on the reaction rate in the earliest stages. If concentrations of acetaldehyde are increased beyond 3 mol %, not only are induction times to oscillations reduced but the amplitude of the first oscillation is also increased.

Kirsch's second point really has to do with the imperfections of trying to categorize chemical reactions in closed vessels in terms that are only properly appropriate to open systems. True limit cycles in these oxidations are possible only if fresh reactants can be constantly supplied. As experimentalists we can recognize, in a closed system, behaviour that is like a limit cycle in the phase plane when we see 5 or 6 successive undamped or weakly damped oscillations; the behaviour is more like a stable focus in the phase plane when we see strongly damped oscillations.

Even these remarks do not apply to very many closed-vessel oxidations—propane is perhaps the most favourable case, giving up to 11 pulses in a closed vessel. Acetaldehyde is the opposite; no experimentalist seems to have observed more than one or two cool flames in a closed vessel. In an open system, of course, oscillations can persist for as long as we wish. Recent studies by B. F. Gray and P. G. Felton working in Leeds have revealed oscillatory oxidation supported indefinitely in a variety of conditions, including acetaldehyde as well as propane.

In our fig. 4, the broken line dividing the regions is correspondingly subjective, and it ought perhaps to be drawn in its lower parts with a finite positive slope instead of as a nearly vertical line. (After all, a single oscillation can hardly be called undamped.) Even so, we do not encounter either a train of (say 4) undamped oscillations followed by another train of (say 4) damped oscillations, or the reverse sequence. It is as though Kirsch's calculations assume that the development in time of the (real) closed system, with a particular initial reactant concentration but suffering continual reactant consumption, can be legitimately imitated by putting together a short sequence of open systems each without any reactant consumption at all but differing from one another by successively diminished initial reactant concentrations.

So far as fig. 4 is concerned, there is a similarity between a vertical traverse of the diagram and the course of events in time but to represent either fuel consumption (or reactant temperature) on it would require a third dimension.

Prof. P. Gray (*Leeds*) (*communicated*): Among gaseous oxidations often said by reviewers of oscillatory reactions to be accompanied by oscillations, is that of hydrogen sulphide. Our work [1] shows it to involve thermal and kinetic feedback by self-heating and chain-branching, but neither we nor any authors save Thompson [2] have so far succeeded in producing oscillations. I think multiple ignitions reported may have arisen from their technique of admitting reactants successively to their relatively narrow cylindrical reaction vessel. At the pressures employed, mixing could have been so imperfect that the first ignition would have been very incomplete and localized. When it was over, time for mixing would be required to elapse before a second partial ignition could occur, and so on. If this is correct, then successive ignitions in H_2S oxidation belong with old observations on phosphorus oxidation in unclosed vessels as originating in slow diffusive processes rather than in the chemistry.

[1] P. Gray and M. E. Sherrington, *J.C.S. Faraday I*, 1974, **70**, 2338.
[2] H. W. Thompson, *J. Chem. Soc.*, 1931, 1809.

Prof. P. Gray (*Leeds*) said: In answer to Shashoua, I would remark that acetaldehyde is added here principally to shorten the induction period before oscillatory cool flames set in—it is of course one of the many intermediate products of propane oxidation—and it also has some secondary effects on the oscillations observed. Rather than being a sink for radicals as in Shashoua's polymerisations, added acetaldehyde in these oxidations is a potential radical source. I agree with Shashoua that it (and other added reactants) offer a means of varying conditions, controlling behaviour and investigating the chemical part of the thermokinetic mechanism. We believe these would be worthwhile investigations though there are other fundamental aspects to clear up first.

Dr. R. G. Gilbert and **Mr. R. Ellis** (*University of Sydney*) said: Concerning Yang's isothermal mechanism, both he and P. Gray have mentioned the possibility of important non-isothermal effects. In a closed adiabatic system, inclusion of temperature as an additional independent variable in numerical integration of the kinetic equations is fairly straightforward. We have written a general program to do this, and can verify that such systems are very sensitive to this temperature variation. This program was written for our study of sound propagation in gas-phase reactions, extending our previous work in this field.[1] As a result of these studies we suggest that acoustic-kinetic interactions in systems which are capable of undergoing homogeneous oscillations would lead to amplification and interference effects which could provide useful data for deriving kinetic information.

Prof. C. H. Yang (*Stonybrook*) said: In general, the non-isothermal effects in the oxidation of CO is not as significant as in the case of hydrogen[2] or hydrocarbon[3, 4] oxidation. For sustained oscillations and long-lasting glows (minutes) in the CO system Dove[5] showed that it is essentially isothermal.

In the case of explosion or glow (seconds) for wet mixtures (greater than 1 % H_2O), the slow reaction rate is very low at sub-critical conditions. Limit pressures or temperatures are unlikely to be affected by the non-isothermal effect. Limits for drier mixture may be influenced by " self-heating " mildly. Only the duration of explosions will almost certainly be drastically shortened by thermokinetic considerations as the temperature difference between the reacting mixture and the bath will be high in such cases. However, the duration of explosion has rarely been used as an important parameter in either theoretical or experimental studies. A unified thermokinetic approach to the problem probably would not add significantly to the understanding of the kinetic mechanism.

Dr. L. J. Kirsch (*Thornton*) said: In the final paragraph of his paper, Yang discusses the effect on his calculations of the rate constant k_2 which describes the recombination rate between atomic oxygen and carbon monoxide:

$$O + CO + M \rightarrow CO_2 + M \qquad \qquad k_2$$

Yang uses the value recommended by Baulch *et al.* (1968) in his computations. Recently a number of studies of this reaction have been made, many employing

[1] R. G. Gilbert, H.-S. Hahn, P. J. Ortoleva and J. Ross, *J. Chem. Phys.*, 1972, **57**, 2672; R. G. Gilbert, P. J. Ortoleva and J. Ross, *J. Chem. Phys.*, 1973, **58**, 3625.

[2] K. K. Foo and C. H. Yang, *Comb. Flame*, 1971, **17**, 223.

[3] C. H. Yang and B. F. Gray, *J. Phys. Chem.*, 1969, **73**, 3395.

[4] C. H. Yang, *J. Phys. Chem.*, 1969, **73**, 3407.

[5] J. E. Dove, *Dissertation* (Oxford, 1956).

direct monitoring of the atomic oxygen decay in a time-resolved system.[1-6] This work has indicated that at room temperature, the rate coefficient k_2 is some orders of magnitude lower than that given by Baulch *et al.* Measurements over a range of temperature [3] give a value of $k_2 = 2.4 \times 10^{15} \exp(-4340(\text{kcal/mol})/RT)$ cm^6 mol^{-2} s^{-1}. The reaction, therefore, appears to exhibit the sizeable positive activation energy that Yang states will improve the fit between calculated results and experimental data. At a temperature of about 700 K the overall value of the rate coefficient is in fact close to the temperature independent value employed by Yang.

Perhaps the most striking fact to emerge from these recent studies of reaction (2) is its extreme sensitivity to traces of impurity, presumed to be metal carbonyls. Thus the most stringent purification procedures must be followed, even with supposedly high purity carbon monoxide, in order that the true value of the rate constant be recorded. For this reason many earlier studies of the reaction must be disregarded. It follows that some doubt must be attached to any quantitative interpretation of more complex reactions in which reaction (2) plays a rate determining role, unless similar precautions regarding purification of CO have been taken.

Prof. C. H. Yang (*Stonybrook*) said: A very interesting point has been raised by Kirsch, namely that the rate constant of reaction (2) is extremely sensitive to traces of impurity, according to recent experimental evidence. In fact, the rate constant for reaction (4) may also be sensitive to impurities. While we have no direct evidence of this at the present time, it appears to be true with some other excited species reported in the literature. Cvetanovic cited the data of Slanger and Welge indicating that the rate constants for the de-excitation reaction of the $O(^1s)$ atom with third bodies of N_2 and H_2O are 3.0×10^7 and 2.1×10^{14} cm^{-3} mol^{-1} s^{-1} respectively. Water would clearly be a very sensitive impurity in this case. It is well known that the oscillation phenomenon in the CO system is difficult to reproduce under apparently identical experimental conditions. Often, this is mysteriously attributed to the role played by the vessel surface. In view of our calculations,[8, 9] it is clear that oscillatory solutions may be completely inhibited by sizeable variations of either of these two rate constants. Perhaps, purity of the reactants may be the most essential quality to strive for in kinetic experiments of this kind.

Prof. R. M. Noyes (*Oregon*) said: The behaviour of oxygen atoms in Yang's mechanism is similar to that of the switched intermediate, X, in the Oregonator. Thus step (1) forms oxygen atoms at a rate independent of their concentration, and the sequence of steps (5) and (6) accomplishes the same net reaction at a rate proportional to the concentration of OH radicals. Because the concentrations of OH and of O are positively coupled through step (7), the sequence of steps (5) and (6) generates oxygen atoms autocatalytically. The sequence (2) + (4) destroys oxygen atoms by a first order process, and the sequence (2)+(3) essentially represents second order destruction of oxygen atoms.

[1] R. J. Donovan, D. Husain and L. J. Kirsch, *Trans. Faraday Soc.*, 1971, **67**, 375.
[2] F. Stuhl and H. Niki, *J. Chem. Phys.*, 1971, **55**, 3943.
[3] T. G. Slanger, B. J. Wood and G. Black, *J. Chem. Phys.*, 1972, **57**, 233.
[4] R. Simonaitis and J. Heicklen, *J. Chem. Phys.*, 1972, **56**, 2004.
[5] W. B. Demore, *J. Phys. Chem.*, 1973, **76**, 3527.
[6] E. C. Y. Inn, *J. Chem. Phys.*, 1973, **59**, 5431.
[7] R. J. Cvetanovic, *Canad. J. Chem.*, 1974, **52**, 1452.
[8] C. H. Yang, *Comb. Flame*, 1974, **23**, 97.
[9] C. H. Yang and A. L. Berlad, *J. C. S. Faraday I*, 1974, **70**, 1661.

As we have pointed out elsewhere,[1] systems with unstable steady states can not be modelled by elementary processes involving only two intermediates. Thus the state of the present system can not be described solely by the concentrations of CO_2^* and O; other species like H, OH, and H_2O_2 undergo coupled variations with delays that are important for the instabilities exhibited.

Although the calculations with this model show impressive similarities to experimental observations; I am disturbed about the postulated CO_2^* intermediate. This excited species reacts with oxygen atoms very efficiently and without activation energy, even though the net reaction involves breaking and forming chemical bonds each much stronger than 100 kcal/mol. However, the same excited species must usually undergo over 10^8 collisions before it loses its excitation energy by step (4). I am not aware of any precedent for an excited species that can so easily react with rearrangement of strong bonds or that is so inert to loss of excitation energy by collision; it would be particularly remarkable to find these unusual and seemingly contradictory tendencies in the same molecule. I am, therefore, unconvinced that the mechanism of this reaction is yet demonstrated.

Prof. C. H. Yang (*Stonybrook*) said: In reply to Noyes I would remark that Tyson and Light[2] have shown that the type of oscillation prescribed by a limit cycle does not exist in a kinetic model which is constructed with two intermediates and involves only first and second order elementary process. This conclusion is, of course, invalid for a kinetic system with more than two intermediates. Under appropriate conditions a kinetic model of multiple components may be reduced to a binary system when the concentrations of some intermediates are eliminated by the steady state assumption. Again, Tyson and Light's conclusion is inapplicable to such a reduced system. In our earlier work[3] we replaced the concentrations of H, OH and OE (where OE represents a vessel wall site which is occupied by an O atom) by their steady state values. The kinetic model was reduced to a binary system containing only the concentrations of CO_2^* and O. A stable limit cycle was shown to exist in that system. In our present work, on the other hand, all intermediates in the proposed scheme have been considered. Clearly, the system is not solely prescribed by the intermediates CO_2^* and O. It is useful to point out after comparing the present results with our earlier work that the assumption of steady states for some intermediates, while greatly simplifying the mathematical complexities, has not impaired the ability of the simpler model to predict all important kinetic features qualitatively.

It is indeed essential to assume in our current work that the excited CO_2^* molecule is quite stable (actually metastable) as far as the quenching reaction (4) is concerned. As indicated in the paper our sole criterion for selecting the rate constants for reactions (3) and (4) is based on a favourable fitting of the calculations to the oscillation and explosion limit data. There are no known values for these rate constants reported in the literature to the knowledge of the author. They probably remain to be independently determined. However, we find it difficult to accept the argument that the stability of the CO_2^* molecule would imply a slow rate for reaction (3). Many bimolecular reactions which involve a stable molecule and a radical are known to be fast. One example is the well known titration reaction $NO + N \rightarrow N_2 + O$, where a bond of 150 kcal is broken and a bond of 225 kcal is formed for which a rate constant of the value 3×10^{13} cm^3 mol^{-1} s^{-1} with nearly zero activation energy is accepted.[4]

[1] R. M. Noyes and R. J. Field, *Ann. Rev. Phys. Chem.*, 1974, **25**, 95.
[2] J. J. Tyson and J. C. Light, *J. Chem. Phys.*, 1973, **59**, 4164.
[3] C. H. Yang, *Comb. Flame*, 1974, **23**, 97.
[4] D. L. Baulch, D. D. Drysdale, D. G. Hoare and A. C. Lloyd, *High Temperature Reaction Rate Data* (Leeds University 1969), no. 4.

This value is almost three orders of magnitude greater than what we proposed for reaction (3). As noted [1] before, the rate constants for reactions (3) and (4) may be changed to 5.0×10^{13} and 1.25×10^8 respectively if only one thousandth of the total excited CO_2^* molecules produced in reaction (2) are effective in reaction (3). All computed results will be invariant for such a change.

Reaction (5) is another example. At $T = 1000$ K the rate constant for reaction (5) is greater than or at least comparable to the one we suggested for reaction (3).

Dr. J. R. Bond (*Leeds*) said: Yang's theoretical paper on carbon monoxide oxidation is a carefully fitted complex of elementary reactions, and quite narrow restrictions must be placed upon kinetic parameters (e.g., relative third body efficiencies of CO_2 to other gases) to ensure reasonable agreement with experiment. It is also an isothermal model, although it is for a highly exothermic reaction. To discover whether temperature changes in real systems are indeed negligible, we have studied both " dry " and " wet " oxidations of carbon monoxide using exceedingly fine thermocouples as probes.

In all but one case,[2] oscillations observed by previous workers have occurred in dry mixtures, and we too have observed many oscillations in such systems. However, when we deliberately add small amounts of hydrogen to the initially " dry " system, we find that oscillations are still readily obtained over a range of temperatures and pressures. The number of cycles observed in a closed vessel is always far less than for the " dry " case but the reactant consumption in each cycle is much greater. In a mixture containing 0.1 mol % of hydrogen, ten or more oscillations can be observed, and this mixture is very " wet " indeed when compared with the low hydrogen content of mixtures used in the investigation of " dry " oxidation phenomena. The most striking feature of the oscillations is the size of the temperature pulse accompanying

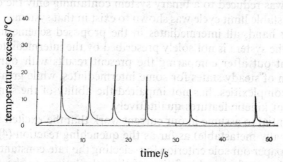

FIG. 1.—Temperature pulses in a " wet " $CO + O_2$ mixture at 480°C and 30 torr.

each cycle. Temperature peaks exceed 30 C for the first few cycles; they reduce to 10°C as reactant consumption nears completion at the end of the sequence. In " dry " mixtures, amplitudes are smaller and the train of oscillations in a closed system is longer. Thus, although thermal effects may be small or even negligible, in " dry " systems it is probably necessary to allow for self-heating in " wet " systems, and the isothermal model may be inappropriate. Moreover, even the small temperature pulses may be more that the casual effects of reaction pulses—they may interact integrally with the kinetics.

[1] C. H. Yang and A. L. Berlad, *J.C.S. Faraday I*, 1974, **70**, 1661.
[2] J. W. Linnett, B. G. Reuben and T. F. Wheatley, *Comb. Flame*, 1968 **12**, 325. These oscillations were observed during first or " glow " limit determination by the heating method.

Prof. C. H. Yang (*Stonybrook, N.Y.*) (*communicated*): Our calculated trajectories by oscillating O atom concentration also compared very well with the recent measurements of the successive emission peaks from the CO and O_2 system by McCaffrey and Berlad.[1] Their results will be published shortly.

The destruction of hydrogen-containing intermediates on the wall will probably produce relatively more stable molecular species H_2 and H_2O. At the present time, the detailed mechanisms of these heterogeneous reactions are far from elucidated. We simply assumed that only H_2O is produced from these reactions to avoid the complication of introducing many additional reactions into our calculation. Eqn (27) represents a conservation statement of the total oxygen in the vessel. In the early phase of the oxidation process, the consumption of water is usually limited to less than a few percent of its initial concentration. Calculated limits of explosion, glow and oscillations are not likely affected even if the products of the heterogeneous reactions contain a small fraction of H_2. For the calculation of a long sustained oscillation or glow, the overall oxidation rate will probably accelerate if H_2 is slowly accumulated to reach a significant level as the rate of the reaction $O + H_2 \rightarrow OH + H$ is undoubtedly faster than the rate of reaction (7). The general kinetic behaviour, however, will remain unchanged.

Dr. A. Perche (*Lille*) said: Yang's paper concerning simulation of periodic carbon monoxide explosions would certainly have been much more informative if a direct

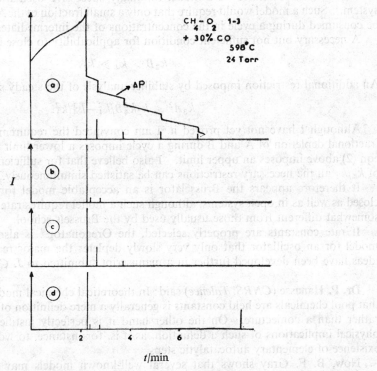

FIG. 1.—Pressure variation (Δp) and luminous intensity (I) versus time. (*a*) Usual evolution—reaction vessel tap opened. (*b*) Tap closed. (*c*) Tap initially opened, closed after the first explosion. (*d*) Tap initially closed, opened after 15 min.

[1] B. McCaffrey, personal communication.

comparison with experiments had been performed. Thus, in the case of high temperature oxidation of methane and carbon monoxide mixtures a similar periodicity was observed by me [1] in Lucquin's laboratory. As is shown in fig. 1, this oscillatory phenomena in our case mainly depends on the diffusion of initial reactants from the external dead volume into the reaction vessel. Another problem is that water formation from heterogeneous radical destruction (reactions 8, 10 and 14), each producing $\frac{1}{2}H_2O$, does not seem very clear. We simply assumed that only H_2O is produced from these reactions to avoid the detailed mechanisms of these heterogeneous reactions.

Prof. P. Gray (*Leeds*) said: Perche makes a valuable experimental point about the carbon monoxide oxidation " lighthouse ", and speaking as someone with both theoretical and experimental interests in the system, I have not the slightest doubt in asserting that however hard the computer calculations may be, they are far less difficult in this system than the experiments. The open tap is clearly important in Perche's study, and the repetitive entry of fresh reactant through it may indeed contribute to repetitive reactions in his particular system. An extreme case is furnished by old Russian work.[2] However, in Ashmore's (1939) study there was no dead space at all, and oscillations were still found.[3] They persisted apparently unchanged when the vessel was reconnected to a " dead space ".

Prof. R. M. Noyes (*Oregon*) (*communicated*): After further consideration, I have concluded the argument of Section 3 of the paper by B. F. Gray and Aarons does *not* invalidate the Brusselator as a model for many successive oscillations in a closed system. Such a model would require that only a small fraction of the A and B reactants be consumed during a cycle in the concentrations of the intermediates X and Y.

A necessary but not sufficient condition for applicability to closed systems is that

$$k_2B > k_4 \gg k_1. \tag{1}$$

An additional restriction imposed by stability analysis of the steady state is that

$$k_3A^2 < k_2k_4^2B/k_1^2 - k_4^3/k_1^2. \tag{2}$$

Although I have not yet proved it, I am convinced the requirement of minimal fractional depletion of A and B during a cycle imposes a lower limit on k_3A^2 just as eqn (2) above imposes an upper limit. I also believe that for sufficiently large values of k_4/k_1 all the necessary restrictions can be satisfied simultaneously.

It therefore appears the Brusselator is an acceptable model for oscillations in closed as well as in open systems, although such a model requires rate constant ratios somewhat different from those usually used by the Brussels school.

If rate constants are properly selected, the Oregonator [4] is also a satisfactory model for an oscillator that only very slowly depletes the major reactants. These ideas have been developed further in a manuscript submitted to *J. Chem. Phys.*

Dr. P. Hanusse (*CNRS, Talence*) said: In theoretical chemical models, postulating that pool chemicals are held constants is generally a mere definition of pool chemicals rather than a conjecture. On the other hand it is perfectly justified to check the physical implications of such a definition, as it is, for instance, to wonder about the existence of elementary autocatalytic steps.

Now, B. F. Gray shows that several well-known models may be structurally

[1] A. Perche, *Thèse de 3ème cycle* (Lille, 1970).
[2] Tokarev and Nekrasov, *Russ. J. Phys. Chem.*, 1936, **8**, 504.
[3] P. G. Ashmore, *Nature*, 1951, **167**, 390.
[4] R. J. Field and R. M. Noyes, *J. Chem. Phys.*, 1974, **60**, 1877.

unstable with respect to the introduction of some new steps. We think that this result depends on the perturbation introduced, or the way it is defined. For instance, one may replace a step of form:

$$A+X \to \ldots \qquad A \text{ constant}$$

by the two following steps:

$$A_0 \to A \qquad \text{diffusion at system limits, } A_0 \text{ constant}$$
$$A+X \to \ldots \qquad \text{and } A \text{ variable,}$$

but other physically meaningful ways are possible, for instance in a steady flow reactor, this becomes

$$A_0 \to A \qquad \text{influx, constant rate}$$
$$A \to \ldots \qquad \text{outflux}$$
$$A+X \to \ldots$$

We have already made experiments which illustrate the control of pool species as well as the effect of a small parameter, namely temperature.

In a steady flow reactor, where influxes are controlled we have studied an oscillating reaction derived from Bray's reaction, as proposed by T. S. Briggs and W. C. Rauscher. The evolution of the system is continuously recorded by electrochemical potential, spectrophotometric absorption, and temperature measurements. Many interesting phenomena occur in this system: oscillations are perfectly sustained, very stable in magnitude and frequency—better than 0.5 %. Studies may be achieved in isothermal or adiabatic conditions. In both conditions, temperature oscillation is observed. In some conditions a double frequency oscillation may be observed as those shown by Sørensen; the high frequency part is due to purely chemical oscillation, and the low frequency oscillation seems to be very drastically dependent on temperature.

Several stationary states have been found, and transitions between them have been studied. Hysteresis phenomena occur also when varying the constraints on the system, namely, mass flows. We think that the investigation of such completely sustained systems is the only way to avoid artefacts [1] and they may lead to interesting new informations on oscillating reactions, since the theoretical prerequisites are fulfilled.

Dr. B. F. Gray (*Leeds*) said: As Hanusse remarks, our results depend on the perturbation of the system which we consider, but our main point is that physically realistic *improvements* in the description of the system (such as trying to say what happens to the " pool " chemicals, whose concentrations are not quite constant) can alter the behaviour of the intermediates, i.e., remove or produce oscillation. We agree with the comment about completely sustained systems and are ourselves operating such a reactor (see comment on paper by Gray, Griffiths and Moule, this Symposium).

In response to Noyes, our section 3 does not invalidate the Brusselator as a model for many successive oscillations in a closed system in *general*, we simply said that it does not oscillate in the region of parameter space chosen by Lefever and Nicolis for their computations. Your condition (1) is stated in the text in Section 3, and its compatibility with the condition for an unstable singularity (your condition 2) has been discussed [2] and shown to be unlikely. Briefly, besides (1) one also needs

[1] P. Hanusse, *Compt. Rend.*, 1975.
[2] B. F. Gray, *Kinetics of Oscillatory Reactions*, (Specialist Periodical Reports, 30, The Chemical Society, London, 1974).

$k_3 \gg k_2$. To be a little more precise, if we take $X_0 = A_0 = 1 = Y_0 = B_0$, for example, then the " low consumption per cycle " requirements become

$$k_3 \sim k_2/\xi, \; k_4 \sim k_1/\xi.$$

Substitution of these into the instability conditions gives $k_1 < 0$, i.e., a contradiction due to the leeway allowed in the $\sim$ signs. In view of the limited physical interest of the Brusselator we decided not to pursue the matter numerically, but the main conclusion from our paper on this general aspect is that in future the onus is on any proposers of new oscillatory shemes to show that it is stable with respect to small parameter stability of this type.

Dr. R. Lefever (*Brussels*) said: 1. The question whether trimolecular steps, as in the Brusselator, are " physically unrealistic " should not be decided in a general *a priori* way. From the paper of Matsuzaki *et al.* at this meeting, it can be seen that sometimes they may appear as a good approximation to the behaviour of realistic systems. Often such a step can be regarded as the overall step of several realistic partial steps, e.g.,

(i) of the isomerization process :

$$2X \rightarrow Z$$
$$Y + Z \rightarrow 3X$$

(ii) of the enzymatic chain

$$E + X \rightarrow (E_1 X)$$
$$(E_1 X) + X \rightarrow (E_1 XX)$$
$$(E_1 XX) + Y \rightarrow (E_1 XXY)$$
$$(E_1 XXY) \rightarrow E^* + 3X.$$

It can also easily be seen that for some values of the parameters the allosteric model of the glycolytic oscillator, which Goldbeter, Boiteux and Hess [1] have mentioned at this meeting, would present a third order molecularity, or even higher if more than two subunits are considered for the enzyme.

2. The criticisms with respect to the effect of small parameters look exaggerated. (*a*) The Volterra–Lotka model is structurally unstable in any event. Thus we know that the smallest modifications will alter significantly even the qualitative behaviour, sometimes in completely opposite ways. For example : if instead of considering diffusion of the initial products, one considers diffusion of the intermediates, and investigates the equations

$$\frac{\partial X}{\partial t} = X - XY + \varepsilon \frac{\partial^2 X}{\partial r^2}$$

$$\frac{\partial Y}{\partial t} = XY - Y + \varepsilon \frac{\partial^2 Y}{\partial r^2}$$

then, however small ε, no sustained oscillatory behaviour is possible. If on the other hand one considers the following slight change in the equation :

$$\frac{dX}{dt} = X - XY^{(1 + \varepsilon)}$$

$$\frac{dY}{dt} = XY^{(1 + \varepsilon)} - Y \qquad 0 < \varepsilon \ll 1$$

then their solution turns out to be a spiral which blows up in the first quadrant.

[1] A. Boiteux, A. Goldbeter and B. Hess, *Proc. Nat. Acad. Sci.*, 1975.

(*b*) For the trimolecular model it is obvious that keeping *A, B* constant is equivalent to infinitely large flows of A and B, or perfect instantaneous stirring inside the system. The main test for the consistency of a model is to have it working in a certain limiting case. This is certainly so for the Brusselator. After all, in physics, the model of a perfect pendulum without friction is generally unrealistic, but quite useful.

Furthermore, our own work in Brussels (see for example the recent papers by Nicolis and Auchmuty,[1] and Kaufman-Herschkowitz [2]) has already shown that by modifying the original Brusselator one can obtain new types of behaviour such as localized structures.

Dr. B. F. Gray (*Leeds*) (*communicated*): 1. I agree with Lefever that a third order reaction may be a representation of a more complex system involving perhaps only first and second order steps, but in making such an approximation one is introducing yet *another* small parameter relating for example [E₁XXY] to X and Y, besides the original one relating X and Y to A and B. It may be possible to do this correctly, but our main point is that this cannot be assumed.

2*a*. We do not agree that our criticism of the method of neglecting small parameters is exaggerated. The Lotka–Volterra model is well known to be structurally unstable (i.e., a characteristic root is zero), but we are not discussing this type of stability here, we are discussing stability with respect to a small parameter, and consequent increase of the degree of the characteristic equation, producing *new* roots. If one or more of these is positive then it is possible for the system to be unstable even though the unperturbed system was structurally stable, i.e., all its characteristic roots were <0. The other examples in our paper treat systems which are not structurally unstable and the structural instability of the Lotka–Volterra system is in this context incidental.

2*b*. The reference to a perfect pendulum without friction is completely misleading in this discussion since including friction does not increase the degree of the characteristic equation of the number of variables in the problem. This is a simple perturbation problem with no possibility of the type of instability we are discussing, which is essentially characterised by non-uniform convergence in the neighbourhood of $\varepsilon \to 0$. On the contrary, the convergence of a damped pendulum to an undamped one is completely uniform as the damping factor tends to zero, i.e., the solution of the damped pendulum equation

$$M \frac{d^2y}{dt^2} + \varepsilon \frac{dy}{dt} + Ky = 0$$

which is $y = \exp\{-\varepsilon t/M\} \sin(K/M)^{\frac{1}{2}}t$ is well approximated by the undamped solution $y_0 = \sin(K/M)^{\frac{1}{2}}t$ provided $t \ll M/\varepsilon$, and as $\varepsilon \to 0$, this interval gets larger and larger. The correct analogy within the realm of damped springs is where the *mass* of the spring tends to zero and the unperturbed differential equation is only first order

$$dy/dt + Ky = 0;$$

hence a spring with zero mass is *not* a sensible approximation to a spring with a small mass, however small this may be. This is a singular perturbation problem, as compared to Lefever's example which is a secular perturbation problem.

Dr. B. L. Clarke (*Alberta*) said: The term " jump point " used in Section 1 of the paper by Gray and Aarons can have two interpretations. The term is used in the

[1] G. Nicolis and A. Auchmuty, *Proc. Nat. Acad. Sci.*, **71**, 1974, 2748.
[2] M. Kaufman-Herschkowitz and G. Nicolis, *J. Chem. Phys.*, 1972, **56**, 5.

paper to refer to the values of y for which the curve $F(x;y) = 0$ has tangents which are perpendicular to the y axis. If eqn (2) moves x to a stable pseudosteady state on this curve, x will essentially be determined by y. As y passes through the "jump point" $\dot{x}$ becomes infinite briefly. However, the trajectory $(x;y)$ through this point may still be a smooth curve.

The sudden changes in the dynamical variables shown in the papers by Field and Noyes and by Franck are usually caused by a situation which is the second interpretation of "jump point". Since the stability of the steady state of eqn (2) depends on y, the curve $F(x;y) = 0$ may have regions of stability and instability for perturbations Δx *off* the curve. The trajectory $(x;y)$ will follow this curve in the regions of stability and jump off the curve at the points which separate the stable and unstable regions of the curve. The rapid evolution of x at essentially constant y which follows can move x into a limit cycle around the curve. If the solutions of $F(x;y) = 0$ are multivalued for a given y the evolution may go to a second branch of the curve which is stable. The "jumping off" points of the curve are found by examining the Hurwitz determinants of the matrix $(\partial F_i/\partial x_j)$.

If the general equations of motion are used instead of eqn (1) the variables can often be treated as slowly evolving variables like y over part of their range and rapidly evolving variables like x over much of the remainder of their range. The variable η of the Oregonator behaves in this fashion. The argument of the preceding paragraph can still be used; however, the stability of points on the possible curves must be determined for fluctuations in all the variables.

Dr. A. Babloyantz (*Brussels*) said: In connection with the paper of Gray and Aarons I would like to mention that structural stability formalism has already been applied to biological problems in the context of prebiotic evolution of informational macromolecules.[1] It can be shown that a favourable mutation giving rise to a new macromolecule, although in a small quantity, may take over and destabilize the originally stable system.

Dr. B. L. Clarke (*Alberta*) said: Regarding the paper by Gray and Aarons: the model reaction systems which become unstable when the pool chemicals are allowed to fluctuate suggest a question. "Which network features are necessary and sufficient to guarantee structural stability with respect to the inclusion of the pool chemicals?" Several theorems given in a paper on the stability of topologically similar chemical networks[2] give conditions which ensure that including the pool chemicals in the dynamics will not alter after the stability. If a pool chemical enters the network through a decomposition reaction of the form A → ..., permitting A to fluctuate is equivalent to adding a reactant X to the network as follows, A → X → .... I call the reactant X a "type B flow through reactant" (FTRB) and from Theorem 5 of my paper it follows that for large steady state concentrations of X the network is stable if and only if the network with X eliminated is stable.

The only other way a pool chemical may enter a network is by reacting directly with other reactants as in A + Y → .... If A is allowed to fluctuate, such a reaction is replaced by the following pair of reactions, A → X and X + Y → .... I call the reactant X a "type A flow through reactant" (FTRA) and from Theorem 8 of my paper the following can be said: if the original network has deficiency zero and if the reactant Y does not appear by itself on the left or right hand side of any reaction

[1] I. Prigogine, G. Nicolis and A. Babloyantz, *Physics Today*, November and December 1972.
[2] B. L. Clarke, *J. Chem. Phys.*, 1975, **62**, 3726.

of the extended network, the extended network also has deficiency zero. This theorem is an extension of work by Horn, Feinberg and Jackson and the significance of the deficiency is explained in the immediately following comment by Feinberg. Roughly speaking, if the original network has deficiency zero it is stable and, because the freeing of the pool species does not change the deficiency, the extended network is also stable.

From the Hurwitz determinants one can understand in general why pool chemicals which enter the network via reactions of the form $A + Y \rightarrow \ldots$ sometimes destabilize the network when they are allowed to fluctuate. The extended network has a new positive feedback loop in which A and Y inhibit one another. Such loops will destabilize a network if they pass through reactants which have marginally strong autocatalysis such as $Z + Y \rightarrow 2Y$ and, in addition, certain other conditions are met. In all the examples of Gray and Aarons' paper the pool chemicals enter the network in this manner.

Dr. M. Feinberg (*Rochester*) said: In the discussion regarding B. F. Gray's paper Clarke made reference to some theorems proved Horn, Jackson and Feinberg which might bear upon the problem of how stability might be affected if a physico-chemical model is structurally perturbed. Horn and I have published a description of the most intriguing of these theorems.[1] Perhaps I should explain this theorem and some newer results (in terms more schematic than precise), and offer my views regarding their utility in answering questions of the type raised by Gray.

According to a rather simple formalism described in the aforementioned article there can be assigned a non-negative integer (called the *deficiency*, δ) to every chemical mechanism according to its algebraic structure. (Some of the reactions of a mechanism might be " pseudo-reactions " incorporated to reflect special physico-chemical effects for a system under study). Thus mechanisms can be classified according to whether they are of deficiency zero, one, two, etc.

Mechanisms of deficiency *zero* are surprisingly common, and it is for these that we have been able to prove what I think is an interesting theorem for open homogeneous reactors. For *arbitrary* positive rate constants (in the context of the usual mass-action kinetics) the existence of an equilibrium for which all species concentrations are positive suffices for *preclusion* of pathological statics and dynamics, e.g., sustained composition cycles. Moreover, there exists such an equilibrium if and only if the reaction " arrows " in the mechanism are directed such that the mechanism is what we call " weakly reversible ".

The weak reversibility constraint is somewhat limiting since most models people have been considering these days do not fall into this category. However, since publication of the article cited above I have been able to prove the following. For mechanisms of deficiency zero which are *not* weakly reversible it is true that for *arbitrary kinetics* (subject only to very weak constraints) the dynamical equations for open homogeneous systems cannot give rise to sustained temporal composition cycles for which at some time all species concentrations are positive. (In fact this can be shown to hold for a class of mechanisms somewhat broader than those of deficiency zero.) Moreover, for mechanisms of deficiency zero governed by *mass-action* kinetics with *arbitrary* positive rate constants, sustained temporal composition cycles cannot be obtained at all *whether or not the mechanism is weakly reversible*. That is *all* mechanisms of deficiency zero taken with mass action kinetics are, loosely speaking, of stable character.

[1] M. Feinberg and F. Horn, *Chem. Eng. Sci.*, 1974, **29**, 775.

It can be shown that if one begins with a mechanism of deficiency δ the addition of reactions to that mechanism results in a new mechanism of deficiency *not less than* δ. Similarly if one begins with a mechanism of deficiency δ and one *removes* reactions the resulting mechanism has deficiency *not greater than* δ. In each case the deficiency may be unchanged. These ideas might in some instances help in deciding the effect of structural perturbations in a model (resulting in addition or removal or reactions) upon stability characteristics.

If one begins with a mechanism of deficiency zero and one *adds* to it new reactions the resulting mechanism may also be of deficiency zero so that it falls within the realm of the theory. Thus, the " perturbed " model will be of stable character.

On the other hand, if one begins with a mechanism of deficiency in excess of zero (perhaps exhibiting static and dynamic " exotica ") and one structurally perturbs the model by *removing* reactions then the resulting mechanism may have deficiency zero, and the modified model will have essentially stable character.

I might add that mechanism deficiency has a surprising bearing upon matters quite divorced from stability considerations (e.g., upon the determinability of complete sets of rate constants from certain classes of dynamic experiments). I have discussed these matters in a chapter of the forthcoming Wilhelm Memorial Volume on Chemical Reactor Theory, several other chapters of which (particularly those by Luss and Bailey) might hold interest for participants in this Symposium.

Dr. H. Tributsch (*Berlin*) said: There have been many theoretical investigations on the Lotka Volterra model but there has been up to now a lack of experimental systems which permit a verification of conclusions derived from it. For this reason I would like to present a new and simple oscillating system which in my opinion corresponds to the first example which has been discussed in the paper by B. F. Gray, namely to that of a Lotka Volterra oscillator, which has to be considered as being perturbed by small parasitic parameters. The system consists of a crystal of Cu_2S or Cu_5FeS_4 in contact with an electrolyte that contains hydrogen peroxide. Oscillations appear during the reduction of H_2O_2 at the sulphide electrode above a certain

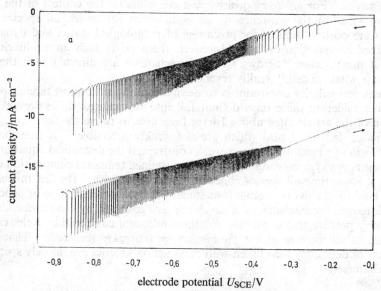

FIG. 1.—Dynamical reduction curves for H_2O_2 on Cu_5FeS_4 electrode.

minimum potential. Fig. 1 shows two reduction curves which have been recorded at 2 mV/s in opposite potential-directions. For this system we could not only derive a reasonable kinetic scheme of the Lotka Volterra type but we were also able to find clear experimental evidence for the behaviour which is considered to be characteristic of this kind of oscillator, especially an unstable frequency which is dependent on the initial conditions of the system as well as a complementary amplitude–frequency correlation which is a result of its conservative character and of a constant entropy production. (Consider the amplitude decrease which coincides with an accidental increase in the oscillation frequency in the central portion of one of the reduction curves (fig. 1)). A closer investigation of the dynamics and shape of oscillations has shown that they are usually very periodic and well formed in the central region of the reduction curve, but that they are often composed of compact spike groups of fast rising amplitude in the border regions where now and then oscillations also fail to occur.

The available data are all consistent with the concept that the investigated system is basically of the Lotka Volterra type. Its instability is, however, so pronounced that it will—depending on the predominant chemical perturbation reactions—either slip into a limit cycle type of oscillation (case discussed in Gray's paper) or result in a gradually rising or damped oscillation (cases discussed by Frank–Kamenetskii and Sal'nikov [1]).

[1] D. A. Frank–Kamenetskii and I. E. Sal'nikov, *Zhur. Fiz. Khim.*, 1943, **17**, 79.

C. Membranes, Heterogeneous and Biological Systems

Factors Controlling the Frequency and Amplitude of the Teorell Oscillator

By Kenneth R. Page and Patrick Meares*

Biophysical Chemistry Unit, Department of Chemistry,
The University of Aberdeen, Old Aberdeen, Scotland

Received 12th July, 1974

The properties of the Teorell oscillator are analysed by considering a membrane containing parallel cylindrical pores with a low surface charge density. Particular attention is given to alterations in the properties produced on varying the pore diameter and surface charge. Brief consideration is given to the relation between the behaviour of the oscillator and of biological mechano-electric transducers.

The Teorell oscillator is a device which makes use of the coupling of ion and water fluxes within a highly porous membrane of low internal charge. It was originally devised as an analogue for the study of biological excitability.[1] It is particularly interesting in relation to the problems of mechano-electric transduction, i.e., the transformation of mechanical stimuli into electrical signals, which is found in organs such as baroreceptors. In this paper, some factors which affect the frequency and amplitude of the oscillations are examined because these may have a bearing on the study of natural mechano-electric transduction. The analysis relies on a theory developed by Meares and Page [2, 3] which has been experimentally verified using Nuclepore filters supplied by General Electric. These filter membranes have an exceptionally well-defined and uniform parallel pore structure.

In the Teorell oscillator the membrane separates two well-stirred electrolyte solutions of different concentrations. Each solution is contained in a compartment with a vertical capillary on the top. A net movement of solution through the membrane therefore generates a hydrostatic pressure opposing the flow. When, as in the case examined here, the membrane surface charge is negative an electrical potential across the membrane generates an electro-osmotic flow in the direction of the cathode. The solutions are arranged so that this flow takes place from the dilute to the concentrated side.

Sustained oscillations of the hydrostatic pressure ΔP and electric potential $\Delta \Psi$ across the membrane may be produced when a constant electric current, greater than a critical minimum, is passed from one compartment to the other. The cycle of events can be visualised as follows. Initially Δp is low and the electro-osmotic flow causes dilute solution to enter the membrane pores. Thus the membrane conductivity is low and the potential $\Delta \Psi$ is large. As the hydrostatic pressure builds up, the flow progressively reduces until a point is reached when the direction of flow is reversed. The membrane then fills with the concentrated solution and its conductance rises. Hence the potential difference decreases and the reversed flow allows the pressure to fall until a point is reached when electro-osmosis once again takes control and a new cycle commences.

166

THEORY

Attention is restricted to the case of a Nuclepore filter membrane separating two solutions of sodium chloride. It has been found empirically that the negative surface charge density on the membrane is directly proportional to the cube root of the salt concentration in contact with it. Owing to the uniform pore structure of the membrane the overall flow may be analysed in terms of that in a single pore. Flows from the dilute to the concentrated solution will be assigned positive values.

The motion of the fluid is governed by a balance between four forces: the hydrostatic pressure P_c, the electro-osmotic force P_E, the viscous drag force F_η and an inertial force caused by accelerations of the fluid. It has already been shown that the inertial force may be neglected in the cases considered here.[3] The viscous drag force F_η is obtained from the Navier–Stokes equation, the result being

$$F_\eta = -8\eta v l/a^2 \tag{1}$$

where v is the volume flow per pore, a the radius and l the length of the pore. η is the coefficient of viscosity of the fluid in the pores. P_c is related to the difference in height between the menisci in the two vertical columns. The geometry of the cell permits the variation of P_c with time to be related to the volume flow by

$$v = (1/\gamma\lambda)(dP_c/dt). \tag{2}$$

Here γ and λ are constants of the system and include the total area of pores open to flow and the cross sectional area of each vertical tube.

The electro-osmotic force also is a function of the volume flow because of convective coupling between the ion fluxes and the flow of fluid. Either the dilute or the concentrated electrolyte solution is swept into the pores to an extent which depends on the magnitude and direction of the volume flow and the thickness of the electrical double layer. Provided the current is held constant the magnitude of the membrane potential varies accordingly with the concentration profile in the pore. Although this concentration profile cannot change instantaneously with changes in v, provided dv/dt is sufficiently small, the difference between the actual profile at any instant and the one appropriate to a stationary state will be negligibly small. When this holds it is possible to estimate P_E by using the stationary-state equations given elsewhere.[2] In order to do this the ion fluxes are expressed by the Nernst–Planck equations, extended to include a term for convective flow, and the Gouy–Chapman theory is employed to describe the electro-osmotic component of the volume flow. The local equations have to be averaged over the pore cross-section and integrated along its length. Allowance must also be made for the presence of unstirred layers of solution immediately adjacent to the membrane faces. The final result can be expressed in the form

$$P_E = f(c_\alpha, c_\beta, i, v) \tag{3}$$

where c_α and c_β are the concentrations on either side of the membrane, and i is the current density in the pore.

It is convenient to regard the forces P_c and P_E as being in opposition and, neglecting the inertial force, to express the force balance on the fluid by

$$P_E - P_c = F_\eta. \tag{4}$$

When eqn (1) to (3) are taken into account, eqn (4) can be expressed

$$P_c = P_E - F_\eta = f(dP_c/dt). \tag{5}$$

The complete functional relationship in eqn (5) is complicated and the detailed formulation, given by Meares and Page,[3] shows that it can be related to the Van der Pol equation. Without going into the details of this relationship, its properties will be discussed here with the aid of numerical solutions.

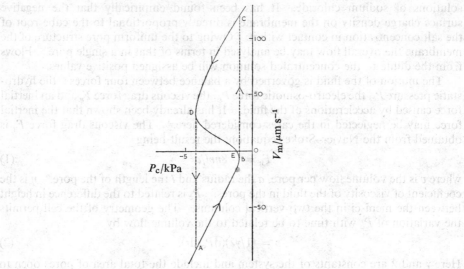

FIG. 1.—(P_c, V_m) limit cycle when $I > I^*$. Plot calculated for membrane 2MA at a current density of 690 A m^{-2}.

Fig. 1 illustrates the relationship between P_c and the total volume flow in the membrane V_m, calculated for membrane 2MA, at a total current density I, greater than the critical minimum I^* required to produce oscillations. V_m and I are related to the corresponding quantities in a single pore by eqn (6) and (7)

$$V_m = \pi a^2 N v \qquad (6)$$
$$I = \pi a^2 N i \qquad (7)$$

where N is the number of pores per unit area of membrane. The curve in fig. 1 represents the solution of eqn (5) calculated by using the surface charge density, pore density and hydrodynamic permeability of membrane 2MA. These quantities were determined in separate experiments.

The system is in a stable state when on curves AB and CD. On curve AB the volume flow is negative, and concentrated solution is being dragged into the pores. In this state the behaviour of the system is controlled chiefly by the hydrostatic pressure. Curve CD represents the opposite condition; dilute solution is being dragged into the pores and the flow is dominated by electro-osmosis. On curve BD the system would be unstable and this region is inaccessible under constant current conditions. Points B and D mark transitions between the two stable states. Point E, at which V_m would be 0, lies on DB and the system cannot therefore achieve a truly stationary state. Instead the state of the system will constantly progress around the closed path ABCD that represents the limit cycle of the oscillations. Provided the times taken to traverse BC and DA are small compared to those taken along AB and CD the period of the oscillations τ is given by

$$\tau = \frac{1}{\lambda \gamma} \left\{ \int_{P_c(A)}^{P_c(B)} \frac{1}{V_m} \, dP_c + \int_{P_c(C)}^{P_c(D)} \frac{1}{V_m} \, dP_c \right\}. \qquad (8)$$

In fig. 2 the relationship between pressure and volume flow is shown for the same membrane at a current density less than the critical value I^*. The point E at which $V_m = 0$ now lies on curve AB and hence represents an accessible state. When I is less than I^* the system will move around the curve BCDA until E is reached at which it will have attained a truly stationary state. If the curve were plotted for the case $I = I^*$ the turning point B and the zero point E would coincide.

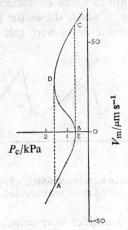

FIG. 2.—(P_c, V_m) limit cycle when $I < I^*$. Plot calculated for membrane 2MA at a current density of 250 A m^{-2}.

DISCUSSION

The theoretical equations permit the calculation of the limit cycle and period of oscillation for a given set of membrane characteristics, electrolyte solutions and electric current density. They also indicate the value of I^*. The theory has been tested under a variety of conditions with several types of Nuclepore membranes.[3] Table 1 lists three of the results in order to indicate the extent of agreement between

TABLE 1.—CHARACTERISTICS OF (P_c, V_m) LIMIT CYCLES AND PERIODS FOR 0.1 M-0.1 M NaCl IN THREE MEMBRANES

[Units: $L_p = $ m Pa^{-1} s^{-1}; $K = $ mC m^{-1} kg^{-1}; $a = \mu$m; $I = $ A m^{-2}; $P_c = $ kPa; $V_m = \mu$m s^{-1}; $\tau = $ s. The letter bracketed after V_m denotes point on limit cycle shown in fig. 1.]

membrane 2MA

$L_p = 2.35 \times 10^{-8}$; $K = -9.84$; $a = 0.22$; $I = 690$

	$P_{c\ max}$	$P_{c\ min}$	$V_m(A)$	$V_m(B)$	$V_m(C)$	$V_m(D)$	τ
observation	3.12	1.30	−78	−2.4	166	27	1194
prediction	4.21	1.06	−84	−3.9	119	33	1463

membrane 6MC

$L_p = 5.62 \times 10^{-8}$; $K = -5.21$; $a = 0.42$; $I = 1070$

observation	1.15	0.396	−73	−3.5	130	50	576
prediction	1.28	0.340	−62	−4.6	91	26	519

membrane 2MD

$L_p = 36.2 \times 10^{-8}$; $K = -4.35$; $a = 0.76$; $I = 1790$

observation	0.262	0.058	−110	−6.1	180	42	90
prediction	0.366	0.075	−120	−6.8	160	42	93

prediction and observation. Each example refers to the same pair of electrolyte solutions, 0.1 M and 0.01 M NaCl.

The values of the hydrodynamic permeability L_p and the proportionality constant relating the surface charge density to the cube root of concentration K for the membranes were measured in separate experiments. The pore radii a were calculated from L_p and the pore density N, which was obtained from photomicrographs of the membrane. The observed quantities which characterise the oscillations were taken from recorder traces such as fig. 3 which shows the pressure oscillations produced with membrane 6MC. The volume flows were calculated from the slopes of the

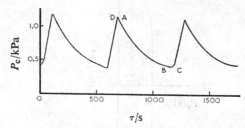

FIG. 3.—Oscillations of P_c observed using membrane 6MC. The positions of the turning points on the (P_c, V_m) limit cycle are indicated ABCD.

pressure traces. The predicted quantities were calculated by using the appropriate values of L_p, K, c_α, c_β, and i. It can be seen from table 1 that the correspondence between prediction and observation is generally good, the worst discrepancy being the overestimation of the maximum pressure P_c for membrane 2MA.

A variety of work has shown that many biological mechano-electric transducers respond to strain rather than stress.[4] In organs such as the Pacinian corpuscle the sensory process appears to involve a two-step mechanism in which a sensory membrane produces a graded potential (generator potential). This in turn triggers off an oscillatory discharge at an associated afferent nerve ending.[4] It is possible, however, that some receptors operate by a single step process. In these the sensory membrane produces the oscillatory discharge directly. The baroreceptors in the carotis sinus and crustacean muscle stretch receptors may be examples of the single step mechanism.[1] It is of interest, therefore, to examine what effects strain might have on the properties of the oscillator studied here.

The tensile strength of the Nuclepore filters precludes any direct experimental study of strain in the present system. As indicated in table 1, however, a series of Nuclepore filters with different pore radii and surface charges was studied. The consequences of these differences in properties were successfully described by the theory and it is possible, therefore, to investigate theoretically the effect of varying either the pore radius or the surface charge. Thus one may infer the effects of radial tension upon the behaviour of an extensible membrane which had otherwise exactly the same properties as a Nuclepore filter. As shown by Burton,[5] a membrane of this type should be highly sensitive to strain, provided its Poisson ratio was not too low. If such a membrane were stretched the pores would act as foci of stress and a small increase in the membrane diameter would cause a disproportionately greater increase in pore diameter.

Fig. 4(i) shows the effect on the limit cycle of varying the pore radius at constant K. The points labelled A, B, C and D correspond exactly with those shown in fig. 1 for membrane 2MA at a current density of 690 A m^{-2}. To clarify the presentation, the pairs of points A–B and C–D are joined by straight lines. The lines labelled (a)

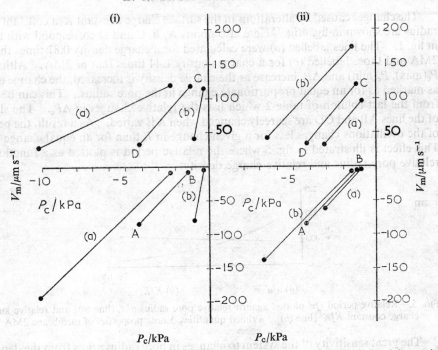

P_c/kPa P_c/kPa

FIG. 4.—(i) Effect of altering pore radius on (P_c, V_m) limit cycle. Lines AB and CD join turning points for membrane 2MA. Lines (a) and (b) correspond to AB and CD when the pore radius is 0.81 and 1.44 respectively times that of membrane 2MA. (ii) Effect of altering the surface charge constant on the (P_c, V_m) limit cycle. Lines AB and CD join turning points for membrane 2MA. Lines (a) and (b) correspond to AB and CD when the surface charge constant is 0.81 and 1.44 respectively times that of membrane 2MA.

TABLE 2.—EFFECTS OF CHANGES IN PORE RADIUS a AND SURFACE CHARGE CONSTANT K UPON TIMES τ_{AB}, τ_{DC} AND THE PRESSURE AMPLITUDE ΔP_c

[Units: $a = \mu m$; $K = mC\ m^{-1}\ kg^{-\frac{1}{3}}$; $\tau_{AB}, \tau_{DC} = s$; $\Delta P_c = kPa$. a', K', $\Delta P_c'$: values appropriate to the experimental run on membrane 2MA.]

a	K	a/a'	K/K'	τ_{AB}	τ_{DC}	τ_{AB}/τ_{DC}	ΔP_c	$\Delta P_c/\Delta P_c'$
1.78	9.84	0.81	1.00	2316	1080	2.14	7.76	2.44
3.17	9.84	1.44	1.00	264	78	3.38	0.63	0.20
2.20	9.84	1.00	1.00	930	420	2.21	3.15	1.00
2.20	14.20	1.00	1.44	1068	552	1.93	5.37	1.69
2.20	7.96	1.00	0.81	912	348	2.62	2.24	0.72

correspond to AB and CD calculated for pores of radius 0.81 times those in membrane 2MA. The lines labelled (b) were similarly calculated for pores of radius 1.44 times those in 2MA. It may be seen that the extremum values of P_c, i.e., $P_c(\text{max})$ and $P_c(\text{min})$, both increase as the radius decreases. The amplitude of the pressure oscillations ΔP_c also increases (see table 2) and the slopes of AB and CD decrease. Table 2 lists the times τ_{AB} and τ_{CD} taken to traverse AB and CD respectively. Clearly the period increases as the radius decreases and τ_{AB} becomes smaller relative to τ_{CD}. As a result of τ_{AB}/τ_{CD} becoming closer to unity the oscillation curves become more nearly symmetric with decreasing pore radius.

The changes caused by alterations in the surface charge constant K at constant pore radius are shown in fig. 4(ii).   Here also points A, B, C and D correspond with those in fig. 1.   The lines labelled (a) were calculated for a charge density 0.81 times that of 2MA and those labelled (c) for a charge density 1.44 times that of 2MA.   Although P_c(max), P_c(min) and ΔP_c increase as the charge density is increased, the change is not as marked as for an equal proportionate change in the pore radius.   This can be seen from the last column of table 2 which lists the relative changes in ΔP_c.   The slopes of the lines AB and CD are scarcely changed when K is varied.   As a result, the period of the oscillations changes less for a given change in K than for an equal change in a. This effect is illustrated in fig. 5 where the relative period is plotted as a function of relative pore radius and relative charge density.

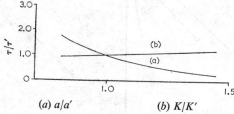

FIG. 5.—Relative period τ/τ' plotted against relative pore radius a/a' {line (a)} and relative surface charge constant K/K' {line (b)}.   Primed quantities denote properties of membrane 2MA.

The great sensitivity of the system to changes in pore radius arises from the dependence of L_p on the fourth power of the pore radius, whereas electro-osmosis is more nearly related to the first power of K.   These considerations explain why dust proved to be a major hazard in early work with this system; repeatable results were obtained only after carefully cleaning the apparatus and filtering all solutions.   During a typical early experiment K was found to remain reasonably constant while L_p changed markedly owing to the progressive accumulation of small particles in the pores. Although it is unlikely that all pores became silted in the same uniform manner the results indicated the great sensitivity of the system to L_p.   Fig. 6 shows the trace

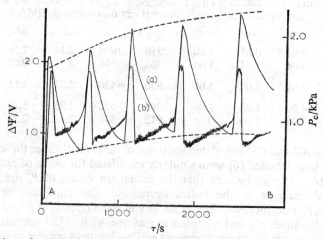

FIG. 6.—Oscillations of pressure P_c line (a) and potential between probe electrodes $\Delta\psi$ line (b) observed using a membrane of pore radius 0.8 μm.   At point A the hydrodynamic permeability was 32×10^{-8} m Pa^{-1} s^{-1} and at point B 17×10^{-8} m Pa^{-1} s^{-1}.

obtained in an experiment during which L_p fell to 54 % of its original value. P_c(max) and P_c(min) increased, as also did ΔP_c, and the oscillation period lengthened as the run progressed. These features are all in agreement with the predictions made in the discussion above.

This investigation has shown that a radial extension would increase the frequency and decrease the amplitude of the oscillations in a Teorell oscillator. When the pore walls are extended the surface charge density is likely either to remain constant or to decrease. If the latter occurred this would reinforce the changes produced by the increase in pore radius. As shown in fig. 4, an increase in pore radius and a decrease in K each move point B closer to the line $V_m = 0$. If B reached this line, oscillations would cease because the current density applied would then be less than the critical current density I^* of the stretched membrane. Too great a stretch would, therefore, stop the system from oscillating and allow a stationary state to be reached, a phenomenon known as " overstretch ".

Lack of space prevents a detailed discussion of the variations in membrane potential but the frequency of its oscillations must match those of pressure. Experimentally, the amplitude of the membrane potential oscillations was found to be less sensitive than ΔP_c to changes in L_p (see fig. 6) because most of the potential drop recorded occurred between the sensing electrodes in the solutions and the membrane surfaces.

There are some similarities between the predicted properties of the membrane oscillator under stretch and the behaviour of natural mechano-electric transducers. An increase in frequency with increasing strain is observed in most such organs, whilst some also show the " overstretch " phenomenon. For example the oscillations in cray-fish stretch receptors cease if the stretch is excessive.[6] This behaviour is reversible and oscillations recommence when the strain is reduced.

It is possible that some biological transducers rely on processes that phenomenologically resemble the mechanism analysed in this paper and this conclusion is independent of the nature of the processes occurring at the molecular level. Structurally the present system differs in a number of important ways from a biological tissue and a detailed correspondence with natural processes should not be sought. The present findings show only that a single step mechanism involving the deformation of a permeable membrane may provide a useful working model with which to guide future investigations in this field.

[1] T. Teorell, *Handbook of Sensory Physiology*, ed. W. R. Loewenstein (Springer Verlag, Berlin, 1971), vol. 1, chap. 10.

[2] P. Meares and K. R. Page, *Phil. Trans. A*, 1972, **272**, 1.

[3] P. Meares and K. R. Page, *Proc. Roy. Soc. A*, 1974, **339**, 513.

[4] W. R. Loewenstein, *Cold Spring Harbour Symp. Quant. Biol.*, 1965, **30**, 29.

[5] A. C. Burton, *Permeability and Function of Biological Membranes*, ed. L. Bolis *et al.* (North Holland Publishing Co., Amsterdam, 1970), p. 1.

[6] C. Terzuolo and Y. Washizu, *J. Neurophysiol.*, 1962, **25**, 56.

Electrical Oscillatory Phenomena in Protein Membranes

By Victor E. Shashoua

McLean Hospital Biological Research Laboratory,
and Department of Biological Chemistry,
Harvard Medical School, Belmont, Mass. 02178, U.S.A.

Received 5th August, 1974

Membranes were prepared by interacting polycations with polyanions at an interface to give a structured system in which a cationic phase was separated from an anionic phase by a neutral polyampholyte zone. Such a membrane system with a cation ⇌ anion junction exhibits electrical oscillations in an electric field. Measurements of the (current, voltage) characteristics of the membranes under current clamp conditions shows a " negative resistance " region coincident with the polarization conditions required for producing electrical oscillations. Both proteins and polynucleic acids can be used as the polyelectrolyte components of the membrane.

Biological membranes exhibit many types of electrical oscillatory properties. These include the relatively slow phenomena characteristic of plant cells and the fast events observed in nerve and muscle cells. Tasaki and Takenaka,[1] in an analysis of the electrical properties of the squid giant axon, demonstrated that substantially all the electrical excitability and conduction properties of axons can be attributed to the cell membrane alone, i.e., these properties remain intact even when practically all cytoplasmic components have been perfused out of the axons. Thus the axonal membrane consisting of a 70 Å thick layer of proteins complexed with lipids can generate the electrical properties of neurons.

A number of model systems have been proposed for simulating various aspects of the excitability properties of axonal membranes. One of the first models was described by Lillie.[2] He showed that an iron wire covered with glass tubing can propagate an electrical impulse in a manner suggestive of the characteristics of myelinated nerve fibers. In 1954, Teorell[3] and subsequently Franck[4] showed that glass membranes can generate slow oscillatory electrical signals. More recently lipid membrane systems, based on the bilipid layer concept of Davson and Danielli[5] have been the subject of many investigations.[6-10] Mueller and Rudin[6] showed that when lipid bilayers were modified by certain macromolecules they then could generate many of the electrical characteristics of neuronal membranes. This type of model suggests that specific macromolecules can convert a lipid bilayer, which has the high electrical resistance characteristics of a good insulator (10^8 ohm/cm^2) into a membrane with a low resistance (10^2 ohm/cm^2) and a capacity to generate electrical oscillations. In addition " semiconductor like " properties are obtained for some protein–lipid interactions. Clearly, no such properties can be predicted from an analysis of the bulk properties of either the lipid or the protein constituents of the membrane. Both classes of these molecules are insulators in the dry state. Lipids behave like detergents and proteins become polyelectrolytes in aqueous solutions.[11]

In an effort to find out if there are any fundamental characteristics of proteins or more generally polyelectrolytes, which may be useful for defining the electrical properties of biological membranes, we explored the possibility that specific changes might occur when polyelectrolyte membranes are organized into layered structures.[12]

In this way we attempted to simulate the interaction of proteins with charged lipid monolayers organized in a smectic phase. The possibility exists that new properties may be obtained following such an interfacial interaction.

ELECTRICALLY ACTIVE POLYELECTROLYTE MEMBRANES

In the initial experiments aqueous solutions of polyacids were layered onto solutions of polybases.[12] The membranes produced at the interface were found to be capable of generating random electrical oscillations in an applied electric field (see fig. 1A). In subsequent work,[13] some membrane systems produced sustained electrical oscillations (see fig. 1B, C, D). Essentially these membranes could simulate a transduction process with the properties of an electrical oscillator circuit to convert a d.c. potential into an a.c. output with " spike-like " characteristics. The amplitude (1-100 mV) and duration (1-10 ms) of the oscillations were of the same order of magnitude as those of the neuronal spikes. These properties were found to be directly attributable to the " sandwich-like " structure of the membrane in which a cation exclusion barrier (polycationic phase) was separated from an anion exclusion barrier (polyanionic phase) by a neutral polyampholyte zone. This type of structure may be called a polycation ↔ polyanion (c ↔ a) junction membrane. The double arrow in the c ↔ a symbol is used to designate the presence of a neutral polyampholyte layer between the two polyelectrolyte phases of the membrane.

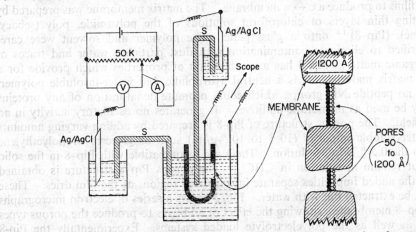

FIG. 2.—Experimental arrangement for study of electrical oscillatory properties of c ↔a junction membranes : S, Agar salt bridges with 0.15 N NaCl ; V and A are a voltmeter and ammeter ; output to oscilloscope is through a d.c. amplifier with high impedance ; insert shows diagram of the structure of a matrix supported c ↔a membrane.

In a typical experiment a c ↔ a membrane separates two compartments containing 0.15 N NaCl (see fig. 2). The current is passed through the system via two agar salt bridges containing 0.15 N NaCl. These connect each compartment to a silver/ silver chloride electrode—immersed in 0.15 N NaCl. The potential across the membrane is detected by means of two Ag/AgCl electrodes connected to a d.c. differential amplifier with a very high input impedance (Metametrics Corp., Cambridge, Mass.). This essentially draws no current from the system. The output of the amplifier is fed into an oscilloscope to display the pattern of the potential changes obtained. When a current is passed through the membrane to drive anions into the polycation phase and cations into the polyanion phase, a sequence of electrical and mechanical events takes place as a function of the applied voltage. At first the recording electrodes show that there is an instability region and that the output voltage recorded across the membrane becomes very sensitive to mechanical vibrations. Further increase in the applied voltage results in a loss of the mechanical instability followed by the generation of electrical transients with " spike-like " characteristics. Higher voltages produce a breakdown of the membrane. Thus there is a critical voltage range at which oscillations take place.

MATRIX SUPPORTED c ↔ a MEMBRANES

The preparation of c ↔ a membranes by the direct interaction of a polyacid with a polybase was found to be difficult to control. The membranes frequently had regions of imperfection and holes where short-circuiting occurred. In order to obtain a more experimentally feasible situation, a matrix supported system was developed. This was achieved by using a matrix membrane to act as a neutral hydrophilic support polymer for the two polyelectrolyte phases. Fig. 2 shows a diagram of the cross-section of such a polymer matrix membrane cemented in place at the 2 mm aperture of a cellulose nitrate tube separating the two electrode compartments. The membrane is generally about 1200 Å thick and contains a distribution of pores of 50-3000 Å in diameter, as shown by electron microscopy. This " sieve-like " structure acts as a support for the two polyelectrolyte components which are electrophoretically loaded into the films to produce a c ↔ a membrane. The matrix membrane was prepared by evaporating thin layers of chloroform solutions of the polyamide, poly (sebacyl piperazine). (Pip–8) [14] onto a glass plate. The polymer and solvent were carefully purified to eliminate contamination with dust particles, water and traces of other organic matter. Pip–8 has a combination of properties which provide for a suitable matrix material. It is a neutral hydrophilic, but water-insoluble polymer, and has no peptide NH groups which could promote denaturation of any proteins that may be used as membrane additives. It generates no oscillatory activity in an electric field. The porous structure of Pip-8 is prepared by adding varying amounts of a water soluble impurity (10^{-7} to 10^{-6} g/ml) such as glycerol or polyethylene glycol to the 0.5 % Pip-8 solution. These are incompatible with Pip–8 in the solid phase but remain in solution in $CHCl_3$. The porous Pip–8 structure is obtained because the added impurities separate into isolated regions as the film dries. These can then be extracted out with water. Fig. 3 shows a series of electron micrographs of the Pip–8 membranes, showing the effect of additives to produce the porous types of films as well as the polyelectrolyte loaded systems. Experimentally the Pip-8 matrix is loaded simultaneously with a polyacid and polybase from opposite sides. The conditions of loading, such as concentrations of the polyelectrolyte, viscosity of the solutions and pH were adjusted so that the mobilities of the polyacid and polybase were the same and that they could interact within the pores to form films and thus the required barrier for a c ↔ a structure. This type of experimental method was used to investigate a variety of synthetic polyelectrolytes, proteins and polynucleic acids as membrane components (see table 1).

EXPERIMENTAL

PREPARATION OF MATRIX MEMBRANES

The polymer for the matrix membrane was synthesized by the interfacial polycondensation method from sebacyl chloride and piperazine (Eastman Organic Chemicals Co.).[14] The polymer was rigorously purified by repeated extraction with 1 M sodium carbonate and water. The wet polymer was then dissolved in chloroform and precipitated by a mixture

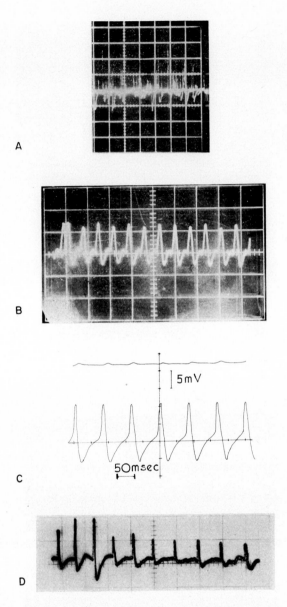

Fig. 1.—Oscilloscope traces of spikes generated by polyelectrolyte membranes in 0.15 N NaCl:
(A) dextran sulphate ↔ polylysine membranes in 0.15 N NaCl: (A) dextran sulphate → poly-
L-sarcosine; scan = 50 ms, amplitude = 20 mV per major division.   (B) RNAse; scan 20 ms and
amplitude 5 mV per major division.   (C) methylacrylate/acrylic acid ↔ Ca^{2+} membrane, upper
record is a d.c. trace at 100 mV/cm.   (D) RNA ↔ Ca^{2+} membrane scan = 0.1 s, amplitude 20 mV
per major division.

[*To face page* 176

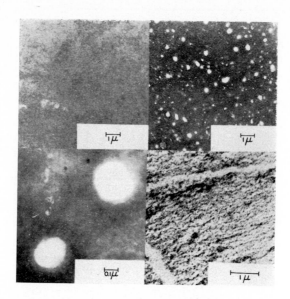

FIG. 3.—Electron micrographs of Pip-8 membranes: upper left shows the nonporous matrix membrane; upper right and lower left show a matrix membrane with pores; lower right shows a matrix membrane loaded with yeast RNA; the deeply stained areas are regions of RNA in the membrane.

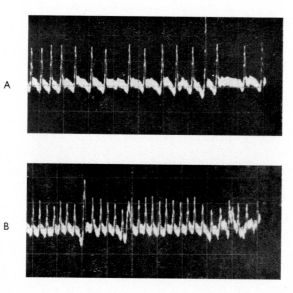

FIG. 4.—Oscillatory patterns from a polylysine ↔ DNA membrane: A—initial output B: stabilized output; scales are 2 mV/cm and 0.5 s/cm for each major division on oscilloscope screen.

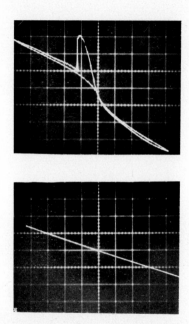

FIG. 8.—Current-clamp data for a polylysine ↔ polyglutamic acid junction membrane; the upper photographs shows an oscilloscope trace of the (I, V) characteristics of a c↔a membrane; lower photograph is for the unloaded matrix membrane. The x and y axes correspond to the current and voltage values. Each major division corresponds to 1 V and 0.1 mA respectively. The zero voltage and current value is at the centre of the photograph.

of ethanol and water. This was repeated three times to remove low molecular weight components. The white fibrous polymer was then dried at 40°C for several days. The matrix membranes were obtained by casting a 0.5 % Pip-8 solution in chloroform over a glass plate coated with a thin layer of a water soluble polymer such as polyethylene glycol (American Cyanamid) or Dextran sulphate (Pharmacia Inc.). The thickness of the chloroform solution was adjusted by a knife edge supported at a distance of 0.025 cm from the glass. After drying for 20 min at room temperature the position of the Pip-8 film on the glass surface was visible only by tilting the glass to produce interference fringes. The film was then cut with a razor blade into 3 cm squares and distilled water was added at the cut edges. The water dissolves the water soluble polymer under the Pip-8 and dislodges the segments, which then float onto the water surface. Next, a cellulose nitrate tube with a 2 mm aperture was used to pick up the matrix membrane. The Pip-8 film was sealed onto the plastic tube with a cement of Pip-8 (10 % in chloroform) at the edges to form a window as shown in fig. 2. The electrophoretic loading with polyelectrolytes was accomplished by placing the polyacid in the outside compartment and the polybase in the inner compartment. Each c ↔ a membrane requires its own experimental parameters. In one example of a polylysine ↔ DNA preparation, the polylysine hydrobromide (polycation phase) at a pH of 3 was electrophoresized against a DNA (polyanion phase) at pH 3. Initially the resistance of the system of salt bridges and electrolytes was 1.5×10^6 ohm. A current of 0.02 mA at 1.5 V was passed through the system. After 15 min the current dropped to less than 0.005 mA. This was considered to be an indication of complete loading of the matrix membrane. Fig. 4 shows the type of electrical oscillations obtained with this membrane when the inner and outer compartment electrolytes were 0.15 N NaCl and when the system was polarized at a potential of 1.2 V. Trace A is the initial type of sustained oscillations obtained when the critical voltage of 1.2 V was first applied. Trace B (fig. 4) shows the pattern after about 1 min of firing. This membrane continued to generate electrical oscillations for a total of 10 min and then abruptly stopped due to the formation of a short circuit.

MEASUREMENT OF (CURRENT, VOLTAGE) CHARACTERISTICS

Two methods were used for measuring the current–voltage characteristics of the c ↔ a membranes. Fig. 5 shows the experimental arrangement for the measurements. The first method used Ag/AgCl electrodes to apply a potential across the membrane M. The current

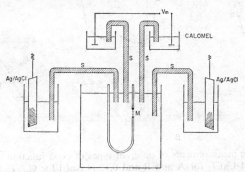

FIG. 5.—Experimental arrangement for measurement of current-voltage characteristics of c ↔ a membranes—S, salt bridges ; V_m, voltage across membrane.

flow in this circuit was detected by a milliammeter. The voltage across the membrane was detected by a voltmeter connected by the two salt bridges S to two Calomel electrodes as shown. The results obtained in this type of measurements are shown in fig. 6. This method did not clamp the current or voltage during the measurement. The second method used a current clamp circuit to apply a given voltage V across the membrane with a nanosecond time constant. Fig. 7 shows the circuit diagram with the operational amplifier in place. The current flowing was detected across a 10 k ohm resistance in series with the membrane.

Both the current and voltage measurements were carried out with Ag/AgCl electrodes connected with salt bridges across the membrane. A characteristic curve for the membrane was directly plotted onto an oscilloscope screen adjusted to display the voltage and current on the x and y axes respectively. All Ag/AgCl electrodes used were converted to chloride form immediately before each measurement. Fig. 8 shows a photograph of the oscilloscope screen for data of a polylysine $\leftrightarrow$ polyglutamic acid membrane system.

RESULTS AND DISCUSSION

Table 1 lists the various c $\leftrightarrow$ a junctions membranes studied to date and the polarizations applied to the membranes to initiate electrical oscillatory phenomena. Fig. 1 and 4 show the types of results obtained. In general random oscillatory patterns were more common. The constant frequency oscillations obtained ranged

TABLE 1.—COMPOSITION OF CATION $\leftrightarrow$ ANION JUNCTION MEMBRANES

polycationic phase	polyanionic phase
polylysine HBr	polyglutamic acid
polydimethylaminoethyl acrylate	polyacrylic acid
Ca^{2+}	yeast RNA
Ba^{2+}	dextran sulphate (m.w. 2×10^6)
cytochrome c	acrylic acid/acrylamide copolymer (50/50)
Ca^{2+}	acrylic acid/methylacrylate copolymer (50/50)
poly L-sarcosine	polyglutamic acid
poly L-lysine HBr	DNA

During the oscillatory mode the positive and negative electrodes were connected to the polycations and polyanionic compartments respectively of the c $\leftrightarrow$ a junction membranes.

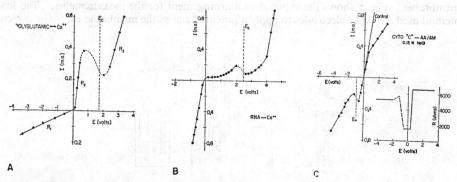

FIG. 6.—Current-voltage measurements of different types of c$\leftrightarrow$a junction membranes; electrolyte was 0.12 N NaCl+1 mM CaCl$_2$ for A and B and 0.12 N NaCl for C. E_c is the initial voltage at which electrical oscillations are observed.

from a low range of 13 Hz for a RNA $\leftrightarrow$ Ca^{2+} to about 100 Hz for a polyglutamic acid $\leftrightarrow$ Ca^{2+} membrane. While it is not yet possible to specify the detailed procedures required for producing constant frequency patterns, we believe that uniform loading and a narrow distribution of pore sizes in the matrix membrane may be among the critical factors.

Fig. 6 shows the current–voltage characteristics of three c $\leftrightarrow$ a membranes, measured under conditions of no current clamping. All three membranes show characteristics of rectifiers with a " negative " resistance region. The polyglutamic acid ↔ Ca⁺⁺ membrane in fig. 6A has three regions of linear–current–voltage properties corresponding to resistance values of 888 ohm/cm², 45 ohm/cm² and 110 ohm/cm². The resistance of the unloaded matrix membrane was 12 ohm/cm² considerably lower than for each of results for loaded membranes. Fig. 6C depicts

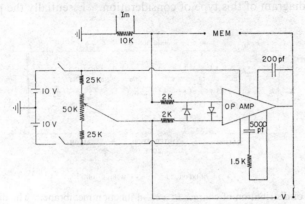

FIG. 7.—Diagram of the current-clamp circuit used for obtaining (I, V) curves for the c↔ membranes.

a more complicated (current, voltage) curve obtained for an acrylic acid/acrylamide (50/50) ↔ cytochrome c membrane. In all the graphs in fig. 6 the oscillatory properties of the membranes are obtained when the membranes are polarized at the critical voltage E_c, i.e., at the negative resistance region of the membrane. The shift from non-oscillatory to oscillatory behavior can actually be seen on an oscilloscope. Experimentally, the scope is set for a.c. recording, then, as the voltage is raised to a critical region there is a rapid transition and the oscilloscope trace shifts from one stable mode to another accompanied by the onset of " spike " generation. This transition in membrane properties occurs within a few seconds and is better illustrated in current–clamp type of measurements.

Fig. 8 shows the current–clamp data for a polylysine ↔ polyglutamic acid membrane. The photograph of the oscilloscope trace was obtained by a point by point setting of each voltage. It is seen that when a voltage of +3 V was applied across the membrane that a sudden decrease of voltage occurred at a constant current of −0.12 mA down to +1.0 V. From then on, the properties of the membrane had ohmic resistance characteristics. The only way to restore the membrane to its original state was to reverse the voltage and obtain (I, V) characteristics as shown in the photograph. It is clear that the membrane behaves like a typical " tunnel–diode " with " negative resistance " characteristics.

Katchalsky [15, 16] has proposed a molecular mechanism for the properties of these c → a membranes. It is based on the concept that polyelectrolytes undergo a phase transition at certain critical salt concentrations. Thus current flow through the membrane causes cations to move into the polyanionic phase, but when they pass into the polyampholyte layer they suddenly encounter the polycation phase. This represents a very highly charged positive layer, so the cations are repelled and they accumulate at the polyampholyte interface. Similarly anions arrive through the polycation phase. The net result is that the NaCl (or electrolyte) concentration builds up at the interface. At a critical concentration there is a sudden shrinkage of the polyelectrolyte and the membrane produces a " breakdown " region. This shrinkage is the result of the well known conformation change of polyelectrolytes at high ionic strength. The result is a conductance change, and electrolyte rushes through to wash out the excess salt to regenerate the original membrane state.

An electrical analogy to the properties of c ↔ a junction membranes can be obtained from comparison of the c ↔ a structure to that of an n–p semiconductor. Fig. 9 shows a diagram of this type of consideration. Essentially the polycation and

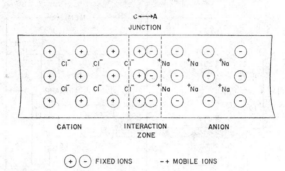

FIG. 9.—Diagram of distribution of charges in a c↔a junction membrane. The diagram indicates the presence of fixed charges provided by the polyelectrolyte and the mobile charges (current carrying) provided by the electrolyte.

polyanion phases represent (in cross-section) regions of fixed ions. The poly-ampholyte zone is the interaction zone and the current carrying species are the mobile ions Na^+ and Cl^-.

Additional similarities are to be found from a consideration of the dynamic properties of c ↔ a junction membranes. Katchalsky and Spangler [16] derived a theoretical equation for the frequency of the oscillations of c ↔ a membranes.

$$v = \frac{RT}{\pi d_0}\sqrt{3LpwC_p}$$

where v = frequency, d_0 = membrane thickness, Lp = the filtration coefficient, w = the salt permeability of the neutral zone and C_p = the concentration of the polymer in mol at the membrane surface. One remarkable aspect of this equation is that it predicts a square root relationship to concentrations of the polymer at the interface zone, i.e., for the concentration of fixed ions at the interaction zone. The equation for the self-resonant frequency of a tunnel diode also has a square root relationship,

$$v = \frac{1}{2\pi R_j C_j}\sqrt{\frac{R_r^2 C_s}{L_s} - 1}$$

where R_j, C_j and L_s are the junction resistance, junction capacitance and series inductance of an equivalent circuit of a tunnel diode. At present we are not able sufficiently to control the parameters for preparing the membranes so as to test various aspects of the " tunnel diode " and the dynamic polyampholyte models.

The general concept of a c ↔ a junction membrane and its analogy to semi-conductors can be applied to a number of membrane models. The requirements are that a two phase system with fixed ions should be present to act as a barrier for the current carrying mobile ions. In the simplest example (see table 1) both the cationic and anionic phases are fixed as polymeric components. It is also possible to have one fixed phase and one " pseudo " fixed phase. For example, Ca^{2+} ions can be used

as the polycationic phase. These ions are introduced into a matrix membrane already loaded with the polyanionic component to give a c ↔ a membrane in which the Ca^{2+} ions cross-link the polyanion to form a graded " polycationic " structure. Thus a preponderance of Ca^{2+} ions are present at the outer surface of the membrane, and a " neutral " phase is developed in the centre of the polyanionic component. In such a system, the polycationic phase is dynamic in nature and it is necessary to have extra Ca^{2+} ions present in the polycationic compartment so that any removal of the Ca^{2+} by ion exchange can be replaced from the electrolyte.

Examples of c ↔ a membranes can also be prepared in which the polyanionic phase is derived from labile components. Thus polylysine can be used as the polycationic phase to provide the fixed cations and a bilayer of lipids (such as lecithin) can act as the anionic phase. If the lipid bilayer is stabilized as a smectic phase then the outer monolayer facing the electrolyte acts as the anionic (negatively charged) component while the inner monolayer interacts with the polycationic component to form the neutral zone of the c ↔ a junction membrane. The integrity of such a membrane can be maintained as long as the bilayer remains intact and extra lipid molecules are available to replace molecules lost by diffusion and electrophoresis.

Such membrane systems have recently been prepared by Montal [10] and earlier in some of the experiments of Mueller et al.[17] using lipid bilayers with the addition of polylysine and protamine respectively. In both these examples the c ↔ a junction model can be used to provide a mechanism for the generation of the excitability and semiconductor properties of the membranes.

This paper is dedicated to the memory of Aharon Katchalsky for his enthusiastic encouragement of the research. Thanks are due to Dr. K Kornacker for the circuit used in the current clamp experiments.

[1] I. Tasaki and T. Takenaka, *Proc. Nat. Acad. Sci. USA*, 1964, **52**, 804.
[2] R. S. Lillie, *J. Gen. Physiol.*, 1925, **7**, 473.
[3] T. Teorell, *Exp. Cell Research Suppl.*, 1954, **3**, 339.
[4] U. F. Franck, *Prog. Biophys.*, 1956, **6**, 171.
[5] H. Davson and J. F. Danielli, In *Permeability of Natural Membranes* (Cambridge University Press, London 2nd Edition, 1952).
[6] P. W. Mueller and D. O. Rudin, *Nature*, 1968, **217**, 713.
[7] C. Huang and T. E. Thompson, *J. Mol. Biol.*, 1965, **13**, 183.
[8] A. M. Monnier, *J. Cell. Comp. Physiol.*, 1965, **66**, 147.
[9] J. Del Castillo, A. Rodriguez, C. A. Romero and V. Sanchez, *Science*, 1966, **153**, 185.
[10] M. Montal, *Biochim. Biophys. Acta*, 1973, **298**, 750.
[11] U. P. Strauss and P. L. Wineman, *J. Amer. Chem. Soc.*, 1953, **75**, 3935; A. Katchalsky and I. R. Miller, *J. Polymer. Sci.*, 1954, **13**, 57.
[12] V. E. Shashoua, *Nature*, 1967, **215**, 846.
[13] V. E. Shashoua, In *The Molecular Basis of Membrane Function*, Ed. D. E. Tosteson (Prentice Hall, N.Y., 1968) p. 147.
[14] P. W. Morgan and S. L. Kwoleck, *J. Polymer Sci.*, 1962, **62**, 33.
[15] A. Katchalsky. In *Neurosciencesv A Study Program*, Ed. G. Quarton, T. Melnichuck and F. O. Schmitt (Rockfeller University Press, N.Y., 1967) p. 335.
[16] A. Katchalsky and R. Spangler, *Quart. Rev. Biophys.*, 1968, **1**, 127.
[17] P. Mueller, D. O. Rudin, H. T. Tien and W. C. Wescott, *Nature*, 1962, **194**, 979.

On the Nature of Certain Oscillations on Bimolecular Lipid Membranes

By Béla Karvaly

Institute of Biophysics, Biological Research Centre, Hungarian Academy of
Sciences, H-6701 Szeged, P.O.B. 521, Hungary

Received 24th July, 1974

The origins and nature of bioelectric oscillations are mostly still obscure. The sustained electro-mechanical oscillations observed by Pant and Rosenberg on bimolecular lipid membranes (BLMs) are, therefore, of great theoretical and practical importance and worthy of detailed investigations. It will be shown that lipid molecules arranged in a bimolecular structure are in a spontaneously excited state (called exciton state) and that these excitons are involved in the charge-transfer processes at the interfaces. The detailed mechanism of the interfacial electrode processes suggests that certain lipid complexes may serve as catalysts promoting interfacial charge transfer and gives further support for the view that a mechanism different from the electrohydraulic one (Teorell's oscillator) may be responsible for the oscillations mentioned above.

1. INTRODUCTION

Pulsating, rhythmic and oscillatory phenomena, at macroscopic, microscopic and submicroscopic (molecular) dimensions alike, are frequently observable in both inanimate nature and the living world. Marked attention has been paid to oscillatory processes in typically biological objects at organ, cellular, subcellular and molecular levels. We think, first of all, of rhythmic reactions in biochemical systems, as well as of spontaneous and induced periodicities of bioelectric phenomena, such as rhythmic changes in membrane potential during stimulus [1]; the propagation of excitation along nerve membranes [2]; the noise-like fluctuations of the membrane potential [3, 4]; the large voltage fluctuations on skeletal muscle membrane [5] and other periodic processes in muscle,[6] etc.

The rhythmic phenomena in biology are comprehensively summarized in Sollenberg's monograph.[7] The thermodynamical bases of oscillatory systems have been laid down by Prigogine and his colleagues.[8, 9] The general mathematical treatment of the kinetics of certain oscillations can be found in ref. (10) and (11). A detailed study of the dynamics and control of cellular reactions has been provided by Higgins [12]; oscillatory properties in chemical systems have been reviewed by Nicolis and Portnow [13]; those in biochemical systems by Hess and Boiteux [14] and the electrical oscillations on porous, fixed charge membranes have been discussed by Teorell,[15, 16] Franck [17] and Meares and Page.[18] Despite the intensive research work done in the field of bio-oscillations, the mechanisms and the physical and physico-chemical background of these phenomena have not yet been cleared up.

The present paper is not concerned with experimental details. It merely summarizes and adopts the most recent results obtained in our laboratory, which furnish further evidence for the electronic nature of the Pant–Rosenberg oscillator.

2. COUPLED ELECTRO-MECHANICAL OSCILLATIONS ON BLMS

Large-amplitude voltage and current oscillations accompanied by periodic change of the Plateau–Gibbs border of BLMs separating inorganic redox electrolytes have

been reported by Pant and Rosenberg.[19]   These phenomena were found to be quite
different from those observed on BLMs in the presence of proteins and proteinaceous
substances,[20, 21] and from spontaneous random fluctuations occurring on unmodified
BLMs.   Since the BLMs can be considered the best models of biomembranes, the
oscillations described by Pant and Rosenberg seem to be of great importance as regards
the understanding of the rhythmicity of bioelectric phenomena, and especially the
membrane oscillations.   The system to be discussed below is outlined in fig. 1.

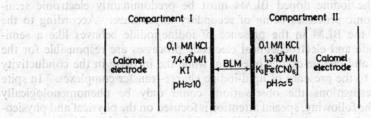

FIG. 1.—The BLM oscillator.

Let us recall the basic properties of this membrane oscillator :

1. The necessary conditions for the appearance of oscillations are the acidity
(pH $\lesssim$ 5) of the potassium ferricyanide solution and the alkalinity (pH $\gtrsim$ 10) of the
potassium iodide compartment.

2. The oscillations are sustained, practically undamped, of constant frequency and
of constant amplitude.

3. The phenomenon is membrane-specific but it is independent of the nature of
the membrane lipids.

4. The amplitude of oscillations is determined by the potassium iodide concentration but it is not affected by pH.

5. The frequency can be altered by changing the pH of the bathing solution, but
it cannot be influenced via the concentrations of the redox components.

6. The periodic electro-mechanical behaviour emerges only in a given range of
the transmembrane potential, i.e., the oscillatory process is voltage-controlled.

7. The voltage and current oscillations can be resolved into at least two sinusoidal
or near-sinusoidal components.

8. The electric and mechanical oscillations are synchronized.

9. Periodic bulging of the membrane is not noticeable under the usual geometry of
observation, but periodic change of the Plateau–Gibbs border is detectable.

10. The system possesses rectifying character, showing that the transport of
negative charge-carriers is more likely from the potassium iodide compartment into
the potassium ferricyanide compartment than in the opposite direction.

Two basically different theories have been proposed to explain the experimental
findings.   Pant and Rosenberg presumed that a mechanism similar to that operating
in the Teorell oscillator,[15-18] or even the Teorell electrohydraulic mechanism itself,
gives rise to the electro-mechanical oscillations.   It was recently suggested by the
present author [22] that electrochemical electrode reactions may be involved in bringing
about the prolonged periodic changes in the electrical properties of the BLM.   Whereas the first interpretation tacitly assumes the existence of ionic conduction in the BLM,
the latter considers rather the electronic properties and processes.

3. CONDUCTIVITY MECHANISM OF LIPIDS AND BLMS

The electrical properties of BLMs in the presence of the iodine/iodide redox couple have been extensively investigated in recent years.[23-31] Nevertheless, both the nature of the charge-carriers and the mechanism of the conduction remain hotly-debated questions. Several scientists [25-27, 31] have arrived at the conclusion that the generation of the electromotive force and the dramatic resistance drop should be attributed to ionic processes rather than to electronic ones, while others [28-30] have concluded that the iodine doped BLMs must be predominantly electronic semi-conductors, the ionic processes being of secondary importance. According to the latter conception the BLM in the presence of iodine/iodide behaves like a semi-conductor electrode and electrochemical electrode processes are responsible for the e.m.f. generation and charge-carrier injection, while the increase in the conductivity can be attributed to the presence of lipid-iodine charge-transfer complexes. In spite of extensive investigations the observations could only be phenomenologically interpreted. In the following, special attention is focused on the physical and physico-chemical aspects of conduction on which the electromechanical oscillatory properties are based.

3.1 CONDUCTIVITY MECHANISM OF LIPIDS AND LIPID-IODINE COMPLEXES

It has been pointed out in very recent investigations [32-37] that the conduction mechanism of wet, oxidized bulk cholesterol samples and unmodified BLMs formed from oxidized cholesterol are very closely related to those of wet, oxidized bulk cholesterol-iodine charge-transfer complexes and iodine-doped BLMs. Thus, all the results concerning the conductivity properties of wet, bulk samples give valuable information on the BLM conductivity too. Therefore, from dielectric studies on oxidized bulk cholesterol and its iodine complexes,[32, 36, 37] as well as from conductivity measurements on BLMs separating iodine/iodide solutions with different concentrations the following could be concluded:

1. The oxidized cholesterol (and probably all lipids) and its iodine complexes come into a very specific interaction with water, thereby leading to the development of Maxwell-Wagner polarization zones.

2. The lipids and their complexes with iodine may become spontaneously excited especially in the presence of water and according to the reaction

$$[L : I_2 : W] \rightleftharpoons [L : I_2 : W]^* \tag{3.1}$$

in both the bulk samples and BLMs spontaneously excited states (molecular excitons) may be present in high density ($[L : I_2 : W]$ denotes the lipid-water-iodine complexes in the ground state and $[L : I_2 : W]^*$ refers to the excited or exciton states).

3. By means of exciton-exciton interaction these excitons are capable of producing free charge-carriers. As a result of the autoionization process

$$[L : I_2 : W]^* + [L : I_2 : W]^* \rightleftharpoons [L : I_2 : W]^{\pm}_{2} + e^{\mp} \tag{3.2}$$

there exist definite concentrations of mobile electronic charge-carriers and of less mobile (occasionally immobile) ionic species.

4. The BLMs (and probably the biomembranes and biomacromolecules too) are covered by an exciton coat, the existence and properties of which are closely related to the stability, dynamic structure and electric and transport characters of the membranes. Consequently, it is practical to make a distinction between the molecular architecture and functional structure of membranes. The molecular architecture (the morpho-logical membrane) includes the lipid constituents, proteins, etc. and the substances

dissolved in these components (fig. 2a). The functional structure of membranes (the functional membrane) consists of two interfaces* and the bulk membrane phase (fig. 2b).

5. The injection of charge-carriers into the BLMs occurs according to the Poole-mechanism, i.e., the injection process at the interfaces is governed by the trans-membrane potential V and/or by the electric field strength E actually present in the interfacial region. In the event of a steady state, the density of the electronic charge-carriers produced by field-enhanced electron/hole emission is proportional to

$$\exp(\bar{\alpha} V) \quad \text{or} \quad \exp(\tilde{\alpha} E). \tag{3.3}$$

Here $\bar{\alpha}$ and $\tilde{\alpha}$ are constants and related to the potential distribution across the membrane (fig. 2c).

6. The electronic charge carriers produced move in the bulk samples and cross the membrane by hopping.

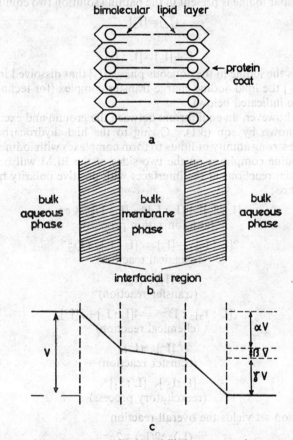

a

b

c

FIG. 2.—The structure of membranes. a. The Danielli–Davson membrane model. b. Schematical representation of the functional structure of membranes. c. The potential distribution across membranes.

* The interfaces are very special parts of the membrane-aqueous phase system, which cannot be delimited on the basis of structural and geometrical considerations. The interface means the space region at the membrane surfaces, the processes taking place in which appear as membrane processes. Of course, the dimensions of the interfaces depend upon the interaction in question and may vary from process to process.

7. If ions are present in the bathing solutions, exciton-ion interactions too may be involved in the generation of charge carriers.

The results summarized above strongly support the idea that the iodine-doped BLMs are predominantly electronic, rather than ionic conductors. It should be noted that the charge carrier generation by exciton–exciton and/or exciton–ion interactions is the explanation of the extremely high resistance-lowering ability of the iodine and water. So far as the hopping conduction is concerned, this accounts for the apparent duality in the interpretation of conductivity data. In the case of such a conduction mechanism the ion-complexes formed as a by-product of autoionization convey their charges to the neighbouring neutral species, this process being reflected as electrodiffusion of ionic charge-carriers.

3.2 MECHANISM OF IODINE EFFECT ON BLMS

If only molecular iodine is present in the bathing solution two equilibria dominate :

$$(I_2) \rightleftharpoons [I_2] \tag{3.4}$$

and

$$L + [I_2] \rightleftharpoons [L : I_2]. \tag{3.5}$$

Here (I_2) denotes the iodine in the aqueous phase, $[I_2]$ that dissolved in the membrane phase and $[L : I_2]$ the lipid–iodine charge-transfer complex (for technical reasons the water will not be indicated below).

There exists, however, an equilibrium between the ground and excited states of the complexes, as shown by eqn (3.1). Owing to the high hydrocarbon-solubility of iodine and to the strong affinity of lipids to form complexes with iodine, the concentrations of lipid–iodine complexes on the two sides of the BLM will be practically the same. The partial reactions at the interfaces with positive polarity take place in the following sequence :

$$[L : I_2]^* + [L : I_2]^* \rightleftharpoons [L : I_2]_2^- + e^+, \tag{3.6}$$
(autoionization process)

$$2[L : I_2]_2^- + [I_2] \rightleftharpoons \langle [L : I_2]_2 : I \rangle_2^{2-}, \tag{3.7}$$
(chemical reaction)

$$(I_2) \rightleftharpoons [I_2], \tag{3.8}$$
(transfer reaction)

$$\langle [L : I_2]_2 : I \rangle_2^{2-} \rightleftharpoons 4[L : I_2] + 2[I^-], \tag{3.9}$$
(chemical reaction)

$$[I^-] \rightleftharpoons (I^-), \tag{3.10}$$
(transfer reaction)

$$[L : I_2] \rightleftharpoons [L : I_2]^*. \tag{3.11}$$
(reexcitatory process)

The above reaction set yields the overall reaction

$$(I_2) \rightleftharpoons 2(I^-) + 2e^+ \tag{3.12}$$

which is assumed to be enacted in the opposite direction at the interface with negative polarity (fig. 3).

3.3 THE MECHANISM OF IODIDE EFFECT ON BLMS

For the sake of simplicity, let the iodide content of the bathing solution be much higher than the iodine content. The equilibria given in eqn (3.4) and (3.5) are valid

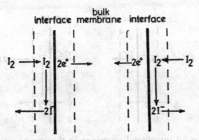

FIG. 3.—Overall interfacial reactions in the presence of iodine.

of course, and the iodide effect can be considered a result of exciton-ion interactions. The consecutive partial reactions are then

$$(I^-) \rightleftharpoons [I^-], \qquad (3.13)$$
(diffusion, transport reaction)

$$[L:I_2]^* + [I^-] \rightleftharpoons [L:I_2]^* : [I^-], \qquad (3.14)$$
(chemical reaction)

$$2[L:I_2]^* : [I^-] \rightleftharpoons \langle [L:I_2] : [I] \rangle_2 + 2e^-, \qquad (3.15)$$
(autoionization)

$$\langle [L:I_2] : [I] \rangle_2 \rightleftharpoons 2[L:I_2] + [I_2] \qquad (3.16)$$
(chemical reaction)

and finally the reexcitation described by eqn (3.11) follows. These partial reactions lead to the overall process (fig. 4)

$$2(I^-) \rightleftharpoons (I_2) + 2e^-. \qquad (3.17)$$

This single electrode process generates a considerable part of the e.m.f.

The basic difference between the effects of molecular iodine and iodide ions is that the interactions involved in liberating electronic charge-carriers are exciton–exciton interactions in the first case and exciton–ion interactions in the second case. As far as the propagation of the electronic charge-carriers is concerned, they pass the membrane by hopping as shown in ref. (32), (36) and (37).

3.4 THE EFFECT OF $K_3[Fe(CN)_6]$

Unfortunately, a detailed mechanism for potassium ferricyanide is not yet available. However, from preliminary studies it is clear that the effect of $K_3[Fe(CN)_6]$ is significantly different from that of either iodine or iodide. The ferricyanide ion appears to be capable of adsorption on the membrane surface, but this adsorption process does not lead to drastic changes in the electrical properties of the BLM. Nevertheless, this adsorbed state of the ferricyanide ion is a favourable one as regards participation in interfacial processes. The electrons supplied by the overall process (3.17) at the interface in iodide compartment may be picked up by the adsorbed ferricyanide ions producing ferrocyanide ions.

4. SOME REMARKS ON THE OSCILLATIONS

The general scheme of the Pant–Rosenberg oscillator has been outlined in an earlier paper.[22] The further results summarized above confirm the suggestion that the interfaces can be considered thermodynamically open systems and that a significant portion of the oscillatory processes is electronic in nature. On the basis of

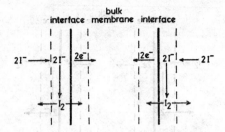

FIG. 4.—Overall interfacial reactions in the presence of iodide.

results presented in ref. (32) and section 3, two simplified versions of the oscillatory system can be given as shown in fig. 5 and 6.*   The importance of the ionic processes arises primarily in the development of mechanical oscillations and in the voltage-controlled nature of the oscillatory behaviour.   The appearance of the mechanical oscillation is thought to be due to the adsorption on the membrane of the products of the overall reaction

$$2[Fe(CN)_6]^{4-} + I_2 \rightleftharpoons 2[Fe(CN)_6]^{3-} + 2I^-, \qquad (4.1)$$

and the excess surface charges result in the periodic change in the membrane surface at the expense of the Plateau–Gibbs border.   Unfortunately, we have no evidence as

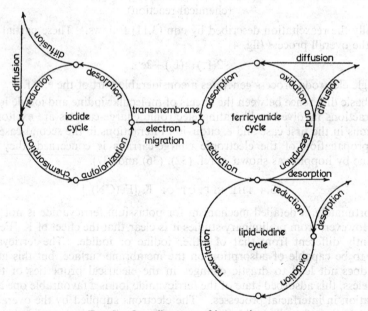

FIG. 5.—Overall system of interactions.

to whether the iodide ion, or the ferrocyanide ion or even both, are involved in creating the mechanical oscillations.   It is not clear either whether the iodide ions originate from the iodine dissolved in the membrane or from the lipid–iodine complexes. Although the latter possibility seems more probable, this question still remains open and is under further examination.

* The details of these branched reaction models will be described elsewhere.[38]

As can be seen from the reaction models in fig. 5 and 6, there are two systems capable of oscillating, these being connected by an electronic charge-transfer process. Franck [39] has treated in detail the problems of coupling between two oscillating

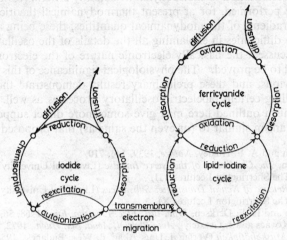

FIG. 6.—Overall system of interactions.

electrochemical metal electrodes. It was pointed out that in the low-frequency region such as ours the capacitive and inductive couplings play a subordinate role. Considering that the bulk membrane phase has a relatively high resistance, independently of the iodine dissolved in it, and that the low-frequency capacity of iodine-doped BLMs is relatively high, it is believed that the system in question may be replaced by two oscillating semiconductor electrodes connected by a parallel RC circuit (fig. 7). Since the transmembrane charge transport is hopping, the possible role of local currents in coupling appears to be excluded.

Finally, some difficulties of a practical and theoretical nature should be noted. The reaction networks proposed in fig. 5 and 6 could be checked in principle by computer simulation, for which, however, the equilibrium constants would be needed. Owing to the complicated structure of the system (diffusion layer, interfacial phenomena, charge-transfer complex forming, etc.), these parameters are still unknown.

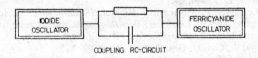

FIG. 7.—Equivalent circuit for the coupled oscillating interfaces.

The theoretical difficulties originate from the fact that a considerable part of the transmembrane potential drops in the interfacial region, thereby causing an extremely high electric potential gradient of the order of 10^6 V/cm or more (fig. 2c). The presence of such a high electric field in the interfacial region would make it difficult to restore the membrane system to the equilibrium state if it were perturbed. Consequently, it seems plausible to assume that the membrane (i.e., the interfacial regions and the bulk membrane phase together) does not and can not exist in the equilibrium state of classical sense. Accordingly, due to the high electrical potential gradient in the interfacial region, at the best a steady-state may prevail in the interfaces, i.e., the natural state of the membrane is non-equilibrium.* Another consequence of this enormously high potential gradient is that a thermodynamic treatment of such a system (e.g., the formulation of thermodynamic stability criteria for non-equilibrium states) cannot be performed, for at present thermodynamical theories operate with space and time gradients of thermodynamical quantities, these being assumed small.

Despite many difficulties in explaining all the details of the oscillatory behaviour of the system discussed, the basically electronic nature of the electronic oscillations can be considered to be proved. The physiological significance of this type of oscillation is fairly obvious, and these preliminary results demonstrate the suitability of BLMs for modelling certain bioelectric oscillatory processes as well. On the other hand the mechanism outlined here may give some more direct support for a nerve conduction mechanism similar to (or even the same as) that proposed by Lillie.[40]

[1] A. L. Hodgkin and A. F. Huxley, *Nature*, 1939, **144**, 710.

[2] A. L. Hodgkin, *The Conduction of the Nerve Impulse* (Liverpool University Press, Liverpool, 1st ed. 1964, The Sherrington Lectures VII).

[3] B. Katz, *The Release of Neural Transmitter Substances* (Liverpool University Press, Liverpool, 1st ed. 1969, The Sherrington Lectures X).

[4] A. A. Verveen and H. E. Derksen, *Proc. Inst. Elec. Electron. Eng.*, 1968, **56**, 906.

[5] E. Varga, L. Kovács and I. Gesztelyi, *Acta Physiol. Acad. Sci. Hung.*, 1972, **41**, 81.

[6] E. Ernst, *Die Muskeltätigkeit* (Verlag d. Ung. Akad. d. Wiss. Budapest, 1958).

[7] A. Sollenberg, *Biological Rhythm Research* (Elsevier, Amsterdam, 1965).

[8] I. Prigogine and G. Nicolis, *Quart. Rev. Biophys.*, 1971, **4**, 107.

[9] P. Glansdorff and I. Prigogine, *Thermodynamic Theory of Structure, Stability and Fluctuations* (Wiley, New York, 1971).

[10] J. I. Gmitro and L. E. Scriven, in *Intracellular Transport*, Symposia of the Int. Soc. for Cell Biology, ed. K. B. Warren (Academic Press, New York-London, 1966), Vol. 5, p. 221.

[11] J. Higgins, *Ind. Eng. Chem.*, 1967, **59**, 19.

[12] J. Higgins, in *Control of Energy Metabolism*, ed. B. Chance, R. W. Estabrook and J. R. Williamson (Academic Press, New York-,1965), p. 13.

[13] G. Nicolis and J. Portnow, *Chem. Rev.*, 1973, **73**, 366.

[14] B. Hess and A. Boiteux, *Ann. Rev. Biochem.*, 1971, **40**, 237.

[15] T. Teorell, *Biophys. J.*, 1962, **2** Suppl. 27.

[16] T. Teorell, in *Laboratory Techniques in Membrane Biophysics*, ed. H. Passow and H. Stämpfli (Springer Verlag, Berlin 1969), p. 130.

[17] U. Franck, *Ber. Bunsenges. phys. Chem.*, 1963, **67**, 657.

[18] P. Meares and K. R. Page, *Phil. Trans. A*, 1972, **272**, 1.

[19] B. Rosenberg and H. C. Pant, *Biochim. Biophys. Acta*, 1971, **225**, 379.

[20] P. Mueller and D. O. Rudin, *J. Theor. Biol.*, 1968, **18**, 222.

[21] P. Mueller and D. O. Rudin, *Nature*, 1968, **217**, 713.

[22] B. Karvaly, *Nature*, 1973, **244**, 24.

[23] P. Läuger, W. Lesslauer, E. Marti and J. Richter, *Biochim. Biophys. Acta*, 1967, **135**, 20.

[24] P. Läuger, J. Richter and W. Lesslauer, *Ber. Bunsenges. phys. Chem.*, 1968, **71**, 906.

[25] P. A. Peshayev and L. M. Tsofina, *Biophys.* (Russ.), 1968, **13**, 360.

[26] A. Finkelstein and A. Cass, *J. Gen. Physiol.*, 1968, **52**, 145.

[27] G. L. Jendrasiak and H. E. Lyon, 13th Annual Meeting of the American Biophysical Society, Los Angeles 1969, *Biophys. Soc. Abstr.*, p. 72a.

[28] V. Ya. Vodyanoj, I. Ya. Vodyanoj and N. A. Fedorovich, *Fiz. Tverd. Tela* (Russ.), 1970, **12**, 3321. (*Soviet Phys. Solid State*).

[29] L. I. Boguslavsky, F. I. Bogolepova and A. V. Lebedev, *Chem. Phys. Lipids*, 1971, **6**, 296.

[30] B. Karvaly, B. Rosenberg, H. C. Pant and G. Kemeny, *Biophysik*, 1973, **10**, 199.

[31] G. Szabó, G. Eisenman, R. Laprade, S. M. Ciani and S. Krasne, in *Membranes*, ed. G. Eisenman (Marcel Dekker, New York, 1973), p. 179.

[32] B. Karvaly, *Thesis for the Academy Degree Candidate of Sciences* (Szeged, 1974).

* The biological macromolecules, as a consequence of their structure, relevant properties and localization, are presumably covered too by an exciton coat, and, therefore, their natural state is plausibly non-equilibrium in most cases for just this reason. This is outside the scope of this paper, however, and will be treated elsewhere.

[33] B. Karvaly, *Bioelectrochem. Bioenergetics* 1975, **2**, 124.

[34] B. Karvaly, I. Szundi and K. Nagy, *Bioelectrochem. Bioenergetics*, 1975, **2**.

[35] I. Szundi and B. Karvaly, *Acta Biochim. Biophys. Acad. Sci. Hung.*, 1973, **8** (Suppl.), 200.

[36] I. Szundi, B. Karvaly and K. Nagy, *M.T.A. Biol. Oszt. közl.*, 1974, **17**, in press.

[37] B. Karvaly, I. Szundi, *2nd Conf. Condensed Matter Division of the European Physical Society*, Budapest 1974.

[38] B. Karvaly (to be published).

[39] U. Franck and L. Meunier, *Z. Naturforsch.*, 1953, **8b**, 396.

[40] R. S. Lillie, *J. Gen. Physiol.*, 1920, **3**, 107, 129.

Current Oscillations in Iodine-doped Polyethylene Film

By G. T. Jones and T. J. Lewis

School of Electronic Engineering Science, University College of North Wales,
Dean Street, Bangor, Gwynedd, North Wales

Received 25th July, 1974

Very low frequency ($\sim 10^{-1}$ Hz) regular current oscillations may be induced in iodine-doped polyethylene films when subjected to electric fields in excess of about 3×10^7 V m^{-1}. The oscillations are similar whether iodine is introduced from aqueous KI electrodes or from the dry vapour. The frequency depends on film thickness and iodine concentration and has an activation energy of ~ 1.2 eV. Most significantly, it decreases with increasing field suggesting that space charge domains are propagating in the films encouraged by a negative differential charge carrier mobility-field characteristic. This is confirmed by direct measurement. The acceptor action of iodine probably generates mobile electron vacancies in the polymer chains, the effective mass of which increases with field to give the negative characteristic.

It has been demonstrated already [1] that the electrical conductivity of thin polyethylene films will increase by several orders of magnitude when they absorb iodine by contact with aqueous sodium iodide solutions. It appears that neutral iodine rather than iodine ions diffuse into the polyethylene and the current growth at constant applied electric field follows a Fickian diffusion law. [2] Neutral iodine is known to be preferentially absorbed in the amorphous regions of the polymer film [3] and at the same time it appears that electron transfer from polymer molecules to vacant acceptor levels in the iodine system generates mobile " holes " in the polymer chain which leads to enhanced conduction. [2]

Swan [4] has also shown that when the field applied exceeded about 4×10^7 V m^{-1}, regular slow current oscillations were superimposed on the steady background current, the frequency of these depending on temperature, iodine concentration and field strength. Swan considered the possibility that the oscillations might be due to the propagation of space charge domains across the film but concluded because the static (current, voltage) characteristics did not appear to show negative differential characteristics, that high field accumulation domains as observed in more conventional semiconductors were not occurring. McCumber and Chynoweth [5] have shown, however, that the absence of such characteristics is not in fact evidence against the existence of high field domain propagation in a solid.

Our present studies of a system similar to that of Swan confirm the existence of oscillations with the same characteristics. We have also found that oscillations may be generated in a dry system in which iodine is diffused directly into the polyethylene from the vapour and also in a polyethylene + sulphuric acid system. Furthermore we have been able to make direct determinations of the charge carrier mobility in such systems and to show that this parameter exhibits a strong negative differential coefficient with respect to the field, such as would be required for the establishment of space charge instabilities in conventional semiconductors. The implications of this for carrier motion in organic systems generally is briefly discussed.

192

EXPERIMENTAL

The arrangement for most of the experiments was essentially similar to that used earlier.[2,4] Samples of low density polyethylene film without additives in the thickness range 50-350 μm were cleaned of surface grease by washing in methanol and sealed between p.t.f.e. cups, each normally containing an aqueous 1 M solution of KI with iodine added and arranged so that a surface area of 0.32 cm^2 of the solution was in contact with the film on either side. Iodine, added to the KI solutions in known concentrations, was allowed to diffuse into the polymer, thereby increasing the conductivity by orders of magnitude. Contact to the aqueous KI solution in each cup was made via tungsten leads and the whole cell could be housed in an oven of which the temperature could be controlled to 0.1°C.

Dry samples were prepared by first outgassing the films for several hours at a pressure of about 10 N m^{-2} and then immersing in iodine vapour until saturated, and excess iodine appeared on the surfaces. The excess was removed by gentle washing with methanol. The sample was then placed between flat copper electrodes, one fitted with an insulated guard ring. This system was difficult to control since high field measurements had to be made under a vacuum in order to avoid surface leakage and, consequently, the iodine content of the sample gradually fell with time. Some improvement in stability was achieved by pre-iodizing the copper electrodes before making contact to the sample. This stabilised the contact and prevented an iodine deficiency adjacent to the electrodes. Gold electrodes were also tried but without success, the surfaces deteriorating rapidly.

A stabilised controlled voltage source was used to apply selected fields to the samples and currents (10^{-11} A to 10^{-6} A) were measured by electrometer.

RESULTS

CURRENT OSCILLATIONS

Provided the film thickness was greater than 50 μm, and less than about 350 μm, sustained slow current oscillations could be obtained for the " wet " system when the field was raised above a threshold value, E_t. A typical example is shown in fig. 1, where the threshold field lay between 3.9 and 4.3×10^7 V m^{-1}. When the sample was thin (~ 50 μm) current oscillations could be induced only at high iodine concentrations (100 g/l. 1 M KI solution). At lower concentrations, oscillations were intermittent and were rapidly damped out. The general chracteristics present in fig. 1 and

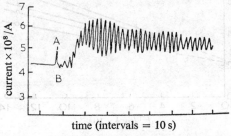

time (intervals = 10 s)

Fig. 1.—Typical (current, time) characteristic at the onset of oscillations. Polyethylene film 127 μm thick. Iodine concentration 10 g in 1 l. of 1 M KI solution at 30°C. At point A the field is changed from below threshold (3.9×10^7 V m^{-1}) to just above (4.3×10^7 V m^{-1}). Note that the oscillations begin by a *downward* swing of current (B).

present in all cases are as follows. On raising the field to the threshold value, a fast capacitative transient is generated which does not appear to have any influence on the subsequent oscillations. Following the transient, oscillations commence by a downward swing of current to a level below that existing before the increase in field. Following this the background steady current, with oscillations superimposed, rises to a peak and then decays slightly to leave a regular oscillating pattern.   The fact that the oscillations begin invariably by a downward swing of current is important for our later discussion of the mechanism.

The amplitudes' of oscillations were not markedly dependent on iodine concentration but were always greatest after a field change and then (as in fig. 1) decreased somewhat with time.   Sometimes, as Swan [4] also found, the oscillations were modulated in a slow way and sometimes there were simultaneous oscillations of slightly different frequency which distorted the sinusoidal form.   Occasionally, as in fig. 2, there would be a gradual transition from one frequency to a slightly different one reflecting presumably some slow change in conditions within the sample.

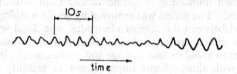

FIG. 2.—Oscillations showing the slow transition from one frequency to another. Note also that there is initially some distortion of the wave due to the presence of other frequencies. Polyethylene film 127 μm thick. Iodine concentration 5 g in 1 l. of 1 M KI solution at 30°C. Field 5.51×10^7 V m^{-1}.

As fig. 3 shows, the threshold field E_t varied inversely as the sample thickness d and also increased as the iodine concentration decreased.   The dry system, for which the iodine concentration was low but not precisely determined, produced a result in agreement with the others.   The characteristics, if extrapolated, converge to a limiting value $E_t \approx 2.2 \times 10^7$ V m^{-1} for infinite thickness.   It is interesting to note that if two films were put together as a single composite film of double thickness, the oscillations were characteristic of a single film of double thickness for both wet and dry systems.

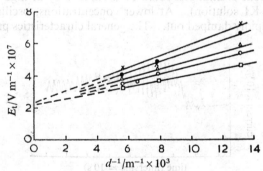

FIG. 3.—Variation of threshold field E_t with reciprocal thickness and with iodine concentration at a temperature of 30°C.   Concentration of iodine (g in 1 l. of 1 M KI solution): □, 25; ○, 10; △, 5; ●, 1; ×, dry system.

The frequency of oscillation f was found to be a maximum at E_t and to decrease as the field was raised above that value as Swan [4] also found.   Ultimately at a sufficiently high field, oscillations become uncertain and periodicity was lost.   By interpreting (arbitrarily at this stage) f^{-1} as a transit time τ of a disturbance across the film, the mobility μ of the disturbance may be written as $d/(\tau E)$ where E is the average applied field ($E \geqslant E_t$).   The results for a range of fields and film thicknesses at a constant concentration lie on a universal curve (fig. 4), μ decreasing with increasing E. Even the result for the double film $(76+76)$ μm fits well on the curve. Arrhenius plots of the frequency against reciprocal temperature produced an activation energy of 1.2 eV in remarkable agreement with Swan.[4] Oscillations could be obtained even when the

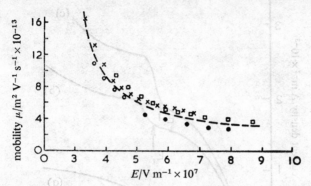

FIG. 4.—Mobility μ determined from frequency of oscillations as a function of field E showing independence of thickness d. Iodine concentration 10 g/l., 1M KI at 30°C. $\times$, 178 μm; $\bigcirc$ $(76+76)$ μm; $\square$ 127 μm; $\bullet$ 76 μm.

temperature was low enough to freeze the electrodes. The period of oscillation decreased or the mobility μ increased with increasing iodine concentration as shown in fig. 5 but the indication is that a limiting value might be reached ultimately.

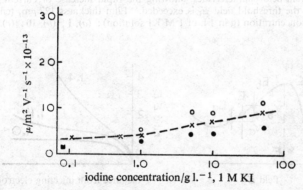

FIG. 5.—Effective mobility μ determined from f as a function of iodine concentration for a 127 μm sample at 30°C. $\bigcirc$, 4.7×10^7 V m^{-1}; $\times$, 5.5×10^7 V m^{-1}; $\bullet$, 7.9×10^7 V m^{-1}; $\blacksquare$, dry system at 5.5×10^7 V m^{-1}.

The background static current has several interesting features which are related to the oscillations. Some (current, field) characteristics are shown in fig. 6. The current first varies supralinearly with the field and then approaches a saturation value which for most samples begins to set in at a field just above 2×10^7 V m^{-1}. Subsequently at the threshold field for oscillations E_t, there is a rapid increase in current and thereafter a variation with field which is less than linear. Again these results are very similar to those of Swan [4] and illustrate how reproducible are the conductivity characteristics of iodine doped polyethylene in contrast to the poor reproducibility of such characteristics for natural polyethylene.

The combination of oscillatory current, the particular form of (current, field)

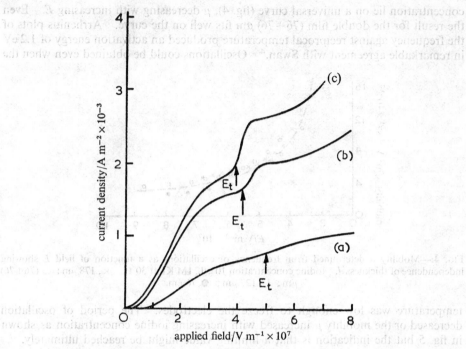

FIG. 6.—Static current-field characteristics, showing the rapid increase of current above a quasi-saturation value as the threshold field E_t is exceeded. Film thickness 127 μm, temperature 30°C. Iodine concentration (g in 1 l. of 1 M KI solution): (a), 1; (b), 10; (c), 50.

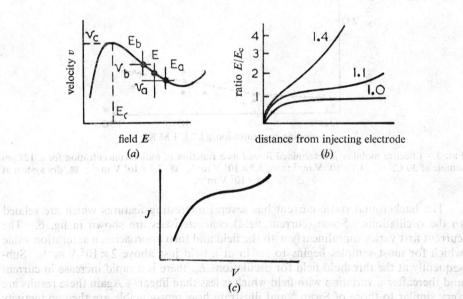

FIG. 7.—Characteristics associated with the onset of space charge domain propagation. (a) (Velocity, field) characteristic with negative slope above the critical field E_c; (b) field distribution in solid for various current ratios J/J_c ($J_c = qnv_c$ where n is the carrier density); (c) general form of (current, average field) characteristic deduced from (b).

characteristic and an apparent mobility which decreases with increasing field (fig. 4) supports a thesis that space charge domains are propagating in the films above the onset field E_t. It is, therefore, worthwhile to review briefly the salient features of space charge domain propagation as developed for conventional semiconductors.[5, 6] The essential requirement is that the mobile charge carrier concerned should have a (velocity, field) characteristic with negative slope over the range of fields of interest (fig. 7a). Assuming this velocity characteristic, a uniform carrier concentration n in the bulk and an efficient carrier injecting electrode, it has been shown that [5] the field distribution will be as in fig. 7b. For current density $J < J_c$ where $J_c = qnv_c$ is the current corresponding to E_c and q is the charge of the carrier, the field approaches a limit $E < E_c$ which extends over most of the sample. For $J > J_c$, the field increases monotonically with distance from the injecting electrode. The (current, field) characteristic may also be found and has the form shown in fig. 7c, the marked upturn in current coming when the ratio J/J_c is such that there is a rapid increase in field away from the carrier injecting electrode. In the negative slope regime (fig. 7a) space charge may accumulate locally and then propagate as a domain.[5, 6] When such an accumulation domain propagates, the field ahead of it, E_a, will tend to increase and that behind, E_b, to decrease (fig. 7a). If $v_a < v_b$, then accumulation continues but when $v_a > v_b$, any accumulation is dissipated. Thus depending on the field distribution and notably on the operating mean bias field E, so domains might propagate fully across the sample, or first grow and then decline part way across or, if the bias field were high enough (or low enough) so that operation moved outside the negative regime in fig. 7a, not be generated at all.

This model thus suggests why there will be a critical field for the onset of oscillations and why at much higher fields they will die out. It also provides a (current, field) characteristic generally in agreement with experiment.

MOBILITY MEASUREMENTS

A central feature of the space charge domain theory is a negative differential (velocity, field) or (mobility, field) characteristic. Thus most convincing for our present argument is that we have been able to demonstrate directly the existence of a negative differential mobility coefficient by extending the method outlined by Davies,[7] Wintle [8] and others to high fields. The experiment utilises essentially the same

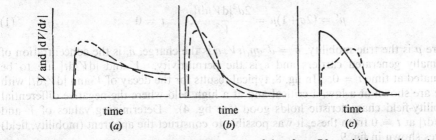

FIG. 8.—Typical results for V and $|dV/dt|$ as functions of time for a 76 μm thick sample containing 10 g iodine/l. of 1 M KI solution at 23°C; ——— V, – – – $|dV/dt|$. (a) 0.79×10^7 V m⁻¹, (b) 3.9×10^7 V m⁻¹, (c) 7.9×10^7 V m⁻¹. In (c) it should be noted how $|dV/dt|$ increases as V decreases with time, which is indicative of the negative differential (mobility, field) relationship.

arrangement as for our other measurements reported above, except that the constant voltage applied to the electrodes is replaced by a fixed charge placed on the high voltage electrode at an initial time and of sufficient magnitude to raise the field in the sample to the value required for the experiment. The electrode is then isolated so that charge can decay only through the sample and the electrode potential V falls according to the mobility of carriers in the sample. The rate of fall is determined by monitoring the potential of the electrode using a rotating vane field-mill electrostatic voltmeter exposed to the field of the electrode and differentiating the output electronically.

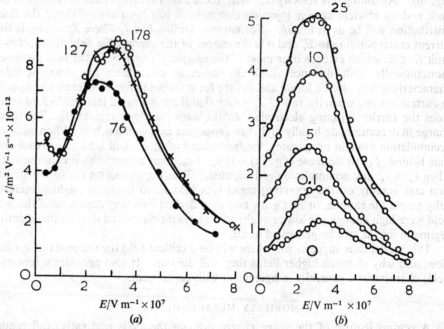

FIG. 9.—Apparent mobility μ' from potential decay curves. (a) Iodine concentration 10 g per litre KI solution at 30°C. Film thickness in μm shown on curves. (b) Film thickness 76 μm, temperature 23°C. Concentration of iodine in g per litre KI solution shown on curves.

It is possible to show that the apparent carrier mobility μ' is given by [9]

$$\mu' = (2\sigma+1)\mu = \frac{2d^2|dV/dt|}{V^2}, \qquad t = 0 \tag{1}$$

where μ is the true mobility, $\sigma = d^2 q n_t / \varepsilon V$, q is the charge, n_t is the concentration of thermally generated carriers and ε is the permittivity. V and $|dV/dt|$ are to be evaluated at time $t = 0$. In fig. 8, typical results for the decay of V and $|dV/dt|$ with time are shown at a low, a critical and at a high field where the negative differential mobility-field characteristic holds good (see fig. 4). Determining values of V and $|dV/dt|$ at $t = 0$ from these, it was possible to construct the apparent (mobility, field) curves shown in fig. 9.

It is seen that μ' reaches a maximum in the range 2-3×10^7 V m^{-1} for all the conditions chosen and, significantly, exhibits a negative slope above this range. The onset of the latter regime coincides with the field range in which oscillations are found and for which there is also a negative slope of the mobility curve (fig. 4) derived from these oscillations. Our values of μ' are in quite reasonable accord with those reported by Davies [10] on iodine-doped polyethylene at low fields. The mobility μ estimated from the period of oscillations (fig. 4) is less than μ' (fig. 9) and it is very likely that the difference is accounted for by the factor $(2\sigma+1)$ in eqn (1). From the data of fig. 4 and 9a and the expression for σ, it is possible to determine n_t, assuming that the permittivity ε of the film is known. We have taken ε to be the value for undoped polyethylene which direct measurement of the permittivity of the films in situ suggested was appropriate. Values of n_t are shown in the table immediately below.

ESTIMATES OF THERMALLY-GENERATED CARRIER DENSITIES n_t BASED ON MOBILITIES μ AND μ' (IODINE, 10 g PER LITRE 1 M KI SOLUTION AT 30°C)

$d=76\mu m$		$d=127\mu m$		$d=178\mu m$	
$E/\text{V m}^{-1}\times 10^7$	$n_t/\text{m}^{-3}\times 10^{20}$	E	n_t	E	n_t
5.25	2.8	4.5	1.4	3.5	0.7
6.0	2.7	5.0	1.7	4.0	0.9
7.0	3.0	6.0	1.9	5.0	1.3
		7.0	1.8	6.0	1.3
		8.0	1.8	7.0	1.3

The value of n_t remains sensibly constant for any given sample at a value of about 10^{20} m^{-3} in close agreement with trap densities estimated by Davies.[10] Choosing a value of n_t, it is possible to calculate the contribution to the static current density from such carriers. For a 127 μm sample, at a field of 4.5×10^7 V m^{-1} and assuming a value of mobility from fig. 4 we find a current density of 8×10^{-4} A m^{-2}. This is less than the measured value from fig. 6 ($\sim 1.5\times 10^{-3}$ A m^{-2}) but in view of the various possible inaccuracies is surprisingly close and suggests that a major part of the static current characteristic comes from thermally generated carriers.

We should note that the onset of the peak in μ' (fig. 9) coincides with the plateau region of the current-field curves (fig. 6) and also with the extrapolated threshold field E_t (fig. 3). Although the magnitude of the peaks (fig. 9b) increases with iodine concentration, the position of the peaks on the field axis is not so affected, in agreement with the fact that the limiting E_t in fig. 3 is also independent of concentration. The situation appears to be similar for the KI solution alone in fig. 9b but there will be some free iodine in this case also.

Since the essential action of the iodine is likely to depend on its electronegativity and strong acceptor action it is important to try other similarly active dopants. In fact, using sulphuric acid in water in a ratio 1 part in 5 in place of the iodine-KI solution produced similar although less repeatable oscillation phenomena and a (current, field) curve with all the characteristic features of fig. 6 but at a lower level of current. The calculated mobility ($\mu \sim 10^{-13}$ m^2 V^{-1} s^{-1}) was in agreement with that for the dry iodine samples (see fig. 5), i.e., those of lowest iodine concentration.

DISCUSSION

The direct evidence of a negative differential coefficient for mobility adds considerable weight to the argument that the current oscillations are associated with domain formation.[5, 6] There may also be significance in the fact that a second general condition for the onset of domain formation, namely that the product of carrier density and specimen length should exceed a certain minimum,[6] is also obeyed for the present system. There is more than one derivation of this condition but the general result for conventional semiconductors is that the product should be in the range 10^{15} to 10^{16} m^{-2}. In our case if we choose values from the table we find the product to be practically constant at 2×10^{16} m^{-2}. We can see also from this condition why oscillations might not be possible in thin specimens ($d = 50$ μm) unless the iodine concentration was very large.

It is now necessary to consider in more detail the action of iodine (or sulphuric acid) and the nature of the carrier whose motion could give rise to the behaviour illustrated in fig. 4 or 9. The iodination action in polyethylene may be twofold. First, the strong electronegativity of iodine will create donor–acceptor complexes with the polymer and the evidence is that this occurs in association with the terminal double bond vinyl groups.[11] Electron transfer to the iodine releases vacancies (" holes ") into the polymer chain which then become mobile in the crystalline region of the polymer structure since, in the close-packed crystallite, hole transfer between chains is also possible.[12, 2] Secondly, neutral iodine as associated complexes I_n[13] might be able to link polymer molecules across the amorphous gap between crystallites. Electron transfer along an I_n chain could occur readily. Thus composite conduction pathways are opened up which, unlike those in conventional semiconductors, may be localised and devious.[2] Vacancies will have an effective mobility which takes account of both drift through the polymer crystallite lattice and trapping at iodine sites with an associated activation energy W.[2] Thus

$$\mu = \mu_1 \theta \varepsilon^{-W/kT} \tag{2}$$

where μ_1 is the lattice mobility and θ is a factor involving the trapping site density, which will probably depend on the density of iodine at low concentrations and on the density of vinyl end groups at higher saturation concentrations of iodine. Eqn (2) is similar to that used by Davies[10] and W may be assigned the value 1.2 eV estimated from the temperature dependence of the frequency of oscillations as well as the static current. The lattice mobility μ_1 can be represented by a relaxation time approximation $\mu_1 = \tau/m^*$ where τ is the relaxation time for collisions within the polymer crystallites themselves and m^* is an effective mass for the carrier moving in an appropriate energy band of the crystallite structure. If, as the field increases, the carrier momentum increases, m^* could increase markedly, particularly if it is moving in a narrow band. Thus μ_1 could exhibit the necessary decrease with increasing field. At fields below the critical value E_t, we find according to fig. 9 that μ', and therefore μ, increases rather than decreases with field. This might be explained by the counter influence of a field lowering of the activation barrier W. Much more work is required to confirm these ideas conclusively but it is certainly plausible that a decrease in effective mass of holes in the polymer chain is responsible for the mobility characteristic, and thus for the current oscillations.

The role of the electrodes has not been questioned so far, but it has been implicit according to the model that the cathode and anode are ready hole and electron acceptors respectively. Diffusion experiments by Taylor and Lewis[2] have indicated that it is certainly necessary to have iodine at the cathode interface for electron injection there, but that under steady state conditions the current is bulk rather than injection limited. The experiments with double films either in the wet or the dry arrangement also lead to the conclusion that interfaces do not play a crucial role in the oscillatory process.

It will be important to establish whether the phenomena observed with the polyethylene system may also be found with other long chain organic structures provided vacancies can be introduced into the chain. Negative differential mobility for carriers moving in narrow energy bands in organic crystals or even for movement along extended chain molecules might be a more common property than hitherto expected and overlooked in the past because of the need to operate at high electric fields and with low currents.

Oscillatory currents have been observed in other organic polymers, for example, by Swaroop and Predecki[14] in dry polyethylene terephthalate and polystyrene as well as polyethylene films without iodine doping. The likelihood in many such cases is simply that a relaxation breakdown phenomenon is being observed [15] especially since fields greater than 10^8 V m^{-1} are required. Recently Toureille, Reboul and Caillon [16] have reported observations of oscillatory current phenomena in the form of regularly spaced peaks in the same three polymers. They find that oscillations coincide with the onset of a negative differential resistance which disappears along with the oscillations at higher fields. From the period of oscillations they estimate a carrier mobility of 10^{-13} m^2 V^{-1} s^{-1} for polyethylene which, since their sample is undoped, is much greater than ours would be in similar circumstances. The model they propose is that above a threshold field the injection capacity of an electrode increases markedly so that excess charge is momentarily injected into the film. While it propagates, this charge serves to lower the field at the electrode and so interrupts the injection. Injection thereby becomes intermittent and the current fluctuates in a quasi-regular manner.

Further work is required to confirm whether this phenomena differs fundamentally from that found for the iodine-doped system but it is encouraging that the subject is receiving continued attention.

The authors thank Dr. D. M. Taylor and Mr. M. G. Jones for much help, especially in earlier stages of this work. Many thanks are also due to Dr. F. J. Smith and the Monsanto Chemical Company for the supply of polyethylene samples and helpful discussions. One of us, G. T. Jones, is grateful for the award of a Science Research Council Research Studentship.

[1] D. W. Swan, *J. Appl. Phys.*, 1967, **38**, 5051.
[2] T. J. Lewis and D. M. Taylor, *J. Phys. D: Appl. Phys.*, 1972, **5**, 1664.
[3] V. A. Marikhin, A. I. Slutsker and A. A. Yastrebinskii, *Sov. Phys.-Solid State*, 1965, **7**, 352.
[4] D. W. Swan, *J. Appl. Phys.*, 1967, **38**, 5058.
[5] D. E. McCumber and A. G. Chynoweth, *Trans. I.E.E.E. Electron Devices*, 1966, **13**, 4.
[6] P. H. Butcher, *Rep. Prog. Phys.*, 1967, **30**, 97.
[7] D. K. Davies, *Static Electrification*, I.P.P.S. Conference, 1967, Ser. No. 4, 29.
[8] H. J. Wintle, *J. Appl. Phys.*, 1970, **41**, 4004.
[9] K. Keiji Kanazawa, I. P. Batra and H. J. Wintle, *J. Appl. Phys.*, 1972, **43**, 719.
[10] D. K. Davies, *J. Phys. D.: Appl. Phys.*, 1972, **5**, 162.
[11] D. K. Davies and P. J. Lock, *J. Electrochem. Soc.*, 1973, **120**, 266.
[12] R. J. Fleming, *Trans. Faraday Soc.*, 1970, **66**, 3090.
[13] E. A. Liberman and V. P. Topaly, *Biochim. Biophys. Acta*, 1968, **163**, 125.
[14] N. Swaroop and P. Predecki, *J. Appl. Phys.*, 1971, **42**, 863.
[15] N. Swaroop, P. Predecki and J. P. Allen, *J. Appl. Phys.*, 1973, **44**, 1943.
[16] A. Toureille, J. P. Reboul and P. Caillon, *Compt. Rend.*, 1974, **278**, 849.

Oscillations in Glycolysis, Cellular Respiration and Communication

By Arnold Boiteux and Benno Hess *

Max-Planck-Institut für Ernährungsphysiologie, Dortmund

Received 3rd October, 1974

In an attempt to understand oscillatory phenomena in chemical, subcellular and cellular systems, the mechanisms of oscillation as well as the general dynamics of glycolysis, cellular respiration and cyclic-AMP-controlled oscillation in the slime mould *Dictyostelium discoideum* have been studied. The enzymic sources of the oscillations have been identified with glycolysis as well as for the slime mould, and appropriate mathematical models have been developed. The primary reaction mechanism of mitochondrial oscillations is not yet known in detail. However, the studies show unequivocally that the conditions for the generation of oscillations can be summarized for all cases in four common statements:

(i) The " primary source " of the oscillations has a nonlinear kinetic characteristic.
(ii) The kinetic structure involves simple and multiple types of feed-back interactions.
(iii) The system moves on a limit cycle.
(iv) The system operates far from equilibrium and is thermodynamically open.

Recently oscillations in biochemical and biological systems have been observed (for summary see ref. (1)-(4)), recognized as results of thermodynamic conditions, and defined as " dissipative structures " by Glansdorff and Prigogine.[5] Indeed, theory and experiment demonstrate a number of dynamic states in nature, which evolve in open systems operating with a critical degree of non-linearity distant from equilibrium. Three types of behaviour have been observed:

(i) The maintenance of multiple steady states with transitions from one to another.

(ii) The maintenance of rotation on a limit cycle around an unstable singular point.

(iii) The maintenance of sustained oscillations coupled to diffusion, resulting in chemical waves.

Recent analyses of biochemical and cellular oscillators provide a remarkable insight into the molecular source of non-linearity although detailed mechanisms might not be at hand in every case. In some cases, promising progress has been made in biochemical studies of the dynamics observed and of the structures involved as well as in the establishment of quantitative mathematical models. This paper summarizes results obtained in our laboratory in the field of bioenergetic processes and of intercellular communication, in partial collaboration with Dr. A. Goldbeter of the Weizmann Institute of Science, Rehovot, and Dr. G. Gerisch of the Friedrich-Miescher-Laboratorium der Max-Planck-Gesellschaft, Tübingen.

OSCILLATIONS IN GLYCOLYSIS

At the present time, oscillating glycolysis is the best understood oscillating biochemical system. Since the original observation of NADH-cycles in yeast cells initiated on the transition from aerobiosis to anaerobiosis (for review see ref. (1,)) this phenomenon has received considerable attention. It has been observed in intact cells, in cell ghosts, in cell extracts, and by a variety of analytical methods it can be reduced to its molecular mechanism and satisfactorily simulated by digital computer techniques.

In a suspension of intact cells following the addition of glucose under anaerobic conditions, a number of cyclic metabolic changes can be observed and directly demonstrated by recording the fluorescence of NADH located inside the cells and part of the metabolic process involved in the cyclic phenomena. A typical experiment is shown in the fluorometric record of fig. 1, which was obtained after a single addition of 100 μmol glucose leading to a series of cyclic changes of NADH-fluorescence.

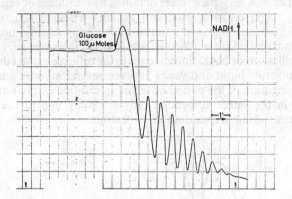

Fig. 1.—Oscillations in yeast cells grown aerobically for 32 h. The arrow indicates a single addition of glucose. (From ref. (7)).

By means of an injection technique allowing the continuous addition of glucose at a steady rate, undamped oscillations of NADH can be maintained as demonstrated in fig. 2. Whenever suitable glycolytic substrates—either glucose, fructose, sucrose, maltose or trehalose—are injected into a suspension of yeast cells, the cells continue to glycolyze their substrates in an oscillatory manner as long as the injection rate is maintained constant within an upper and lower limit.[6, 7] The period of the cycles varies with the experimental conditions between 15 and 75 s.[7]

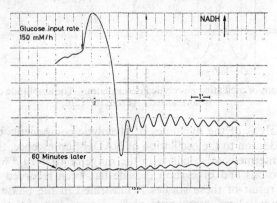

Fig. 2.—Oscillations in yeast cells grown anaerobically for 32 h. Glucose is injected at a constant rate. (From ref. (7)).

By a short treatment with toluene the cellular membrane of yeast can be made permeable to low molecular weight substances, whereas the enzymic equipment of the cells is retained within the cellular volume.[8] This technique allows a study of the functjon of the low molecular weight substances in the mechanism of glycolytic oscillation. It has been shown that as soon as a lower limit of NAD and adenine nucleotide concentrations is maintained, substrate addition leads to glycolytic oscillations with a period between 30 s and a few minutes, comparable to what is observed in intact cells. To a first approximation the NAD concentration determines the amplitude, while the phosphate/nucleotide system strongly influences period and waveform of the oscillation of NADH.

The complexity of the cellular system can be reduced further without losing essential properties. A cell-free extract obtained by treatment of cells with ultrasonic waves and high speed centrifugation still exhibits self-sustained oscillations when supplied with glucose or fructose at a limiting rate, using the injection technique or a controlled enzymic generation of glucose from disaccharides (see ref. (1) and (7)). The period of the oscillations of NAD fluorescence in the cell-free extract depends primarily on the rate of substrate input and varies between several minutes and approximately 1 h.[6] The prolongation of the period observed in cell-free extracts as compared with those observed in the intact suspended cells is mainly due to the dilution of the glycolytic enzymes during preparation of the extract.

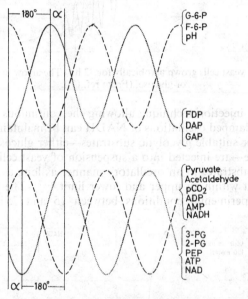

FIG. 3.—Phase relations of oscillating glycolytic intermediates.   The amplitude of the concentrations is normalized.   (From ref. (2)).

A detailed analysis of the dynamics of glycolytic components during oscillation showed that, in addition to the oscillation of NADH fluorescence, the concentrations of all glycolytic intermediates oscillate in the range of 10^{-5}-10^{-3} M.[6] Also, pulsed production of protons and of carbon dioxide has been recorded by suitable techniques.[2, 9] The result of these studies is illustrated in the phase pattern given in fig. 3, where the normalized concentration changes per unit of time are summarized. The figure shows that the concentrations of metabolites oscillate with equal frequency

but different phase angles, relative to each other.   According to their time dependence
they can be classified into two groups in which maxima and minima of the concentrations coincide in time.   The two groups differ by the phase angle α, which depends on
the experimental conditions.[6]

Phase angle analysis of the concentration changes of glycolytic intermediates
during oscillation allows location of the enzymic steps controlling the oscillatory state.
The cross-over diagram given in fig. 4 is obtained by plotting the phase angles between
adjacent glycolytic intermediates according to their reaction sequence along the
glycolytic pathway.   The phase shift of 180° between fructose-6-phosphate and
fructose-1,6-bisphosphate, as well as between phosphoenolpyruvate and pyruvate,
indicates the enzymes phosphofructokinase and pyruvate kinase to be essential
control points.

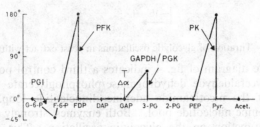

Fig. 4.—Phase angles of glycolytic intermediate concentrations during oscillations in a cross-over
plot.   (From ref. (2)).

The question, which of the kinases is the " primary oscillophore " could be
answered by a demonstration that the substrate of phosphofructokinase fructose-6-
phosphate does indeed induce glycolytic oscillation, whereas an injection of fructose-1,
6-bisphosphate or even phosphoenolpyruvate, the substrate of pyruvate kinase, does
not.   Furthermore, the role of phosphofructokinase as the generator of glycolytic
oscillations is stressed by the fact that its substrate fructose-6-phosphate clearly
exhibits the relatively largest concentration amplitude of all oscillating glycolytic
intermediates.[6, 9]

The function of the control point attributed to pyruvate kinase can be described
as control of pulse propagation along the chain of enzymically catalyzed reactions of
glycolysis.   Indeed, it can be shown that ADP, one of the products of the phosphofructokinase reaction, strongly interacts with pyruvate kinase, whereas the other
product fructose-1,6-bisphosphate does not influence the latter enzyme under
conditions observed during glycolytic oscillation.   Since ATP, the product of the
pyruvate kinase reaction reacts again with phosphofructokinase, feed-forward (ADP)
and feed-backward (ATP) components couple the primary oscillator phosphofructokinase with pyruvate kinase and propagate metabolic pulses via the nucleotide
system.

The function of the adenine nucleotides can be directly demonstrated by phase
shifting experiments as shown in fig. 5 (see also ref. (10)).   The addition of ADP at
the NAD-minimum has no influence on the oscillation, whereas addition of ADP at
the NAD-maximum—corresponding to the lowest level of ADP—shifts the phase
of the oscillation by 180°, because of the rapid acceleration of the reactions catalyzed
by phosphoglycerate kinase and pyruvate kinase, both being susceptible to adenine
nucleotide control.   In addition to the feed-forward control by ADP, the adenine
nucleotide system imposes a strong feed-back activation on phosphofructokinase
via AMP, the allosteric activator of the latter.   As should be expected, ATP and

AMP also shift the phase of glycolytic oscillation if added at appropriate time intervals.

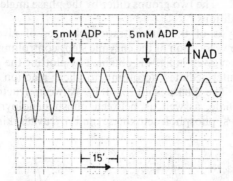

FIG. 5.—Titration of glycolytic oscillations in yeast extract with ADP.

The phase angle diagram of fig. 4 indicates a third control point located at the enzyme couple glyceraldehyde dehydrogenase/phosphoglycerate kinase. The first enzyme is part of the nicotinamide adenine dinucleotide loop. The latter one reacts with the adenine nucleotide pool. Both enzymes, strongly cooperating, exert a duplex control function on the glycolytic oscillation. This duplex control is responsible for the variation of the phase angle α by any change in the concentration of the nucleotide or dinucleotide systems, compare ref. (6).

The actual rates of the different enzymic reactions are coupled via a simple conservation equation to the time dependent concentration changes of the respective metabolites as follows :

$$\dot{c}_{(i)}(t) = v_{(i)+}(t) - v_{(i)-}(t)$$

indicating that the observed time course of a metabolite concentration reflects the imbalance between the source term $(v_{(i)+})$ and sink term $(v_{(i)-})$ for this metabolite. With given restrictions, from the known input rates and the measured time course of the glycolytic intermediates, the time dependences of the respective enzymatic reactions have been calculated.* The pattern obtained resembles a simplified version of fig. 3. The enzymic flux rates split into two groups with coinciding maxima and minima in each group. The maximum activity of enzymes in the " upper " part of the glycolytic reaction sequence precede the maxima of the residual enzymes by the angle α. Since α is in the range of 15°-60° it can be concluded that all enzymes operate almost synchronously, which corroborates the experimentally observed temporal coincidence in the production of fructose-1,6-bisphosphate and carbon dioxide. Again the amplitude of the phosphofructokinase activity, which is the largest, and a low amplitude of aldolase point to the generation of metabolic pulses by phosphofructokinase and to their propagation via the adenine nucleotide system (see also ref. (11)).

In fact, the oscillatory dynamics of glycolysis can be reduced to the operation of the " oscillophore " phosphofructokinase. An analysis of its structure and activity revealed complex properties. The molecular weight of the enzyme was found to be 720 000, the enzyme can be dissociated into 8 subunits of approximately 86 000 and 94 000 Daltons.† The kinetics of the enzyme have been analyzed in the yeast extract under conditions observed during glycolytic oscillations as shown in fig. 6. The

* in collaboration with Dr. H.-G. Busse.
† In collaboration with Dr. N. Tamaki.

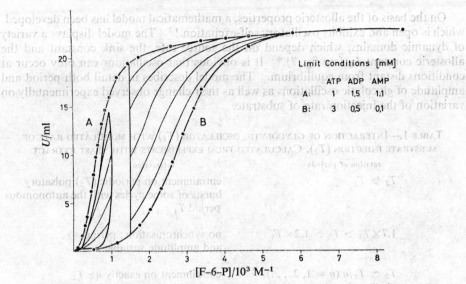

FIG. 6.—Activity of phosphofructokinase under conditions of oscillation.

enzyme activity is strongly dependent on the state of the adenine nucleotide system as well as on the concentration of fructose-6-phosphate. In the oscillatory state the activity can change up to a factor of 80-90. The area on the left hand side of the diagram describes the range of activities which can be observed during oscillations. This kinetic behaviour fits an allosteric model with the homotropic effectors fructose-6-phosphate and ATP and the strong heterotropic activator AMP.

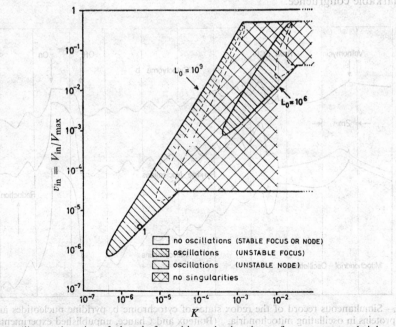

FIG. 7.—Oscillatory domains of phosphofructokinase in the plane of source rate and sink constant calculated for the oscillophore model with two different allosteric constants (L_0).

On the basis of the allosteric properties, a mathematical model has been developed, which is open and exhibits oscillatory self-excitation.[12]   The model displays a variety of dynamic domains, which depend on the source rate, the sink constant and the allosteric constant L_0 (see fig. 7).*   It is obvious that oscillations can only occur at conditions distant from equilibrium.   The model describes in detail both period and amplitude of glycolytic oscillations as well as their change observed experimentally on variation of the injection rate of substrate.

TABLE 1.—INTERACTION OF GLYCOLYTIC OSCILLATOR (T_1) WITH MODULATED RATE OF SUBSTRATE INJECTION (T_2). CALCULATED FROM EXPERIMENTS WITH YEAST EXTRACT

relation of periods	interaction
$T_2 \gg T_1$	entrainment on periodic (T_2) pulsatory bursts of some cycles with the autonomous period T_1
$1.7 \times T_1 > T_2 > 1.2 \times T_1$	no synchronisation; period and amplitude variable
$T_2 \simeq T_1/n \ (n = 1, 2, \ldots)$	entrainment on exactly $n \times T_2$

Recently, entrainment of the glycolytic oscillophore to a periodically varied injection rate has been demonstrated experimentally.   This behaviour is predicted from the common properties of non-linear oscillators.   Whether, and what type of, entrainment is observed depends on the relation of the periods of the oscillophore T_1 and of the input rate T_2 as shown in table 1.   Given a stochastic variation of this rate the oscillophore settles at its autonomous frequency.   Indeed, a comparison between the biochemical experiments and the properties of the model reveal a remarkable congruence.

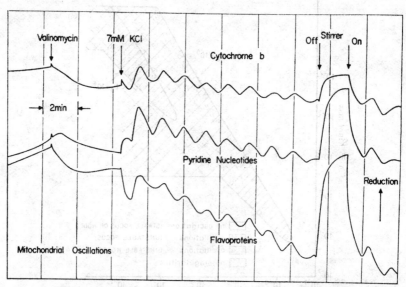

FIG. 8.—Simultaneous record of the redox states of cytochrome b, pyridine nucleotides and flavoproteins in oscillating mitochondria.   (Boiteux and Chance, unpublished experiments.)

* In collaboration with Dr. T. Plesser and V. Schwarzmann.

In partial summary, one now notes that analysis and simulation of the oscillating glycolysis reveal a number of basic principles, which obviously are necessary conditions for this phenomenon :

(i) The " primary source " of the oscillations has a non-linear kinetic characteristic.

(ii) The kinetic structure involves some type of feed-back interaction.

(iii) The system moves on a limit cycle.

(iv) The system operates far from equilibrium and is thermodynamically open.

OSCILLATION IN MITOCHONDRIAL RESPIRATION

The control system of cellular respiration possesses an inherent potential to break into self-sustained oscillations. In a number of laboratories, this property has been observed during studies of intact cells as well as of isolated mitochondria.[1] In a typical experiment, fig. 8 demonstrates the dynamics of some components of the respiratory chain in suspensions of pigeon heart mitochondria, supplied with substrate and oxygen. Since an integral part of the oscillatory system is the rate of ion transport across the mitochondrial membrane, which must match the rate of electron flow through the respiratory chain, the oscillation can be influenced by interaction with the mechanism of ion transport. Therefore, in this experiment valinomycin, an ionophoric antibiotic, is used to activate the potassium transport. It is shown, that the oscillations start right away, as soon as potassium ions are added. The use of an ionophore, however, is not obligatory. Oscillations can be induced as well by other methods critically matching the rate of ion transport and electron flow. As shown in the figure, the components of the respiratory chain, cytochrome b, pyridine nucleotides and flavoproteins are oxidized and reduced simultaneously with a period of about 70 s, all three responding with identical kinetics to the change of oxygen potential, which is maintained by vigorous stirring. The mitochondrial system is complex enough to show a variety of dynamic states. In a phase plane plot of the redox

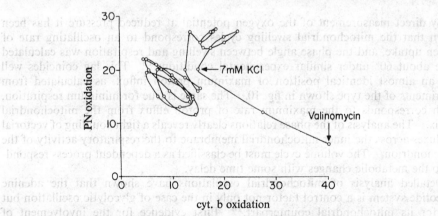

FIG. 9.—Phase plane plot of the redox states of pyridine nucleotides and of cytochrome b in oscillating mitochondria. (Calculated from ref. (14).)

states of pyridine nucleotides and cytochrome b, as given in fig. 9, the transition from an unstable spiral to a stable limit cycle can be demonstrated on addition of potassium chloride to valinomycin treated mitochondria.

The oscillatory changes in the redox state of the respiratory chain are accompanied by periodic changes of the concentrations of protons and of potassium ions, and by concomitant shrinking and swelling of the mitochondrial matrix volume. Fig. 10 is a cutout obtained from a long train of virtually undamped cycles, demonstrating the phase relations between the recorded parameters. Obviously, the oscillations of redox components, ion fluxes and volume changes are not synchronous. First of all, the directions of the fluxes of protons and potassium ions are opposite. This ion movement, however, is not a simple one-to-one exchange reaction. The amplitude of the ion fluxes is up to 15 μ equiv. g^{-1} of mitochondrial protein in the case of protons, and up to ten times more for potassium ions, depending on the experimental conditions. This electrical imbalance must be counteracted by a concomitant flux of ions, predominantly phosphate and substrates, which in turn implies a simultaneous transport of water by osmotic forces. In fact, the maximum of the potassium concentration inside the mitochondria slightly precedes maximum swelling, indicating that the volume change is a secondary process.

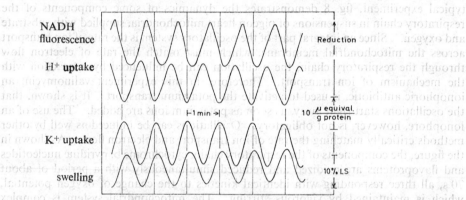

FIG. 10.—Simultaneous record of NADH fluorescence, H⁺ uptake, K⁺ uptake, and swelling in oscillating mitochondria. (A. Boiteux and B. Chance, unpublished experiments.)

By direct measurement of the oxygen potential at reduced pressure it has been shown that the mitochondrial swelling cycles correspond to an oscillating rate of oxygen uptake, and the phase angle between swelling and respiration was calculated to be about 60° under similar experimental conditions.[13] This lag coincides well with an almost identical position of maximum proton influx, as calculated from experiments of the type shown in fig. 10. The same is true for minimum respiration, which corresponds to the maximum rate of proton efflux from the mitochondrial matrix. The analysis of the phase relations clearly reveals a tight coupling of vectorial ion fluxes across the inner mitochondrial membrane to the respiratory activity of the mitochondrion. The volume cycle must be classified as a dependent process responding to the metabolic changes with some time delay.

Detailed analysis of mitochondrial oscillations have shown that the adenine nucleotide system is a control factor not only in the case of glycolytic oscillation but also for its mitochondrial counterpart.[14] First evidence for the involvement of adenosine triphosphate in the control mechanism was the observation that mitochondrial oscillation can be suppressed by oligomycin,* an antibiotic inhibitor, which prevents the interconversion of ADP and ATP at the mitochondrial phosphorylation sites The internal ATP-level in oscillating mitochondria clearly exhibits periodic

* In collaboration with Dr. B. Chance.

cycles, as shown in fig. 11. The ATP-minimum succeeds maximum swelling with a lag of approximately 50°, coinciding with the phase angle for maximum respiration rate. Accordingly, the highest ATP-level corresponds to the lowest respiration rate.

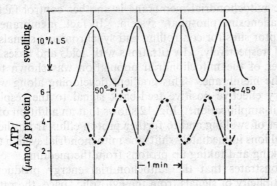

FIG. 11.—Phase relations between swelling and endogenous level of ATP in oscillating mitochondria. (Boiteux and Chance, unpublished experiments.)

Final proof for adenine nucleotide control in mitochondrial oscillations comes from titration experiments. Addition of ADP at any phase angle of the oscillation results in a new synchronization of the mitochondrial oscillations at highest respiratory rate. Titrations with ATP lead to a synchronization at minimum respiration. Within a limited range of concentrations the synchronization by adenine nucleotides is completed within fractions of a second, and the oscillation continues without time delay from the newly obtained position. The phase relations between volume change, respiration rate, redox state of cytochrome b, and ion fluxes are given in the outer shells of fig. 12. In addition, the diagram shows in its centre sections the phase angles of the cycle obtained by titrations with ADP, ATP, potassium ions, acids and bases, respectively.

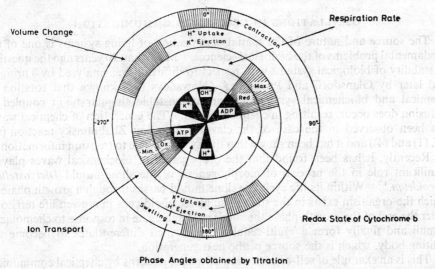

FIG. 12.—Phase relations in mitochondrial oscillations and shift of phase angles by titration. (From ref (14).)

From these experiments it is obvious that the ADP/ATP-ratio in the mitochondrial matrix is a control factor of the oscillations, though not the only one. Experiments with valinomycin clearly demonstrate that the activity of the potassium carrier determines both amplitude and period of the oscillation.[14] Thus, the rate of ion transport across the mitochondrial membrane is another control factor in addition to the endogenous adenosine phosphate system. In fact, membrane-bound carrier systems are the receptor sites for controlling and synchronizing signals in suspending medium and cytosol, respectively. By titrations with acids and bases (cf. fig. 12) the external synchronizer of the mitochondrial population was shown to be a proton gradient [14] across the membrane. This gradient developing along with the proton flux serves as a very effective positive feed-back signal to the respiratory system, promoting strong self-amplification. Fig. 12 shows that an addition of protons shifts the cycle to the onset of swelling and to further proton efflux from the matrix space. Addition of hydroxyl ions immediately shifts the mitochondrial system to the position, where it starts shrinking and taking up protons from the medium.

The study demonstrates that the mitochondrial energy production is tightly controlled by a multiplicity of signals from the cytosolic space, the rate of ion fluxes across the inner membrane being a critical controlling parameter. The detection of the very effective feed-back action of a proton gradient raises the question of the relation between the membrane charge and the structural and functional changes at critical control sites. Though the mechanism of generation of mitochondrial oscillations is not yet known in detail, it is obvious, that the statements from points (i)-(iv) of the previous section also hold for this case. (1) The involvement of vectorial transport processes for controlling reaction steps leads to non-linear kinetics. (2) The self-exciting proton gradient is the basis of the feed-back structure of the membrane system. (3) Fig. 9 shows that the system moves on a limit cycle. (4) Approximation towards the equilibrium state by substrate exhaustion or oxygen limitation leads to rapid damping of the oscillation, which can be maintained only if a minimal flux is provided. The system is open for the source, namely oxygen, and for a number of sink reactions, leading to the production of water, carbon dioxide, and heat.

OSCILLATIONS IN CELLULAR COMMUNICATION

The source and nature of the spatial organization of living systems is one of the fundamental problems of the biological sciences. More than 20 years ago the question of stability of biological systems with respect to diffusion was analyzed by Turing [15] and later by Glansdorff and Prigogine.[5] Nowadays it is known that rotation of chemical and biochemical systems around an unstable singular point coupled to diffusion does occur, resulting in chemical waves. The generation of chemical waves has been observed in the case of the classical Belousov–Zhabotinsky reaction (see ref. (1) and (4)) and it has been shown that this system is able to transmit information.[16]

Recently, it has been found that the propagation of biochemical waves plays a significant role in the process of morphogenesis of the slime mould *Dictyostelium discoideum*.[17] Within its life cycle the slime mould passes through a growth phase in which the organism exists in the state of single ameboid cells in a given life territory. After the end of the growth phase the single cells aggregate in response to chemotactic stimuli and finally form a multi-cellular body which differentiates to become the fruiting body, which is the source of the next generation.

This is an example of self-organization of spatial patterns by chemical communication starting with a layer of randomly distributed identical cells. The aggregation territories are controlled by the centres which release chemical pulses of cyclic AMP with a frequency of 0.2-0.3 min⁻¹. The pulses are propagated from cell to cell as excitation waves which spread in a uniform layer with a constant speed of 40-50 μm min⁻¹. The waves are either concentric or spiral shaped. Formally the system can be treated as a set of diffusion coupled oscillators in which the *Dictyostelium* cells operate as the oscillating element by receiving, amplifying and ejecting periodically concentration gradients of cyclic AMP.

The periodic motion of the cells as well as the response of the cells to cyclic AMP can be recorded by light scattering, indicating changes in cellular shape, volume and degree of agglutination.[17, 18] A typical example is given in fig. 13, where periodic spike formation, and later on, sinusoidal oscillations are recovered. The system is sensitive to cyclic AMP addition.[18] A half maximal reaction is obtained with 1×10^4 molecules of cyclic AMP per cell, and the response is still detectable at a molarity of 10^{-9} with a molecule to cell ratio of 3000. Cyclic AMP pulses interact also with the oscillating cellular system, resulting in phase shift or suppression of spike formation.

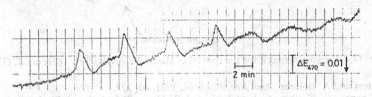

$\Delta E_{470} = 0.01$

2 min

FIG. 13.—Oscillations of light scattering in a suspension of *Dictyostelium discoideum*. The drift of the curve indicates shift towards larger aggregates. (From ref. (18).)

Biochemical analysis identified the function of a membrane-bound adenyl cyclase which is responsible for the production of cyclic AMP. The activity of a phosphodiesterase controls the upper threshold level of cyclic AMP at the membrane as well as in the intercellular space into which a soluble diesterase is secreted. The binding of cylic AMP to a receptor site is transitory, which should be expected for an oscillating binding function.[17-20] It is interesting to note, that the receptor system by which the slime mould aggregates in response to periodic cyclic AMP pulses is similar to the acetylcholine receptor/cholinesterase system by which nerve impulses are transmitted through the synaptic cleft.[17] It is beyond the scope of this paper to discuss the coupling mechanism by which waves of cyclic AMP are received and transduced, resulting in an orientated chemotactic motion towards the aggregation centre.

An analysis of the enzymic system involved in the production of cyclic AMP reveals an interesting network of interactions as demonstrated in fig. 13 on the basis of recent experimental data.[19, 20] The cyclase and pyrophosphohydrolase form a bienzyme system which is autocatalytically controlled by feedback. The heterotropic interactions of both enzymes with cyclic AMP and 5′ AMP respectively are highly cooperative. Goldbeter recently[21] analyzed the overall dynamics of the system, applying the concerted transition theory of allosteric enzymes. A computer solution demonstrates that under suitable activation by 5′ AMP the system readily settles on a limit cycle around a non-equilibrium unstable stationary state with a period independent of initial conditions. The model satisfactorily describes the experimental observations.

The experimental results and model analysis demonstrate that intercellular communication transmitted by chemical waves plays a significant part in the morphogenesis of highly complex species. Furthermore, one of its essential elements, namely the pulsewise production of the transmitting chemicals, could be reduced and identified as the periodic function of the cyclic AMP producing enzyme in the cellular wall. Indeed, nonlinear chemical and biochemical oscillators are well suitable to serve as devices to measure time, to memorize chemicals and to synchronize chemical activities.

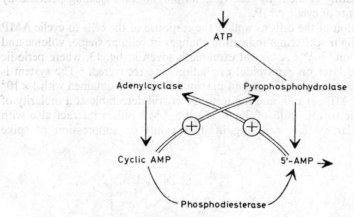

FIG. 14.—Enzyme network for the control of the cyclic AMP-level.   (From ref. (17)).

[1] B. Hess and A. Boiteux, *Ann. Rev. Biochem.*, 1971, **40**, 237.
[2] B. Chance, E. K. Pye, A. K. Ghosh and B. Hess, *Biological and Biochemical Oscillators* (Academic Press, New York and London, 1973).
[3] B. Hess, A. Boiteux, H.-G. Busse and G. Gerisch, *Membranes, Dissipative Structures in Evolution* (Wiley-Interscience New York, 1974) in press.
[4] G. Nicolis and J. Portnow, *Chem. Rev.*, 1973, **73**, 365.
[5] P. Glansdorff and I. Prigogine, *Thermodynamic Theory of Structure, Stability and Fluctuations* (Wiley-Interscience, N.Y., 1971).
[6] B. Hess, A. Boiteux and J. Krüger, *Adv. Enzyme Regul.*, 1969, **7**, 149.
[7] B. Bess and A. Boiteux, *Hoppe-Seyler's Z. Physiol. Chem.*, 1968, **349**, 1567.
[8] R. E. Reeves and A. Sols, *Biochem. Biophys. Res. Comm.* 1973, **50**, 459.
[9] B. Hess and A. Boiteux, *Biochim. Biophys. Acta Library*, 1968, **11**, 148.
[10] B. Chance, R. Schöner and S. Elsaesser, *Proc. Natl. Acad. Sci. USA*, 1964, **52**, 335.
[11] B. Hess, *Nova Acta Leopoldina*, 1968, **33**, 195.
[12] A. Goldbeter and R. Lefever, *Biophys.*, 1972, **12**, 1302.
[13] A. Boiteux and H. Degn, *Hoppe-Seyler's Z. Physiol. Chem.*, 1972, **353**, 696.
[14] A. Boiteux, *Ergebn. exp. Medizin* 9 (Volk und Gesundheit, Berlin, 1972), p. 347.
[15] A. M. Turing, *Phil. Trans. B*, 1952, **237**, 37.
[16] H.-G. Busse and B. Hess, *Nature*, 1973, **244**, 203.
[17] G. Gerisch, D. Malchow and B. Hess, *Cell Communication and cyclic AMP Regulation during Aggregation of the Slime Mould Dictyostelium Discoideum* (Biochemistry of Sensory Functions) (Springer Verlag, Heidelberg, Berlin, New York) in press.
[18] F. Gerisch and B. Hess, *Proc. Natl. Acad. Sci. USA*, 1974, **71**, 2118.
[19] E. F. Rossomando and M. Sussman, *Proc. Natl. Acad. Sci. USA*, 1973, **70**, 1254.
[20] D. Malchow and G. Gerisch, *Proc. Natl. Acad. Sci. USA*, 1974, **71**, 2423.
[21] A. Goldbeter, *Nature*, 1974.

GENERAL DISCUSSION

Dr. B. L. Clarke (*Alberta*) said: The cooperation between positive and negative feedback effects discussed by Franck shows up clearly in the linear steady state stability problem for chemical reaction networks. The necessary and sufficient conditions for the asymptotic stability of a linearized network are that the Hurwitz determinants are all positive. I have shown [1] that each term in the Hurwizt polynomials can be interpreted as a *product of feedback loops*. Negative terms are combinations of feedback loops that cooperate to destabilize and positive terms are combinations which stabilize. The Hurwitz determinant can, therefore, be viewed as a mathematical algorithm for constructing all the combinations of feedback loops which are significant for stability and weighting them with coefficients and algebraic signs according to their importance and stabilizing or destabilizing character. Only one product of feedback loops is usually important at any one time.

Franck has illustrated that the waveforms of oscillatory systems have portions which are dominated by various feedback loops. The reason for this can be understood from steady state stability analysis. Fast changing variables attain pseudo-steady states for given values of the slowly changing variables. The latter variables move slowly on a manifold which passes through regions of instability of the pseudo-steady state. In these regions the trajectory moves rapidly in a manner determined by the product of feedback loops which destabilize the pseudosteady state.

Dr. R. Lefever (*Brussels*) said: I must say that the necessity of two counteracting feedbacks, positive and negative, in order for a chemical system to produce an oscillatory behaviour, is not obvious to me, or else that the definitions of negative and positive feedback are not so clear as it may seem. In the Brusselator, there is indeed a step, usually viewed as a positive (autocatalytic) feedback on the production of X, but then no step can be associated with what would usually be called a negative feedback. Indeed, what brings the concentration of X down during an oscillation, is the fact that, at the same time, the autocatalytic step produces X and consumes Y. This consumption occurs at a rate proportional to $X^2 Y$ which cannot be matched by the linear Y production step ($B + X \rightarrow Y + D$), if the concentration of X goes beyond a certain level. One thus sees that in a way the autocatalytic step is self inhibitory. On the other hand, following the author's rules since here both X and Y may have sharp breaks in the course of their oscillation, both variables should be considered of the positive feedback type.

As another illustration of the ambiguity between negative and positive feedback concepts, I may perhaps mention the Yates–Pardee type of enzymatic chain regulation;

$$A \xrightarrow{E_0} X_1 \xrightarrow{E_1} X_2 \xrightarrow{E_2} \ldots \rightarrow X_n \xrightarrow{E_{n-1}} B$$

In this chain of reaction, the first enzyme of the chain E_0 is inactivated by the end of chain product X_n. This is generally considered as a purely negative feedback type of control. Nevertheless limit cycle behaviour can be observed in such systems.

Dr. P. Rapp (*Cambridge*) said: In his paper, Franck notes the importance of time delays in allowing nonlinear feedback systems to oscillate. I would like to confirm

[1] B. L. Clarke, *J. Chem. Phys.*, 1974, **60**, 1481.

the importance of this observation and also connect points raised by Auchmuty and Lefever in the discussion of this paper. Auchmuty has pointed out that if the differential equation contains terms with retarded arguments, then it is possible for a system with a single variable to possess a stable periodic solution. As an example he cites the well known equation proposed by Hutchinson [1] as a model for a fluctuating population

$$\frac{dx(t)}{dt} = x(t)[1 - Kx(t-T)]. \tag{1}$$

This equation has been investigated by Cunningham [2] and Jones.[3] I would like to point out that the presence of time delays has an important effect on the stability properties of differential equations which describe phase shift oscillators mentioned by Lefever. Systems of this type consist of a sequence of reactions in which the first step is inhibited by the product of the last reaction. The most familiar class of differential equations describing these oscillations takes the following form (Goodwin,[4] and Walter [5]).

$$\dot{x}_1 = h(x_n) - b_1 x_1$$
$$\dot{x}_j = g_{j-1} x_{j-1} - b_j x_j \,;\, j = 2 \ldots n. \tag{2}$$

For eqn (2) it can be shown that for an arbitrary nonlinear function $h(x_n)$, $n \geqslant 3$ is required for periodic solutions (Higgins [6] and Rapp [7]). Eqn (3) is the governing equation corresponding to eqn (2), where delays occur between each reaction. The constants T_j are positive real numbers.

$$\left.\begin{array}{l} \dot{y}_1 = h(x_n) - b_1 y_1 \\ x_1 = y_1(t - T_1) \\ \\ \vdots \\ \\ \dot{y}_j = g_{j-1} x_{j-1} - b_j y_j \\ x_j = y_j(t - T_j) \quad ; \qquad j = 2 \ldots n \end{array}\right\} \tag{3}$$

Here it is possible to construct a one dimensional system which possesses a stable periodic solution. For example ;

$$\frac{dx}{dt} = \frac{k}{1 + \alpha x(t-T)^2} - bx$$

when $k = 30\,000$, $\alpha = 1$, $b = 1$, $T = 2$ oscillates from $x_{min} = 11.01$ to $x_{max} = 156.90$ with period $= 5.99$. Clearly this is an extreme case which is valuable primarily as an object lesson. However, such considerations are important when selecting between theoretical models of oscillatory processes. For example, it can be shown

[1] G. E. Hutchinson, *Ann, N.Y. Acad. Sci.*, 1948, **50**, 221.

[2] W. J. Cunningham, *Proc. Nat. Acad. Sci.*, 1954, **40**, 708.

[3] G. S. Jones, *Interlinear Symposium on Nonlinear Differential Equations and Nonlinear Mechanics*, ed. J. LaSalle and S. Lefschetz (Academic Press, New York, 1963).

[4] B. C. Goodwin, *Temporal Organization in Cells* (Academic Press, New York, 1963).

[5] C. Walter, *Biochemical Regulatory Mechanisms in Eukaryotic Cells*, ed. E. Kun and S. Grisolia, (Wiley Interscience, New York, 1971).

[6] J. Higgins, R. Frenkel, E. Hulme, A. Lucas and G. Rangazas, *Biological and Biochemical Oscillators*, ed. B. Chance, E. Pye, A. Ghosh, B. Hess (Academic Press, New York, 1973).

[7] P. Rapp, FEBS (Federation European Biochemical Societies) General Meeting Budapest, August, 1974.

that it is impossible to find a set of physically realizable reactions constants such that
the following system oscillates:

$$\dot{x}_1 = \frac{k}{1+\alpha x_4^4} - b_1 x_1$$

$$\dot{x}_2 = g_1 x_1 - b_2 x_2$$

$$\dot{x}_3 = g_2 x_2 - b_3 x_3 \qquad\qquad (4)$$

$$\dot{x}_4 = g_3 x_3 - b_4 x_4.$$

However, in the corresponding delayed system if $\varepsilon = T_1 + T_2 + T_3 + T_4 > 0$, then it
is possible to find reaction constants which give an oscillatory system.

Dr. H. Tributsch (*Berlin*) said: In his interesting review article Franck has demonstrated a remarkable similarity between artificial oscillating systems and a variety of
non linear biological phenomena. In this connection it is interesting to note that there
are apparently no simple model systems known which can simulate the property of
biological receptors of modulating the frequency of excitable membranes according
to the intensity of external physical and chemical stimuli.

The transformation of continuous signals into spike patterns (pulse signals) is an
important prerequisite for advanced data acquisition and processing methods. It
would, therefore, be of considerable interest to develop technical detectors with
digital response.

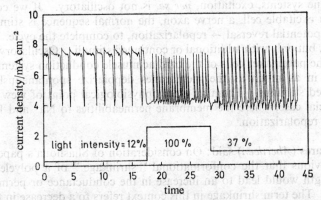

FIG. 1.—Influence of visible light on the oscillation behaviour of the system Cu_2S-electrode/elec-
trolyte + H_2O_2. Electrode potential : -0.5 V against S.C.E.

Our own experience with the development of a digital photodetector indicates that
efforts in this direction might be promising. To achieve our aim it was necessary to
find a suitable electrochemical oscillator with a light sensitive feedback mechanism.
This condition was met by the new oscillating system :

$$Cu_2S - \text{electrode/electrode} + H_2O_2.$$

The light sensitive reactant (absorption between 400 and 600 nm) turned out to be an
oxide bond on the electrode surface which is formed through the oxidizing action of
hydrogen peroxide and which can be destroyed by a photo-electrochemical reaction.
It is participating in an autocatalytical reaction which is crucial for the occurrence of
the electrode current oscillations. Light can consequently be used to modulate the
frequency of these oscillations (fig. 1).

Detectors of this type might become examples for a useful technical application
of oscillating mechanism.

Dr. M. Blank (*ONR, London*) said: The ionic mechanism proposed by Shashoua, to account for the properties of a polycation ↔ polyanion junction membrane, can also explain the oscillations we have observed in a much thinner bilayer system. We formed the bilayers from decane solutions of cholesterol in contact with two identical aqueous phases containing NaCl and cetyltrimethylammonium bromide (CTAB). The resistance of these bilayers depended strongly upon the concentration of CTAB.[1] At low CTAB concentration the bilayer behaves as if it were negatively charged (probably as a result of adsorbed anions), and as the CTAB concentration increases the bilayer becomes positively charged. The variation of the resistance with surface charge is in line with earlier experiments on monolayers.[2] On passing current across bilayers formed in a range of CTAB concentrations around the point of zero surface charge, we produced potential oscillations, which we believe are due to the variations in CTA^+ ion concentration that cause the bilayer to alternate between positive and negative surface charge. The concentration changes arise because of differences between the transport numbers of the aqueous phases and the bilayer. It should be possible to analyze the oscillatory behaviour of this bilayer system in terms of the physical properties of the components and the known behaviour of monolayer and interfacial systems. It may, therefore, serve as a model to elucidate the mechanism of oscillatory bioelectric phenomena involving ion flow in more complex biological systems.

I would also like to point out that although oscillatory phenomena exist in biological membrane systems, excitation, *per se*, is not oscillatory. If we consider the prototype of an excitable cell, a nerve axon, the normal sequence is stimulus → depolarization → potential reversal → repolarization, to complete the cycle. The cycle can be repeated, but only with additional or continued stimuli. Oscillatory behaviour can be seen in the pacemaker cells of the cardiac muscle conduction system, but these are specialized in as much as the membranes depolarize spontaneously with no externally applied stimulus. From an electrophysiological point of view these cells represent a special case, where the membrane permeabilities to Na^+ and K^+ ions do not stabilize on repolarization.

Prof. P. Meares (*Aberdeen*) said: On consideration of Shashoua's paper, it is not immediately obvious that the conformational " shrinkage " of a polyelectrolyte at high ionic strength would lead to an increase in the conductance or permeability of the membrane. The term shrinkage in this context refers to a decrease in the end-to-end vector of the polymer chains. Such a shrinkage does not require a decrease in the volume of the polymeric component such as would create pores. It is also uncertain whether the polyelectrolyte complex in the interaction zone of the $c \rightleftharpoons a$ junction would respond to changes in ionic strength in the same way as a purely polycationic or purely polyanionic material.

For these reasons I would like to offer for consideration an alternative mechanism of the periodic changes in conductance of the polyelectrolyte membranes. The material in the interaction zone is held together primarily by electrostatic attractions between polyanions and polycations (in this respect Ca^{2+} and Ba^{2+} can be regarded as polycations crosslinking polyanions by electrostatic forces). When, during the passage of a current, the salt concentration in the interaction zone is increased by the simultaneous arrival of Na^+ and Cl^- from opposite sides the osmotic pressure in the interaction zone must increase. A point will be reached when this osmotic pressure is sufficient to overcome the electrostatic binding forces between the polyanions and

[1] G. W. Sweeney and M. Blank, *J. Colloid Interface Sci.*, 1973, **42**, 410.
[2] I. R. Miller and M. Blank, *J. Colloid Interface Sci.*, 1968, **26**, 34.

polycations. These will then be forced apart, adopting the Na^+ and Cl^- ions respectively as their counterions, by the incoming water i.e. the polyelectrolyte complex will be dissociated. The resulting conductance increase would cause the membrane to depolarize rapidly. Subsequent reformation of the polyelectrolyte complex by diffusion of NaCl away from the interaction zone would lead to a relatively slow recovery of polarization and the cycle of events would then be repeated.

Dr. T. Keleti (*Budapest*) said: I would like to make a comment concerning one of the possible mechanisms of oscillations in biological systems. We are interested in the analysis of " three-body systems ", in which at least one substrate and two modifiers interact on the enzyme. These modifiers can be either one inhibitor and one liberator or two inhibitors.[1, 2]

The liberator is a modifier which has no effect on the free enzyme, but liberates the enzyme from the action of an inhibitor or activator.[3] A peculiar effect can be shown in the case of certain types of inhibitions and liberations. One of these types is illustrated in fig. 1: as a function of liberator concentration its effect changes from partial liberation (i.e., inhibition), through complete liberation to activation.[1, 3]

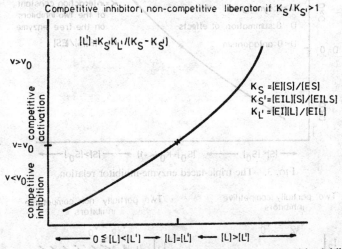

FIG. 1.—Interaction of a non-competitive liberator with a competitive inhibitor.

Similar effects can be shown in the case of double inhibitions, where different interactions between the substrate and/or inhibitors may appear (fig. 2).[4] One peculiar effect is the triple-faced enzyme-inhibitor relation. As a function of substrate concentration the antagonistic effect of the two inhibitors changes through the simple summation of their action to a synergetic effect (fig. 3).[2] Moreover, in the case of two partial inhibitors, if the concentrations of the inhibitors and the substrate are higher than the " characteristic " one, we obtain the inhibition paradox (fig. 4). In this case the initial rate of the enzyme in the presence of two inhibitors may be higher than in the absence of inhibitors.[1, 2]

[1] T. Keleti, in Proc. 9th FEBS Meeting. Vol. 32. *Symp. on Mechanism of Action and Regulation of Enzymes*, (Akadémiai Kiadó, Budapest and North–Holland, Amsterdam, 1975), ed. T. Keleti.

[2] Cs. Fajszi and T. Keleti, in *Mathematical Models of Metabolic Regulation*, ed. T. Krlei and S. Lakatos (Akademiai Kiado, Budapest, 1975).

[3] T. Keleti, *J. Theoret. Biol.*, 1967, **16**, 337–355.

[4] T. Keleti and Cs. Fajszi, *Mathemat. Biosci.*, 1971, **12**, 197.

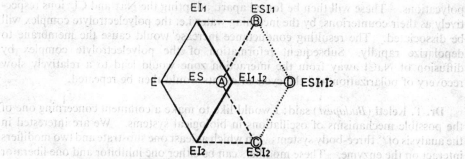

FIG. 2.—Scheme of the general mechanism of double inhibitions.

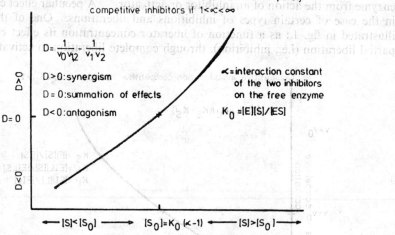

FIG. 3.—The triple-faced enzyme-inhibitor relation.

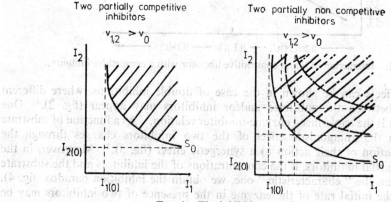

FIG. 4.—The inhibition paradox.

Assuming a linear chain of three enzymes, where the substrate of the first enzyme is an inhibitor or liberator and the product of the third enzyme is the inhibitor of the second one, we can obtain oscillations in the above-mentioned cases. This means that oscillations do not require the presence of an allosteric enzyme in the enzyme system.

Dr. H. Tributsch (*Berlin*) said: As most researchers in the field are aware, one must sometimes resist the temptation to interpret biological oscillations as true chemical oscillations. Periodic phenomena in complicated molecular structures such as sub-cellular and cellular systems may also be the consequence of a nonlinear energy conversion mechanism known as parametric energy coupling or variable parameter energy conversion. Parametric energy conversion proceeds through the periodical variation of an energy storing quantity (e.g., an electrical capacitance or an elastic constant) and is known to occur in all fields of physics both on the macroscopic and atomic levels (e.g., mechanism of a swing, movements of a pendulum on a spring, vibrating capacitor amplifiers or parametric amplifiers in electronics, nonlinear (laser) optics, the Raman effect, the parametric motor) (cf. N. Minorsky, ref. (1)).

An example of a parametric mechanism which could be operative in subcellular systems can easily be demonstrated with a physical model (fig. 1). If a pendulum is constructed with a piece of dielectric material and placed within the plates of a capacitor to which an alternating voltage is applied, it will—under certain conditions—start oscillating and reach a stationary state of constant amplitude. A periodical

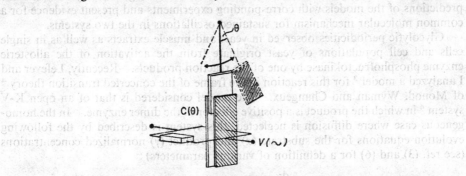

FIG. 1.—Dielectric model of phenomenon of Bethenod.

perturbation of a variable (energy storing) quantity (the capacitance) by a high frequency low amplitude oscillation can thus lead to the generation of a low frequency high amplitude (capacity) oscillation. This type of parametric phenomenon, which is difficult to understand in a purely intuitive way, was first described by Bethenod [2] with a model consisting of a coil carrying alternating current and a pendulum made of a piece of soft iron placed above it and has theoretically been investigated by Minorsky.[1] By analogy with the phenomenon of Bethenod, low amplitude relatively high frequency chemical oscillations (e.g., of the type reported by Shashoua in protein membranes) could be the actual cause for the occurrence of pronounced, low frequency (parametric) oscillations in subcellular (e.g., mitochondria) systems.

It is suggested that a search for these faster oscillations of much lower amplitude should be made since parametric energy conversion has recently been proposed [3] as playing a major role in membrane bound bioenergetical mechanisms.

Dr. A. Boiteux (*Dortmund*) (*communicated*): Tributsch's initial comment is perhaps based on a misunderstanding. We do not propose a sequence of simple chemical reactions as in glycolysis to cause oscillations in the mitochondrial system. On the

[1] N. Minorsky, in *Nonlinear Oscillations*, (D. van Nostrand Inc., Princeton, New York, 1962), p. 390 and p. 438.

[2] J. Bethenod, *Compt. Rend.*, 1938, 207.

[3] H. Tributsch, *J. Theor. biol.*, 1975, **52**, 847.

contrary, we have pointed out in our paper that the involvement of membrane permeability and vectorial transport processes provide the necessary non-linearity for this parametrically controlled system. Compare also ref. (14) p. 354, cit. : " The state of energy charge in the capacitor for the storage of chemical energy therefore is a controlling variable for the mitochondrial oscillation. "

Dr. A. Goldbeter (*Rehovot*) said: Metabolic oscillations controlled by cyclic AMP in the slime mould *Dictyostelium discoideum* [1] present a striking similarity to those observed in yeast glycolysis [2] (see the communication of Boiteux and Hess at this Symposium for a detailed presentation of these oscillatory systems). In both cases, models based on the molecular properties of the enzymes involved in the oscillatory mechanism indicate that periodic behaviour corresponds to a temporal dissipative structure, i.e., to sustained oscillations of unique amplitude and frequency around a nonequilibrium unstable stationary state (see the communication of Nicolis and Prigogine at this Symposium). The models account for several experimental observations in yeast and in the slime mould. In the following, I briefly compare the predictions of the models with corresponding experiments and present evidence for a common molecular mechanism for sustained oscillations in the two systems.

Glycolytic periodicities observed in yeast and muscle extracts as well as in single cells and cell populations of yeast originate from the activation of the allosteric enzyme phosphofructokinase by one of the reaction products. Recently, Lefever and I analyzed a model [3] for this reaction in the frame of the concerted transition theory [4] of Monod, Wyman and Changeux. The model considered is that of an open K-V system [4] in which the product is a positive effector of the dimer enzyme. In the homogeneous case where diffusion is neglected, this system is described by the following evolution equations for the substrate (α) and product (γ) normalized concentrations (see ref. (3) and (6) for a definition of various parameters) :

$$\frac{d\alpha}{dt} = \sigma_1 - \sigma_M \Phi$$

$$\frac{d\gamma}{dt} = \sigma_M \Phi - k_s \gamma$$

with

$$\Phi = \frac{\left(\dfrac{\alpha}{\varepsilon+1}\right)\left(1+\dfrac{\alpha}{\varepsilon+1}\right)(1+\gamma)^2 + L\theta\left(\dfrac{\alpha c}{\varepsilon'+1}\right)\left(1+\dfrac{\alpha c}{\varepsilon'+1}\right)}{L\left(1+\dfrac{\alpha c}{\varepsilon'+1}\right)^2 + (1+\gamma)^2\left(1+\dfrac{\alpha}{\varepsilon+1}\right)^2}.$$

The limit cycle behaviour of the model matches the oscillations observed in yeast extracts with a constant, periodic or stochastic source of substrate.[5] Qualitative and quantitative agreement is obtained in the model for the variation of period and amplitude in the oscillatory range of substrate injection rates which extends over one order of magnitude of parameter σ_1, for the phase-shift exerted by the reaction product ADP, and for the periodic change in enzyme activity.[3, 5] Thus in the middle of the

[1] G. Gerisch and B. Hess, *Proc. Nat. Acad. Sci.*, 1974, **71**, 2118.
[2] B. Hess, A. Boiteux and J. Kruger, *Adv. Enzyme Regul.*, 1969, **7**, 149.
[3] A. Goldbeter and R. Lefever, *Biophys. J*, 1972, **12**, 1302.
[4] J. Monod, J. Wyman and J. P. Changeux, *J. Mol. Biol.*, 1965, **12**, 88.
[5] A. Boiteux, A. Goldbeter and B. Hess, *Proc. Nat. Acad. Sci.*, 1975, **72** (10), in the press.
[6] A. Goldbeter and G. Nicolis, *Progr. Theoret. Biol.*, 1975, **4**, in the press.

unstable domain of σ_1 values the enzyme reaction rate v oscillates between 0.95 % and 73 % of the maximum rate V_M, with a mean value of 17.5 % V_M (fig. 1). Experimentally, Hess *et al.*[1] have reported a periodic variation of phosphofructokinase activity between 1 % and 80 % V_M, with a mean value of 16 % V_M and an activation factor of 80 comparable to the factor 77 obtained theoretically. Further analysis of the model shows that the coupling between limit cycle behaviour and diffusion can give rise to propagating concentration waves at the supracellular level.[2]

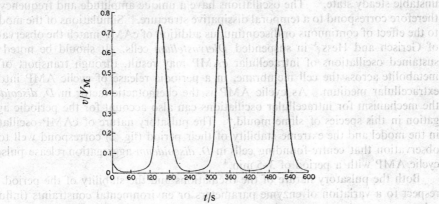

Fig. 1.—Periodic variation of enzyme activity in the concerted allosteric model for the phosphofructokinase reaction. The curve is obtained by integration of the evolution equations of α and γ for $\sigma_1 = 0.7/s$, $k_s = 0.1/s$, $\sigma_M = 4/s$, $L = 10^6$, $c = 10^{-5}$, $\varepsilon = \varepsilon' = 10^{-3}$, $\theta = 1$ (see text and ref. 6).

The model[3] for metabolic oscillations in *Dictyostelium discoideum*[4] is based on the regulation of two membrane-bound enzymes involved in the synthesis of cyclic AMP, namely ATP pyrophosphohydrolase and adenyl cyclase, which transform ATP into 5'AMP and cAMP, respectively (see ref. (8) and fig. 14 in the communication of Boiteux and Hess at this Symposium). The system is further coupled by a membrane-bound phosphodiesterase which transforms cyclic AMP into 5'AMP. Rossomando

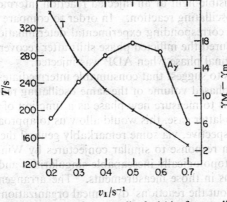

Fig. 2.—Variation of the period (T) and of the amplitude (A_γ) of normalized cAMP concentration in the oscillatory domain of ATP injection rates v_1, in the model for the oscillatory synthesis of cAMP in *D. discoideum* (see text and ref. (8)).

[1] B. Hess. A. Boiteaux and J. Kruger, *Adv. Enzyme Regul.*, 1969, **7**, 149.
[2] A. Goldbeter, *Proc. Nat. Acad. Sci.*, 1973, **70**, 3255.
[3] A. Goldbeter, *Nature*, 1975, **253**, 540.
[4] G. Gerisch and B. Hess, *Proc. Nat. Acad. Sci.*, 1974, **71**, 2118.

and Sussman have shown [1] that cAMP activates ATP pyrophosphohydrolase whereas 5'AMP activates adenyl cyclase. In both cases regulatory interactions are highly cooperative, pointing to the oligomeric structure of the enzymes.

The variables considered in the model are the intracellular concentrations of ATP, 5'AMP and cAMP. A stability analysis indicates that the cooperative and regulatory properties of adenyl cyclase and ATP pyrophosphohydrolase can give rise to sustained oscillations in the synthesis of cyclic AMP around a nonequilibrium unstable steady state.[2] The oscillations have a unique amplitude and frequency and therefore correspond to a temporal dissipative structure. Simulations of the model as to the effect of continuous or discontinuous addition of cAMP match the observations of Gerisch and Hess [3] in suspended *Dictyostelium* cells. It should be noted that sustained oscillations of intracellular cAMP may result, through transport of this metabolite across the cell membrane, in a periodic release of cyclic AMP into the extracellular medium. As cyclic AMP is the chemotactic factor in *D. discoideum*, the mechanism for intracellular oscillations can also account for the periodic aggregation in this species of slime mould.[4] The pulsatory nature of cAMP oscillations in the model and the extreme stability of their period (fig. 2) correspond well to the observation that centre-founding cells in *D. discoideum* aggregation release pulses of cyclic AMP with a period of 3–5 min.[4]

Both the pulsatory nature of the oscillations and the stability of the period with respect to a variation of enzyme parameters or environmental constraints (influx of substrate, protein concentration, etc.) result from the allosteric properties of the enzymes involved in the oscillatory mechanism, as in the case of phosphofructokinase for glycolytic oscillations. A further similarity between these oscillatory systems becomes apparent when noting that, in the absence of ATP pyrophosphohydrolase, the mechanism of cAMP-controlled oscillations in the slime mould reduces to that of oscillating glycolysis of yeast and muscle.[2] Thus the molecular basis of periodic behaviour is identical in the two metabolic systems : both phosphofructokinase and adenyl cyclase are allosteric enzymes under positive feedback control.

Dr. A. Winfree (*Indiana*) said: Hanusse, Noyes, Tributsch and Goldbeter have independently used visible light or an injected reaction intermediate to test the limit-cycle stability of an oscillating reaction. In order to compare his analytic model for glycolysis with Pye's corresponding experimental determinations, Goldbeter went a step further : he measured the inflicted phase shift after recovery to the limit cycle, as a function of the original phase when ADP was injected.

Is it too ambitious to suggest that consumable intermediates, light, thermal shock, or even a differently-phased volume of the same oscillating reaction could be used as transient perturbations to measure new phase as a function of old phase?

Particularly in the latter case, this would allow us to approach chemical dynamics from a quite new perspective, via some remarkably general theorems recently proved by Guckenheimer [5] in response to similar conjectures by Winfree. These theorems describe a variety of topologically-inescapable singularities and discontinuities which should be conspicuous in those measurements. The arrangement of these features reveals some thing about the reactions' dynamical organization : about the number of critical reactants, and of stationary states, about their stability properties, etc. Such

[1] E. F. Rossomando and M. Sussman, *Proc. Nat. Acad. Sci.*, 1973, **70**, 1254.
[2] A. Goldbeter, *Nature*, 1975, **253**, 540.
[3] G. Gerisch and B. Hess, *Proc. Nat. Acad. Sci.*, 1974, **71**, 2118.
[4] G. Gerisch, *Curr. Top. Devel. Biol.*, 1968, **3**, 157.
[5] J. Guckenheimer, *J. Math. Biol.*, 1975, **1**, 259.

measurements have begun to bear fruit in oscillating *biological* systems such as glycolysis [1] the circadian " clock " [2], and the mitotic cell cycle.[3]

Why not in physical chemistry?

Dr. Th. Plesser (*Dortmund*) said: Goldbeter's description of biochemical oscillations by the concerted model of Monod, Wyman and Changeux can be generalized very easily to the case of n promoters.

The linear stability analysis gives the following results: For all n there is only one stationary state in the positive quadrant of the phase plane. If the product is an allosteric inhibitor of the enzyme, then the stationary state is always stable. If the product is an allosteric activator, then the stationary state may be stable or unstable and give rise to self excited oscillations. The stationary state in the positive quadrant can never be a saddle point, if the input rate of substrate is lower than the maximum velocity of the enzyme. If the number of protomers is odd, there is only one stationary state. For even n, there is a second one for negative and therefore unphysical concentration. Numerical analysis shows, that the oscillatory domain is not very sensitive to the number of protomers.

[1] A. T. Winfree, *Arch. Biochem. Biophys.*, 1972, **149**, 388.
[2] A. T. Winfree, *Nature*, 1975, **253**, 315.
[3] S. A. Kauffman, private communication.

D. Theory of Excitable Media

Excitability and Spatial Order in Membranes of Developing Systems

By B. C. Goodwin

School of Biological Sciences, University of Sussex, Brighton, Sussex BN1 9QG

Received 15*th July*, 1974

A simple biological model of metabolic activities on membranes is explored with respect to excitability and spatial gradient formation. The model is shown to have the property of undergoing a phase transition from a non-excitable to an excitable condition. Once excitable, it is shown that a spatially-stable gradient can be formed on the membrane. Periodic gradients can be formed by a slightly modified process.

---

1. INTRODUCTION

The primary concern of this paper is the exploration of some temporal and spatial consequences of certain very elementary properties of biological membranes. The term "elementary" is employed in a biological, not a physico-chemical, sense. Thus it is taken as axiomatic that biological membranes are complex lipo-protein structures and that the proteins are as important as the lipids. This is the emphasis that has emerged in recent years as a result of the molecular biological approach to membrane structure and activity.[1] It is, therefore, assumed from the outset that membranes contain enzymes, and proteins capable of reacting with specific ligands. Then, on the basis of some very simple postulates about interactions between these proteins a model of metabolic excitability as a phase transition is presented; ultimately it is shown how the postulates give rise to processes which result in either monotonic or periodic gradients of "morphogens" on the membrane.

The reason for concentrating attention on biological membranes as the sites of spatial order in developing organisms is provided by experimental evidence which points very strongly in this direction.[2-6] Furthermore, unicellular organisms show essentially the same morphogenetic capacities and behaviour as do the multicellulars, so that a basic understanding of morphogenetic mechanisms must be arrived at independently of assumptions about cellular partitions. These points have been argued out more fully in a paper by Goodwin and McLaren[7] in which the basic features of the model considered in this paper have been presented, together with applications to particular developmental processes. The present goal is to look at membrane excitability from a particular perspective, and to explore some further consequences of the basic model, thus extending its range of applicability.

2. EXCITABILITY AS A PHASE TRANSITION

The type of phenomenon now considered is the rather sudden appearance of spontaneous action potentials during the regeneration of the unicellular marine alga, *Acetabularia*, as reported by Novak and Bentrup.[8] Other developing systems show similar types of state transition also, but the electrophysiological behaviour of *Acetabularia* is particularly clearly defined. Whether this change of membrane state is correctly described as a phase transition of the type here described requires closer experimental study, but the model draws attention to this possibility, and hence to the eventuality that such transitions are an important aspect of development.

OUTSIDE

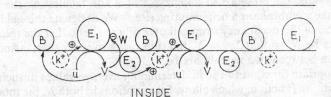

INSIDE

FIG. 1.—Model of interactions between enzymes, E_1 and E_2, metabolites, U, V, and W, and the ionic species K^+, which can result in propagating metabolic activity waves and the formation of a morphogenetic gradient on a membrane. For explanation see text.

It is convenient first to describe the molecular organization within the membrane which characterizes the metabolically excitable condition underlying the model. This is shown in fig. 1. The circles labelled E_1, lying within the membrane, represent enzymes whose active sites face the interior of the cell. They catalyze the conversion of a metabolite U into another metabolite, V. A second enzyme, E_2, which is either a soluble enzyme or is loosely associated with the membrane, converts V into W. It is assumed that an ionic species, represented in the figure by K^+, is in equilibrium between a membrane bound state and a free state, and that V has the effect of displacing this equilibrium towards the free state. The ion is an activator of E_1, so that a positive feed-back loop is created which introduces an instability into the kinetics. W, on the other hand, is an inhibitor of E_1, so that the control circuit can be stabilized. A kinetic system of this type belongs to the same category of reaction system as the glycolytic oscillator, extensively studied by Hess and his colleagues,[9,10] and to the general class of kinetic processes to be considered by Franck [11] in this symposium. Since there are many possible kinetic realizations of such a system, there is no need to make particular assumptions; one simply supposes that propagating waves of metabolic activity can occur providing certain constraints are satisfied. Several contributions to this Symposium are concerned with the definition of these constraints.

It is evident that one condition which must be satisfied in order for there to be wave propagation in such a system is that the density of enzymes, E_1, within the membrane must be sufficiently large. The exact value will depend upon the stoichiometries of the reactions, the diffusion constants of the reacting species and of the activating ion, the temperature, etc. The density or the concentration of E_1 in the membrane will depend upon the balance between the rate of its incorporation into the membrane and its rate of loss or dilution (if new membrane is being formed). All these factors will be represented by means of a biochemical potential, μ, so that the concentration of enzyme is determined by this quantity. We may assume that E_1 is effectively inactive until it takes the spatial configuration resulting from incorporation into the membrane, so that the cytoplasm is not excitable.

It is assumed that there are sites for enzyme E_1 on the membrane and that their mean number per unit area is constant. The binary variable ε_i is used to designate the state of site i on the membrane. This variable takes the value 0 or 1 according as the site is unoccupied or occupied respectively. The model now follows closely the lattice gas description of the process of condensation, which is in turn based upon the Ising model of ferromagnetism, using the mean field approximation.[12, 13] This assumes that each unit in the system under consideration (in our case, an enzyme) experiences a mean field of force due to the presence of other units, and the essence of the procedure is to find an equation which expresses this field in terms of itself (the self-consistent field equation). The field under consideration in this model is assumed to arise from interactions between enzymes occupying membrane sites and those entering the membrane, the presence of an enzyme at a site facilitating the entry of an enzyme into the membrane at a neighbouring site. We designate this field as $\rho = \langle \varepsilon_i \rangle$, the mean value of the occupancy averaged over all the sites. Let the total number of occupied sites be N, so that $\Sigma \varepsilon_i = N$, the sum being over all sites. And finally, let the interaction between two neighbours be u_{ij}.

The appropriate function to use in determining average values in such a system is the grand partition function, which allows for variations in both N, the total number of active enzymes on the membrane, and in the energy of the system resulting from interactions. The Hamiltonian for such a system is then

$$H = -\tfrac{1}{2}\Sigma u_{ij}\varepsilon_i\varepsilon_j - N\mu. \tag{2.1}$$

The self-consistent equation for ρ which is obtained from this expression via the grand partition function is

$$(2p-1) = \tanh[\beta/2(\rho u(0) + \mu] \tag{2.2}$$

where $u(0)$ is the Fourier transform of Σu_{ij} taken at the point 0. The assumption here is that the interaction field is everywhere the same, $u_{ij}(R)$ being translation invariant. The parameter β is $1/kT$. Thus it is assumed that the system is at a defined temperature. This temperature, in fact, implies that the relevant processes inolved are quasi-equilibrium ones, in accordance with the use of a biochemical potential to represent the factors which determine the concentration of enzyme E_1 in the membrane. Other processes such as those involved in the maintenance of the membrane and of a constant pool of precursor, U, are taken as constant parameters of the system.

The consequences of this model are as follows. There is a region of parameter space in which eqn (2.2) has a single root and another region where it has three. The latter condition, which is the interesting one for phase transitions, requires that $\mu = -u(0)/2$, whence eqn (2.2) becomes

$$(2\rho-1) = \tanh[\tfrac{1}{4}\beta u(0)(2\rho-1)]. \tag{2.3}$$

For sufficiently large β this equation has three roots in ρ, one of which is always $\rho = \tfrac{1}{2}$ and the other two are symmetric about this. The negative value of the potential in this region has no physical significance, since this quantity is defined relative to an arbitrary constant.

Plotting ρ as a function of T, one gets a curve such as that shown in fig. 2. The region of triple roots lies below T_c, the critical temperature, and the curve shown in this region is called the coexistence curve. Above T_c, no excitability is possible, but below it, a transition from the non-excitable to the excitable state can occur. Thus if the membrane is in the state shown by P and ρ is then increased (e.g., more enzyme is incorporated into the membrane) then the system can move to P', an unstable point on the coexistence curve. A sudden state transition to Q' can then occur, so that the membrane becomes excitable. As the temperature is decreased, this transition can occur at smaller values of ρ; i.e., thermal noise interferes less and less with the processes resulting in excitation.

Now clearly in a biological system there will also be a lower temperature bound to the domain of excitability as well. Thus the coexistence curve should close on itself again. This could be modelled by making further assumptions about, say, the temperature-dependence of the quantities $u(0)$ and μ at lower values of T. However, it is far from clear as yet that we are dealing with true phase transitions in excitability phenomena, so a further analysis does not seem worthwhile. The point of this model is to demonstrate the possibility of such behaviour in membranes. The analysis could also be extended to a consideration of metastable regions of the coexistence domain, corresponding to super-cooled and super-heated states. These would be conditions where very small fluctuations cause either excitation or the loss of excitation, respectively. The former might describe the state of membranes which show latent excitability, responding to small stimuli, but are not recurrently active as in a pacemaker system.

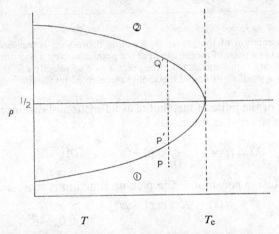

FIG. 2.—Coexistence curve for non-excitable (1) and excitable (2) states on a membrane with molecular interactions of the type shown in fig. 1. The region above T_c corresponds to a condition in which no transition to excitability can occur, but below T_c such transitions are possible. The dotted line from P' to Q' describes such a transition, which involves a rapid change in the variable ρ, the fraction of the total number of sites which is occupied.

3. EXCITABLE MEMBRANES AND SPATIAL GRADIENTS

Attention is now directed to certain spatial consequences of the model described in fig. 1. It is of interest to ask under what conditions activity waves can leave " memory " traces on membranes in the form of gradients of a metabolite. Such a process would clearly be of significance in relation to both morphogenetic and neural activities. A simple mechanism for achieving this is to assume the existence in the membrane of proteins with binding sites for the metabolite V. These are shown as circles marked B in fig. 1. In order to get a graded distribution of bound V, hereinafter referred to as morphogen, we may assume that these sites are activated after the production of V by E_1. Thus, for example, we could suppose that W is an activator of B, as well as an inhibitor of E_1. This temporal asymmetry of the processes which constitute the propagating wave of activity on the membrane results in a graded distribution of morphogen with its maximum at the point of origin of the wave.[7] To illustrate the principles of the model, consider the following local representation. A pulse of metabolite of magnitude M is produced at the point $x = \rho$ on a one-dimensional membrane, as shown in fig. 3. The wave of activation of the binding sites is represented by a pick-up function which starts at $x = \rho$ and propagates with velocity v in the positive direction. We may then ask what the distribution of morphogen will be after the passage of the wave. The correct representation of the process involves the propagation of the pulse, M, as well, as is done in another publication,[7] but its essence is illustrated by the local model.

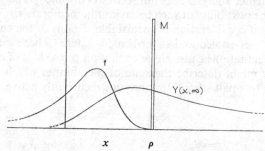

FIG. 3.—Local representation of the gradient-generating process on a membrane. The diffusible metabolite is produced in a pulse of magnitude M at the point $x = \rho$ on a one-dimensional membrane and the pick-up function (f) propagates from the point $x = \rho$ at velocity v in the positive direction. The result is a gradient of bound metabolite represented by the function $Y(x, \infty)$.

The behaviour of the pulse of metabolite as a function of distance from the origin, x, and time, t, is given by the function

$$X(x, t) = \frac{M}{\sqrt{4\pi Dt}} \exp(-(\rho-x)^2/4Dt) \qquad (3.1)$$

where D is the diffusion constant. The pick-up function is represented by

$$
\begin{aligned}
f(s) &= K\alpha s \exp(-\alpha s) && s \geqslant 0 \\
&= 0 && s < 0
\end{aligned}
\qquad (3.2)
$$

where s is the distance from the wave-front, α is a parameter determining the distance from the wave-front to the maximum of $f(s)$ which is Ke^{-1} at $s = 1/\alpha$, and K is the product of the concentration of binding sites and a rate constant for bound complex formation. The expression for the morphogen concentration, $Y(x, t)$, assuming that the bound complex is stable and that $Y(x, 0) = 0$ is

$$Y(x, t) = \frac{M\alpha K}{\sqrt{4\pi D}} \int_{\tau_0}^{t} d\tau (v\tau + \rho - x)\tau^{-\frac{3}{2}} \exp(-\alpha(v\tau + \rho - x) - (\rho - x)^2/4D\tau). \qquad (3.3)$$

where the lower limit

$$
\begin{aligned}
\tau_0 &= 0 && \text{for } x \leqslant \rho \\
&= x - \rho && \text{for } x > \rho.
\end{aligned}
$$

Defining $\beta(\rho - x)/v$ and $\gamma = (\rho - x)^2/4D$, this becomes

$$Y(x, t) = C \int_{\tau_0}^{t} d\tau (\tau + \beta)\tau^{-\frac{3}{2}} \exp(-\alpha v(\tau + \beta) - \gamma/\tau)$$

where

$$C = \frac{M\alpha v K}{\sqrt{4\pi D}}.$$

For the case $\tau_0 = 0$, letting $t \to \infty$ this integral is [14]

$$Y(\xi, \infty) = \frac{MK}{4\sqrt{\alpha v D}}\left[1 + \xi\left(2\alpha + \sqrt{\frac{\alpha v}{D}}\right)\right] \exp\left(-\left(\alpha + \sqrt{\frac{\alpha v}{D}}\right)\xi\right) \qquad (3.4)$$

where $\xi = \rho - x \geqslant 0$. For $x > \rho$ the integral does not permit a simple closed form, but it is easily shown that $Y(x, \infty)$ decreases monotonically. The peak of morphogen occurs at a distance $\xi_0 = \alpha/\sigma(\alpha + \sigma)$ (where $\sigma = \alpha + \sqrt{\alpha v/D}$) to the left of the origin of the pulse, as shown in fig. 3. Biologically plausible values of the parameters are $v = 8 \ \mu m \ s^{-1}$, $\alpha = 0.025 \ \mu m^{-1}$ and $D = 200 \ (\mu m)^2 \ s^{-1}$, giving $\xi_0 = 5.3 \ \mu m$.

Now if the temporal asymmetry is reversed so that the activation of the binding sites precedes the activation of enzyme and production of V, then instead of (3.3) one obtains for Y the expression

$$Y(x, t) = \frac{M\alpha K}{\sqrt{4\pi D}} \int_{\tau_0}^{t} d\tau (v\tau + \rho - x)\tau^{-\frac{3}{2}} \exp\left(-\alpha(v\tau + \rho - x) - \frac{x^2}{4D\tau}\right).$$

This represents a situation in which the pulse of metabolite occurs at the origin, $x = 0$, while the pick-up function starts as before at $x = \rho$. Using the same procedure as previously, one finds for the case $\tau_0 = 0 \ (x < \rho)$

$$Y(x, \infty) = \frac{MK}{4}\sqrt{\frac{\alpha}{vD}}\left[\frac{1}{\sqrt{2\alpha}} + (\rho - x) + x\sqrt{\frac{v}{2D}}\right] \exp\left(-\alpha\rho + x\left(\alpha - \sqrt{\frac{\alpha v}{D}}\right)\right). \tag{3.5}$$

This function increases monotonically up to ρ, so that the maximum occurs for a value of $x > \rho$. Thus we see that the temporal relationships between metabolite production and pick-up determine which way the gradient of morphogen is formed.

4. PERIODIC DISTRIBUTION OF MORPHOGEN

In order to have spatially stable distributions of morphogen on membranes, it is necessary to have well-defined, self-stabilizing origins of wave initiation. This requires a further postulate in the model, but this is easily provided by a consideration of how the morphogen is likely to act. It is reasonable to suppose that, directly or indirectly, it affects membrane properties, as discussed more fully elsewhere.[15] One such effect could be on the concentration of free K^+ itself. If the morphogen increases this concentration, then the wave will tend to recur at the wave origin, where the morphogen concentration is maximal. The kinetics of the system described in fig. 1 can then result in regularly recurring activity waves, which will maintain dynamically a spatial gradient of morphogen.

A spatially periodic distribution of morphogen can result from the recurrent propagation of metabolic activation waves and binding-site activation waves if these propagate at different velocities from a common origin. For the waves will then pass one another at defined points on the membrane, where their temporal relationships will be reversed. Thus the gradient will form in opposite directions on either side of these passing points. However, in order for the waves to propagate independently of one another at different velocities, it is necessary that the processes involved in metabolite production be different from those involved in binding site activation. Therefore, the picture of fig. 1 must be modified so that each process has kinetics of the general type described for metabolite production. This could be achieved by assuming another pair of enzymes coupled as are E_1 and E_2, with an ion activator distinct from K^+. Then both these processes could be initiated by a spatially-localized pacemaker region, established by a gradient formed by a previous wave process. This complication of membrane-localized activities in no way overburdens this organelle with specific structure, since a mitochondrial membrane, for example, is very much more complex than this.

The biological attraction of the above model is that it provides a plausible mechanism for both aperiodic and periodic gradient formation in either unicellular or multi-cellular organisms. Furthermore, having established a wave origin, the spatially periodic process is a reliable one, unlike the difficulties encountered by the Turing model in this respect. However, although aperiodic gradients generated by the model regulate (are size-independent),[7] the periodic gradients do not; so here the model fails biologically, as does also the Turing process.

This paper was written while the author held a Leverhulme Visiting Professorship at the Department of Theoretical Physics, National University of Mexico, and grateful acknowledgement is made to the Royal Society and to Professor M. Moshinsky, Head of the Department, for assistance and hospitality.

[1] C. Gitler, *Ann. Rev. Biophys. Bioeng.*, 1972, **1**, 51.
[2] A. S. G. Curtis, *The Cell Surface* (Academic Press, London, 1967).
[3] H. F. Stumpf, *Dev. Biol.*, 1967, **16**, 144.
[4] D. L. Nanney, *Science*, 1968, **160**, 496.
[5] L. S. Sandakhchiev, L. I. Puchkova and A. V. Pikalov, *Biology and Radiobiology of Anucleate Systems*, ed. S. Bonnotto, 1972, p. 297.
[6] W. Herth and K. Sander, *Wilhelm Roux' Arch. Ent. Org.*, 1973, **172**, 1.
[7] B. C. Goodwin and D. I. McLaren, *J. Theor. Biol.*, 1975.
[8] B. Novak and F. W. Bentrup, *Planta* (Berl), 1972, **108**, 227.
[9] B. Hess, H. Kleinhaus and D. Kuschmitz, *Biological and Biochemical Oscillators*, ed. B. Chance, and K. Pye (Academic Press, New York, 1973), p. 253.
[10] A. Boiteux and B. Hess, paper at this Symposium.
[11] U. F. Franck, paper at this Symposium.
[12] R. Brout, *Phase Transitions* (Benjamin, New York, 1965), Chap. 2 and 3.
[13] H. E. Stanley, *Introduction to Phase Transitions and Critical Phenomena* (Clarendon, Oxford, 1971), chap. 6 and appendix A.
[14] I. S. Gradsteyn and I. M. Ryzhik, *Tables of Integrals, Series, and Products* (Academic Press, New York, 1965), p. 340.
[15] B. C. Goodwin, *Analytical Physiology of Cells and Developing Organisms* (Academic Press, 1976), chap. 5.

Unified Theory of Temporal and Spatial Instabilities

By I. Balslev

Institute of Physics

AND

H. Degn

Institute of Biochemistry,
Odense University, Denmark

Received 29th July, 1974

Chemical reaction-diffusion systems are discussed with the view of clarifying the relation between spatial and temporal concentration variations. The predicted absence of oscillating or propagating structures in closed two component systems, which are not oscillatory when stirred, points to the necessity of studying systems with more than two components.

1. INTRODUCTION

The occurrence of concentration oscillations in homogeneous chemical systems was a matter of controversy for several decades until the sixties, when there came a boom in experimental and theoretical work on the subject. The existence of homogeneous chemical oscillations then became firmly established. This had hardly happened before the logical extension of temporal periodicity, namely spatial periodicity, took over the role of the flying saucers of chemistry.

The possibility of the emergence of a spatial concentration pattern in an initially homogeneous chemical system seems to have first occurred to Turing.[1] He was led to the idea when searching for an explanation of the stage in morphogenesis when a small spherical array of identical cells assigns different parts of its surface to develop into different organs. Turing's work, published in 1952, only became famous after 1967 when the Brussels school began the publication of a series of papers on spatial instability in homogeneous chemical systems. The report by Zhabotinskii[2] on spatial phenomena in the Belousov reaction at the Prague Symposium on Oscillating Reactions in 1968 inspired experimental work on spatial patterns in many laboratories. Despite the parallel courses of the theoretical and experimental approaches, it is not always evident that the theoreticians and the experimentalists talk about the same thing.

The pretentious title of this paper reflects the fact that temporal and spatial instabilities are closely related phenomena as they both require reaction schemes with feedback. The aim of the paper is to clarify this relationship and to give operational interpretations of theoretical results regarding spatial instabilities and spatial periodicity.

The concept of concentration oscillations in a homogeneous phase is so straightforward that it needs no special introduction. Spatial periodicity, on the other hand, is an equivocal expression covering different phenomena which may be easily confused. Before we proceed to theoretical analyses we shall therefore classify these phenomena in operational terms.

Imagine a solution in an ordinary vessel made of non-reactive material. We stir the solution and it does not exhibit time periodicity as long as it is stirred. When we stop the stirring, conceivably we may observe that concentration gradients form spontaneously in the solution. These gradients may tend to form a time-invariant structure or they may be periodic in time either as propagating or as oscillating standing concentration waves.

A reaction solution which forms spatial structures after it was homogenized by stirring and did not exhibit concentration oscillations while stirred, is said to be *spatially unstable*. Current theory predicts spatial instability in reaction systems which comprise diffusion of chemical components and certain types of chemical reactions with feed-back loops. In the discussion below systems which produce time-invariant structures are said to have *aperiodic spatial instability* and systems giving rise to concentration waves are said to have *oscillatory spatial instability*.

Let us now return to an imaginary experiment where a solution in an ordinary vessel with non-reacting walls exhibits concentration oscillations when stirred. When we stop the stirring, conceivably we may observe the emergence of concentration structures which are oscillating with the same period as the previous homogeneous oscillation. In this case spatial instability is not involved.[3] Time-dependent structure can be explained by small local heterogeneities or externally induced gradients causing different parts of the oscillating solution to come out of phase. Such patterns in oscillating systems will tend to be destroyed by diffusion of the reaction components of the chemical system.

This paper is concerned with structures and waves which result from spatial instability and thus grow and are maintained with the assistance of diffusion.

2. TWO-COMPONENT SYSTEMS

We consider a system in which two substances A and B are present with concentrations $A(r, t)$ and $B(r, t)$ as function of space r and time t. The relevant reactions are assumed to be fully described by two rate equations

$$\frac{\partial A}{\partial t} = f_A(A, B) + D_A \nabla^2 A$$

$$\frac{\partial B}{\partial t} = f_B(A, B) + D_B \nabla^2 B$$

(1)

where f_A and f_B describe the chemical reaction rates, D_A and D_B are diffusion coefficients and ∇ is the gradient operator. We shall first neglect the spatial dependence of A and B by assuming stirring or large diffusion coefficients.

As shown in several works a linear theory provides important information on the dynamics near singular points (A_s, B_s) in the A-B phase plane where

$$f_A(A_s, B_s) = f_B(A_s, B_s) = 0.$$

For small deviations from the singular point

$$a(t) = A(t) - A_s$$
$$b(t) = B(t) - B_s$$

one obtains

$$\frac{da}{dt} = m_{AA}a + m_{AB}b$$

$$\frac{db}{dt} = m_{BA}a + m_{BB}b$$

(2)

where the coefficients m are partial derivatives

$$m_{PQ} = \frac{\partial f_P}{\partial Q} \tag{3}$$

in the singular point.   Solutions to eqn (2) may be written

$$a = a_0 \exp((\mu - i\omega)t)$$
$$b = b_0 \exp((\mu - i\omega)t) \tag{4}$$

where μ is the growth rate and ω is the angular frequency.   The complex growth rate $\lambda = \mu - i\omega$ is then determined as the roots of the secular equation

$$(m_{AA} - \lambda)(m_{BB} - \lambda) - m_{AB}m_{BA} = 0. \tag{5}$$

Let us denote these roots by $\lambda_1 = \mu_1 + i\omega$ and $\lambda_2 = \mu_2 - i\omega$.   For $\omega \neq 0$ we have $\mu_1 = \mu_2$ and we shall in this case denote the singular point as *oscillatory unstable* if $\mu_1 > 0$ and *oscillatory stable* if $\mu_1 < 0$.   For $\omega = 0$ the singular point is *aperiodically stable* if both μ_1 and μ_2 are negative.   Two positive values correspond to *aperiodical instability* and different signs lead to a *saddle type instability*.   In the later discussion of spatial effects this notation is slightly more suitable than the conventional terms : focus, node and saddle.

As it turns out that the case $m_{AB}m_{BA} \geqslant 0$ is uninteresting in connection with spatial effects included later, we shall assume $m_{AB}m_{BA} \leqslant 0$.   In a qualitative discussion it is convenient to scale the time unit and the unit for the concentrations a and b.   In such rationalized units the complex growth rate is obtained from solutions to

$$(\alpha - \lambda)(\beta - \lambda) + 1 = 0 \tag{6}$$

where

$$\alpha = m_{AA}/(-m_{AB}m_{BA})^{\frac{1}{2}}$$
$$\beta = m_{BB}/(-m_{AB}m_{BA})^{\frac{1}{2}}. \tag{7}$$

The solutions of eqn (6) are conveniently represented in the $\alpha\beta$ parameter space with domains for different types of dynamics according to the values of λ_1 and λ_2.   The domains are shown in the $\alpha\beta$ diagram in fig. 1.   This graphical representation is useful in the analysis of specific chemical reaction schemes.   For fixed rate constants the scheme defines a single point (α, β).

For example, in the reaction scheme of Lotka [4] (1910) the points obtained for different values of rate constants fall on the negative β-axis and the Volterra–Lotka scheme [5] has always $(\alpha, \beta) = (0, 0)$.   In the reaction scheme of Prigogine and Lefever [6] it is possible to obtain points in the region between the hyperbola branches and the axes.   This latter scheme and reaction systems involving forward inhibition or an enzymatic step were analyzed previously [7] by transforming the $\alpha\beta$ diagram into a parameter space directly related to the chemical rate constants.

Including the spatial dependence of A and B of eqn (1) and assuming a spatially periodic solution we find in the linear theory

$$a(r, t) = a_0 \exp(\mu t - i\omega t + i\mathbf{k}.\mathbf{r})$$
$$b(r, t) = b_0 \exp(\mu t - i\omega t + i\mathbf{k}.\mathbf{r}) \tag{8}$$

where k is the wave vector defining a propagation direction and a wave length of $2\pi/|\mathbf{k}|$.   Introducing a scaled unit for length we obtain the secular equation

$$(\alpha - D_A k^2 - \lambda(k))(\beta - D_B k^2 - \lambda(k)) + 1 = 0 \tag{9}$$

where k is the magnitude of k and $\lambda(k) = \mu(k) - i\omega(k)$ is the scaled complex growth rate for wave vector k.

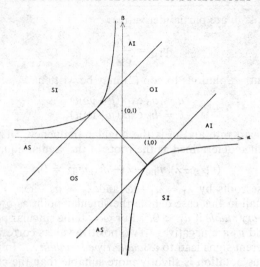

FIG. 1.—Domains in the $\alpha\beta$-plane for different types of dynamics near singular points in two.
component systems. OS and AS correspond to oscillatory and aperiodic stability, respectively-
The character of instabilities are abbreviated to OI, AI and SI corresponding to oscillatory, aperiodic
and saddle type instability, respectively.

Comparing eqn (6) and (9) we note that the dynamics of periodic patterns are
obtained by subtracting $D_A k^2$ and $D_B k^2$ from α and β, respectively, corresponding to a
displacement in fig. 1 towards negative parameter values along a line with slope D_B/D_A.
These considerations lead to a classification of spatio-temporal behaviour near singular points, shown as domains I-III in fig. 2. On the left hand side of these domains
there is global stability and on the right hand side, the long wave limit is aperiodically
unstable. In the latter case, the non-linear terms are generally unable to prevent a

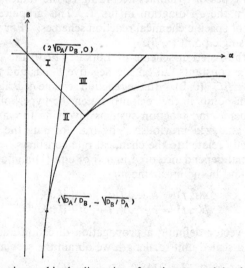

FIG. 2.—Instability domains used in the discussion of spatio-temporal development beyond the linear
approximation. The slope of the straight line is D_B/D_A. Consequently no lines with this slope
through domain I intersects the domain for saddle type instability to the left of the hyperbola. In the
figure $D_B/D_A \approx 9$.

global drift away from the singular point. In domains I and III the long wave limit* is oscillatory unstable while domains II and III have aperiodically growing solutions in a finite wave vector interval. Domain II and III vanish unless the self-activating component (here the A-component) is diffusing slower than the other component. The necessity of having $D_B/D_A > 1$ for obtaining aperiodical growth for finite wave vectors has been clearly stated by Edelstein [8] and by Segel and Jackson.[9]

The discussion of the further development of unstable systems beyond the linear approximation is limited to systems which do not drift away to other singular points and to closed systems having zero gradient of all concentrations perpendicular to the to the surface of the system. Such boundary conditions allow all space-independent solutions to the rate equations.

Under these conditions, the nature of the final state is closely related to the position in the parameter space of the point (α, β) (see fig. 2).

In Domain I-systems the homogeneous solution ($k = 0$) has the largest growth rate and is allowed by the boundary conditions. The final state of such systems is a homogeneous limit cycle.

In Domain II the long wave limit is stable and there is saddle type instability in a limited wave vector range $k_1 < k < k_2$. If the boundary conditions allow the presence of Fourier components in this range, the system develops into states with large deviations of the steady state. So far the ultimate destiny of such systems has not been analysed in general. Numerical integrations [1, 6, 10] on specific reaction schemes seem to indicate that the above situation generally leads to stable structures. However, at present it is not established, if a system with specific spatial and chemical parameters allows a multiplicity of different stable patterns, and to what extent these are accessible from a a slightly perturbed homogeneous steady state. From the linear theory it is obvious that below a critical size $L_c \approx \pi/k_2$ of the largest linear dimension there is spatial stability, k_2 being the largest wave vector with positive growth rate. In the case of much larger systems it would be interesting to search for general trends concerning the favoured spatial wave length of the final pattern.

Oscillatory instability in the long wave limit and aperiodic growth of patterns is simultaneously present in Domain III (see fig. 2). Numerical integration in one dimension on such systems indicates that homogeneous temporal oscillations are strongly favoured if the concentration gradients are zero at the boundaries.[10] In fact, the existence of any form of stable patterns in closed Domain III systems is questionable.

We can summarize that in closed two-component systems spontaneous oscillations tend to be homogeneous and spatial instabilities will develop into time-independent structures.

For the experimentalist the most pertinent result of the above analysis is first that the formation of stable spatial structures may occur from an initially homogeneous solution. Secondly, the interpretation of observed propagating structures in systems which are not oscillating when stirred cannot be based on two-component models. Thirdly, in order to obtain spatial instability the self-activating component must diffuse considerably slower than the other one.

In the search for systems capable of forming time-independent patterns one should look for reaction systems where an autocatalytic component is considerably more bulky than its precursor. This condition may not be fulfilled in the Belousov reaction where the autocatalytic species almost certainly is a fragment such as BrO_2 of the bromate ion.[11] Since bromate is present in excess it should not be considered as a variable component. The functional precursor of the autocatalytic species in the

---

* The long wave limit corresponds to homogeneous solutions.

Belousov reaction is cerous ion. We have no data on the diffusion coefficients of these substances. However, cerous ion is probably the bulkier one because of its tendency to form complexes. In the Bray reaction the autocatalytic substance is probably a fragment of the iodate ion and its functional precursor is molecular iodine. The large mass of molecular iodine and, possibly, its tendency to form complexes suggest a large diffusion coefficient compared to that of an IO_n species. Thus, neither of the systems mentioned above seem to have ratio of diffusion coefficients required for spatial instability in a two-component model.

3. THREE- AND FOUR-COMPONENT SYSTEMS

The absence from two-component systems of oscillatory spatial instabilities is believed to be a consequence of their simplicity rather than a general trend in reaction-diffusion systems. This suggests that models with more than two components should be investigated.

We have studied a special class of three- and four-component systems, namely those containing a chain of first order reactions with a feed-back on the first reaction from the last intermediate in the chain. Such systems can be symbolized as follows

$$\begin{array}{c} \downarrow \hspace{1em} | \\ \rightarrow A \rightarrow B \rightarrow C \rightarrow \end{array}$$

$$\begin{array}{c} \downarrow \hspace{2em} | \\ \rightarrow A \rightarrow B \rightarrow C \rightarrow D \rightarrow \end{array}$$

where the feed-back is activation or inhibition.

The structure of the matrix of partial derivatives (cf. eqn (3)) for a chain of three components is given by

$$\begin{pmatrix} -M_A & m_A & 0 \\ 0 & -M_B & m_B \\ -m_0 & m_0 & -M_C \end{pmatrix} \tag{11}$$

where

$$M_P = m_P + D_P k^2 \tag{12}$$

and m_A, m_B and m_C are positive (no self-activation). This system can be shown to be globally stable if $m_0 < 0$ (backward inhibition). For $m_0 > 0$ the system may have oscillatory or aperiodic instabilities. However, as shown in the Appendix the oscillatory solutions have increasing growth rate for increasing wave lengths in the vicinity of marginal stability $(d\mu(k^2)/d(k^2) < 0$ for $\lambda(k) = \pm i\omega)$ whereas aperiodic spatial instability may occur. Thus this three-component system does not offer any spatial phenomena unknown in two-component systems.

Preliminary studies [10] of four-component systems by the methods shown in the Appendix indicate that backward inhibition may lead to oscillatory spatial instability. This requires a stable long wave limit (all four roots of the secular equation have negative real parts for $k = 0$) and the existence of wave vectors for which two roots with $\mu(k) = 0$ and $\omega(k) \neq 0$ have $d\mu(k)/d(k^2) > 0$. From eqn (12) we have

$$\frac{d\mu}{d(k^2)} = \sum_P D_P \frac{\partial\mu}{\partial M_P}. \tag{13}$$

Thus, oscillatory spatial instability requires that at least one of the derivatives $\partial\mu/\partial M_P$ is positive for a marginally stable, oscillatory solution. This is the case for large diffusion coefficients of the A-component and a large rate constant of the reaction which is subject to inhibition.

The above preliminary results for models with more than two components indicate that such models should be explored in more detail. The results should be related to chemical reaction schemes involving backward inhibition, such as the model for the Belousov reaction.[12]

If models predicting oscillatory spatial instabilities turn out to be chemically realistic then a quite new field is opened not only in connection with instabilities. It is known in general linear wave mechanics that globally stable steady states in systems which are close to oscillatory spatial instability allow wave propagation with little spatial attenuation.

APPENDIX

Considering the matrix in eqn (11) of a three component system with backward inhibition we find that the complex growth rate $\lambda = \mu - i\omega$ is the solution to

$$(M_A+\lambda)(M_B+\lambda)(M_C+\lambda)-m_0m_BM_A+m_0m_Am_B = 0. \tag{14}$$

With reference to eqn (13) we concentrate on the signs of $\partial\mu/\partial M_P$.

Investigating first the behaviour near marginally stable, aperiodic solutions ($\lambda \simeq 0$) we require for a finite wave vector k_0

$$M_AM_BM_C-(M_A-m_A)m_0m_B = 0. \tag{15}$$

In the vicinity of k_0

$$\lambda = \mu = -\frac{M_AM_BM_C-(M_A-m_A)m_0m_B}{M_AM_B+M_AM_C+M_BM_C-m_0m_B}. \tag{16}$$

Evaluating $\partial\mu/\partial M_P$ and inserting in eqn (12) and (15) we find

$$\partial\mu/\partial M_A = (M_BM_Cm_A)/(D_Ak^2N), \quad \partial\mu/\partial M_B = -M_AM_C/N \tag{17}$$

and

$$\partial\mu/\partial M_C = -M_AM_B/N$$

where

$$N = M_AM_B+M_AM_C-M_BM_Cm_A/D_Ak^2. \tag{18}$$

The different signs of the derivatives $\partial\mu/\partial M_P$ allow the existence of aperiodic spatial instability.

For oscillatory solutions we find in the case of marginal stability

$$-\lambda^2 = \omega^2 = \frac{M_AM_BM_C-m_0(M_A-m_A)}{M_A+M_B+M_C} = M_AM_B+M_AM_C+M_BM_C-m_0m_B > 0.$$

In the evaluation of $\partial\mu/\partial M_P$ we consider the left hand side of eqn (14) as a function $f(M_A, M_B, M_C, \lambda)$ and use the fact that

$$\frac{\partial\mu}{\partial M_P} = -\mathrm{Re}\,\frac{\partial f/\partial M_P}{\partial f/\partial\lambda}.$$

Inserting in this expression the above value of $\lambda = \pm i\omega$ we find that all derivatives $\partial\mu/\partial M_P$ are negative. Consequently, oscillatory spatial instability is absent in this three-component system.

[1] A. M. Turing, *Phil. Trans. B*, 1952, **237**, 37.
[2] A. M. Zhabotinskii in B. Chance, E. K. Pye, A. K. Ghosh and B. Hess (eds.), *Biological and Biochemical Oscillators* (Academic Press, New York, 1973), p. 89.
[3] P. Ortoleva and J. Ross, *J. Chem. Phys.*, 1973, **58**, 5673.
[4] A. Lotka, *J. Phys. Chem.*, 1910, **14**, 271.

[5] V. Volterra, *Leçon sur la Theorie Mathematique de la Lutte pour la Vie* (Gautier-Villars, Paris, 1931).

[6] I. Prigogine and R. Lefever, *J. Chem. Phys.*, 1968, **48**, 1695.

[7] I. Balslev and H. Degn, *J. Theor. Biol.*, 1974.

[8] B. B. Edelstein, *J. Theor. Biol.*, 1970, **26**, 227.

[9] L. A. Segel and J. L. Jackson, *J. Theor. Biol.*, 1972, **37**, 545.

[10] I. Balslev, unpublished work.

[11] R. J. Field, E. Körös and R. M. Noyes, *J. Amer. Chem. Soc.*, 1972, **25**, 8649.

[12] H. Degn, *J. Chem. Ed.*, 1972, **49**, 302.

Nucleation in Systems with Multiple Stationary States

By A. Nitzan, P. Ortoleva and J. Ross

Department of Chemistry, Massachusetts Institute of Technology,
Cambridge, Massachusetts 02139

Received 26th July, 1974

We consider a reaction diffusion system, far from equilibrium, which has multiple stationary states (phases) for given ranges of external constraints. If two stable phases are put in contact, then in general one phase annihilates the other and in that process there occurs a single front propagation (soliton). We investigate the macroscopic dynamics of the front structure and velocity for two model systems analytically and numerically, and for general reaction-diffusion systems by a suitable perturbation method. The vanishing of the soliton velocity establishes the analogue of the Maxwell construction used in equilibrium thermodynamics. The problem of nucleation of one phase imbedded in another is studied by a stochastic theory. We show that if the reaction dynamics is derived from a generalized potential function then the macroscopic steady states are extrema of the probability distribution. We use this result to obtain an expression for the critical radius of a nucleating phase and confirm the prediction of the stochastic theory by numerical solution of the deterministic macroscopic kinetics for a model system.

1. INTRODUCTION

Chemical reaction mechanisms with macroscopic rate laws of sufficient non-linearity in systems maintained far from equilibrium may have multiple stationary stable states [1-6] for given external constraints. We refer to each such state as a phase and transitions between phases are possible. The analogy of the theory of phase transitions and critical phenomena to transitions between stable states and critical (marginal stability) points has been discussed in some detail, both at the macroscopic and statistical mechanical level of analysis.[7-9] In this paper we investigate the nucleation of one phase within another phase, as well as the conditions of coexistence of two phases (the analogue of the Maxwell construction). These problems have been considered by: Kobatake [7] who showed for a particular case the similarity of the behaviour of the generalized entropy production [3] in transitions between stable branches of steady states, and the Gibbs free energy in equilibrium phase transitions; by Schlögl,[8] who took into account reaction and diffusion and treated coexistence of phases and the analogue of the Maxwell construction for one variable systems with the help of a mechanical analogy; and by Nicolis, Malek-Mansour, van Nypelseer and Kitahara,[10] who analyzed the onset of instability as a nucleation process, derived a non-linear master equation for that purpose, and applied that result to some examples.

We approach these problems from two different points of view. First, in section 2 we investigate the macroscopic dynamics of two stable stationary states of semi-infinite extent, at given external constraints, placed in contact with each other. Except for one value of the constraints, one phase annihilates the other and in that process there occurs a single-front propagating wave, a soliton. We investigate the behaviour of the concentration profile and velocity of such a wave for two model systems analytically and numerically and show that the values of the external constraints for zero soliton velocity establish the Maxwell construction.

Further we study these quantities for general reaction-diffusion systems by a suitable perturbation method and find that the soliton velocity vanishes linearly in the deviation of the external constraint from its value at the Maxwell construction.

Next, in section 3, we discuss some aspects of a stochastic theory of instability phenomena on the basis of a Fokker–Planck equation assumed for a reaction-diffusion system. That equation can be solved for the steady state probability distribution for which the reaction dynamics is obtainable from a generalized potential function. In that case we show that the macroscopic steady states are extrema of the probability distribution, maxima for stable steady states. We apply this theory to analyze the behaviour of a system consisting initially of a nucleus of one phase embedded in an infinite bulk of the other phase; we do so in a manner similar to the conventional treatment of nucleation in first order phase transitions.[11, 12] We thus obtain an expression for the critical radius for such a nucleus above which the nucleus grows, and below which it disappears in time. This prediction agrees with numerical solutions of the deterministic diffusion-reaction equation for a simple model system.

2. MACROSCOPIC DYNAMICAL THEORY

We consider a diffusion-reaction system for which the macroscopic deterministic equation of motion is

$$\frac{\partial \psi}{\partial t}(r, t) = \mathbf{D}\nabla^2\psi + F[\psi, \lambda]. \tag{2.1}$$

The symbols denote : ψ, a column vector of concentrations and possibly other state variables such as temperature; r, spatial coordinates; t, time; $\mathbf{D}$, a matrix of diffusion coefficients; F, the variations due to chemical reaction; and λ, the set of external constraints (boundary concentrations of species, light intensity in an illuminated system,[5] etc.). We assume that : 1. the diffusion matrix is constant and symmetric and the diffusion process is stable in that in the absence of reaction eqn (2.1) with $F = 0$) the steady state solution is stable; 2. F is analytic in ψ and λ. We consider systems for which there exist two stable stationary states (and one unstable stationary state) in a given range of λ. Thus if, in a one-dimensional system (coordinate x) we prepare two systems under identical external constraints, one system in one stable state (one phase) and the other system in the other stable state (the other phase), then the boundary at $x = 0$ between the two systems will move in one direction or another depending on λ. There occurs a single front propagation,[13] a soliton, in which one phase annihilates the other. We shall find the condition for coexistence of two such systems (phases) to be zero soliton velocity, or a standing front at a given λ. We therefore seek wave solutions of (2.1) which are most readily studied in a reference frame moving with the front. We define the phase

$$\phi = x - vt, \tag{2.2}$$

where the soliton velocity v is a function of λ. With (2.2) the partial differential eqn (2.1) is converted into an ordinary differential equation

$$\mathbf{D}\frac{d^2\psi}{d\phi^2} + v\frac{d\psi}{d\phi} + F[\psi, \lambda] = 0, \tag{2.3}$$

which is a nonlinear eigenvalue equation for v at a given λ. Coexistence occurs when $v = 0$ at $\lambda = \lambda_M$. The value of λ_M is obtained from eqn (2.4) which may be interpreted as an eigenvalue equation for λ_M,

$$\mathbf{D}\frac{d^2\psi_0}{d\phi^2} + F[\psi_0, \lambda_M] = 0. \tag{2.4}$$

We first consider a soluble model system and then present an analysis for more general $\{\mathbf{D}, \mathbf{F}\}$ systems.

The model system consists of a single species with third-order kinetics which obey the equation

$$\frac{\partial \psi}{\partial t} = D \frac{\partial^2 \psi}{\partial x^2} - q(\psi - a)(\psi - b)(\psi - c). \tag{2.5}$$

Homogeneous steady states occur at $\psi^* = a, b, c$, where these parameters are arbitrary functions of λ. We take $b \geqslant c \geqslant a, q > 0$, and thus only states a and b are stable to small homogeneous perturbations. If we define a new variable u such that

$$u \equiv \frac{\psi - a}{b - a},$$

then we obtain

$$\frac{\partial u}{\partial t} = D \frac{\partial^2 u}{\partial x^2} - q(b-a)^2 u(u-1)\left(u + \frac{(c-a)}{(b-c)}\right). \tag{2.6}$$

This is identical in form to a model considered by Montroll.[12] In terms of our notation his solutions for soliton fronts are

$$\psi(\phi) = a + (b-a)(1 + \exp(\pm \beta \phi))^{-1} \tag{2.7}$$

where

$$\beta = \left(\frac{q}{2D}\right)^{\frac{1}{2}} (b-a). \tag{2.8}$$

The front velocity is given by

$$v = (\tfrac{1}{2}qD)^{\frac{1}{2}}(a+b-2c). \tag{2.9}$$

Coexistence occurs at $v = 0$, i.e.,

$$c = \tfrac{1}{2}(a+b) \text{ at } \lambda = \lambda_M, \tag{2.10}$$

which is obvious from the symmetry of the trilinear kinetics for which the two stable states a and b are located symmetrically about the unstable state c.

It is interesting to take a particular case and analyze it further. Let

$$F = -q\psi(\psi^2 - 2\psi + 1/\lambda). \tag{2.11}$$

Then $v = (\tfrac{1}{2}qD)^{\frac{1}{2}}[3(1 - 1/\lambda)^{\frac{1}{2}} - 1]$ and $\beta = (q/2D)^{\frac{1}{2}}[1 + (1 - 1/\lambda)^{\frac{1}{2}}]$, from (2.8, 9). The uniform steady states and the soliton velocity are shown in fig. 1(a) and (b). Coexistence is found at $\lambda_M = 9/8$. With this we obtain the structure of the coexistence region to be

$$\psi_{\text{coex.}}(x) = \tfrac{4}{3}(1 + \exp(\pm \beta_M x))^{-1}, \tag{2.12}$$

where $\beta_M = \tfrac{1}{3}(8q/D)^{\frac{1}{2}}$; the structure is shown in fig. 2.

We note from fig. 1(b) that the velocity passes through the origin at λ_M with a finite but non-zero slope $|(dv/d\lambda)_{\lambda_M}| \neq \infty$. This general property provides a useful aid in determining λ_M experimentally.

Next we consider the general system (2.1) by analyzing the solutions of the front eqn (2.3) in the vicinity of λ_M. To do so we introduce a parameter A to be defined at a later stage in the development, which measures the deviation of the system from coexistence conditions. In general, A is a function of λ,

$$\lambda = \sum_{n=0}^{\infty} \lambda_n A^n, \tag{2.13}$$

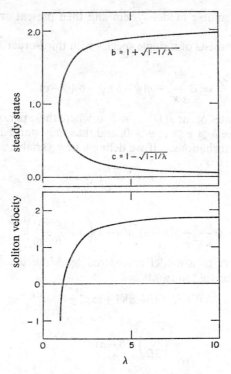

FIG. 1.—Steady states and soliton velocity versus the parameter λ for the model system eqn (2.11).

and we expect that A approaches zero proportional to the factor $(\lambda - \lambda_M)$ raised to some positive power. Furthermore ψ and v depend on A,

$$\psi = \sum_{n=0}^{\infty} \psi_n(\phi) A^n \tag{2.14}$$

$$v = \sum_{n=0}^{\infty} v_n A^n. \tag{2.15}$$

Such expansions have been employed elsewhere for the study of chemical waves.[15] On substitution of the expansions (2.13-15) into (2.3) we recover to zero'th order in A,

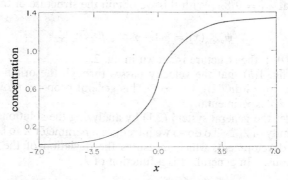

FIG. 2.—Structure of the coexistence region versus distance, eqn (2.12), for the model system eqn (2.11).

eqn (2.4), with the boundary conditions of the system being one stable stationary state at $x = -\infty$, and the other stationary stable state at $x = +\infty$. We also have by definition in this order $v_0 = v(\lambda_M) = 0$.

To first order in the parameter A we find

$$-\mathscr{L}\psi_1 = v_1 \frac{d\psi_0}{d\phi} + \left(\frac{\partial F}{\partial\lambda}\right)_0 \lambda_1 \qquad (2.16)$$

with the operator $\mathscr{L}$ defined by

$$\mathscr{L} \equiv D \frac{d^2}{d^2\phi^2} + \Omega \qquad (2.17a)$$

$$\Omega \equiv (\partial F/\partial\psi)_0 \qquad (2.17b)$$

and the subscript 0 following differentiation implies evaluation in lowest order. We note that differentiation of (2.4) with respect to ϕ yields

$$\mathscr{L} \frac{d\psi_0}{d\phi} = 0. \qquad (2.18)$$

We assume that $d\psi/d\phi$ is the only eigenfunction of $\mathscr{L}$ with zero eigenvalue. $\mathscr{L}^{-1}(d\psi/d\phi)$ does not exist and hence the solution of (2.8) for ψ_1 cannot be found for arbitrary (v_1, λ_1). The problem is overcome by a proper choice of these coefficients. We then define a vector Θ such that

$$\mathscr{L}^+\Theta = 0, \qquad (2.19)$$

where $\mathscr{L}^+$ is the adjoint operator of $\mathscr{L}$. If we define an inner product

$$(A|B) = \sum_i \int_{-\infty}^{\infty} d\phi\, A_i^* B_i, \qquad (2.20)$$

where the sum on i runs over all species, then from (2.17, 20) $\mathscr{L}^+$ is given by

$$\mathscr{L}^+ = D^+ \frac{d^2}{d\phi^2} + \Omega^+. \qquad (2.21)$$

(By assumption we have $D^+ = D$, Onsager's relations).

We can be assured of the solubility of the first order eqn (2.16) if we can choose v_1 and λ_1 such that the r.h.s. of (2.16) is orthogonal to the zero eigenvalue function $d\psi_0/d\phi$. Thus we obtain by use of the adjoint eigenfunction to $d\psi_0/d\phi$, that is Θ, see (2.19), that

$$v_1(\Theta|(d\psi_0/d\phi)) + \lambda_1(\Theta|(\partial F/\partial\lambda)_0) = 0. \qquad (2.22)$$

The coefficient of v_1 may be chosen to be unity by proper normalization of Θ. Hence if the coefficient of λ_1 is finite, then this relation can be satisfied for finite v_1, λ_1. Finally $(\partial F/\partial\lambda)_0$ approaches constant values as $|\phi| \to \infty$. A necessary condition for solubility is $\int_{-\infty}^{\infty} d\phi |\Theta(\phi)| < \infty$ in order to find $\lambda_1, v_1 \neq 0$. For systems where $\Omega^+ = \Omega$, $\mathscr{L}$ is self adjoint and therefore $\Theta = d\psi_0/d\phi$. Since $d\psi_0/d\phi$ is localized to the region of the soliton front, it satisfies the necessary condition of solubility. However we have not investigated this condition for systems with $\Omega^+ \neq \Omega$.

It is sufficient to all orders to choose $\lambda_{n \geq 2} = 0$; thus with the convenient choice $\lambda_1 = 1$ we have

$$A \equiv \lambda - \lambda_M. \qquad (2.23)$$

To first order the perturbation solution is

$$\psi(\phi)_{\lambda \to \lambda_M} \approx \psi_M(\phi) - \mathscr{L}^{-1}\left\{\left(\frac{dv}{d\lambda}\right)_M \frac{d\psi_0}{d\phi} + \left(\frac{\partial F}{\partial \lambda}\right)_M\right\}(\lambda - \lambda_M) + \cdots \qquad (2.24)$$

$$v_{\lambda \to \lambda_M} \approx -[(\Theta|(\partial F/\partial \lambda)_M)/(\Theta|(d\psi_0/d\phi))](\lambda - \lambda_M) + \cdots \qquad (2.25)$$

For the one variable system (2.5-12) we have verified these relations using the equation $\Theta = d\psi_0/d\phi$.

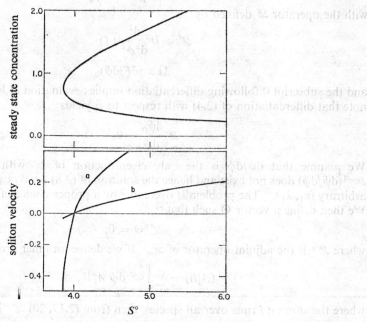

FIG. 3.—Steady state concentration and soliton velocity versus $S°$ for the model system described in Appendix A.   For curve (a) the diffusion coefficients D_S, D_P are both 1 and for curve (b) $D_S = 1$ and $D_P = 0.05$.

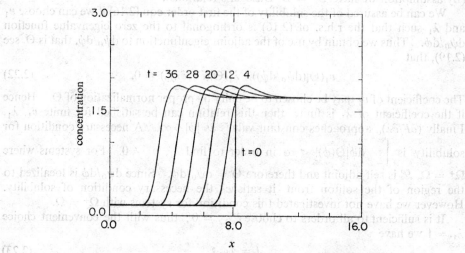

FIG. 4.—Soliton structure against distance at different times for the model system, Appendix A.   The diffusion coefficients are $D_S = 1$, $D_P = 0.05$.   The external concentration $S° = 5$.

Finally we note that the structure and velocity of the soliton depends in general not only on the reaction term F but also on the diffusion matrix D. If D is equal and diagonal then it may always be eliminated from the equation for the soliton front by a suitable scaling of the length, $\phi \rightarrow (\sqrt{D})\phi'$. However for more complex diffusion matrices this simple scaling no longer holds. In Appendix A we consider a two species system with rate mechanism

$$S^\circ \rightleftharpoons S \xrightarrow{\varepsilon} P \rightleftharpoons P^\circ \tag{2.26}$$

where S°, P° are maintained constant, and the activity of the enzyme $\mathscr{E}$ depends on P.

In fig. 3 we show the steady state concentration P^* as a function of S° and the results of the numerical integration of the partial differential equations yielding the variation of the soliton velocity against the concentration of S° for various choices of diffusion coefficients of S and P. For this system the value of S° at coexistence (λ_M of the general theory) does not appear to depend on the choice of diffusion coefficients to within the accuracy of the numerical computations; the results of the stochastic theory, section 3, indicate no dependence on that choice for systems for which the kinetics is derivable from a generalized potential function.

In fig. 4 we show the soliton stucture for the same system (Appendix A) for the choice of diffusion coefficients $D_S = 1$, $D_P = 0.05$, and $S^\circ = 5$.

3. STOCHASTIC THEORY

Our treatment is based on a model in which to the set of reaction diffusion rate equations we add a stochastic source term to account for fluctuations in the system. The set of concentrations at each spatial point becomes then our set of stochastic variables. It is convenient to make this set discrete by dividing the system into cells of homogeneous composition. We thus consider the system of volume V to be partitioned into N cells each of volume $\bar{v} = V/N$. The macroscopic reaction kinetics is again given by

$$\left(\frac{d\psi}{dt}\right)_{rx} = F[\psi], \tag{3.1}$$

with a similar equation for ψ_n in the nth cell. The rate of change of numbers of particles of any species, $\chi_n = \bar{v}\psi_n$, due to diffusion is

$$\left(\frac{d\chi_n}{dt}\right)_{diff.} = \bar{v}^{1-(2/d)} \mathbf{D} \sum_i (\psi_{n+i} - \psi_n) \tag{3.2}$$

where the sum extends over the $2d$ nearest neighbour cells (d is the dimensionality of the system). The factor of $\bar{v}^{1-2/d}$ comes from multiplying by the surface area $\bar{v}^{1-1/d}$ of the cell and dividing by the characteristic length ($\bar{v}^{1/d}$) of a cell such that $(\psi_{n+i} - \psi_n)/\bar{v}^{1/d}$ is an approximation to the concentration gradient. Now we write a Langevin equation for this system as

$$\frac{d\chi_n}{dt} = \bar{v}F[\psi_n] + \bar{v}^{1-2/d}\mathbf{D} \sum_i (\psi_{n+i} - \psi_n) + (2K\bar{v})^{\frac{1}{2}} \tag{3.3a}$$

or

$$\frac{d\psi_n}{dt} = F[\psi_n] + \bar{v}^{-2/d}\mathbf{D} \sum_i (\psi_{n+i} - \psi_n) + \left(\frac{2K}{\bar{v}}\right)^{\frac{1}{2}} \tag{3.3b}$$

For simplicity we take the stochastic term to be Gaussian and thus to satisfy the usual relations $\langle f_n(t) \rangle = 0$ and $\langle f_{\alpha n}(t')f_{\beta m}(t) \rangle = \delta_{\alpha\beta}\delta_{nm}\delta(t-t')$ (α, β denote different chemical

components).   Eqn (3.3b) is equivalent to the following Fokker–Planck equation for
the probability density of finding the concentration distribution ψ_n, $P[\Psi]$, in the system

$$\frac{\partial P[\Psi]}{\partial t} = -\sum_n \sum_\alpha \left\{ \frac{\partial}{\partial \psi_{\alpha n}} (M_{\alpha n}[\Psi]P) - \frac{K}{\bar{v}} \frac{\partial^2 P}{\partial \psi_{\alpha n}^2} \right\} \tag{3.4}$$

where $M_n(\Psi)$ (with components $M_{\alpha n}[\Psi]$) is given by

$$M_n[\Psi] = F[\psi_n] + \bar{v}^{-2/d} \, \mathbf{D} \sum_{i=1}^{2d} [\psi_{n+i} - \psi_n]. \tag{3.5}$$

and where Ψ denotes the "supervector" ψ_n $(n = 1 \ldots N)$ $(= \psi_{\alpha n}$ $(n = 1 \ldots N$,
α goes over the different chemical species)).   In (3.4) K was taken for simplicity as a
scalar matrix, $K = KI$, where I is the unit matrix.   The integration of this many-
variable Fokker–Planck equation for steady state conditions, $(\partial P/\partial t) = 0$, is possible
provided the reaction kinetics are such that the curl of M is zero,[16]

$$\frac{\partial M_{\alpha n}}{\partial \psi_\beta} = \frac{\partial M_{\beta n}}{\partial \psi_\alpha} \tag{3.6}$$

This condition should be modified if $\mathbf{D}$ is not a symmetric matrix or if K is not a
scalar matrix.   If (3.6) holds then we obtain

$$P[\Psi] = C \exp(-U[\Psi]) \tag{3.7}$$

where C is a normalization constant.   For simplicity we assume now that we have a
one-component system, for which (3.6) always holds.   The results for a more general
case with a symmetric diffusion matrix are given in Appendix C.   We write $U[\psi]$ as

$$U[\psi] = -\frac{1}{K} \left[\bar{v} \sum_n \left\{ G[\psi_n] - (\bar{v}^{-2/d}) \frac{D}{2} \sum_{i=1}^d (\psi_n - \psi_{n+i})^2 \right\} \right], \tag{3.8}$$

which in the continuum limit

$$\left(\bar{v} \sum_n \to \int d^d r, \ (\psi_n - \psi_{n+i})/\bar{v}^{1/d} \to \frac{\partial \psi(r)}{\partial r_i} \right)$$

becomes

$$U[\psi(r)] = -\frac{1}{K} \int dr \left\{ G[\psi(r)] - \frac{D}{2} [\nabla \psi(r)]^2 \right\}. \tag{3.9}$$

The function $G[\psi(r)]$ is defined by

$$G[\psi(r)] = \int_0^\psi dy \, F[y] \tag{3.10}$$

which in case of many chemical species should be understood as a line integral in ψ
space.   Condition (3.6) ensures that this line integral is well defined and does not
depend on the integration path.   It is interesting to note that (3.9) has the same form
as the free energy in the Landau–Ginzburg theory [11] (there $G(\psi)$ is usually taken as a
quartic form).   Also critical phenomena have been discussed with Hamiltonians of
that form.[17]

Having arrived at an expression for the steady state probability distribution $P[\psi]$
we show next that extrema in that distribution are given by the steady states of the
deterministic equation.   Differentiation of (3.7) yields

$$\delta P = -P \frac{\delta U}{\delta \psi} \delta \psi \tag{3.11}$$

or

$$\delta P = -\frac{P}{K} \int d\mathbf{r}\{F[\psi]\delta\psi - D\nabla\psi \cdot \nabla\delta\psi\} \qquad (3.12)$$

which on partial integration becomes

$$\delta P = -\frac{P}{K} \int d\mathbf{r}\{F[\psi] + D\nabla^2\psi\}\delta\psi. \qquad (3.13)$$

For P to be an extremum for the arbitrary variation $\delta\psi$ we must have the integrand vanish; that is the macroscopic condition for steady states. It can be further shown that stable steady states correspond to maxima of the probability distributions (see Appendix B). The system may be in a single stable stationary state (although there exist other such states at the same external constraints) and in that case the system is homogeneous, i.e., $F[\psi] = 0$. However, two (or more) stable stationary states may coexist and the distribution of concentrations is the solution of $D\nabla^2\psi + F[\psi] = 0$ for given boundary conditions. Note that this is exactly the equation for the zero velocity soliton structure (eqn (2.3) with $v = 0$).

The stochastic theory for the probability distribution of a system in a steady state may be applied to the problem of nucleation. Consider a spherical nucleus of radius R of one stable stationary state, labelled phase B, immersed in an infinite bulk of another stable stationary state, labelled phase A, both under the same external constraints. Thus we have $F[\psi_A] = F[\psi_B] = 0$ for the individual, homogeneous phase. For phase B immersed in A, however, we may inquire about the expected stability of the spherical nucleus, that is the nucleus is expected to evolve in the direction of increasing the probability of the overall structure (nucleus + bulk). The logarithm of the ratio of the probability distribution for B phase immersed in A phase, P_{BA}, to that of pure A phase, P_A is

$$K \ln \frac{P_{BA}}{P_A} = -\left\{\frac{2\pi^{d/2}}{d\Gamma(d/2)}R^d[G(B)-G(A)] - \frac{2\pi^{d/2}R^{d-1}D}{\Gamma(d/2)l}(\psi_B-\psi_A)^2\right\} \qquad (3.14)$$

where l is the width of the interface region, and where we have approximated the gradient by $(\psi_B - \psi_A)/l$. In eqn (3.14) we have used the following expression for the surface area S and the volume V of a d-dimensional sphere (Γ is the Γ-function)

$$S = \frac{2\pi^{d/2}}{\Gamma(d/2)}R^{d-1}$$

$$V = \frac{2\pi^{d/2}}{d\Gamma(d/2)}R^d.$$

The derivative (with respect to R) of the r.h.s. of (3.14), set equal to zero, gives the radius of the sphere of phase B for which the l.h.s. of (3.14) is a minimum, that is (for a given l)

$$R_c = \frac{D(\psi_B-\psi_A)^2(d-1)}{l(G[\psi_B]-G[\psi_A])}. \qquad (3.15)$$

If $G[\psi_B] > G[\psi_A]$, then there exists a distribution (characterized by R_c) of maximum "free energy" (minimum probability) such that for $R > R_c$ the nucleus will grow (as this is the direction of increasing the probability) and for $R < R_c$ it will disappear in time. If $G[\psi_B] < G[\psi_A]$ then no positive R_c exists. Finally, if $G[\psi_B] = G[\psi_A]$ then $R_c = \infty$, and phases A and B coexist with a planar (zero curvature) surface of contact. This condition is clearly the analogue of the Maxwell construction in equilibrium phase transitions.   Furthermore, we expect that this condition is equivalent to the one of zero soliton velocity and have confirmed the expectation for the example given by (2.5).

In a many component curl-free system with a symmetric non-scalar diffusion matrix, the result (see eqn (C.1), Appendix C) shows that the Maxwell construction is again given by the condition $G[\psi_B] = G[\psi_A]$.

We stress that the arguments presented here are based on probabilistic considerations, especially on the postulate that a system will evolve in the direction of increasing the probability.   The result (3.15) is similar in many respects to that for the radius of critical nucleus in equilibrium first order phase transitions.[11, 12]   The contribution of the diffusion process to the " free energy " U, eqn (3.9), takes here the place of the surface tension in the equilibrium nucleation treatment.   The appearance of the parameter l in eqn (3.15) takes into account the non-zero thickness of the interphase layer.   It should be remembered that l is not an independent parameter but is determined by the external parameters.   Also, the simplifying assumption of linear variation of the concentration ψ across the interphase layer limits the applicability of the result (3.15).

We compare the results of the stochastic theory to the solution of the deterministic dynamical equations.   To this end we numerically integrated the one variable partial differential equation

$$\frac{\partial \psi}{\partial t} = D\nabla^2\psi - \psi^3 + b\psi + c \tag{3.16}$$

in which the cubic polynomial $F(\psi) = -\psi^3 + b\psi + c$ plays the role of a chemical source term.   For $b > 0$ and $|c| < (2b/3)(\sqrt{b/3})$ this equation has two stable and one

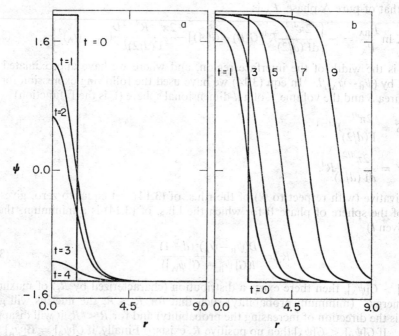

FIG. 5.—Evolution of a nucleus as a function of radial coordinate for various times, as obtained from numerical solution of eqn (3.16) with $b = 3$, $c = 1.5$ and $D = 1$.   Figures at left and right are for initial radius smaller and larger than R_c respectively.

unstable homogeneous steady states.   Denoting the two stable homogeneous steady states by ψ_A (phase A) and ψ_B (phase B) we put our initial condition to be a sphere of radius R centred around the origin and consisting of phase B, imbedded in a spherical region of phase A which in the computation process has to be taken finite but with radius R_1 much larger than R.   Since the initial condition is spherically symmetric and the coefficients of (3.16) are constants, it is sufficient to consider the radial coordinate only.   The boundary conditions are taken as $\partial\psi/\partial r = 0$ at $r = 0$ and at the larger bounding sphere.

Fig. 5 shows the time and space dependent solution to eqn (3.16) in three dimensions for the choice of coefficients $D = 1, b = 3, c = 1.5$ and for two different choices of the initial radius R: $R < R_c$ and $R > R_c$.   The nucleus shrinks and disappears for $R < R_c$ and grows in time for $R > R_c$.   It is interesting to note that the growth process (for $R > R_c$) accelerates for small R but the velocity approaches a constant for larger radii as the front becomes more planar.   We calculate the velocity of a planar front (from eqn (2.9) using the steady state solutions $\psi_a = -1.38$, $\psi_b = 1.94$ (stable) and $\psi_c = -0.558$ (unstable)) to be 1.2, while from fig. 5b, the velocity for the largest radius obtained is 1.0.

TABLE 1.—CRITICAL RADIUS FOR NUCLEATION FOR THE SYSTEM GIVEN BY EQN (3.16)

c	R_c from numerical computations		R_c calculated from eqn (3.15) with $l = 2$	
	2 dimensions	3 dimensions	2 dimensions	3 dimensions
0.50	$2.6 < R_c < 3.2$		3.4	
0.75		$3.5 < R_c < 5.5$		4.6
1.00	$1.5 < R_c < 1.75$	$2.7 < R_c < 3.7$	1.7	3.4
1.25		$2.1 < R_c < 2.3$		2.6
1.50	$0.85 < R_c < 1.0$	$1.65 < R_c < 1.95$	1.05	2.1
1.75	$0.8 < R_c < 1.0$	$1.35 < R_c < 1.5$	0.9	1.75
1.875		$1.2 < R_c < 1.4$		1.6
2.0	$R_c < 0.5$	$R_c < 0.8$	0.7	1.35

In table 1 we present the results of numerical computation of R_c, as well as results based on eqn (3.15), for $b = 3$ and for different values of c in two and three dimensions. The calculations based on eqn (3.15) were done with $l = 2$ which is an estimate based on the numerical results such as given by fig. 5b.   The agreement between the values of R_c as obtained from (3.15) and from numerical integration of (3.16) is fairly good for all the values of the parameter c that were tried, excluding the bifurcation value $c = 2$.   (Our failure to provide a more accurate numerical answer for R_c near the bifurcation point is caused by the slow time evolution of the system near that point, which makes the numerical integration very expensive.)   This agreement is remarkable as the two ways of obtaining R_c are different; one is based on solutions of the time-dependent deterministic equation, and the other on the steady state probability distribution.

To conclude this section we should note that even though the dynamics of nucleus growth as given by eqn (3.16) or fig. 5 give important information on the rate of nucleation of one phase within another, the rate of this process depends also on the rate of formation of nuclei by spontaneous fluctuations which is not discussed in the present work.

We thank John M. Deutch for helpful discussions.   This work was supported in part by the National Science Foundation and Project SQUID, Office of Naval Research.

APPENDIX A

For the reaction mechanism (2.26) the reaction-diffusion equations are

$$\frac{\partial S}{\partial t} = D\frac{\partial^2 S}{\partial x^2} + (S^\circ - S) - \mathscr{E}(P)S \tag{A.1}$$

$$\frac{\partial P}{\partial t} = D\frac{\partial^2 P}{\partial x^2} + (P^\circ - P) + \mathscr{E}(P)S \tag{A.2}$$

with all rate coefficients taken to be unity. The assumed form for the dependence of $\mathscr{E}$ on P is

$$\mathscr{E}(P) = \frac{P^2}{1+P+P^2} \tag{A.3}$$

for which chemical examples are available.[18] The steady state solutions of (A.1, 2) are

$$P^* = 0; \ \tfrac{1}{4}(S^\circ - 1) \pm \tfrac{1}{4}[(S^\circ - 1)^2 - 8]^{\frac{1}{2}}, \tag{A.4}$$

which are shown in fig. 3a.

APPENDIX B

Here we show for the probability distribution given by eqn (3.7) and (3.9), that stationary points which correspond to stable macroscopic steady states are local maxima of the distribution, or minima of the potential $U[\psi]$ given by (see eqn (C.1))

$$U[\psi] = -1/K \int dr\{G[\psi(r)] - \tfrac{1}{2}\mathbf{D} : (\nabla\psi(r))^2\}. \tag{B.1}$$

Let the steady state variable $\psi^*(r)$ satisfy the equation

$$F[\psi^*(r)] + \mathbf{D}\nabla^2\psi^*(r) = 0, \tag{B.2}$$

and let $\psi(r) = \psi^*(r) + \delta\psi(r)$. We assume that $\psi^*(r)$ corresponds to a stable steady state of the system. The first order (in $\delta\psi$) term of U then vanishes and the second order term is given by

$$\delta^2 U = -\tfrac{1}{2}\int dr\delta\psi\{\Omega + \mathbf{D}\nabla^2\}\delta\psi \tag{B.3}$$

where $\Omega = (\partial F/\partial\psi)_{\psi*}$. We have used the identity

$$\int dr\nabla W\cdot\nabla V = -\int dr W\nabla^2 V + \int Wn\cdot\nabla V \, d\sigma \tag{B.4}$$

and the fact that $\psi^*(r)$ and $\psi(r)$ must satisfy the same boundary conditions and therefore $(\delta\psi)_\Sigma = 0$. In eqn (B.4), Σ is the boundary of the system, $d\sigma$ is an element of this boundary and n is a unit vector perpendicular to the boundary. Since we assume that $\psi^*(r)$ is a stable steady state, the operator $\Omega + \mathbf{D}\nabla^2$ must be negative definite and therefore $\delta^2 U$ in eqn (B.3) is positive. This establishes the fact that $U[\psi]$ is a minimum (and $P[\psi]$ is a maximum) for $\psi = \psi^*$.

APPENDIX C

Here we write the modified relations of section 3 for a many variable system and for a general symmetric diffusion matrix. The potential $U[\psi]$ (eqn (3.9)) is given by

$$U[\psi(r)] = -1/K \int dr\{G[\Psi(r)] - \tfrac{1}{2}\mathbf{D} : (\nabla\psi(r))^2\} \tag{C.1}$$

where

$$\mathbf{D} : (\nabla \psi(r))^2 \equiv \sum_{ij} D_{ij} \sum_k \frac{\partial \psi_i}{\partial r_k} \frac{\partial \psi_j}{\partial r_k}. \tag{C.2}$$

The expression for the critical radius R_c is now replaced by

$$R_c = \frac{(\psi_B - \psi_A) \cdot \mathbf{D} \cdot (\psi_B - \psi_A)(d-1)}{l(G(\psi_B) - G(\psi_A))}. \tag{C.3}$$

[1] N. Rashevsky, *Mathematical Biophysics* (Dover, New York, 1960), Vol. I.
[2] B. B. Edelstein, *J. Theoret. Biol.*, 1970, **29**, 57.
[3] P. Glansdorff and I. Prigogine, *Thermodynamic Theory of Structure, Stability and Fluctuations* (Wiley-Interscience, New York, 1971).
[4] G. Nicolis, *Adv. Chem. Phys.*, 1971, **19**, 209.
[5] A. Nitzan and J. Ross, *J. Chem. Phys.*, 1973, **59**, 241.
[6] H. Hahn, P. Ortoleva and J. Ross, *J. Theoret. Biol.*, 1973, **41**, 503.
[7] Y. Kobatake, *Physica*, 1970, **48**, 301.
[8] F. Schlogl, *Z. Phys.*, 1972, **253**, 147.
[9] A. Nitzan, P. Ortoleva, J. Deutch and J. Ross, *J. Chem. Phys.*, 1974, **61**, 1056, see also other references therein.
[10] G. Nicolis, M. Malek-Mansour, A. van Nypolseer and K. Kitahara, private communication (preprint).
[11] L. D. Landau and E. M. Lifshitz, *Statistical Physics* (Addison-Wesley, Reading, Massachusetts, 1958.)
[12] see, for example, B. Widom in *Phase Transitions and Critical Phenomena*, Vol. 2, ed. C. Domb and M. S. Green (Academic Press, New York, 1972).
[13] R. Davidson, *Methods in Nonlinear Plasma Theory* (Academic Press, New York, 1972).
[14] E. W. Montroll in *Statistical Mechanics*, ed. S. A. Rice, K. F. Freed and J. C. Light (Univ. of Chicago Press, 1972), p. 69.
[15] P. Ortoleva and J. Ross, *J. Chem. Phys.*, 1974, **60**, 5090.
[16] R. L. Stratonovich, *Topics in the Theory of Random Noise* (Gordon and Breach, New York, 1963).
[17] see, for example, S. Ma, *Rev. Mod. Phys.*, 1973, **45**, 589.
[18] H. Hahn, A. Nitzan, P. Ortoleva and J. Ross, *Proc. Nat. Acad. Sci.* (submitted).

Limit Cycles in the Plane

An Equivalence Class of Homogeneous Systems

By John Texter

Department of Chemistry and Center for Surface and Coatings Research,
Lehigh University, Bethlehem, Pennsylvania 18015, U.S.A.

Received 30th July, 1974

A set theoretic presentation of the necessary and sufficient conditions for the occurrence of limit cycle behaviour in open homogeneous systems is given for the specific case of two time-dependent components. The results are obtained by using several theorems due to Zubov in the framework of Lyapunov's stability theory. The starting point in the analysis is the definition of a closed invariant set which depicts the hypothesized or experimentally observed limit cycle. Classes of kinetics which evolve to the defined limit cycle may then be generated. When kinetics of a particular polynomic form are presumed, necessary conditions on the rate constants and external constraints are generated. Time reversal generates an equivalence class of kinetics which has the previously defined limit cycle set as an asymptotic stability domain boundary. This concept may be of use in the consideration of the turning on and off of biological clocks. Introductory results regarding the admissible kinetics giving rise to elliptical limit cycles are presented, and the use of the theory in treating experimentally observed limit cycles is discussed.

1. INTRODUCTION

Renewed and increasing interest has, over the past decade, been shown in the study of physicochemical systems which give rise to sustained oscillatory behaviour. The number of such systems examined experimentally seems to be steadily increasing.[1-4] In formulating a firm physical basis for such oscillatory phenomena, it is required that the observed behaviour be reproduced quantitatively in terms of descriptive mathematical models. Attention herein will be focussed on open homogeneous systems with time-independent boundary conditions (external constraints) and two time-dependent components. The dynamics of such systems may be suitably modelled in terms of two coupled nonlinear ordinary differential equations. Since the sustained oscillations which have been reported in the literature are usually preceded by an induction period, attention herein is further restricted to oscillations of the limit cycle type.

It should be noted that many hypothetical chemical mechanisms have been formulated which (as proven by numerical integration procedures) admit limit cycle behaviour.[4-14] Although numerical integration of the rate equations for a proposed mechanism may certainly be sufficient to prove that a given mechanism admits limit cycle behaviour, such models of real systems are often stiff.[14-15] In these circumstances numerical integration becomes prohibitive. The problem at hand then is to develop necessary and sufficient criteria (exclusive of numerical integration) by which it may be judged whether or not a given model will admit limit cycle oscillations. The ultimate test of any proposed model remains one of numerical simulation, however.

Several necessary conditions for chemical limit cycle oscillations have been proposed. Physically these include the requirement of autocatalytic or crosscatalytic interactions among the chemical species. These requirements follow naturally from the necessary condition that such systems be described by nonlinear ordinary differential equations. Another necessary condition usually considered is the requirement that the linear part of a system's rate equations (as obtained by an expansion of the original rate expressions about the critical point or steady state whose stability character is being investigated) have eigenvalues with positive real parts. This guarantees that the critical point (any limit cycle surrounds at least one critical point) will be repulsive to small perturbations. This convention will be adopted herein, in that the describing differential equations (in perturbed coordinates) associated with a given limit cycle will be presumed to have a linear part which is positive definite. The important point here is that the critical point surrounded by the limit cycle (for the sake of discussion, cases involving multiple critical points surrounded by a limit cycle will be excluded herein) must be repulsive with respect to infinitesimal perturbations. To the best of this writer's knowledge no firm physical basis has so far been established to exclude the possibility of limit cycles occurring for systems whose linear part is negative or positive semidefinite. The theory of indices provides another (qualitative) necessary condition for the occurrence of limit cycles and Bendixson's negative criterion gives sufficient conditions for the non-existence of limit cycles.[16] The Poincare–Bendixson theorem gives both necessary and sufficient conditions for the occurrence of limit cycles, but these criteria are not easily implemented.

In this paper it will be attempted to implement a more quantitative theory of limit cycle oscillations than has heretofore been utilized in considering physico-chemical oscillations. Two general questions (not unrelated to one another) are investigated. First, does a given pair of coupled rate expressions admit a limit cycle as an asymptotic solution? Secondly, what is the general class of kinetics which will give rise to an *a priori* defined limit cycle solution? The basic theory utilized herein is that of Zubov [17, 18] in the framework of Lyapunov's [19] theory of stability of motion. Both necessary and sufficient conditions for limit cycles are included. A set theoretic presentation of Zubov's theory is given in sections 2 and 3. In section 2 the general class $\mathscr{E}_s$ of kinetics is constructed which has an *a priori* defined asymptotic stability domain. Usage of Zubov's theory in this context has been noted previously [20] and has successfully been applied to the characterization of bistability in two-mode lasers [21] and the characterization of thermal explosion limits.[22] In section 3 the construction of the general class $\mathscr{E}_1$ of kinetics having a given *a priori* defined limit cycle solution is discussed. Various applications of the theory are illustrated and discussed in section 4. The results reported herein are only to be considered introductory and do not provide means for complete system identification; however, sufficient criteria for constructing physicochemical limit cycle oscillations are obtained.

2. CONSTRUCTION OF $\mathscr{E}_s$

In this section the construction of the general class $\mathscr{E}_s$ of two-dimensional open homogeneous reaction kinetics

$$\dot{X} = F_1(X, Y)$$
$$\dot{Y} = F_2(X, Y)$$

(2.1)

is considered where the system (2.1) admits a finite domain of asymptotic stability about a critical point (X_c, Y_c) contained in this domain. For the sake of what follows the righthand sides of (2.1) are assumed to be holomorphic (i.e., expressible as a convergent Taylor series about any point) in some domain embedding the domain of asymptotic stability. To facilitate discussion, the following definitions are introduced.

Definition 1. The coordinates (X, Y) are restricted to some open (simply connected) set Ω of the real euclidean plane according to their type. For example, if X and Y are both chemical concentrations or activity variables, Ω is the positive quadrant of the plane.

Definition 2. The critical point (X_c, Y_c) comprises an invariant set $\mathcal{M}_s$, $\mathcal{M}_s \subset \Omega$ and $\mathcal{M}_s = \{(X_c, Y_c) \mid F_i(X_c, Y_c) = 0, i = 1, 2\}$. $\mathcal{M}_s$ is assumed to be unique in its domain of attraction $\mathcal{A}_s$.

Definition 3. The domain of attraction of $\mathcal{M}_s$ is an open, bounded, and simply connected set $\mathcal{A}_s$ where $\mathcal{A}_s \subset \Omega$ and $\mathcal{M}_s \subset \mathcal{A}_s$. $\mathcal{M}_s$ is further assumed to be an interior point of $\mathcal{A}_s$.

Definition 4. The boundary of the asymptotic stability domain $\mathcal{A}_s$ in a closed invariant set $\mathcal{M}_1 \subset \Omega$ which is a limit set of $\mathcal{A}_s$. Unless noted otherwise, $\mathcal{M}_1$ constitutes a closed curve in Ω.

Definition 5. The set $\mathcal{A}_1$ is a closed set where $\mathcal{A}_1 = \mathcal{A}_s \cup \mathcal{M}_1$. Note that every neighbourhood of every point in $\mathcal{M}_1$ contains points of $\mathcal{A}_s$.

Definition 6. The set of *a priori* admissibility constraints is denoted by $G = \{G_i, i = 1, 2, \ldots\}$ where the G_i are related to the parameters which appear in the righthand sides of (2.1). For example, if X and Y denote chemical concentrations, certain of the G_i would ensure the positivity of X and Y. Others of the G_i may reflect what is already known about or presumed for the system (2.1). For example, if the righthand sides of (2.1) are presumed to have a given polynomic form, certain of the G_i may denote values of certain of the rate constants or pseudo rate constants and some of the G_i may denote the sign appearing in front of certain of the terms accordingly as to whether or not restrictions are imposed on the molecularities of the interactions between X and Y.

The transformation is now made to perturbed coordinates $(x, y) = (X - X_c, Y - Y_c)$ so that (2.1) becomes

$$\dot{x} = f_1(x, y)$$
$$\dot{y} = f_2(x, y) \tag{2.2}$$

where $f_i(0, 0) = 0$, $i = 1, 2$. The linear part of (2.2) will be assumed to be negative definite. As this exposition is fundamentally based on Zubov's results, his principal theorem for the cases considered herein is stated as follows:

Theorem. The system (2.2) has $\mathcal{A}_s$ as its domain of asymptotic stability. A Lyapunov function $v(x, y)$ exists for (2.2) such that for all $(x, y) \in \mathcal{A}_s$, $0 \leqslant v < 1$, and for all $(x, y) \in \mathcal{M}_1$, $v = 1$. A continuous positive definite function $\phi(x, y)$ exists for all $(x, y) \in \Omega$ such that v is defined by the condition $v(0, 0) = 0$ and the equation

$$f_1 \frac{\partial v}{\partial x} + f_2 \frac{\partial v}{\partial y} = -\phi(1-v)(1+f_1^2+f_2^2)^{\frac{1}{2}}. \tag{2.3}$$

Proofs may be found in Zubov,[17, 18] Hahn,[23] or in Bhatia and Szëgo [24] where $\mathcal{A}_1$ is an open invariant set. The construction now proceeds by defining $\mathcal{M}_1$ through a choice of v such that $v(x, y) = 1$ for all $(x, y) \in \mathcal{M}_1$ and such that $\mathcal{M}_1$ describes a closed curve. $\mathcal{E}_s$ is thus obtained as the solution (f_1, f_2) to the inhomogeneous system

$$\begin{pmatrix} \partial v/\partial x & \partial v/\partial y \\ a & b \end{pmatrix} \begin{pmatrix} f_1 \\ f_2 \end{pmatrix} = \begin{pmatrix} -\phi(1-v)(1+f_1^2+f_2^2)^{\frac{1}{2}} \\ c \end{pmatrix} \tag{2.4}$$

where ϕ (satisfies the hypothesis of the Theorem), a, b and c (all scalar functions of x and y or constants) are selected in conformity with the desired spectral properties required of the linear part of (2.2) and such that the determinant of the coefficient matrix of the lefthand side of (2.4) may not vanish in $\mathcal{A}_1$ except possibly at the origin (in which case the righthand side of (2.4) vanishes identically at the origin also). The a, b and c may be restricted in such a manner that the solutions have a particular polynomic or nonpolynomic form (i.e., they must be chosen in conformity with the admissibility constraints G). As will be subsequently illustrated, the solution of (2.4) is simplified somewhat if the substitution $\phi = \phi(1+f_1^2+f_2^2)^{-\frac{1}{2}}$ is made. It should also be noted that the set $\mathcal{M}_1$ may or may not (depending on the choices of a, b, c and G) contain additional critical points of the system (2.2) obtained from (2.4).

To illustrate the flexibility of the method, the class $\mathcal{E}_g$ of kinetics giving rise to global asymptotic stability is constructed. The construction proceeds analogously as that for $\mathcal{E}_s$ except that $\mathcal{A}_s = \Omega$ and the solutions are obtained from the inhomogeneous system

$$\begin{pmatrix} \partial v/\partial x & \partial v/\partial y \\ a & b \end{pmatrix} \begin{pmatrix} f_1 \\ f_2 \end{pmatrix} = \begin{pmatrix} -\phi \\ c \end{pmatrix} \tag{2.5}$$

rather than from (2.4). In addition, v must be chosen positive definite such that $v(x, y) \to \infty$ as $x^2+y^2 \to \infty$. The extension of the construction to higher dimensional systems is straightforward. Both polynomial and nonpolynomial kinetics are generated depending on the selection of the functions v, ϕ, a, b and c. Nazarea[25-27] has illustrated a different generating function method for obtaining these results.

3. CONSTRUCTION OF $\mathcal{E}_1$

In this section the construction of the general class $\mathcal{E}_1$ of admissible kinetics having an invariant set $\mathcal{M}_1$ as an asymptotic stability boundary is outlined. Proceeding as before in section 2, the class $\mathcal{E}_s$ is constructed under the additional constraint that $\mathcal{M}_1$ contain no critical points of the solutions f_1 and f_2 of (2.4), i.e., f_1 and f_2 do not vanish identically for all $(x, y) \in \mathcal{M}_1$. Under the previously presumed condition that the righthand side of (2.1) (and hence of (2.2)) is holomorphic in some domain embedding $\mathcal{A}_1$, the continuation of integral curves from points in $\mathcal{A}_1$ to the negative semi-trajectory, $t \in (0, -\infty)$, is assured as is their boundedness. As a practical consideration it may additionally be required that there exist some sufficiently large open set $\mathcal{C}$ embedding $\mathcal{A}_1$ such that there exist no critical points of (2.2) in the set $\mathcal{D}$ where $\mathcal{D} = \mathcal{C} - \mathcal{A}_s$. This can be an especially important consideration in systems admitting multiple steady states. The class $\mathcal{E}_1$ is then obtained by the substitution of $t \to -t$ in the solutions (2.2) obtained from (2.4). The added restrictions of this section are, where necessary, implemented in the selection of the functions ϕ, a, b and c and the admissibility constraint G. Under such considerations as above, $\mathcal{E}_s$ and $\mathcal{E}_1$ are seen to be equivalent classes under time reversal.

4. APPLICATIONS AND EXAMPLES

The procedures outlined in sections 2 and 3 have several foreseeable applications in the identification and analysis of temporal ordering in nonlinear open systems. The generation of the classes $\mathcal{E}_s$ and $\mathcal{E}_1$ is a very general procedure and therefore highly flexible. The flexibility is inherent in the selection of the functions v, ϕ, a, b and c (cf. (2.4)) which are chosen in accordance with various constraints. Since Zubov's theory provides necessary and sufficient stability criteria, it may be used as a double-edged sword in practical applications. As the ensuing discussions will indicate, these procedures provide us with a powerful analytical tool. Implementation of the theory is not devoid of difficulties and restrictions however.

ANALYSIS OF STIFF SYSTEMS

In cases where a proposed model of a limit cycle is sufficiently stiff to make numerical dynamical simulation prohibitive, the following procedure may be used to establish the existence of a limit cycle. If the limit cycle exists, its phase portrait is also obtained. The substitution $t \to -t$ is first made in the differential equations suspected of admitting a limit cycle. The equations are then inserted into (2.3) where $\phi(1 + f_1^2 + f_2^2)^{\frac{1}{2}}$ is replaced by a quadratic form in conformity with the linear parts of f_1 and f_2. The solution $v(x, y)$ is then obtained necessarily by the method of undetermined coefficients as a power series of homogeneous forms

$$^n v(x, y) = \sum_{k=2}^{n} v_k(x, y) \tag{4.1}$$

until convergence is achieved. Numerical details of this method have been given elsewhere.[28, 29] After convergence has been obtained, the indicated boundary set $\mathcal{M}_1$ is checked to see if it contains any critical points of the system f_1 and f_2. If there are no critical points in $\mathcal{M}_1$ the existence of the cycle is proven.

Unfortunately, the methods of section 3 are restricted to two-dimensional systems so that higher-dimensional limit cycles cannot be constructed in the manner for two-dimensional cases. However, three-dimensional cases may be treated in the following manner, but only necessary conditions are generated. The three-dimensional limit cycle model is first transformed as before $(t \to -t)$. One of the coordinates is treated as a parameter and two-dimensional cross-sections of the asymptotic stability domain boundary are generated in a step-wise manner proceeding along both the positive and negative semiaxes of the parameterized coordinate. The three-dimensional stability surface is thus generated sequentially. If the original three-dimensional system considered admits a limit cycle, it will necessarily describe a closed curve on the three dimensional asymptotic stability surface. The locus of points observed experimentally may be compared with the numerically generated stability surface.

TIME REVERSAL AND BIOLOGICAL CLOCKS

The mapping of time reversal has been illustrated to generate an equivalence class of kinetics. This concept may be of some use in developing a fuller appreciation of biological clocks and rhythms which exhibit intermittent periodic behaviour. For the sake of discussion consider the equivalence class of kinetics which admits the invariant set $\mathcal{M}_1$ illustrated in fig. 1 respectively as a domain of asymptotic stability and as a limit cycle. The subspaces k and k' denote the external constraints for the respective situations. The desired time reversal may thus be realized as an invertible transformation $\hat{T}$ operating on the external constraints where $\hat{T}k = k'$ and $\hat{T}^{-1}k' = k$. The case illustrated in fig. 1 is highly idealized. From a practical standpoint, the respective invariant sets $\mathcal{M}_s$ need not coincide with each other and, similarly, the invariant sets $\mathcal{M}_1$ may also be disjoint. A physical example of a system which admits this kind of switching is the Teorell membrane oscillator [30-33] which exhibits damped oscillatory behaviour below a given current threshold, and sustained oscillations above the threshold.

SYSTEM IDENTIFICATION

The present theory should prove to be useful in furthering our understanding of the dynamics of physicochemical systems exhibiting limit cycle behaviour. It will complement the more physical methods used in studying oscillations. Its ultimate utility will only be realized when considered in conjunctjon with exhaustive experimental data. The reason for this is that the theory focusses on invariant sets. A single application of the theory to an isolated data set describing an observed limit cycle would probably be of marginal utility as far as system identification is concerned. This is because an infinity of kinetics may exist which admits a given invariant set $\mathcal{M}_1$ as a limit cycle. In other words, induction period behaviour is not considered explicitly in the theory although certain implicit constraints may be included in the construction as indicated in section 2.

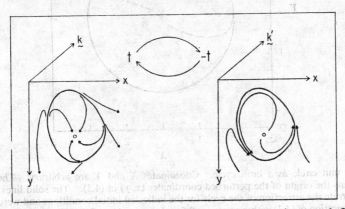

FIG. 1.—Schematic representation of the concept of time reversal in the turning on and off of biological clocks. The small open circles depict the critical point associated with the illustrated invariant sets $\mathcal{M}_1$. In the lefthand figure $\mathcal{M}_1$ is an asymptotic stability domain boundary and in the right figure $\mathcal{M}_1$ is a limit cycle. The coordinates k and k' denote the external constraints associated with the respective situations.

Best use of the theory is made when an appreciable amount of information is already known (or suspected) about the physical system under investigation. Thus if kinetics of a given polynomic or nonpolynomic form are suspected, necessary conditions on the rate constants, proportionality coefficients, and external constraints may be generated. These considerations then provide a severe test of admissibility that must be satisfied by any finally accepted model for a given system.

EXAMPLE CONSTRUCTIONS

To illustrate the methods of section 3, a class of kinetics which admits the unit circle as a limit cycle is constructed. The invariant set $\mathcal{M}_1 = \{(x, y) \mid v(x, y) = x^2 + y^2 = 1\}$ is illustrated in fig. 2 in arbitrary nonperturbed coordinates. Polynomial kinetics of the form

$$\dot{x} = a_1 x + a_2 y + a_3 x^2 + a_4 xy + a_5 y^2 + a_6 x^3 + a_7 x^2 y + a_8 xy^2 + a_9 y^3$$
$$\dot{y} = b_1 x + b_2 y + b_3 x^2 + b_4 xy + b_5 y^2 + b_6 x^3 + b_7 x^2 y + b_8 xy^2 + b_9 y^3 \qquad (4.2)$$

are presumed where the variables x and y are perturbed coordinates and the coefficients a_i, b_i are to be determined. A particular solution is obtained by solving (2.4) with $\phi(1 + f_1^2 + f_2^2)^{\frac{1}{2}} = d_1 x^2 + d_2 y^2 (d_1, d_2 > 0)$, $a = 0$, $b = 1$, and $c = \dot{y}$. The result is obtained as three inhomogeneous equations in the unknowns a_1, b_1, a_2 and b_2, four homogeneous equations in the unknowns a_3, b_3, a_4, b_4, a_5 and b_5, and five inhomogeneous equations in the unknowns $a_6, b_6, a_7, b_7, a_8, b_8, a_9$ and b_9. As all three sets of equations are undetermined, an infinity of solutions exists. Unfortunately, these solutions will not necessarily admit limit cycle behaviour following time reversal as

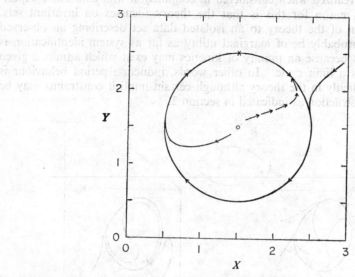

Fig. 2.—The unit circle as a limit cycle. Coordinates X and Y are arbitrary. The small open circle represents the origin of the perturbed coordinates (x, y) of (4.3). The solid lines approaching the cycle illustrate the attraction of the cycle for the indicated initial conditions and were obtained by numerical integration of (4.3) with $a_1 = b_1 = 0.5$ and $a_2 = -b_2 = 0.1$. The dashed curve illustrates the type of behaviour that may be obtained when critical points are not explicitly excluded from the proposed limit cycle set. The dashed curve was obtained by numerical integration of the system (4.2) with $a_1 = -a_6 = 1.0$, $b_2 = -b_9 = 0.5$, $a_8 = 1.5$ and the remaining coefficients set equal to zero and is seen to come to rest at the critical point "X" at the top of the circle.

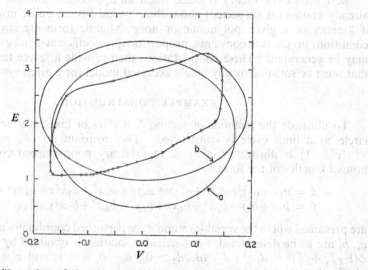

Fig. 3.—Qualitative illustration of the treatment of experimental data reported for the Teorell membrane oscillator. E is the potential in V and V is the net volume rate in ml min^{-1}. Curves a and b were defined by making a least squares fit of the function $\alpha e^2 + \gamma v^2 = 1$ to the experimental data (open circles) where e $= E - E_c$ and v $= V$ for two different choices of E_c (a: $E_c = 1.85$ V; b: $E_c = 2.25$ V). The curves a and b were generated by numerical integration of the system (4.4) with $d_1 = d_2 = 1.0$, $l = 1.0$, $(\alpha, \gamma) = (0.545, 35.3)$ for a and $(\alpha, \gamma) = (0.639, 27.7)$ for b.

critical points have not necessarily been excluded from the boundary. In fact, ten arbitrarily initiated distinct solutions were found to have critical points on the boundary. To obviate this difficulty the following less general kinetics were considered:

$$\dot{x} = a_1 x(1 - x^2 - y^2) + a_2 y \qquad (4.3)$$
$$\dot{y} = b_1 y(1 - x^2 - y^2) + b_2 x.$$

The general solution ($d_1 = d_2 = 1$) is easily demonstrated to be given as $a_1 = b_1 = 0.5$ and $a_2 = -b_2 \neq 0$. The form of (4.3) was selected so as to exclude critical points from the unit circle. In view of the fact that the determinant

$$\begin{vmatrix} 0.5(1 - x^2 - y^2) & -b_2 \\ b_2 & 0.5(1 - x^2 - y^2) \end{vmatrix}$$

is positive for all (x, y) and finite b_2, it is assured that the origin is the only critical point admitted by (4.3) so that the limit cycle is globally attractive. It should be noted that (4.3) is not admissible as a purely chemical model if interactions among the species are limited to trimolecular and lower order interactions (i.e., certain of the admissibility constraints G_i for chemical reaction kinetics are never satisfied by (4.3)).

The results for (4.3) may immediately be extended to the case where $v(x, y) = \alpha x^2 + \gamma y^2 = 1$ describes the general ellipse in standard position and $\phi(1 + f_1^2 + f_2^2)^{\frac{1}{2}}$ is chosen as the positive definite quadratic form $d_1 x^2 + d_2 y^2$. The general solution is obtained as

$$\dot{x} = \frac{d_1}{2\alpha} x(1 - \alpha x^2 - \gamma y^2) + \frac{\gamma}{\alpha} l y$$

$$\dot{y} = \frac{d_2}{2\gamma} y(1 - \alpha x^2 - \gamma y^2) - l x \qquad (4.4)$$

where l is finite.

For illustrative purposes the results (4.4) were generated ($d_1 = d_2 = 1$) for two ellipses obtained by a least squares fit to the limit cycle (in the potential—net volume rate plane) reported by Teorell[34] for his membrane oscillator. The results are depicted in fig. 3 where the open circles are the data to which the ellipses were fitted. The limit cycles a and b were defined by arbitrarily presuming the location of the critical point to be given by $(E_c, V_c) = (1.85, 0)$ and $(2.25, 0)$ respectively and then carrying out the least squares fit in perturbed coordinates (e, v) where $e = E - E_c$ and $v = V$. The cycles a and b are respectively defined by the relations $v(e, v) = 0.545e^2 + 35.3v^2 = 1$ and $v(e, v) = 0.639e^2 + 27.7v^2 = 1$; the illustrated curves were generated by numerical integration of (4.4). The purpose here is not to confirm or dispute the empirical model proposed by Teorell for this system as the experimental data do not depict an ellipse, but rather to stress the importance of having at least limited *a priori* knowledge of the model being investigated. For example, what is the actual location of the critical point encircled by the experimental cycle? In this instance a good initial guess might be obtained by extrapolation of the observed sequence of critical points observed in the relaxation of the system at various driving currents below threshold.

5. CONCLUSIONS

Necessary and sufficient conditions for the admissibility of limit cycle oscillations in physicochemical systems have been presented. Hypothetical reaction kinetics giving rise to an *a priori* limit cycle solution are constructed for the first time in a deterministic non-trial-and-error fashion, and use of the theory in treating experimental results has been outlined. Extensions of the current results are currently being attempted with the hope of determining whether or not purely chemical kinetics (with the accompanying constraints G) will admit a quadratic form (as well as higher order forms) as a limit cycle.

The author would like to thank Prof. G. Stengle for several criticisms and suggestions concerning this paper.

[1] B. Hess and A. Boiteux, *Ann. Rev. Biochem.*, 1971, **40**, 237.
[2] G. Nicolis and J. Portnow, *Chem. Rev.*, 1973, **73**, 365.
[3] B. Chance, E. K. Pye, A. K. Ghosh and B. Hess (eds.), *Biological and Biochemical Oscillators* (Academic Press, New York, 1973).
[4] D. E. F. Harrison and H. H. Topiwala, *Adv. Biochem. Eng.*, 1974, **3**, 167.
[5] I. Prigogine and R. Lefever, *J. Chem. Phys.*, 1968, **48**, 1695.
[6] R. Lefever, *J. Chem. Phys.*, 1968, **49**, 4977.
[7] R. Lefever and G. Nicolis, *J. Theor. Biol.*, 1971, **30**, 267.
[8] G. Nicolis, *Adv. Chem. Phys.*, 1971, **19**, 209.
[9] B. Lavenda, G. Nicolis and M. Herschkowitz-Kaufman, *J. Theor. Biol.*, 1971, **32**, 283.
[10] M. Herschkowitz-Kaufman and G. Nicolis, *J. Chem. Phys.*, 1972, **56**, 1890.
[11] J. J. Tyson, *J. Chem. Phys.*, 1973, **58**, 3919.
[12] J. J. Tyson and J. C. Light, *J. Chem. Phys.*, 1973, **59**, 4164.
[13] M. L. Smoes and J. Dreitlin, *J. Chem. Phys.*, 1973, **59**, 6277.
[14] R. J. Field and R. M. Noyes, *J. Chem. Phys.*, 1974, **60**, 1877.
[15] E. M. Chance, ref. (4), p. 177.
[16] V. V. Nemytskii and V. V. Stepanov, *Qualitative Theory of Differential Equations* (Moscow, 1949, English translation: Princeton University, Princeton, 1960).
[17] V. I. Zubov, *Prikl. Mat. Mekh.*, 1955, **19**, 179.
[18] V. I. Zubov, *Methods of A. M. Lyapunov and Their Application* (Noordhoff, Groningen, Netherlands: 1964).
[19] A. M. Lyapunov, *General Problem of Stability of Motion, Comm. Soc. Math., Kharkov*, 1892 (*Problème Generale de la Stabilité de Mouvement, Ann. Math. Studies* 17, Princeton, 1949).
[20] J. Texter, *J. Chem. Phys.*, 1973, **58**, 4025.
[21] J. Texter and E. E. Bergmann, *Phys. Rev. A*, 1974, **9**, 2649.
[22] D. Dwyer and J. Texter, unpublished.
[23] W. Hahn, *Stability of Motion* (Springer-Verlag, Berlin, 1967).
[24] N. P. Bhatia and G. P. Szëgo, *Stability Theory of Dynamical Systems* (Springer-Verlag, Berlin, 1970).
[25] A. D. Nazarea, *Biophys.*, 1971, **7**, 85.
[26] A. D. Nazarea, *Biophys.*, 1972, **8**, 96.
[27] A. D. Nazarea, *Biophys.*, 1973, **9**, 93.
[28] S. G. Margolis and W. G. Vogt, *IEEE Trans. Automat. Contr.*, 1963, **AC-8**, 104.
[29] J. J. Rodden, in *5th Joint Automatic Control Conf.* (Stanford Electronic Laboratories, Stanford, California, 1969), p. 261.
[30] T. Teorell, *Disc. Faraday Soc.*, 1956, **21**, 9.
[31] T. Teorell, *J. Gen. Physiol.*, 1959, **42**, 831.
[32] T. Teorell, *J. Gen. Physiol.*, 1959, **42**, 847.
[33] T. Teorell, *Biophys. J.*, 1962, **2**, 27.
[34] See ref. (31), fig. 3.

GENERAL DISCUSSION

Dr. P. Ortoleva and **Prof. J. Ross** (*MIT*) said: We have investigated the problem of the establishment of gradients of concentrations in a chemical system far from equilibrium which is originally homogeneous.[1]   For this purpose we chose a volume surrounded by a membrane and that composite system is immersed in a bath of fixed concentrations.   Note that the boundary condition across the membrane for each

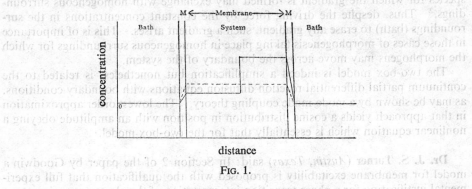

distance

FIG. 1.

concentration is a flux proportional to the concentration difference across the membrane.   For cells this is likely to be a physically better condition than one of constant concentration or constant flux.   We formulated a general theory for symmetric and asymmetric perturbations and then derived for a model system the equations for the symmetric and asymmetric stable states.   Thus we showed that from an initially stable symmetric state in a system immersed in a bath of constant concentrations (dotted line in the figure) an asymmetric stable state (solid line) evolves in certain ranges of kinetic and transport coefficients and bath conditions.

A different problem of spatial order is that of Liesegang rings, referred to earlier in this Discussion.   A theory and some experiments on this topic have been presented by Flicker and Ross.[2]

Dr. A. Babloyantz and **Prof. G. Nicolis** (*Brussels*) (*partly communicated*): To our knowledge, the paper by Ortoleva and Ross refers to a *discontinuous* system of two compartments subject to a flow of matter from the outside.   One cannot really refer in terms of " boundary condition " to the flux across the external " membrane ", as this flux is *incorporated* into the kinetic equations themselves.

Asymmetric states of concentration under similar conditions were first obtained by Prigogine and Lefever.[3]   In spite of its interest and its impact as the first model for dissipative structures, a discontinuous two-box model suffers from the inconvenience of *imposing* artificially a wavelength on the system.   This point is discussed in detail in Prigogine's paper at the second Versailles conference *From Theoretical Physics to Biology*, (Versailles, 1969).   When diffusion is taken into consideration, either in a

[1] P. Ortoleva and J. Ross, *Biophys. Chem.*, 1973, **1**, 87.
[2] M. Flicker and J. Ross, *J. Chem. Phys.*, 1974, **60**, 3458.
[3] I. Prigogine and R. Lefever, *J. Chem. Phys.*, 1968, **48**, 1695.

continuous formalism or by dividing the space into at least 3 cells, then the system can evolve to a structure whose wavelength is no longer imposed but is instead determined by the system's parameters, the boundary conditions and the size. Under these conditions it has recently been shown that polarity can appear spontaneously in an initially homogeneous morphogenetic field.

Dr. P. Ortoleva and **Prof. J. Ross** (*MIT*) (*communicated*) : Prigogine and Lefever[1] showed with a model reaction in a two-box enclosure that gradients of chemical species may be established under the conditions that these species (X, Y) do not interchange with the surroundings. We showed that in a similar system (eqn A. 7, 8) a gradient of species can be established under more restrictive conditions such that the species for which the gradient is formed may exchange with homogeneous surroundings.[2] Thus, despite the driving force by the constant concentrations in the surroundings (bath) to erase any gradient, such a gradient arises. This is of importance in those cases of morphogenesis taking place in homogeneous surroundings for which the morphogens may move across the boundary of the system.

The two-box model is indeed a simplification but nonetheless is related to the continuum partial differential reaction diffusion equations with boundary conditions, as may be shown by a mode-mode coupling theory. The lowest order approximation in that approach yields a cosine distribution in position with an amplitude obeying a nonlinear equation which is essentially that for the two-box model.

Dr. J. S. Turner (*Austin, Texas*) said: In Section 2 of the paper by Goodwin a model for membrane excitability is proposed with the qualification that full experimental justification for a phase transition interpretation of the phenomenon is not yet available. Insofar as the underlying chemical kinetics of an appropriate model for membrane excitability is found to exhibit multistationary states, then interpretation as a phase transition may well be accurate. The question of phase transitions in systems for which macroscopic chemical kinetics predicts several stable states has been studied recently by Lefever, Prigogine, and myself. Analyzing spontaneous localized fluctuations in a simple chemical model, we have verified the existence of metastable states in the coexistence region and have obtained the minimum coherence length L_c for spontaneous fluctuations necessary to induce a transition from such a state. (For details see my discussion remark following the paper of Nicolis and Prigogine in this Symposium.) In this view, for example, the phenomenon of latent excitability referred to at the end of Section 2 would correspond to naturally occurring fluctuations having a characteristic spatial extent $L < L_c$, a consequence possibly of physical constraints on membrane-related processes.

Dr. A. Babloyantz (*Brussels*) said: In his paper Goodwin considers the sudden appearance of spontaneous action potential during the regeneration of the unicellular marine alga: *Acetabularia*. But the differential equations constructed to describe his model are linear ones and cannot give a spontaneous generation of inhomogeneity in a given physico-chemical parameter starting from a homogeneous situation. The inhomogeneous result found by the author is due to his special choice of the so-called pick-up function. In particular for the special case of Acetabularia any theory has to explain how in anucleated fragments one can have a generation of polarity giving rise either to one cap formation at one end of the fragment or two caps at either end

[1] I. Prigogine and R. Lefever, *J. Chem. Phys.*, 1968, **48**, 1695.
[2] P. Ortoleva and J. Ross, *Biophys. Chem.*, 1973, **1**, 87.

of the fragment. Can the author account for these experimental facts? And if so, how is the number of caps determined within his linear theory?

Dr. B. C. Goodwin (*Sussex*) said: Babloyantz and Auchmuty questioned whether the model of wave propagation and gradient formation on a membrane was spatially uniform initially, and whether it is dynamically stable. It is evident from fig. 1 that all the molecular components of the model are distributed uniformly in the membrane so that there is no initial asymmetry. Once a wave begins to propagate from some point, however, there is, first dynamical asymmetry (i.e., at any point in space one process is followed by another in a defined manner, as determined by the pattern of interactions described by the kinetics of the reaction scheme) and, subsequently, spatial asymmetry in the bound metabolite arising from the dynamical asymmetry together with wave propagation. This leaves open the question how the wave origin is determined, which was done deliberately in order to cover the various cases to which the model could be applied. These include (1) an external, environmental event determining where the symmetry-breaking wave is initiated, as in the case of light acting upon a fertilized *Fucus* egg; (2) an initial gradient of some kind which establishes the wave origin, as in an amphibian embryo or a decapitated *Hydra*; (3) a system, with neither initial polar order nor an environmental asymmetry, which can break symmetry spontaneously as a result of spontaneous fluctuations which exceed the level for initiation of the wave. In this instance, more than one wave origin can arise, so that competing organising centres may result as occurs in bipolar *Acetabularia* segments. The model is applicable to all these instances, but it does not, of course, explain them.

Regarding stability, the basic equations of the model as described, for example, by Goodwin [1] are necessarily stable when $P(x,t)$ is a propagating delta function (the case analysed) because a finite amount of substance is produced over any finite spatial region of the membrane. Thus there cannot, in this instance, be any question of instability and the problem considered is simply the spatial distribution of the morphogen resulting from the passage of a wave along a membrane.

Dr. J. F. G. Auchmuty and **Prof. G. Nicolis** (*Brussels*) (*partly communicated*): In their paper Balslev and Degn appear to overlook the effect of boundary conditions on the solutions of reaction-diffusion equations. Their conclusions that $(D_B/D_A) > 1$ is a necessary condition for obtaining an aperiodic instability and that in domain I of fig. 2, the system will necessarily evolve to a uniform limit cycle, only hold under special boundary conditions.

An analysis of the one-dimensional trimolecular model in the case where the concentrations are held fixed and equal to their uniform steady state values at the boundaries (see the paper of Nicolis and Prigogine at this Symposium) shows that the first instability may give rise to a *space-dependent and time-periodic* structure. Moreover such structures arise before aperiodic solutions provided

$$\left(\frac{D_X}{D_Y}\right)^{\frac{1}{2}} > \sqrt{1 + \frac{1 + (D_X + D_Y)\pi^2}{A^2} - \frac{1}{A}} \tag{1}$$

where we are using the notation of Nicolis and Prigogine. Here D_Y, D_X are the diffusion coefficients of X (the " self-activating " component) and of Y. In particular D_Y/D_X need not be greater than 1 to lead to an aperiodic stability when A is sufficiently small.

[1] B. C. Goodwin, *Theoretical Physiology*, in *Selected Topics in Physics, Astrophysics and Biophysics*. ed. Laredo and Juristic, (D. Reidel Publ. Co., 1973) pp. 381–420.

The reason for this is the fact that linear stability analysis, for one dimensional systems of length l with fixed concentrations at the boundaries, allows modes of the form

$$x \propto \exp(\omega t) \sin \frac{n\pi r}{l} \qquad (0 \leqslant r \leqslant l) \qquad (2)$$

for $n = 1, 2, \ldots$. The state with $n = 0$ is now excluded because of the boundary conditions and so one gets these different results. Fixed concentration boundary conditions may arise if the reaction takes place inside semipermeable membranes and there is a reservoir of reactants outside. Prof. D. Thomas at Compiègne has been conducting experiments with these boundary conditions.

Dr. A. Winfree (*Indiana*) said: Balslev and Degn point out that in the linear approximation near a homogeneous stationary state, propagating waves cannot occur in a 2-component reaction/diffusion system unless it oscillates while well stirred (i.e., with ∇^2 term omitted, eigenvalues are now real) and that initiation of such *spatial* instability requires that the diffusion coefficients be unequal.

As usual with linear systems, one must enquire how much non-linearity or how great an excursion from the stationary state is required to admit the forbidden phenomena. I submit by way of answer only an isolated example, but it is an instructive one, I think:

$$\frac{\partial A}{\partial t} = \nabla^2 A - A - B + H(A - 0.05)$$

$$\frac{\partial B}{\partial t} = \nabla^2 B + A/5$$

where $H(x) = 1$ if $x > 0$, or 0 otherwise. It will be noted that the diffusion coefficients are equal and the equation is piece-wise linear, the two pieces meeting along $A = 0.05$. In the well-stirred case (∇^2 terms omitted), the stationary state (0,0) is a unique global attractor with negative real eigenvalues -0.72, -0.27.

Yet this equation's stable solutions (depending on initial conditions and boundary conditions) include not only the homogeneous stationary state, but also propagating plane waves and concentric circular waves, and spiral waves polarized to rotate clockwise or counterclockwise (unpublished numerical computations).

Prof. I. Balslev and **Dr. H. Degn** (*Denmark*) said: In answer to Auchmuty and Nicolis we remark that in our paper we have emphasized that the boundary conditions we wish to consider are the ones a chemist will always assume if nothing else is explicitly stated, namely the ones of " an ordinary vessel with non-reacting walls ". We have abstained from discussing the consequences of other boundary conditions such as fixed concentrations at the walls because we would feel obliged to discuss also the operational meaning of such boundary conditions. We have not been able to think of any operations which will assure fixed concentrations at the walls.

In answer to Winfree we affirm that the hypothesis of the existence, in certain reaction systems, of spatial instability towards *infinitesimal* perturbations is a fundamental one. The aim of our paper is to contribute to the clarification of the conditions required for such spatial instability. The interesting observations of propagating patterns presented by Winfree in this symposium seem to be dependent on *finite* perturbations. Likewise, Winfree's example, given in his comment, has a globally stable steady state. Therefore, a departure from the corresponding homogenous state can be brought about only by a *finite* perturbation.

Dr. R. G. Gilbert and **Mr. S. McPhail** (*University of Sydney*) said: In theoretical discussions of reactions which produce spatial oscillations, an obvious point which is frequently overlooked is that a stable final stationary inhomogeneous state, $\psi^1(x)$, is required in addition to an unstable initial state. In a one-dimensional system described by

$$\partial\psi/\delta t = F(\psi) + D\partial^2\psi/\partial x^2 \tag{1}$$

(where ψ is the usual vector of concentrations, etc.), then ψ^1 is defined by putting the right-hand side of (1) equal to zero. An indication of the stability of ψ^1 is that the real parts of the eigenvalues of the matrix $\partial F/\partial\psi^1$ be negative, for all x. We have written a general program for numerically determining ψ^1, requiring the solution of coupled differential equations with double-ended boundary conditions (which, as the equations are very badly conditioned, poses considerable numerical difficulty).

esting this program over a small range of x on the Brusselator mechanism, we found not only the solution derived by Lefever [1] by complete numerical solution of (1) (which was of course stable) but also several unstable solutions, one of which was asymmetric and hence relevant to Babloyantz' previous remark. We feel that the search for stable stationary inhomogeneous states will prove most useful in understanding morphogenesis.

Dr. A. Babloyantz and **Dr. J. Hiernaux** (*Brussels*) (*partly communicated*): We wish to present a mechanism explaining the spontaneous onset of polarity in an initially homogeneous morphogenetic field. Ever since Turing's [2] important work it has been known that a reaction diffusion system of two or several morphogens may account for many phenomena of morphogenesis. In particular, the spontaneous formation of inhomogeneous patterns was viewed as the result of the instability of the homogeneous state triggered by random disturbances.

However, due to the particular boundary conditions used by Turing the patterns he obtained did not present any polarity. This was also the case for all subsequent work on dissipative structures.[3-5]

Consider a reaction-diffusion system of two morphogens m_1 and m_2 described by the following kinetic equations:

$$\frac{\partial m_1}{\partial t} = f(m_1, m_2) + D_{m_1}\frac{\partial^2 m_1}{\partial r^2}$$

$$\frac{\partial m_2}{\partial t} = g(m_1, m_2) + D_{m_2}\frac{\partial^2 m_2}{\partial r^2}$$

f and g describe chemical reactions which must obey kinetics similar to those proposed by Turing.

We impose zero flux at the boundaries of the system. Then, it can be shown [6, 7] that for a proper choice of parameters there is a critical length L_0 below which the system remains homogeneous.

[1] R. Lefever, *J. Chem. Phys.*, 1968, **49**, 4977.
[2] A. M. Turing, *Phil. Trans. R. Soc. B*, 1958, **237**, 37.
[3] P. Glansdorff and I.– Prigogine, *Thermodynamics of Structure, Stability and Fluctuations*, (Wiley–Interscience, New York, 1971).
[4] I. Prigogine and R. Lefever, *J. Chem. Phys.*, 1968, **48**, 1965.
[5] J. F. G. Auchmuty and G. Nicolis, *Bull. Math. Biol.*, 1975.
[6] A. Babloyantz and J. Hiernaux, *Science*, 1975, to appear.
[7] A. Babloyantz and J. Hiernaux, *Bull. Math. Biol.*, 1975.

When the length of the system is greater than L_0, a one-wavelength structure appears in the system, and is of the form seen in fig. 1 which presents a polarity. For higher values of L, there appear structures of several wavelengths.

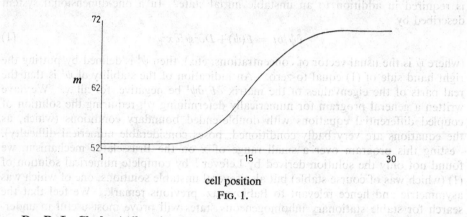

FIG. 1.

Dr. B. L. Clarke (*Alberta*) said: Balslev and Degn have demonstrated the desirability of being able to study reaction-diffusion instabilities which cannot occur in systems with too few reactants. I would like to indicate how the steady state stability of large reaction-diffusion systems can be determined.[1]

The stability analysis of large realistic chemical systems is greatly simplified by the fact that the rate constants usually range over many orders of magnitude. The matrix elements of eqn (2) and (11) (of the paper by Balslev and Degn) are then also of different orders of magnitude and the equations in the Appendix can be simplified to a few dominant terms. For large reaction systems this simplification makes an otherwise unwieldy algebraic problem tractable. Since the stability domains are determined by the signs of various polynomials, the curves separating the stability domains have approximate equations which state that two possible dominant terms of opposite sign are equal. Such equations are always hyperplanes when plotted logarithmically in the parameters. From this it follows that the stability domains of a reaction-diffusion system are asymptotically convex polyhedral cones in the logarithmic parameter space. This approach is only valid when the parameters cannot change sign. If each of the quadrants in fig. 1 and 2 is mapped into the space with coordinates $(\log|\alpha|, \log|\beta|)$ the hyperbolas and lines through the origin will be mapped into straight lines and all the stability domains (except region II of fig. 2) will appear as convex wedges with distortions near their apexes. The cone-like structure of the stability domains breaks down near $\log|\alpha| = \log|\beta| = 0$ because the parameters in eqn (2) are then all of similar magnitude. For chemical networks it can be shown that a complete breakdown of this approximation only occurs when all reactions have the same steady state currents and all evolving reactants have the same steady state concentration. The conditions for even a partial breakdown of this approximate approach are not expected to be encountered often.

If the stability analysis is carried out for a particular reaction network rather than for the general matrix as in eqn (2), the parameters in the matrix will all be of the form nj/c where c is a steady state concentration and j is a current and n is a small integer which comes from kinetics and stoichiometry. If diffusion is included, the parameters Dk^2 also occur. Since these parameters are all positive, the stability

[1] B. L. Clarke, *J. Chem. Phys.*, 1974, **60**, 1481.

domains in the analogue of fig. 1 are only of interest in the positive orthant and only one mapping into the logarithmic space is necessary. Furthermore, each convex cone of instability can be related to the network features which produce it. The dominant term in the stability polynomials which is responsible for destabilizing a network is a product of matrix elements and can be factored into a set of feedback loops which are important in the instability. Thus, each cone of instability is caused by recognizable combinations of feedback loops and the stable regions adjacent to each cone are stabilized by other recognizable network features.

Dr. M. Herschkowitz-Kaufman (*Brussels*) said: In the paper by Balsev and Degn it is stated that the existence of multiple stable patterns is not yet established for a reaction-diffusion system. I should like to emphasize that the multiplicity of stable steady state structures has now been firmly established, both by computer simulation and analytically. Consider the trimolecular reaction scheme. In a bounded medium subjected to zero flux or fixed boundary conditions, computer simulations have shown the existence, beyond a certain distance from the instability point, of a *finite*

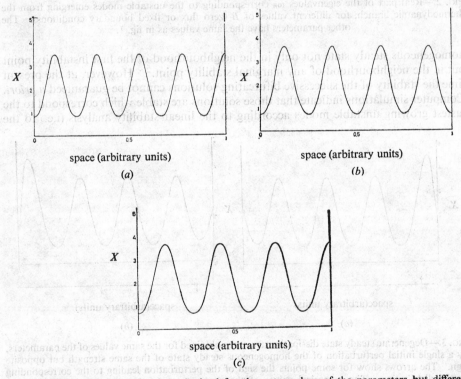

space (arbitrary units)   space (arbitrary units)

(a)   (b)

space (arbitrary units)

(c)

Fig. 1.—Steady state profiles of X obtained for the same values of the parameters but different initial conditions. The curves have been established for zero flux boundary conditions. Numerical values used: $A = 2$, $L = 1$, $B = 4.6$, $D_x = 1.6 \times 10^{-3}$, $D_y = 8.0 \times 10^{-3}$.

multiplicity of spatial steady state dissipative structures, each one having its own domain of attraction as a function of the initial conditions.[1] This is illustrated on fig. 1. Each one of the three different stable structures observed corresponds to a different

[1] G. Nicolis, *Proc. AMS-SIAM, Symposium on Appl. Math.*, vol. 8 (American Math. Soc., Providence, 1974).

wavelength and amplitude. They emerge from the homogeneous steady state for the same values of the parameters of the system but different initial conditions, i.e., depending on the sign, location and number of initial perturbations. This phenomenon can be inderstood on the basis of a bifurcation analysis which can provide an analytical approximation for the new steady state solutions bifurcating from the

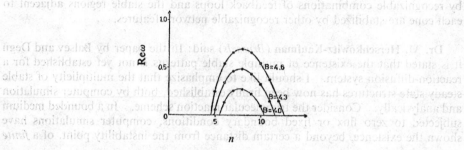

FIG. 2.—Real part of the eigenvalues ω_n corresponding to the unstable modes emerging from the thermodynamic branch, for different values of B (zero flux or fixed boundary conditions). The other parameters have the same values as in fig. 1.

homogeneous steady state not only in the neighbourhood of the first instability point but in the neighbourhood of any marginal stability point.[2]   However, at the present time the stability of the successive bifurcating solutions cannot be guaranteed *a priori*. Computer simulations indicate that those solutions are stable which correspond to the fastest growing unstable modes according to the linear stability analysis (i.e., to the

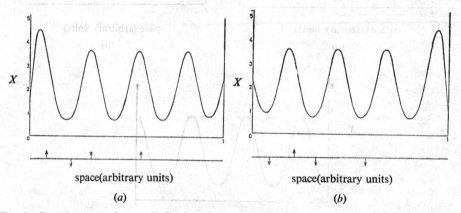

space(arbitrary units)   space(arbitrary units)

(a)   (b)

FIG. 3.—Degenerate steady state dissipative structures obtained for the same values of the parameters, by a single initial perturbation of the homogeneous steady state of the same strength but opposite sign.   The arrows show for some points the sign of the perturbation leading to the corresponding spatial distribution.   The numerical values of the parameters are the same as in fig. 1.   The homogeneous steady state concentrations $X = A$, $Y = B/A$ are maintained at the boundaries.

first few successive bifurcations).   This is seen explicitly in fig. 2 where $n = 7, 8$ and 9 appear indeed as the leading modes for the numerical values corresponding to fig. 1.

An additional mechanism of multiplicity, operating even near the first bifurcation point, is degeneracy as appears from the analytical expression (3.4) presented in the

[1] G. Nicolis and J. F. C. Auchmuty, *Proc. Nat. Acad. Sci.*, 1974, **71**, 2748.

paper by Nicolis and Prigogine. Fig. 3 gives a typical example of such degenerate states obtained by computer simulations.

Dr. P. Hanusse (*CNRS, Talence*) said: I wish to mention to Balslev that we have established several theorems on the occurrence of instability, especially in two component systems.[1]

The main result is that in a system with two components, when only mono and bi-molecular steps are involved, whatever the form of reaction scheme,

1 no unstable node or focus may occur, so no limit cycle type oscillation,

2 when the system is stable with respect to homogeneous perturbations, that is stable when stirred, it can not be spatially unstable.

Consequently, models with two components, can lead to oscillations only if they contain tri-molecular steps, as does the Brussellator.

Dr. J. F. G. Auchmuty (*Brussels*) said: I would wish just to mention to Ross that for single component reaction-diffusion systems, one often gets the single front propagation described in his paper, in which one phase annihilates the other. In multicomponent systems, however, other phenomena besides coexistence and propagating fronts can occur.

These phenomena include the existence of fronts whose propagation velocity is not constant and of unstable propagating fronts as well as the nonexistence of such fronts. In such examples the boundaries between the states can become very complicated (as for example, often occurs in the Stefan problem) and for certain model systems, one can even get explosions when two phases are mixed in the manner described in this paper.

Even the very concept of boundary between two phases is very difficult to define when the steady state solutions are space-dependent.

Prof. J. Ross (*MIT*) said: In the concluding paragraphs of our paper we comment on the agreement of two methods for analyzing aspects of nucleation, one based on solutions of the time-dependent macroscopic (deterministic) equations, the other on the steady state probability distribution obtained from a master equation. Since submission of this article we (K. Kithara, H. Metiu and J. Ross) have found reasons for such agreement. In brief we have derived by path-integral methods solutions of the master equation, for " curl-free " reaction systems and a certain form of the transition probability, which show that the deterministic equations yield the trajectory for the most probable dynamic evolution. Hence the evolution of such non-linear system in the direction of increasing probability is, on the average, given by the deterministic equations.

Dr. P. Fife (*Tucson, Arizona*) said: These remarks extend the results given by Nitzan, Ortoleva, and Ross in their Section 2.

1. Consider the case of a single dependent variable ψ. Let $\psi_0(\lambda)$ and $\psi_1(\lambda)$ be the two stable stationary states, and

$$J(\lambda) = \int_{\psi_0(\lambda)}^{\psi_1(\lambda)} F(\psi, \lambda) \, d\psi.$$

[1] P. Hanusse, *Compt. Rend.*, 1973, **277C**, 263 ; 1972, **247C**, 1247.

If there is only one stationary state (unstable) between ψ_0 and ψ_1, then the critical values $\lambda_M(2.4)$ can be found from the simple relation [1, 4]

$$J(\lambda_M) = 0.$$

If there is more than one stationary state between ψ_0 and ψ_1, then the extra condition

$$\int_{\psi_0(\lambda_M)}^{k} F(\psi, \lambda_M)\,\mathrm{d}\psi > 0 \quad \text{for} \quad \psi_0(\lambda_M) < k < \psi_1(\lambda_M)$$

has to be imposed. In the case of more than one dependent variable, apparently there is no such simple criterion available for determining λ_M.

2. Again consider one dependent variable, but an inhomogeneous medium with the constraints λ held fixed. Thus dependence on x replaces dependence on λ. Define $\psi_0(x)$, $\psi_1(x)$, and $J(x)$ as above. The coexistence problem is to find a solution of

$$D\frac{\mathrm{d}^2\psi}{\mathrm{d}x^2} + F(\psi, x) = 0,$$

with ψ exhibiting a phase transition from values near $\psi_0(x)$ to values near $\psi_1(x)$ as x crosses some transition region. If D is small, a singular perturbation analysis [1, 4] may be applied to determine the conditions for such a solution to exist. Suppose there is a value x^* such that $J(x^*) = 0$ and $J(x)$ changes sign as x passes through x^* (also assume an extra condition like that above, if there is more than one stationary state between ψ_0 and ψ_1). Then a phase transition may occur in an interval centred at x^*, with width of order $D^{\frac{1}{2}}$.

Furthermore, a heuristic argument shows this coexistence solution to be stable, provided that

$$J(x) > 0 \text{ for values of } x \text{ `` in phase } \psi_0 \text{ ''};$$
$$J(x) < 0 \text{ for values of } x \text{ `` in phase } \psi_1 \text{ ''}.$$

This contrasts with the homogeneous case treated in Ross's paper, in which a small perturbation of λ will start the front moving, with consequent annihilation of one of the phases.

If J is negative for some finite interval $a < x < b$ only, then ψ_1 can exist in that same interval, with transitions to ψ_0 at both ends.

3. Consider now the case of two dependent variables u and v, and a homogenous medium.[2] Assume the diffusion coefficient D of u to be much smaller than that of v, which we set equal to 1. The steady-state equations are

$$D\frac{\mathrm{d}^2u}{\mathrm{d}x^2} = f(u, v)$$

$$\frac{\mathrm{d}^2v}{\mathrm{d}x^2} = g(u, v).$$

Suppose the curve $f = 0$ is S-shaped, with ψ_0 and ψ_1 the two stable branches:

[1] P. Fife, *J. Diff. Eqn.*, 1974, **15**, 77. See esp. pp. 102 and 103.
[2] P. Fife, *J. Math. Anal. Appl.*, 1975, to appear.
[3] P. Fife and W. M. Greenlee, *Uspeki Matem. Nauk SSSR*, 1974, **49**, 103; also to appear in *Russ. Math. Surveys*.
[4] A. B. Vasil'eva and V. A. Butusov, *Asymptotic Expansions of Solutions of Singularly Perturbed Equations* (in Russian) (Nauka, Moscow, 1973).

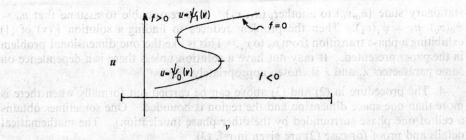

Again, $J(v)$ is defined, and there will be a value v^* at which $J(v^*) = 0$ and J changes sign. Let

$$H(v) = \begin{cases} \psi_0(v), & v < v^* \\ \psi_1(v), & v > v^*. \end{cases}$$

If the reacting mixture is confined to a finite region, say $0 \leqslant x \leqslant 1$, with boundary conditions imposed on u and v at both ends, we proceed by solving

$$\frac{d^2V}{dx^2} = g(H(V),V) \tag{1}$$

with the given boundary conditions for v imposed upon V. This problem is easily analyzed by phase-plane arguments, despite the discontinuity on the right. If there exists a solution $V(x)$ crossing the value v^* at some point x^*, then one can expect a solution (u,v) of the original problem when D is small, for which $v(x)$ is close to $V(x)$, and for which u exhibits a transition in a small interval of width of the order $D^{\frac{1}{2}}$ at $x = x^*$. On one side of the interval, u is approximated by $\psi_0(V(x))$, and on the other side, by $\psi_1(V(x))$. The proof needs some extra minor assumptions. Again, the solution appears to be stable if $J > 0$ in phase ψ_0, and $J < 0$ in ψ_1.

The following example is chosen for clarity rather than reference to any physical system. The function f is not S-shaped, but the above results hold in any event. The fact that negative concentrations are assumed can easily be remedied. Primes denote differentiation in x.

$$\text{Example}: Du'' = (u-v)(u^2-1)$$
$$v'' = -au$$
$$u' = v' = 0 \text{ at } x = 0 \text{ and } 1.$$

Here $\psi_0(v) = -1$, $\psi_1(v) = 1$, $v^* = 0$, and $V(x)$ is piecewise parabolic.

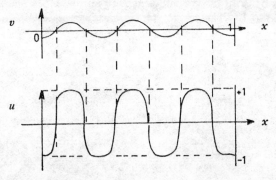

On the other hand if the medium is infinite (or large compared to the square root of v's diffusion coefficient), we may seek a solution passing from one stable stationary state (u_0, v_0) to another, (u_1, v_1). It is reasonable to assume that $u_0 = \psi_0(v_0)$, $u_1 = \psi_1(v_1)$. Then the problem reduces to finding a solution $V(x)$ of (1) exhibiting a phase transition from v_0 to v_1. This is like the one dimensional problem in the paper presented. It may not have a solution unless there is a dependence on some parameters λ, and λ is chosen appropriately.

4. The procedure in (2) and (3) above can be carried out formally when there is more than one space dimension and the region is bounded. One sometimes obtains a cell of one phase surrounded by the other phase (nucleation). The mathematical details and proof for case (2) are given in ref. (3).

5. The nucleation problem discussed in the presented paper is such that when D is small, the radius of the cell R_C is of the order $D^{\frac{1}{2}}$, as is the transition layer width l. The cells constructed as in (4), however, have size independent of D for small D.

Dr. J. S. Turner (*Austin, Texas*) said: I would like to point out that nucleation by *spontaneous* localized fluctuations in systems exhibiting multiple steady states has been studied by Lefever, Prigogine, and myself. We have discussed the occurrence of *metastable* states in a simple model, and have obtained a critical coherence length for spontaneous fluctuations necessary to form an unstable nucleus in an initially homogeneous metastable phase. For details see my discussion remark following the paper of Nicolis and Prigogine in this Symposium.

AUTHOR INDEX*

*The references in heavy type indicate papers submitted for discussion.

The references in heavy type indicate pages submitted for discussion.

Confessing Christ as Lord:

The Urbana 81 Compendium

Edited by
John W. Alexander

InterVarsity Press
Downers Grove
Illinois 60515

InterVarsity Press is the book-publishing division of Inter-Varsity Christian Fellowship, a student movement active on campus at hundreds of universities, colleges and schools of nursing. For information about local and regional activities, write IVCF, 233 Langdon St., Madison, WI 53703.

Distributed in Canada through InterVarsity Press, 1875 Leslie St., Unit 10, Don Mills, Ontario M3B 2M5, Canada.

All Scripture quotations, unless otherwise indicated, are from the Holy Bible: New International Version © 1978 by the New York International Bible Society. Used by permission of Zondervan Bible Publishers. All Scripture quotations marked RSV are from the Revised Standard Version of the Bible, copyrighted 1946, 1952, © 1971, 1973.

ISBN 0-87784-499-2

Printed in the United States of America

Library of Congress Cataloging in Publication Data
Main entry under title:

Confessing Christ as Lord.

   "Messages given at the thirteenth Inter-Varsity Student Missions Convention"–P. [4] of cover.
   1. Missions–Congresses. 2. Christian life–
1960-   –Congresses. I. Alexander, John W.
II. Inter-varsity Student Missions Convention (13th : 1981 : Urbana, Ill.)

| BV2390.C66 | 266 | 82-251 |
| ISBN 0-87784-499-2 | | AACR2 |

| 16 | 15 | 14 | 13 | 12 | 11 | 10 | 9 | 8 | 7 | 6 | 5 | 4 | 3 | 2 | 1 |
| 95 | 94 | 93 | 92 | 91 | 90 | 89 | 88 | 87 | 86 | 85 | 84 | 83 | 82 | | |

FOREWORD

If you are concerned to learn more about the crucial issues in world missions today, you will find assistance in this volume.

These messages were shared at the Urbana 81 Student Missions Convention, a convention strategically planned to bring before thousands of delegates the major issues in world missions confronting the Christian church today.

Although the speakers came from diverse backgrounds, they all have one thing in common. They have an intense personal love for Jesus Christ and a profound burden for the three billion people who do not yet know Jesus Christ as their Lord and Savior.

The theme of Urbana 81 was "Let every tongue confess that Jesus Christ is Lord" (Phil 2:11). Each speaker's message undergirded that theme in a unique manner. It was the desire of those who planned the convention and those who spoke to make Jesus Christ pre-eminent.

As you read these messages, be prepared to have your heart tremendously stirred. God may speak to you afresh about your personal commitment to the cause of world missions. The task ahead is immense. However, it is possible that the total evangelization of our world could be a reality by the year 2000 if each Christian would faithfully carry out his or her share in completing Christ's Great Commission as given in Mark 16:15: "Go into all the world and preach the good news to all creatures."

John E. Kyle

PREFACE

It was the frigid winter of 1981. For five days, December 27-31, nearly 14,000 people occupied the campus of the University of Illinois in Urbana for the 13th national missions convention sponsored by Inter-Varsity Christian Fellowship of the United States and Canada.

The single overriding purpose of Urbana 81 was to glorify the Lord Jesus Christ by helping students find God's place for them in world missions and thus to serve the church by strengthening its ministry in missions.

Within that primary purpose the convention had four major objectives:

1. To help each student consider seriously what the Bible says about God's eternal and unchangeable plan for the world-wide proclamation and demonstration of the gospel of Jesus Christ.

2. To help each student consider seriously the situation in the world today as the church of Jesus Christ proclaims and demonstrates the message of redemption.

3. To help each student consider seriously what these two

things mean, individually and corporately, in obeying God's will for his or her life in the context of the body of Christ and his or her relationship to the world, both on campus and beyond.

4. To help each student to strengthen the missions impact of the student group at his or her school this year.

Along with students, those attending Urbana included missionaries, pastors, Inter-Varsity staff and Urbana volunteers.

Those were trying days on the world scene. In Poland, martial law had recently been enacted, and Polish Christians asked for prayer for their distressed country. Strife in El Salvador continued along with conflicting reports about its causes. At Urbana 81 we prayed that God's will would prevail in the affairs of humanity. The convention was aimed to help students discern God's will as to how to be involved in the witness for Christ in such a world.

The theme of the convention was: "Let every tongue confess that Jesus Christ is Lord." Each day's schedule began with time for a private morning watch (quiet time) in which delegates were encouraged to use a series of studies (prepared by Yvonne Vinkemulder) in Luke and Acts.

From 8:00 to 9:00 each morning the delegates were in small group Bible studies averaging from eight to ten members. Dispersed throughout the university, these small groups were each led by a student or staff member who had been prepared specifically for these daily ministries. All of these small groups also worked their way through studies based on passages from Luke and Acts.

From 10:00 until 11:00 each morning the delegates filled the huge Assembly Hall for an hour of daily Bible exposition by Eric Alexander (from St. George's-Tron Parish Church in Glasgow, Scotland) on chapters of Acts. From 11:00 until noon the assembly sang, prayed and listened to messages on diverse topics: rural missions, urban missions, qualifications for missions, and finding the will of God.

Afternoons offered a host of options. The university Armory and Huff Gym were open for delegates to confer with missionaries stationed at scores of booths. Workshops, seminars and discussions led by missionaries were dispersed throughout classrooms and lecture halls from one end of campus to the other.

Each evening from 7:00 to 9:00 delegates again jammed the Assembly Hall to sing, pray and hear messages from missionaries and mission leaders. The day closed with the small groups meeting once more, this time for sharing and prayer.

There were specialized Bible study groups for pastors each morning from 8:00 to 9:00 led by Gordon MacDonald of Grace Chapel in Lexington, Massachusetts. Another session, for missionaries, was led by Frank Barker of Briarwood Presbyterian Church in Birmingham, Alabama.

Each day the excellent multimedia presentations by Inter-Varsity's TWENTYONEHUNDRED PRODUCTIONS powerfully presented, through music and visuals as well as the spoken word, key issues facing us in missions.

What were the results? God alone can answer that question in full, but a few results are known even as this volume goes to press. About 7,500 students submitted signed decision cards indicating a commitment to greater missions awareness and an openness to a mission call from God. Numerous mission agencies said they received an overwhelming response from students who submitted names and addresses and asked for assistance in discerning opportunities for mission service. An offering of almost $250,000 was taken the evening of December 30 to support student work worldwide. Another offering to help people in distressed areas exceeded $40,000. Most of the delegates also fasted during the noon meal of December 29, enabling the University of Illinois food service to refund $15,000 which was divided up into gifts of $5,000 and awarded to three agencies engaged in poverty relief.

God worked during Urbana 81. In the words of one student, "God opened my eyes to new things at Urbana. I have a new vision of God's work around the world."

May Jesus Christ be praised! May his message be proclaimed around the world so that every tongue will confess that Jesus Christ is Lord.

John W. Alexander

Part 1

Introduction

1
Why Are We Here?

John W. Alexander

Welcome to Urbana 81. It is a privilege to be with you at the start of this missions convention.

At the very outset let us remind ourselves that we are assembled here in the name of Christ Jesus. It is our desire that he will have pre-eminence in all that we think, speak, write and do. May Jesus Christ be praised!

Let's take a look at Acts 13:1-4. The setting for this passage is a local church in Antioch. The Holy Spirit moved in that church and triggered a historic missions endeavor:

In the church at Antioch there were prophets and teachers: Barnabas, Simeon called Niger, Lucius of Cyrene, Manaen (who had been brought up with Herod the tetrarch) and Saul. While they were worshiping the Lord and fasting, the Holy Spirit said, "Set apart for me Barnabas and Saul for the work to which I have called them." So after they had fasted and prayed, they placed their hands on them and sent them off.

The two of them, sent on their way by the Holy Spirit, went . . .

For the rest of that chapter and into the succeeding chapters, we read of how Paul and Barnabas traveled around on their missionary journey. And then in Acts 14:26-27 we read about what happened after they came back. "They sailed back to Antioch, where they had been committed to the grace of God for the work they had now completed. On arriving there, they gathered the church together and reported all that God had done through them and how he had opened the door of faith to the Gentiles."

There are four points to be made from this paragraph. First, we see the missionary-sending church; second, the missionary-sending authority; third, the pair of missionaries who were sent; fourth, the mission field to which they went. Let's look at each of these four topics. They can be likened to the four factors in "the missionary equation."

Missionary-sending church. We are not given much information about the local church at Antioch. In Acts 13 there is no indication of the church's size, the age of its members, its style. We are not told how many men and women comprised it. On the other hand, we are given several facts. The church contained both prophets and teachers. It had a high view of worship and fasting. It was a praying church. It was open and obedient to the movement of the Holy Spirit. Thank God for local churches which today manifest those same characteristics. May the day come when more and more local churches share these same attributes.

Missionary-sending authority. The Holy Spirit is identified in this passage from Acts as the missionary-sending authority. He had a plan by which the Antioch church would serve as the sending agency, sending two of her members to other places as missionaries. The Holy Spirit interrupted the normal life of the church with this message: "Set apart for me Barnabas and Saul for the work to which I have called them."

We are not given much detail on how the church reacted to this interruption, but it must have been jolted. The Spirit's plan called for detachment of forty per cent of the leadership team! Five men are identified as leaders in that local fellowship:

Barnabas, Simeon, Lucius, Manaen and Saul. The Spirit said he wanted two of those five to be set free from local church responsibilities in order to go traveling as missionaries. Imagine your own local church for a moment. Imagine the impact if almost half of her leaders were suddenly detached for mission work elsewhere. No wonder the Antioch church called a fast prayer meeting!

After fasting and praying they laid their hands on Barnabas and Saul and sent them off. I wish I could have been there for that laying-on-of-hands ceremony. Obviously I do not know exactly what the ceremony meant to the various participants. But I have been in a few local churches in recent years when church leaders conducted such a ceremony, and it seems to me that in those sacred moments the leaders were communicating at least two messages to the ones being sent out.

First, it is the Holy Spirit who sends the witnesses out. By laying on hands church members symbolize the fact that it is the Spirit who is moving to send them. And as they go, they represent the church. They are being detached from their church, but they are not being severed spiritually. They are an arm reaching out from a local fellowship to the mission field.

The second message is this. The church is going to stand behind the witnesses. It will support them with its prayers. It will support them with its money. The church is not cutting the missionaries adrift.

Missionaries: the sent-ones. Barnabas and Saul were obedient to the call. I have often wondered if they had received that call *first*, before the church, if they were caught by surprise. Perhaps they were. But at any rate, they obeyed. The Holy Spirit asked the Antioch church to obey—and they did. Likewise, the Holy Spirit asked Barnabas and Saul to obey—and they did. Is not this pattern still operative today? The Holy Spirit has his plan for proclaiming the gospel of Christ around the world, and in that plan he is both calling missionaries to go and calling local churches to send. Verse 3 in the passage says that the two missionaries were sent off by the local church. Verse 4 says they were sent out by the Holy Spirit. Both statements are true. It is certain that there are some today who will be like Barnabas and

Saul—individuals whom the Holy Spirit will be sending out. And then there are others who will be like Simeon, Lucius and Manaen—pillars in a local church whom the Spirit will be calling to lay hands on the sent-ones and send them off.

The mission field. The fourth point in this message deals with the mission fields to which Barnabas and Saul went, where they found men and women ready to listen. Here again the Holy Spirit was at work, preparing the minds and hearts of people for contact with the missionaries. Wherever Barnabas and Saul went they found that the Holy Spirit had already gone before them. Surely the same procedure continues to operate; namely, when the Holy Spirit moves upon a local church to send forth missionaries, he will go before them and make ready the mission fields to which they go. In some parts of the world it appears he is not preparing quite the quantity of receiving hearts at quite the pace he is doing it in other parts of the country. But we'll leave that to the mystery of his will. Our obligation is to obey him as we go. And for the rest of us who don't go abroad, we will obey him in supporting those who do go.

These then are the four major factors in the missionary equation. The missionary-sending authority moves upon the missionary-sending agency to send forth the missionaries to the mission fields signaled by God himself.

One way to answer the question, What is the purpose of Urbana 81? is to reply: to help us all understand more fully the missionary equation and the factors in it and to discern God's will as to how he wants us to respond and to participate.

The stated purposes of this convention are as follows:

1. To consider seriously what the Bible says about God's eternal and unchangeable plans for the worldwide proclamation and demonstration of the gospel.

2. To consider seriously the situation in the world today as the church of Jesus Christ proclaims and demonstrates the message of redemption.

3. To consider seriously what these two things mean, individually and corporately, in seeking God's will for our lives in the context of the body of Christ and our relationship to the world, both on campus and beyond.

4. To strengthen the missions impact of local student groups at their schools this year.

Finally, I want us together to confess our sins, asking God to forgive and cleanse us. We cannot expect him to do much in us if we fail to humble our hearts before him, asking forgiveness and cleansing. The prayer I will ask you all to pray aloud with me is an old one. It is a prayer of confession from *The Book of Common Prayer*, a prayer which some of our forefathers have prayed over the past four hundred years:

Almighty and most merciful Father, we have erred and strayed from Thy ways like lost sheep. We have offended against Thy holy laws. We have left undone those things which we ought to have done, and we have done those things which we ought not to have done. And there is no health in us. But Thou, O Lord, have mercy upon us, miserable offenders. Spare Thou those, O God, who confess their faults. Restore Thou those who are penitent, according to Thy promises declared unto mankind in Christ Jesus our Lord. And grant, most merciful Father, for His sake that we may hereafter live a godly, righteous, and sober life to the glory of Thy holy name. Amen.

2
Urbana: An Ongoing Experience
Edgar S. Beach

Hidden in the rifts and hollows of the Great Continental Divide where it crosses from Guatemala into Mexico is the municipality of Tectitán. You won't find it on many maps, and, as a matter of fact, few Guatemalans have ever heard of the place. One reason for that is the terribly rugged terrain that discourages travel through the area. So most of the world passes way around Tectitán on modern highways or high overhead in sleek jets. But there is a road that goes there. Most of you would not call it a road, though, because it is too narrow, rutty, rocky and slippery, and too precariously hung on the side of the mountain. But nevertheless it is a road and for me it is the road home. Tonight I would like to share with you how I got to that road and a few of the things that have happened as the Lord has graciously led me there.

When I attended Urbana 70 as a college sophomore, I honestly felt not a single hang-up about missions, since I *knew* I was never going to be a missionary. After all, I had long since decided to be an architect. Moreover, I had never had any interest whatsoever in learning a foreign language, and I only took such courses because I was forced to. Being a missionary was great if you were called to it, but I wasn't, and so I would just cheer them on and support them. That's how I went to Urbana 70, and that's how I left Urbana 70—but with one small change. I was so tremendously impressed with the lordship of Jesus Christ and all that he had done for me that I felt I had to make a certain kind of commitment to him. If he were to call me to go overseas to some "uttermost part of the earth," then I would go. Of course I knew he wouldn't do that. But, just in case, I wanted him to know that he was indeed my Lord and that I'd go *anywhere* to do *anything* for him.

The Lord takes care of first things first, though, and so for the next year I really didn't think much more about missions. He was working on other things in my life like having a daily devotional time, learning how to lead a small group Bible study and sharing my faith with other students. But as time went by I kept having this nagging doubt as to whether or not I was really doing the right thing by so confidently pursuing a career in architecture. So I began to read some missionary biographies and some information about missions and to pray about my doubts.

The Big Jolt

My biggest jolt toward seeking the Lord on the matter occurred when my girlfriend confided in me that she felt God was calling her to go overseas to translate the Bible for some group of people whose language had never even been written down. That was a shock! Here was the gal I thought I loved all of a sudden receiving a mysterious "missionary call" and making plans to go off to deepest, darkest who-knows-where. I came to realize that I had not been called by God to any vocation in particular, let alone to architecture. I just wanted to follow that path because I was good at it and I enjoyed it.

I remembered what someone had said at Urbana 70 about finding God's will for your life: you cannot steer a parked car.

As far as my future with the Lord was concerned, I didn't want to be found with the engine off and the keys pulled out. Nor did I want to be a backseat driver. So I decided to pray and to look more carefully into what God might have me do even to considering what seemed to be the most far-out possibility of all, which was becoming a missionary. But I decided that if God wanted me to change direction then he was going to have to indicate that to me in some pretty clear-cut ways, because I was not going to latch on to Elenore's call, claiming it for my own. He would have to call me regardless of his plans for Elenore because I was not going to go off to some steaming, mosquito-infested jungle unless I had to.

After several months seeking the Lord, he finally answered my prayers. The answer came when I saw a film about a Wycliffe couple who had gone to live with the Aguacatec people of Guatemala. The film showed how they had learned to speak the Aguacatec language and how over the years the Lord had enabled them to translate the New Testament. In the process of helping to translate and repeatedly reading and studying God's Word, first one man became a Christian, then his family, then several families, and finally a thriving indigenous church developed.

When I saw that, I knew beyond the shadow of a doubt that that was what the Lord wanted me to do. How? That's the big question, isn't it? How does one know positively that God is calling him or her to the mission field, or anyplace else for that matter? First, I'll tell you what I thought at that time. Later I'll tell you what I've thought since then.

When I saw that film, I saw an unmistakable need and the clear possibility of ordinary people obtaining the specialized training necessary to meet that need. So I said to myself, "Hey! I could do that." But there are lots of needs in the world that one can meet. What really gripped my heart was that I too had become a Christian through reading the Bible on my own, and as I watched the film I saw that exact same thing happening in far-off Guatemala. "Surely," I thought, "God did not lead me to himself in that particular way merely for my own benefit. Surely he must have wanted to instill in me a vision for reaching others in a like manner. I believe that God is calling me to Bible translation work." That was not an impulsive decision. It was the way the Lord had been drawing me through the previous months of prayer and thought and investigation.

Since God had given us the same goals in life and because our love for one another kept growing, Elenore and I decided that God was leading us together, and so a year later we committed ourselves to following him as one. During our first year of marriage we made application to join Wycliffe and to take the first semester of linguistics training that they required of translators.

I honestly expected that the course would be something akin to a Christian camp session, but it wasn't. As a matter of fact, it was the hardest thing I had ever done in my life! Of course everybody is different. We had two friends there that summer who found it all so easy that they had to look for extra work to keep from getting bored. But that was no comfort to me! Here I was, supposedly called by God to dedicate the next decade or two or three of my life to translation, and I was having a tremendous struggle to understand the phonetics, phonology and grammar we were studying. I half wondered if I hadn't made a mistake after all.

By the grace of God I made it through the course. I learned to analyze the sound systems and grammars of languages that I had never even heard of before, and I found that those so-called exotic, impossible-to-learn sounds were quite possible to learn and even fun. And I came to understand something of the magnitude of crossing cultural barriers to translate the Bible into languages that had never been written before. So it was with joy and trembling that at the end of the course I learned we were accepted as members. Our home churches were tremendously excited over how God was leading us, and they were very supportive. So with that kind of reassurance and confirmation of God's leading, we decided to press on with our training.

Laugh at Yourself

Eventually we felt led of the Lord to request assignment to Central America. Since living there would necessitate knowing some Spanish, we headed to San José, Costa Rica, for two semesters in a language school there. Learning a second language in its cultural context *can* be extremely stressful because it takes time to gain your social and linguistic graces. Can you imagine, for instance, a minister or an evangelist, or even a linguist for that matter, who suddenly finds that he can not express himself as well as the five-year-old kid next door. That's very traumatic. But if you're going to learn another language, you've got to learn to laugh at yourself, to enjoy the people in whose country you're a guest and to do things their way without having a defensive and critical spirit. And you've got to make up your mind that you're going to learn that language as well and as fast as you can by all means possible.

I remember the day I learned the Spanish word for *door*. I had just come out of the supermarket and was waiting at the bus stop with two large bags of groceries. When the bus finally came, it was already full twice over. But one of the fun things about traveling on public transportation in Latin America is learning that there's always room for one more. So I squeezed onto the bottom step at the rear door. The bus took off and the automatic door clamped shut—right on my thumb! I was speechless. Here I was jammed against the door embracing two gigantic bags of groceries with the bus whizzing down the boulevard, and, though I was squirming and going through all manner of contortions, I could not get my thumb out of the door. Soon I became aware of at least a dozen pairs of eyes, staring at me and wondering what in the world my problem was, when this fellow with an incredulous look on his face shouted out at the top of his lungs, "*¡Puerta! ¡Puerta!*" Suddenly the door flew open, I pulled my thumb back and a second later it slammed shut again. I had learned the Spanish word for door, *puerta*, and when I finally got home, Elenore and I were able to have a good laugh about it.

With language school over, we packed our bags once again and headed for Guatemala. You can imagine our excitement when the time came to make our move out to the village of Tectitán. On a previous trip there the Lord had miraculously provided a house for us to rent. Since it was just in the process of being built, we made arrangements for it to be finished while we were gone.

When we arrived on our scheduled move-in day, the house was still roofless and far from ready. The adobe walls were up, but somehow the whole thing was a bit smaller than we had had in mind. It measured a petite seven by seventeen-and-a-half feet which was small but okay. The real problem was the height of the finished walls—a mere five-and-a-half feet. "It's safer in case of an earthquake," our landlord said. "But I'm six inches higher than the walls are," I told him. "Well," he said, "you don't have to rent the place if you don't want to." Since we had already heard rumblings that quite a few people were upset about our coming there, and since we knew that there were no other houses for rent, I pretended not to hear him. "That's okay," I said. "I just happen to have some extra wood with me, and so we'll just jack up the beams a bit and scrap out the dirt floor a little more, and it'll be beautiful." Reluctantly he agreed and we set to work.

Finally the day came when we moved into our little tin-roofed house. It seemed rather romantic in a way, and we hardly noticed the lack of electricity and running water, or the dirt floor, or how cramped we were. But what really did come to bother us was that after the three or four years and all the hard work we had put into preparing ourselves and then after our traveling so far to come to help these people by giving them God's Word in their own language, they didn't want it. They didn't want *anything* that we had to offer them. They just wanted us to go away and never come back.

Why? Because they were afraid of us. It was said that we had come to steal their land, to steal their language, to poison their water, to bring disease and to eat their children. One day one of our neighbors came over and asked us point blank, "Do you eat children?" And Elenore said, "Oh, no! Do you?"

And of course no one would talk to us in the Tectitec language since they thought we had come to steal it. "No one here talks that language," they told us in faltering Spanish. So when we asked where the people lived who did speak it, they said, "We don't know but it must be very far away."

Some people started saying that we were only the first of hundreds of foreigners who were going to come and take over their land. So they talked of sending a committee to the governor so that he would get rid of us. But there are always some folk who don't like committees, and they said that it would never work and that the only sure thing was to get a gun and shoot us.

I remember lying awake one night thinking about all this when we began to hear slow, cautious footsteps outside. Did we ever tremble! After praying and waiting what seemed like an eternity, I threw open the door and jumped outside with my flashlight blazing. And there, slowly stalking across the field toward me, was the intruder—our neighbor's mule.

Of course all this tension had a very real effect on us, and as the situation went on for six months and then for a year, I began to feel rejected and depressed. You see, our technical training had prepared us for everything except a spiritual battle. And we were in a spiritual battle the likes of which we had never before experienced and which I frankly did not want to experience. My reaction was not to thank the Lord as we need to in such circumstances but rather to get angry. I was angry with the people who wanted to get rid of us, I was angry with God who just didn't seem to have a very good hand on things, and I was angry with myself for being angry in the first place.

I remember one day when the woman next door left her sheep to feed on a pile of our landlord's freshly harvested corn. Now since our landlord is a very poor man, he couldn't afford to lose any of that corn so I tried to chase the sheep away. They didn't chase too easily, so I decided to do what I had seen the Tectitecs do and tossed a few small stones at the flock. When in Rome, do as the Romans do, right? Perhaps not because the woman came over and started bawling me out for trying to kill her sheep. I just couldn't take that kind of thing any longer, so I started bawling her out. Of course she didn't understand a word of it because it was all in English. But she got the idea.

I smoldered on that for a few days before I just broke down and cried. It seemed to me that surely the Lord must never have had to deal with as bad a case as me. How much bigger of a failure could I be? Why had God ever brought me to this place? I had thought that I knew, but I wasn't sure any longer.

Called to God-sized Tasks

As I searched God's Word to see what he would say to me, I eventually came to understand this: God had *not* called me to an Ed Beach-sized job but rather to a *God*-sized job. All my life I had wanted to do just what *I* was good at, but all my life *God* has wanted to do what *he* is good at. And what is God good at? Things like giving life to the dead and creating something beautiful out of nothing or out of a mess. He is good at loving his enemies, and he's good at planning the course of history and bringing it to *his* conclusion. He is good at molding so-called pots of clay in which to store the treasure of his all-surpassing power and glory. And he is good at guiding our feet along paths which we don't yet know, but which he does. And that is why when he leads, we must follow and not give up. You can't imagine how many times I have, in my heart, left Tectitán and never wanted to hear of the place again. But every time, the Lord brings me back, and he keeps leading me on. So even though at that time I could not see why the Lord had led us there, I decided that I had to keep going, to keep following him.

The Lord eventually gave us *many* friends in Tectitán. As a matter of fact, we now have a fairly warm relationship with the woman next door at whom I blew up. And we have more people wanting to teach us their language than we can hire. It would be impossible to tell you of all those who have come to our house to chat and who have told us, "I *like* talking with you." Or of all those whose lives we've been able to save with a few common medicines for parasites or diarrhea or nausea. Two of our good friends have had four children, the first three of which died before we ever got to Tectitán. The last one is alive because God has repeatedly used us to save the child's life when no other medical help has been available. And there would never be enough time to tell of all those who have thanked us for things we've written down in their language, because now they know that their language can indeed be written and that it's *beautiful* and *not* second rate as some people have told them.

In many ways we're still outsiders in the Tectitec community,

but praise God because he is helping us to build solid friendships based on love and trust and commitment. It takes time,
but it's worth it because it's the context in which we can best
serve people and help them to come to know the Lord and to
walk with him. Let me illustrate.

When we first began to get to know our landlord, we quickly
found out a number of things about him, not the least of which
was that he was an alcoholic. As our friendship deepened, he
eventually shared with us one day how once he had tried to
commit suicide but had failed and how another time he had
fallen through an old footbridge and had miraculously not been
killed. Well, we had just been discussing how to translate the
idea of the lordship of Christ into Tectitec and had decided that
the best way to express this concept was through the notion of
spiritual ownership. In other words, when we say in English
that Jesus is our Lord, if we want to communicate the very same
thing in Tectitec, we have to say that Jesus is *Kajawil*, most
literally meaning "our Owner." With this in mind, Elenore said
to him, "Don't you understand how much God wants to be your
Owner? When you fell through that bridge, you should have
been killed, but he protected you. And even when you tried to
kill yourself, he wouldn't let you. That's because he has been
waiting for you to give the ownership of your life over to him."

We talked this through for some time and later that day he
gave the ownership of his life over to Jesus. Things went along
pretty well for about a year after that as we continued on with
our language learning and translation. But then there was a crisis
in our landlord's life which God used to test him and to confirm
again to us the wisdom of his leading us to Tectitán.

One of the old men of Tectitán died. Since he had been a
friend, our landlord decided to go to the wake. One of the unfortunate parts of traditional funerals in Tectitán is that there
are always some people there who drink a lot of homebrewed
liquor. This is viewed as part of their respect for their ancestors.

Around midnight a fellow came up to our landlord and said,
"Say there, wasn't your honorable deceased father a good friend
of this honorable fellow who just died? If so, you had better
offer the oldest son a drink out of respect." So our fifteen-month-

old babe in Christ found himself put on the spot and he gave in. Of course he who offers a drink must accept the return offer, and with two shots our alcoholic friend was gone.

I learned about the whole thing the next morning when his wife came down and told us about how he had narrowly missed her with his machete when she had tried to stop him.

As I ran off down the trail towards town to look for him, I was thinking of how pleased Satan must have been with the whole mess. "Lord," I prayed, "I don't know what I can do to help this man. But I want to do something because he is one of the closest friends we have in this place. If you want to use me in his life now, then help me to find him and to minister to him."

Who should I meet but a minute later coming slowly down the road towards me but our landlord. "Hi," I said. "What happened?" "Nothing," he said, standing there with eyes so bloodshot that they looked like traffic lights. But he continued and told me what had happened. Then he said, "Now I can never be good enough for God."

Good enough? I didn't want to believe my ears, but what I had feared all along was true. Deep down inside he was still so steeped in the Mayan theology of trying to appease God that no matter how many times we had gone over the point, he still did not understand that we can never make ourselves good enough for God.

We spent some four or five hours that day together. We went back to his house where we talked, cried, prayed and looked over Scriptures that we had been studying and translating and memorizing together. And as we did so, I found that I had an amazing ability to communicate with him in clear, understandable Tectitec in a way that I had never before experienced. I found the words to express every idea about sin, grace and the love of God. And as we considered afresh the love and mercy of God for us sinful, never-good-enough people, it dawned on me that there was nothing, nothing else in all the world, that I would rather be doing than sitting right there in that tiny, thatch-roofed hut, talking about Jesus to a friend with an aching soul and helping him to understand the Word of Life in the language of his heart.

Later, as I left his house and walked down the road home, I thought back on all that had happened in the last ten years, and I couldn't help but be full of praise to God as I said, "Thank you, Lord. Thank you, because this is what you've made me for. Thank you for leading me here. Keep on leading, Lord Jesus, because I want to follow you wherever you lead me."

3
The Message We Confess
John E. Kyle

Today we inhabit an earth that is torn asunder by the discord of hate, threat of nuclear warfare, assassinations, terrorism, selfishness, hunger, sickness, evil and sin. You might respond, "What's new? That's the very sort of world in which Jesus Christ ministered to two thousand years ago."

The Message
As we read the Gospel accounts describing Christ's life, we quickly conclude that wherever he was present there was immediately created an island of love, healing, justice, understanding, truth and peace. Why? He was God incarnate who knew no sin. Christ had a message of good news to proclaim which today we call the gospel. It was this gospel, shared by his disciples, that turned the world upside down.

I believe that today God has created islands of peace built through Christians like us. After receiving Jesus Christ into their lives and knowing the forgiveness of their sins, they proceeded to share the good news with others. Can you envision what this world would be like without these islands of Christian witness? They are indeed God's restraining hand on the power of satanic influence. Yet today there are three billion people who have yet to hear the gospel in a manner to which they can yes or no to Christ's offer of new life.

What is this gospel that can check the power of Satan's evil? The apostle Paul described it this way:

Now I make known to you, brethren, the gospel which I preached to you, which also you received, in which also you stand, by which also you are saved, if you hold fast the word which I preached to you, unless you believed in vain. For I delivered to you as of first importance what I also received, that Christ died for our sins according to the Scriptures, and that He was buried, and that He was raised on the third day according to the Scriptures. (1 Cor 15:1-4 NASB)

Therefore it is clear that the gospel is the message we shared; that the Bible can be trusted in its declaration that sin can be forgiven because of Jesus Christ, God in the flesh; that he died on the cross in Jerusalem almost two thousand years ago; and that his blood shed on that cruel cross was for our sins. Moreover, Christ conquered death by coming back to life from the grave and lives again as attested by many witnesses. Today he is at God's right hand, right now, praying for all of us. This is the good news, the gospel that we are privileged to share here and around the world. In this day of numerous cults and easy answers to salvation, let us affirm that Jesus Christ attested clearly that there is only one way to be assured of eternal life—through him. Remember that Christ said, "I am the way, and the truth, and the life; no one comes to the Father, but through me" (Jn 14:6 NASB).

We must realize that the message we share is not a popular message. The apostle Paul described the personal cost of proclaiming Christ as a missionary when he said,

Five times I received from the Jews thirty-nine lashes. Three times I was beaten with rods, once I was stoned, three times I was shipwrecked, a night and a day have I spent in the deep.

I have been on frequent journeys, in danger from rivers, dan-
gers from robbers, dangers from my countrymen, dangers
from the Gentiles, dangers in the city, dangers in the wilder-
ness, dangers on the sea, dangers among false brethren; I
have been in labor and hardship, through many sleepless
nights, in hunger and thirst, often without food, in cold and
exposure." (2 Cor 11:24-27 NASB)

Paul lived out his ministry in the words of Jesus Christ, "If
anyone wishes to come after Me, let him deny himself, and take
up his cross, and follow Me. For whoever wishes to save his
life shall lose it; but whoever loses his life for My sake and
the gospel's shall save it. For what does it profit a man to gain the
whole world, and forfeit his soul?" (Mk 8:34-36 NASB).

Obviously, the proclamation of the gospel does not necessarily mean a success story. But what is success? A close friend
had just returned to the United States on furlough from missions
service in Latin America. He had earned two undergraduate
degrees, a masters degree and a doctorate from the same university. He had taken his family to Colombia to live in a jungle
for ten years among a forgotten, isolated group of people. My
friend told a group of businessmen that his alma mater was
about to celebrate its 100th anniversary and wanted to publish
a book to share the work of its outstanding graduates. One day
an airplane circled his little jungle clearing and delivered his
"air mail" by throwing out a packet attached to a small para-
chute that floated down into the jungle. He found in the packet
a letter from his university with a series of questions.

The first was: "Do you own your own home?" He thought
back to the days when he and his friends had built this two-
room house out of the jungle for $100. So he checked "yes."
Question two: "Do you own two homes?" "No" was his first
reaction but then he recalled that Jesus said, "I go to prepare a
place for you that where I am you may be also." So he checked
"yes." Question three was: "Do you own a boat?" He looked
out to the river below his house and saw the forty-foot canoe
he and friends had hollowed out of a log with the little outboard
motor on it and answered "yes." Question four asked: "Do you
plan to travel abroad this year?" He called to his wife in the

other room, "Margaret, are we going home on furlough this year?"

She answered, "Yes." So again he answered affirmatively. The final question was: "What is your annual salary?" He searched down the list of suggested salary figures which began at $250,000 and could not find a figure to match his missionary quota. So he drew a line across the bottom of the list and wrote the very small amount of salary he was receiving. After telling this story, my friend leaned over and with a sparkle in his eye said, "When I leave you I am heading out to the West Coast and will visit my alma mater. I just can't wait to get out there and see what I did to the computers!"

We must realize, therefore, that giving ourselves for the sharing of the gospel is not a sign of success in the world's eyes. Yet Paul was convinced of the strategic importance of sharing as follows: Whoever will call upon the name of the Lord will be saved. How then shall they hear without one who proclaims it? And how shall they proclaim it unless they are sent? Just as it is written, "How beautiful are the feet of those who bring glad tidings of good things" (Rom 10:13-15 NASB). We must also remember that the proclamation of the gospel worldwide has been greatly blessed by God. Warren Webster has said well, "The success of the preaching of the Gospel is best illustrated by the fact that the sun never sets on the Church of Jesus Christ worldwide."

My missionary friend from Colombia knew well the cost of his going out as one determined to share the gospel! He is still there. In fact, he is the missionary leader whom the terrorists in Bogota were trying to kidnap in March of 1981. They could not find him so instead they took captive Chester Bitterman, whom they eventually killed. The apostle Paul and my friend knew all the dangers, sorrows and tribulations they would face. Why then do we believers in Christ have this seemingly insane desire in the eyes of the world to share this good news worldwide? I would like to suggest a very simple answer. It is out of obedience to Christ's command as described in Mark 16:15: Go into all the world and preach the gospel to every creature!

Sharing the Message

The great desire of Christ's heart is now called the Great Commission or the missionary mandate of the Church. We Christians have been commanded to proclaim the gospel worldwide so that every man, woman and child of every tongue, nation and tribe might hear of Christ's sacrificial love, that they might have the opportunity to accept or reject the gospel. Are we involved in that which is very close to the heart of Christ?

How about your campus? Are you sharing Christ with your fellow students who come from other nations? There are over 300,000 international students on the doorsteps of North American college campuses today. Or, have you considered going overseas to apply your God-given gifts of sharing the gospel in a cross-cultural setting? Get involved in an evangelistic mission on your campus. Luis Palau, the Latin American evangelist, will be at the University of Wisconsin at the invitation of students involved with campus ministries to help them with an evangelistic mission. If you are faithfully witnessing for Christ, I trust you are also putting your finances to work for the proclamation of the gospel. Start supporting a missionary now. Consider supporting national Christians ministering for Christ overseas. I would encourage you to become acquainted with the book *Unreached Peoples '82* that will open up new horizons for your prayers and service. I would urge you as you graduate from college to immediately tithe your salary to your local church. Consider going beyond that and set aside $5,000 of your income to give to the cause of missions. My guess is that you would never miss it. Dennis De Haan has said, "When we say go into all the world to our missionaries we must say the same to our dollars!"

The Tensions of Sharing

As I have studied the Scriptures I find that fellow Christians are in a tension between two eternal poles established by our loving God. On the one hand we have our marching orders, Jesus' last recorded words to his disciples before he left this earth: "But you shall receive power when the Holy Spirit has come upon you; and you shall be My witnesses both in Jeru-

salem, and in all Judea and Samaria, and even to the remotest
part of the earth" (Acts 1:8 NASB). That's the one pole. The
tension comes, however, when we consider how much time
we have left to march. Jesus Christ is coming back to earth because that's what the Bible teaches. We do not know when he
will return—that's the other pole that puts us in tension. Our
marching orders were no sooner off Jesus Christ's lips than he
was visibly taken up into heaven, and the disciples were left
standing there with their mouths open. Acts 1:10-11 says: And
while they were gazing into heaven as he went, behold, two men
stood by them in white robes, and said, "Men of Galilee, why
do you stand looking into heaven? This Jesus, who was taken
up from you into heaven, will come in the same way as you
saw him go into heaven." Jesus Christ is coming again. Remember, though, these solemn words of Jesus Christ about the
good news he wants us to share as they relate to his glorious
return to earth. "And this gospel of the kingdom shall be
preached in the whole world for a witness to all the nations,
and then the end shall come" (Mt 24:14 NASB).

There is a direct relationship between the proclaiming of
the gospel and the return of Christ. That's the tension. We have
a message to proclaim and this message must be shared before
Christ's return. It is an eternal adventure story. Imagine that,
as we are here today empowered by the Holy Spirit, so Christ
is in heaven awaiting his Father's touch on his shoulder and
the resounding command, "Return to earth, my Son!" What
a message we have had entrusted to us.

Let us go forth with this message. Just because you have a
chemical-engineering degree or whatever does not mean you
could not serve Christ overseas as an evangelist. Don't let your
training pin you down. Don't say, "Hey, I'm a chemical engi-
neer. How can you use my training?" Ask, "I'm willing to go
overseas to be used. How can I help?" Be willing to listen to the
counsel of missionaries who have proven experience and valued
advice for you. Perhaps you will find that you need to get more
training such as a degree from a Bible college or a seminary to
better prepare yourself to join the missionary ranks. Be open!
Learn how you might be used. There are needs for 120,000

more people to take the gospel into cross-cultural settings.

The need for workers is great. I will never forget the first field trip I made as a new missionary to the Philippines. I went with a young missionary who had just completed the Gospel of Mark in the heart language of a particular group of people. The one Christian in the group, Daniel, had come to Christ through helping the missionary with the translation of the Scriptures. One morning we were giving out copies of the Gospel of Mark to people in the villages. Five elderly men clothed only in their colorful G strings were squatting on the ground in a circle, talking together. Each held his copy of the Gospel of Mark. Because I could not understand their language I asked the missionary to tell me what they were saying.

He told me that they were discussing how their ancestors, throughout history, had taught that there were spirits in the trees, rocks, mountains and fields that must always be appeased and kept happy; that various sacrifices of animals were necessary to appease all of the spirits; that according to their customs they, too, continued to carry out these prescribed rituals. Yet Daniel had told them that there is but one God and that he sent his own Son, Jesus Christ, to die for their sins and relieve all of their fears of the spirits. They kept saying over and over, "Why has it taken so long for people to send this young couple to tell us this good news? Why has it taken so long?" I ask you, my brothers and sisters in Christ, why has it taken us so long?

Indeed, we have a life-redeeming gospel to share, a message of eternal importance to tell people everywhere in the world. I urge you to get on with your marching orders! To what end? "That every tongue should confess that Jesus Christ is Lord!"

Part II

God's Witness in Acts

4
Pentecost and Mission: Acts 2

Eric J. Alexander

The book of Acts is the New Testament account of the establishment and expansion of the early church. It is therefore the classical New Testament book on mission.

After much thought, I have concluded that there are two approaches to Acts which we ought to avoid. One would be an attempt to survey the entire contents of the book in four addresses. That I think would be something of a weariness to you. It would also be something of a problem to me, since the gift of brevity is not one with which I am liberally endowed. The other wrong approach would be to focus on a detailed exposition of a comparatively short section of the book, which I think could lead to an unbalanced view of its message.

Between these two approaches, I have sought a compromise (in true British fashion), and I want to suggest that we approach our study in this way. After some words of general introduction, I want first to look with you mainly at chapter 2 and to consider the theme of *Pentecost and mission.* In chapter 5 we will turn to the theme of *mission and opposition* in Acts 4—6. Then, our focus will be on Acts 6—7 with the theme of *mission and the people God calls.* Finally, in chapter 7 we shall focus on *mission and vision,* God's missionary strategy as seen in Acts 13 and 16.

Introduction to the Book: Pentecost

In the very first verse, Acts is described by its author, Luke, as the second of a two-volume treatise on the mighty works of Jesus, the Incarnate, crucified and resurrected Son of God who has now been taken up to heaven and exalted at the right hand of the Father. It is this ascended Lord who continues to do in Acts what he began to do and teach in the Gospels (Acts 1:1). So the acts of which we read are properly not so much the acts of the apostles, as the acts of the exalted Christ who by his Spirit is building his church in the world. It is important to have that basic concept behind all our thinking.

You may have noticed that Luke's Gospel ends and Acts begins in the same place (Jerusalem: Lk 24:52, Acts 1:4), with the same promise (that they will be clothed with power: Lk 24:49 or that the Holy Spirit will come upon them: Acts 1:8) and with the same prospect (of preaching repentance and forgiveness in his name to all nations, beginning with Jerusalem: Lk 24:47, Acts 1:8).

So Jesus is setting out a program for the cause of the gospel. Far from restoring the kingdom in a nationalistic sense to Israel, his purpose is to spread it to the uttermost parts of the earth. And that great statement in Acts 1:8 about the universal extension of the kingdom serves to introduce the theme of Acts, which is the spread of the gospel in Jerusalem (chaps. 3—7), in Judea and Samaria (chaps. 8—12), and to the limits of the known world (chaps. 13 to the end).

The day of Pentecost occupies a very significant place in the book of Acts as the first fulfillment of Jesus' promise and as heralding the future unfolding of his purpose.

When the day of Pentecost came, they were all together in one place. Suddenly a sound like the blowing of a violent wind came from heaven and filled the whole house where they were sitting. They saw what seemed to be tongues of fire that separated and came to rest on each of them. All of them were filled with the Holy Spirit and began to speak in other tongues as the Spirit enabled them.

Now there were staying in Jerusalem God-fearing Jews from every nation under heaven. When they heard this sound, a crowd came together in bewilderment, because each one heard them speaking in his own language. Utterly amazed, they asked: "Are not all these men who are speaking Galileans? Then how is it that each of us hears them in his own native language? Amazed and perplexed, they asked one another, "What does this mean?"

Some, however, made fun of them and said, "They have had too much wine." (Acts 2:1-8, 12-13)

Pentecost was a Jewish feast. Jews came to it (as they did to the Passover) from all over the world, as is indicated in Acts 2:5 and 20:16. The word for Pentecost literally means "fiftieth," and it was the fiftieth day after the Passover. The feast was called by various names and each of these is of great importance for understanding its significance.

First, it was called the *Feast of Weeks*. That was because the fiftieth day after the Passover marked the end of a week of weeks (that is, seven weeks of seven days). It is significant that Pentecost and the Passover are linked in this way. The Passover was the great moment of Israel's history, celebrating their redemption through the shedding of blood from the bondage of Egypt.

This particular Pentecost in Acts 2 came fifty days after that day to which the Old Testament Passover had pointed, when the Lord Jesus was lifted up and offered himself as a full and sufficient sacrifice for sin. We cannot understand the significance of Pentecost apart from the offering up of Christ as our sin bearer. In this sense, the day of Pentecost is a vital part of the redeeming work of Christ, because the coming of the Holy Spirit was to accomplish in us what Christ had accomplished by his death for us. Thus, we could think of the giving of the Spirit in Acts 2 as part of Jesus' saving work which consists in his leaving heaven's glory, his lowly birth of a virgin, his perfect life of obedience, his atoning death on the cross, his resurrection, ascension and establishment at God's right hand, his gift of the Holy Spirit, his present intercession, and his coming again in power and great glory.

In this context, the coming of the Holy Spirit at Pentecost was a once-for-all event, no more repeatable than Jesus' birth, death or resurrection. It ushered in the last days (Acts 2:17). But the gift of the Holy Spirit is given to every believer who repents and receives the forgiveness of sins (2:38). In this sense, every true believer has a personal pentecost at the time of regeneration. Then the Holy Spirit comes to indwell us, to apply the benefits of Christ's death to us, and to raise us into newness of life so that Christ's life may be lived out in us. No more does God only dwell in the Temple at Jerusalem. From Pentecost on, the apostles say to every believer, "your body is the temple of the Holy Spirit."

Second, Pentecost was *a celebration of the giving of the Law*. Some scholars think that the giving of the Law on Mount Sinai was fifty days after the Passover. Certainly in later Judaism there was a close connection between Pentecost and the giving of the Law. The association is significant, because the giving of the Law was designed, by an external standard, to form the lives and characters of God's redeemed people. And God came down upon Sinai in mighty power with thunder, lightning and fire. But God's ultimate purpose was to write his Law in the hearts of men and women: "I will put my law in their minds and write it on their hearts" (Jer 31:33). It is this promise which is being fulfilled at Pentecost, as Paul makes clear in 2 Corinthians 3:7-8.

It is well for us to remember in all our thinking about Pentecost and the coming of the Spirit that the central ministry of the Holy Spirit is to write the law of God in the hearts of men and women.

Third, Pentecost was also called the *Feast of First Fruits*. It was the time when the first ears of ripe corn were offered to God. The first fruits were part of the harvest, as well as the promise of its fullness. Do you see how perfectly God chooses times and seasons? *This* Pentecost was to see the first-fruits of the harvest of the gospel, and that harvest is still being gathered all over the world today. The end of the harvest will not be until the return of Christ in glory.

So Pentecost is not only a *saving* event—bringing the application of the death of Christ to our lives; it is not only a *moral* event—designed to change our character; it is also a missionary event. That is the significance of the reference in Acts 2:5 to people from every nation under heaven gathered in Jerusalem. It seems clear that their cosmopolitan origins are a symbol of God's purpose to spread the gospel throughout the whole earth. There were people from Parthia and Mesopotamia in the north, from Egypt and Libya in the south, from Arabia in the east and as far as Rome in the west. With a view to declaring to them the wonderful works of God, the Holy Spirit descended upon the disciples in fiery tongues (v. 3) which became foreign tongues (vv. 4, 6, 11) so that every person could hear in his own language.

Peter's explanation of the phenomenon is that it is a fulfillment of the words of the prophet Joel in Joel 2:28-32.

> These men are not drunk, as you suppose. It's only nine in the morning! No, this is what was spoken by the prophet Joel:
>> In the last days, God says,
>> I will pour out my Spirit on all people.
>> Your sons and daughters will prophesy,
>>> your young men will see visions,
>>> your old men will dream dreams.
>> Even on my servants, both men and women,
>>> I will pour out my Spirit in those days,
>>> and they will prophesy....
>> And everyone who calls on the name of the Lord will be
>>> saved. (Acts 2:15-18, 21)

The first and main theme of that prophecy is that God is going to pour out his spirit upon all *flesh*: that is, upon all kinds of people, not just prophets, priests and kings, as had been the case in the Old Testament, but upon all nations and races, both sexes (sons and daughters), all ages (young men and old men), all classes (servants, both men and women). Indeed, says Peter, "Everyone who calls on the name of the Lord will be saved."

The missionary nature of Pentecost is further emphasized in Peter's final appeal in verse 39, where he makes it clear that the promise is "for you and your children and for all who are far off." That last phrase is very striking, because that is a reference to the Gentiles (see Eph 2:13). So here God is fulfilling his promise to Abraham, that in him all the families of the earth would be blessed.

Pentecost is a missionary event. And it is mission in the last days, which have been ushered in by Pentecost (Acts 2:17) and will be terminated by the great and glorious day of the Lord (v. 20). These last days are to be marked by the outpouring of the Spirit upon all kinds of men and women that they might bear witness to the wonderful works of God.

Now what did Pentecost produce? Let me suggest three things from Acts 2.

Powerful Biblical Preaching

Pentecost produced powerful preaching, such as the sermon Peter preached in Acts 2:22-39:

> "Men of Israel, listen to this: Jesus of Nazareth was a man accredited by God to you by miracles, wonders and signs, which God did among you through him, as you yourselves know. This man was handed over to you by God's set purpose and foreknowledge; and you, with the help of wicked men, put him to death by nailing him to the cross. But God raised him from the dead, freeing him from the agony of death, because it was impossible for death to keep its hold on him. David said about him:
>
> I saw the Lord always before me.
> Because he is at my right hand,
> I will not be shaken.
> Therefore my heart is glad and my tongue rejoices;
> my body also will live in hope,
> because you will not abandon me to the grave,
> nor will you let your Holy One see decay.
> You have made known to me the paths of life;
> you will fill me with joy in your presence.
> Brothers, I can tell you confidently that the patriarch David

died and was buried, and his tomb is here to this day. But
he was a prophet and knew that God had promised him on
oath that he would place one of his descendants on his
throne. Seeing what was ahead, he spoke of the resurrection
of the Christ, that he was not abandoned to the grave, nor did
his body see decay. God has raised this Jesus to life, and we
are all witnesses of the fact. Exalted to the right hand of God,
he has received from the Father the promised Holy Spirit
and has poured out what you now see and hear. For David
did not ascend to heaven, and yet he said,

    The Lord said to my Lord:
      'Sit at my right hand
    until I make your enemies
      a footstool for your feet.'

Therefore let all Israel be assured of this: God has made this
Jesus, whom you crucified, both Lord and Christ."

When the people heard this, they were cut to the heart and
said to Peter and the other apostles, "Brothers, what shall
we do?"

Peter replied, "Repent and be baptized, every one of you,
in the name of Jesus Christ so that your sins may be forgiven.
And you will receive the gift of the Holy Spirit. The promise
is for you and your children and for all who are far off–for
all whom the Lord our God will call."

What did Peter preach about? We have an immediate answer
in verse 22 "Men of Israel, listen to this: Jesus of Nazareth . . ."
From there until verse 36, when he is interrupted by a cry from
the audience, it is Jesus Christ who is the center and the substance of Peter's preaching. He is that center in three senses.

First, Jesus is a man accredited by God (v. 22). That phrase
would seem to have reference to his humanity and to his divinity. He is the man Jesus of Nazareth, but he is also accredited
by God as the divine Messiah through the mighty wonders and
signs which God did through him.

Second, he is a savior provided by God. You will notice from
verse 23 that behind the death of Jesus as our Savior there lies
both the hand of God and the hands of men. The cross is not
an accident of history but the plan of God from all eternity. And

yet Peter lays the responsibility for Jesus' death without equivocation at the feet of lawless, guilty men. He is saying "it was our sin that nailed him to the tree."

Finally, he is *Lord,* exalted by God (vv. 32-36). The essence of this gospel preaching was Christ in his humanity, his deity, his humiliation by death on the cross, his resurrection and exaltation to the Father's right hand. The gospel contains this vital element that the Jesus we preach is exalted by God to be Lord over all. And Christ may not be received or encountered in any other guise. Peter preached Christ as Lord.

But then we need to go on to ask *from where did he preach Christ?* The answer is inescapable. He preached Christ from the Scriptures. He quoted at length from Psalms 16 and 110 to show that the Christ of Scripture is the Jesus of history. If in interpreting Pentecost from Scripture he says, "This is that which was spoken of the prophets," in proclaiming Christ from Scripture he points to Jesus of Nazareth and says, "This is he." In verses 25-28 he carefully expounds from Psalm 16 that God did not abandon his holy One to death and corruption.

Like Paul, Peter reasons with his hearers out of the Scriptures. He puts several propositions to them. First, he points out that David was not referring to himself in Psalm 16, since he later died and was buried. David was speaking prophetically, knowing God's promise to set one of his descendants on his throne. Peter argues that what was prophesied by David has been fulfilled in Jesus whom God raised from the dead. Now having been exalted to the Father's right hand, it is he who has poured out the Holy Spirit, the evidence of whose presence they are now witnessing.

But Peter's ultimate concern is to demonstrate from the Scriptures that this Jesus has not only been raised from the dead and poured out his Spirit but that he is both Lord and Christ. From Psalm 110 Peter demonstrates that it is of the Messiah David is speaking when he says, "The LORD says to my Lord: 'Sit at my right hand/until I make your enemies/a footstool for your feet.' " Now it is of great significance that in Acts 2:32 Peter adds the apostolic testimony ("we all are witnesses of the fact") to the Old Testament testimony, since he is uniting the apostolic witness with the Old Testament witness as the ground on which truth is presented to the multitude.

So the Jesus Peter presents is Jesus according to the testimony of the Old Testament and the apostles. But the apostolic testimony came to be written down in the New Testament, so we could speak of the Jesus they preached as the Jesus of Scripture.

The significance of this for our age is quite monumental. The Jesus we are to preach and present is the biblical Jesus and no other. There is a fascination in our day with Jesus. But it is often a Jesus of human imagination: Jesus Christ the failed superstar or the stage clown or the Che Guavara of the revolutionary world or the demythologized figure of a liberal theology or the Jesus of my own subjective experience. But the Jesus we are to preach and proclaim is the Jesus of *Scripture:* not "Jesus as I like to think about him" but Jesus as Scripture portrays him.

What did Peter offer and demand? The gospel offer is two-fold: First, forgiveness of sin (v. 38). Second, the gift of the Holy Spirit (v. 38b). Forgiveness is what Jesus had commanded them to preach to all nations, on the basis of his name (Lk 24:47). There is no doubt in Acts that this is what the apostles did. You will notice that in apostolic terms man's primary need is for forgiveness. The primary problem is not that people are unhappy. Their primary problem is not that they are frustrated and confused, suffering from a feeling of emptiness or meaninglessness. That may be the symptom of their problem, but it is important to distinguish the symptom from the actual disease. We can alleviate people's symptoms without ever really touching their disease. The disease from which men and women suffer is that they are under the judgment and wrath of a holy God and need to be put right with him. It remains as true today as it was in Peter's day that however people may dislike it, forgiveness is their primary need and the gospel's primary offer.

But the other thing the gospel brings is the gift of the Holy Spirit. And that bears witness to the fact that Jesus gives us more than forgiveness for the past. He gives us new life in the present, through the regenerating power and indwelling presence of the Holy Spirit, and hope for the future, since the Holy Spirit is the guarantee of our future inheritance in heaven. It is important to note that this gift of the Holy Spirit is part of the gospel promise and is inherent in the very essence of biblical salvation.

What about the demands the apostle makes? They are to "repent and be baptized" (v. 38). Repentance implies a change of direction in one's life as well as a change in one's thinking. It signifies a turning from a sinful and godless way of life to a God-centered way of life. In one sense a person alone is incapable of doing this. That is why repentance is described as something God *grants* (Acts 11:18). But it is also described as something God *demands,* as in verse 38. It is true that faith is not mentioned in Peter's exhortation, but those who repented and were baptized are described in verse 44 as believers. Repentance and faith are two sides of the same coin, and the one includes the other. Baptism was the outward and public sign of this inward and private reality of repentance and faith.

I want to emphasize this apostolic call to repentance. We tend to distort the message by preaching faith without repentance. The sad result of that, I think, is the ocean of nominalism which is the curse of the church of God not only in the Western world but even on the missionary field. This nominalism has been produced by divorcing repentance from faith. The call of the gospel is a call to repentance *and* faith.

So the first thing Pentecost produced was powerful, Christ-centered, biblical preaching. But that preaching in turn produced *deep conviction of heart* (v. 37). Notice the contrast between the result produced by unusual spiritual phenomena (vv. 12-13) and that produced by powerful biblical preaching (v. 37). The first produced curiosity ("What does this mean?"). The second produced conviction leading to repentance, faith and baptism ("Brothers, what shall we do?"). It is often said today that the church is always answering questions no one is asking, and scratching where people are not itching. I would want to plead as much as anybody for relevance to the modern world in our preaching. But is not the deeper problem that people are not asking the right questions, not itching in the right places? What is the answer to that? It is spirit-anointed, Christ-centered, Bible-based preaching in the power and unction of the

Holy Spirit. It is he who brings conviction, as the history of true revival down the ages has shown. But powerful preaching was not the only thing Pentecost produced. It also produced an apostolic church.

An Apostolic Church

The Holy Spirit is not just an evangelist drawing people to Christ. He is a church planter, creating communities of obedient, teachable, worshiping saints who are learning to love God's truth and God's people. That is the message of verses 42-47, and we badly need to learn it:

> They devoted themselves to the apostles' teaching and to the fellowship, to the breaking of bread and to prayer. Everyone was filled with awe, and many wonders and miraculous signs were done by the apostles. All the believers were together and had everything in common. Selling their possessions and goods, they gave to anyone as he had need. Every day they continued to meet together in the temple courts. They broke bread in their homes and ate together with glad and sincere hearts, praising God and enjoying the favor of all the people. And the Lord added to their number daily those who were being saved.

Paul was a church planter, and that is what most missionaries are called by God to be.

What then were the marks of this church which was created on the day of Pentecost? There are four which are easily distinguished. First, it was devoted to *apostolic teaching* (v. 42). It is striking and significant that the first mark of this church of Christ was that it was devoted to the study of apostolic doctrine. Notice that just as the deity of Jesus was attested by mighty works and wonders and signs (v. 22), so the apostles' authority was attested by many wonders and signs (v. 43). The early Christians submitted to that authority in the realm of apostolic doctrine. Every church which is ruled and filled by the Holy Spirit, in every generation and in whatever area of the world, will likewise be submitted to apostolic truth as it is found in Holy Scripture and will devote itself to the study of apostolic teaching.

Second, the church was distinguished by *God-honoring wor-ship*. Notice in verse 42 that they gave themselves not only to apostolic doctrine, but to the breaking of bread and the prayers. These seem to be references to services of worship. There is further reference to this in verses 46 and 47. Their worship seems to have been God-centered and joyful, bringing a spirit of holy reverence upon every soul (v. 43).

So the spirit-filled church in Acts 2 is a worshiping com-munity. So it must be with us. The constant activity of the church in heaven and the chief business of the church on earth is worship. Thus the early church was devoted to apostolic teaching, and it was distinguished by God-honoring worship.

Third, it was *united by Christlike love* (vv. 44-45). These early Christians not only devoted themselves to the apostles' teaching, they also devoted themselves to fellowship (v. 42). The nature of this fellowship is amplified in verses 44 and 45. The very familiar word *koinonia* means a sharing or participa-tion, and its primary reference would of course be to our sharing or participating in God—Father, Son and Holy Spirit. Then it also carries the meaning of sharing out. So the believers in Acts 2 shared out their goods and their very lives with each other. A spirit-filled church is one where the Holy Spirit releases people from their self-centeredness and gives them deep and genuine care for the needs of others within the fellowship.

Finally, the church was *increased by God-centered evan-gelism*. The last verse of chapter 2 tells us that "the Lord added to their number daily those who were being saved." Notice the clear testimony here to the sovereignty of God in salvation. It was the Lord who added to their number. We greatly need to be recalled to this biblical, God-centered understanding of evangelism. There are so many things that human talents and influence can do by itself. We can arouse people emotionally. We can indoctrinate them intellectually. But only God the Holy Spirit can regenerate them spiritually.

Furthermore, this evangelism was church-based. By that I mean that what God did when he saved people was to add them to the church.

And this evangelism was a day-by-day experience. "And the

Lord added to their number daily those who were being saved"
(v. 47). What the Lord did among them he did day by day. Their
evangelism was not a sporadic experience or an occasional
venture into another realm. It was the daily experience of God
calling his people out of darkness into light.

How greatly the church of Jesus Christ throughout the whole
world needs this God-centered, church-based continuous evangelism! So the day of Pentecost produced spirit-anointed, biblical preaching and a spirit-filled, biblical church. May God
pour out such pentecostal blessing upon us in our generation.

5
Mission and Opposition: Acts 4—6
Eric J. Alexander

There is a sense in which the book of Acts is the account of *two* movements, not just one. There is the movement of the Holy Spirit to build the church and the movement of the powers of darkness to destroy it. This is part of a wider biblical principle which we could enunciate in this way: wherever a true work of God is established, that work will be opposed. Everywhere in Scripture you have evidence of this, and it is borne out also in Christian history and in contemporary experience.

What is happening throughout the book of Acts is that Jesus is building his church, and the gates of hell are trying to prevail against it. Now you have an example of this in the section of Acts beginning at chapter 4. In Acts 4 there is *opposition from persecution,* and that is met by the church's giving itself to prayer and finding a new vision of God's sovereignty. Thereafter God pours out his blessing, and there is a new advance in the work of the gospel.

Then in chapter 5 there is *opposition from hypocrisy* within the church, and that is met by a purifying of the church through judgment in the most solemn circumstances.

Finally, in chapter 6 there is a new form of opposition, from *division and diversion* within the church. This is met by a re-ordering of priorities and a restructuring of ministries so that first things might be kept indisputably first. It is at these movements that I want us to look at in more detail.

Opposition from Persecution

Acts 4 records the first persecution which the church suffered. That persecution was, as Dr. Griffith Thomas points out, the commencement of a long chain of worldly opposition. It is the primeval collision of light and darkness, of the church and the world, of God and Satan:

> The priests and the captain of the temple guard and the Sadducees came up to Peter and John while they were speaking to the people. They were greatly disturbed because the apostles were teaching the people and proclaiming in Jesus the resurrection of the dead. They seized Peter and John, and because it was evening, they put them in jail until the next day.

Now one of the immediate lessons we need to learn from this is that there is no ideal place and there are no ideal circumstances in which to serve God or to live for him in the world. We are all prone to say, "Ah, if I only lived in *his* circumstances or in *her* situation everything would be different." But the testimony of Scripture is that there is no ideal place in which to serve God except the place to which he has called you, and where he has set you down to be his servant.

In Acts 4 the significant thing is that this opposition and persecution came from worldly religion. In verse 1 the Sadducees are mentioned as leaders in this persecution (also 5:17). They were the religious rationalists of their day, the antisupernaturalists, and the miracle recorded in chapter 3 of the healing of the lame man really upset them. Their logic was simple if stupid; it had three stages. Stage 1: miracles cannot happen. Stage 2: so miracles don't happen. Stage 3: so this miracle hasn't happened. As Professor Richard Coulson points out in *Science and Christian Belief*, this is the kind of narrow-mindedness which the Christian must deplore.

But the Sadducees were not only antisupernatural, they were anti-Scripture. They relegated much of the Old Testament to a place of secondary importance, venerating only the authority of the Pentateuch. It is significant that Jesus confronts them with these issues in Mark 12:24: "Are you not in error because you do not know the Scriptures or the power of God?" The two things go together: the Scriptures and the power of God.

Professor James Denney, who graced the Scottish theological scene last century, made this comment on the Sadducees in *Jesus and the Gospel*:

> Religion for the Sadducees was an institution, not an inspiration ... living religion the Sadducees dreaded.... When Christianity began to put forth its irrepressible expansive power after the Resurrection, we are told that "they doubted whereunto this would grow." They did not want growing things at all in that sphere. A religion that grew, that operated as a creative or re-creative power, that initiated new movements in the soul or in society—a religion that gave men new and infinite conceptions of duty, making them capable of self-dedication and martyrdom, so that you could never tell what mad, disturbing thing they would do or try—a religion which for anything they could tell might explode the social order—such a religion the Sadducees could only regard as the enemy.

Not bad for a nineteenth-century Presbyterian theologian!

I believe there is a warning for us in the 1980s from the fact that the opposition to the gospel came from a dead and disemboweled religion: dead because it knew nothing of the power of God; and disemboweled because it had tampered with holy Scripture.

It does not matter what name a religion has, or what particular brand of Christianity it is. If it is dead because it knows nothing of the power of God, if it is disemboweled because it has tampered with holy Scripture, it will put us in precisely the position of the Sadducees. I want to underline particularly that when we talk of the authority, inerrancy and sufficiency of holy Scripture, we are not talking about matters that belong only to the remote world of academia; we are talking about issues which concern the future prosperity of the work of the gospel in the world. We are speaking about missionary expansion and evangelism.

When we speak about scriptural authority and inerrancy, we are speaking about experiencing the power of God in our lives. And so I say we should not leave these issues of the authority of Scripture to the theologians—they are much too serious to leave it to them. We need to deal with this ourselves and to come to some solid, well-thought-out position on holy Scripture. This will determine a great deal about your future usefulness to God in the world.

So Peter and John were arrested (v. 3) and interrogated (v. 7) and charged not to preach or teach at all in the name of Jesus (v. 18). Now how did they respond? First, but less importantly, they responded with boldness (v. 19), asserting the principle repeated in Acts 5:29, "We must obey God rather than men!"

Second, and more importantly, Peter and John met their situation with believing prayers (4:23-30):

> On their release, Peter and John went back to their own people and reported all that the chief priests and elders had said to them. When they heard this, they raised their voices together in prayer to God. "Sovereign Lord," they said, "you made the heaven and the earth and the sea, and everything in them. You spoke by the Holy Spirit through the mouth of your servant, our father David:
>
> > Why do the nations rage
> > > and the peoples plot in vain?
> > The kings of the earth take their stand
> > > and the rulers gather together
> > against the Lord
> > > and against his Anointed One.
>
> Indeed Herod and Pontius Pilate met together with the Gentiles and the people of Israel in this city to conspire against your holy servant Jesus, whom you anointed. They did what your power and will had decided beforehand should happen.

Now, Lord, consider their threats and enable your servants to speak your word with great boldness. Stretch out your hand to heal and perform miraculous signs and wonders through the name of your holy servant Jesus.''

This prayer shows not only that the early church prayed, but how they prayed.

The key words here are the first two: "Sovereign Lord." This prayer is really an appeal to God in his sovereign greatness. It carries the conviction that God is sovereign in four areas. First, he is the *sovereign Lord of creation* (v. 24). This is the first thing which persuades them of God's adequacy for their situation. They are threatened by forces, influences and authority far greater than their own. But they remind themselves, "You are the God who made all that is. The very men who are threatening us owe their existence to you, so we gladly bring you this situation." From God's sovereignty in creation they have a new vision of his character, and true prayer depends on that.

This argument from creation to the trustworthiness of God is used two ways in Scripture. It is used by people to plead with God, as in Jeremiah 32:17: "Ah, Sovereign LORD, you have made the heavens and the earth by your great power and outstretched arm. Nothing is too hard for you." It is also used by God to plead with people to trust him:

"To whom will you compare me?
Or who is my equal?" says the Holy One.
Lift your eyes and look to the heavens:
Who created all these?
He who brings out the starry host one by one,
and calls them each by name.
Because of his great power and mighty strength,
not one of them is missing. . . . Do you not know?
Have you not heard?
The LORD is the everlasting God,
the Creator of the ends of the earth.
He will not grow tired or weary,
and his understanding no one can fathom.
He gives strength to the weary
and increases the power of the weak. (Is 40:25-26, 28-29)

This trust is vital in our generation, which has a unique capacity (even within the evangelical church) for thinking great thoughts of men and paltry thoughts of God. We are weak in the doctrine of God as our Creator, and it is to the great impoverishment of our prayer life.

Second, God is the *sovereign Lord of history* (vv. 25-26). Verses 25 and 26 are a quotation from Psalm 2. Artur Weiser says that Psalm 2 "bears witness to a God who is present and active in world history, and who knows how to make himself respected by those who do not want to give heed to Him, and who accomplishes His purpose even though men rebel against Him."

The immediate picture here is of the accession and anointing of a new king and the rebellion of the kings and rulers of the earth against him. The psalmist sees this rebellion as a revolt against the Lord and against *his* anointed king. But the point of the psalm is that the very idea of the nations vaunting themselves against God is ludicrous: "The One enthroned in heaven laughs; the Lord scoffs at them" (Ps 2:4).

Weiser says here: "What the Psalmist sees is that at the centre of history is no longer the struggle of the great powers, but God, whose relationship with earthly powers will determine their destiny." In other words, the vision the psalmist has of history is a theocentric vision, and it is this view of history that the evangelical church desperately needs in our time. God has not abdicated his throne. The center of power today is neither in Moscow or Washington, Peking or London. It is with the Lord God omnipotent.

Third, God is our *sovereign Lord in redemption*. The apostles certainly saw Psalm 2 as messianic. The great watershed of history, the event which bisects it and is therefore the greatest display of God's sovereign power, is the redemption God has accomplished in Jesus Christ. Here they see him turning human wrath to praise and to blessing. So in verse 27 the believers pray to God, in essence: "It has happened in this very city. . . . And yet *they* were under *your* sovereign control." When human wickedness seemed to have reached its nadir, the sovereign controlling hand was God's. "You could have no power," says

Jesus to Pilate, "except it were given to you by my Father."

This leads us to the final clause in this prayer. It is the conviction that God is *sovereign Lord over the contemporary scene*. In Acts 4:29 we read: "Now, Lord, consider their threats and enable your servants to speak your word with great boldness." They recognized that the character of God does not change. He is not the great I WAS, he is the great I AM. And so they spread their contemporary situation before him.

This is the kind of vision we need in our generation: this theocentric vision of life, history, time, redemption and the future. It is such a vision that leads to this kind of believing prayer that rests on a sovereign God. It was meeting the opposition at this level, recognizing that prayer was not supplemental but fundamental in the work of God, which brought down upon their testimony the powerful attestation of God that is described in verses 33-35.

My favorite Old Testament story is recorded in 2 Kings 6. Ben-Hadad, the king of Syria, had been sending his armies out against Israel, and each time he failed to surprise his enemies. It seemed that all of his plans of attack were being leaked somehow to the king of Israel so the Israelites were always one step ahead of him and his troops. So Ben-Hadad gathered his people around and asked, "Which of us is the traitor? Which of us has been leaking information to the enemy?"

They replied, "None of us, my lord. But that little prophet, Elisha, down in Israel seems to know your plans even before you do."

"Where is he?" he cried.

"Dothan."

"Send an army," the king commanded. So they got out the cavalry, the artillery, the infantry and started out from the capital of Syria. You can imagine what a sight they must have been, going out with great armies after the little prophet Elisha in Dothan.

So they neared Dothan, and Elisha's servant went out and saw them coming in the distance—a vast army surrounding Dothan. He went to Elisha and he said, "My lord, what shall we do? Look at the hordes of the enemy."

Elisha replied, "Don't worry. Don't be afraid. They that are with us are more than they that are with them."

So the servant went back out and looked again at the huge army and then he looked at Elisha and he must have been thinking, "This old fellow is out of touch with reality. He should have retired a long time ago." But the fact was that it was the old man who saw reality. And Elisha was getting tired of the young man's fears, so he asked God, "Open his eyes so he may see."

And the Lord opened the young man's eyes and he saw that the mountains were full of the horses and chariots of fire belonging to the Lord. Elisha was surrounded by God's armies.

Greater is he that is in you than he that is in the world. Do you really believe that? It does not matter where God may send you, God is still greater than the forces which oppose you.

Opposition from Hypocrisy within the Church

This second form of opposition the church experienced is described at the beginning of Acts 5. It is the story of Ananias and Sapphira, and their deceitful behavior:

> Now a man named Ananias, together with his wife Sapphira, also sold a piece of property. With his wife's full knowledge he kept back part of the money for himself, but brought the rest and put it at the apostles' feet.
>
> Then Peter said, "Ananias, how is it that Satan has so filled your heart that you have lied to the Holy Spirit and have kept for yourself some of the money you received for the land? Didn't it belong to you before it was sold? And after it was sold, wasn't the money at your disposal? What made you think of doing such a thing? You have not lied to men but to God."
>
> When Ananias heard this, he fell down and died. . . . About three hours later his wife came in, not knowing what had happened. Peter asked her, "Tell me, is this the price you and Ananias got for the land?"
>
> "Yes," she said, "that is the price."
>
> Peter said to her, "How could you agree to test the Spirit of the Lord? Look! The feet of the men who buried your hus-

band are at the door, and they will carry you out also."

At that moment she fell down at his feet and died. (Acts 5:1-5, 7-10)

This is not a case of persecution from outside the church, but corruption from within it, and it is quickly identified by Peter in verse 3 as the work of Satan. Having failed to halt the progress of the gospel by opposition from outside, the evil one now turns to the task of disabling the church by corruption from within.

I think it is significant that throughout history the church of Christ has seldom been permanently harmed by opposition or persecution of an external kind (the news coming out of China reinforces that). But the church has often been grievously harmed by corruption within, when it has gone unchecked or when it has not been taken sufficiently seriously. I want to spell out some of the warnings which derive from this opposition of Satan from within the church.

The first warning note of danger in the church is seen in *Ananias and Sapphira's marriage*. It is essentially a warning about marriage (even Christian marriage) where something is kept back from God. (Cf. 5:2, "with his wife's full knowledge" and 5:9, "how could you agree to test the Spirit of the Lord?")

It seems, then, as if Ananias and Sapphira had sat down together and discussed the whole issue. They were really discussing the level at which they proposed to live before God and the extent of their consecration together to him and to the cause of the gospel. And in the process, they made a travesty of the marriage bond, which in the providence of God was intended to enable a man and woman to be helpers to each other, that they might live for his glory. But Ananias and Sapphira made it a concordat for abetting one another in robbing God of his glory, lying to the Holy Spirit and putting the whole church in jeopardy.

Will you allow me to say to those of you who are married or contemplating it that here is a whole area which you need to bring before God, and the earlier the better. On the last occasion when I was in Malaysia and Singapore, I was greatly struck by the large number of young people in their twenties in the churches. When I remarked on this to a Christian leader, he said to me, "Yes, we are heartened by that, but it has always been so. Our question is, where are the older, mature, middle-aged believers? The problem seems to be that when they marry and settle down, materialism takes over and we lose them."

I also remember someone saying when I was a theological student, "The marriage of a Christian worker either doubles or halves his usefulness: it seldom leaves him the same." Our marriages must be brought under the lordship of Christ.

The second danger within the church is the *peril of hypocrisy and unreality*. The background to Ananias and Sapphira's sin is the account in Acts 4:32-37 of the mutual love and generosity which was a fruit of the Holy Spirit within the church. We are told that Barnabas, for instance, sold a field and brought the money and laid it at the apostles' feet. Barnabas did that in an unaffected way. And doubtless many people recognized the man's spiritual stature. But Ananias saw this, and he may have said, "I would like them to think that of me too."

So Ananias pretended that the same was true of him as of Barnabas. But the hidden fact was that he had kept from God some of the money from the sale of property. The money was only a token. It was really part of himself he had kept back.

The sin here was not that he had not given the whole price of the land. Nobody, and certainly not God, had required that of him. His sin and Sapphira's was in pretending to a godliness to which they were strangers. Their sin was in being more interested in reputation than in reality. And that was like a dreadful blight which could have killed the early church. So God rooted it out so vigorously. It was said of Fred Mitchell, the great missionary statesman in the Overseas Missionary Fellowship, "You never caught Fred Mitchell off his guard, because he never needed to be on it." People should be able to say the same of us.

The third danger in the church is the danger of *religious emotion without moral content*. Acts 5:1-11 has a great deal to teach us about the person and ministry of the Holy Spirit. For example, this passage points to the personality of the Holy Spirit. He is someone to whom you can lie (v. 3). It also points to the divinity of the Holy Spirit, for the Holy Spirit to whom they lied in verse 3 is described as God in verse 4.

We need to be reminded that the adjective given to the Spirit's name describes his nature. He is the Holy Spirit, or the Spirit of Holiness. Wherever he is truly at work, he produces a deep awareness of God's holiness, a deep sensitivity to sin, and creates a reflection of the moral glory of God in the believer. Ananias and Sapphira seem to have been caught up in a work of the Holy Spirit, and no doubt they were deeply moved by it all. I would not be surprised if there were even tears in their eyes at times. But they were careless of moral issues like honesty and integrity. And so grieving was this to God, that he reacted against it in holy judgment.

So the church was cleansed by God and that led to a new authority and power in the apostles' ministry (v. 12) and to a new growth in the church (v. 14).

But there was a third form of opposition which Satan contrived for the church.

Opposition from Division and Diversion in the Church

In Acts 6:1-7 we read:

In those days when the number of disciples was increasing, the Grecian Jews among them complained against those of the Aramaic-speaking community because their widows were being overlooked in the daily distribution of food. So the Twelve gathered all the disciples together and said, "It would not be right for us to neglect the ministry of the word of God in order to wait on tables. Brothers, choose seven men from among you who are known to be full of the Spirit and wisdom. We will turn this responsibility over to them and will give our attention to prayer and the ministry of the word."

This proposal pleased the whole group. They chose Stephen, a man full of faith and of the Holy Spirit; also Philip, Procorus, Nicanor, Timon, Parmenas, and Nicolas from Antioch, a convert to Judaism. They presented these men to the apostles, who prayed and laid their hands on them.

So the word of God spread. The number of disciples in Jerusalem increased rapidly, and a large number of priests became obedient to the faith.

In these first seven verses there was dispute and murmuring between the Hellenists and the Hebrews. The Hellenists felt their widows were being neglected by the church. But their complaint was used by Satan in a sinister and subtle manner. It was an opportunity for the evil one to divert the apostles from their God-given calling into something else. The whole future of the church at Jerusalem was in the balance, as it had been twice before. In chapter 4, Satan was seeking to destroy their confidence in God. In chapter 5, he was trying to dilute their consecration to God, and in chapter 6 he was seeking to divert them from the call of God.

The apostles summoned the church together, and they said, "This is a crucial issue. It is not that the welfare of the widows is unimportant. It is of great importance to God, who calls himself the judge of the widows, and therefore to us." So they described this social ministry as a responsibility (v. 3). But the point was that the devil could so easily have employed this situation to divert the apostles from the great business of their life, which was prayer and the ministry of the Word.

The apostles' response to this crisis is of great significance. They said that they must reassess their priorities and restructure the church's life, so that first things were indisputably first. They therefore appointed seven men to be responsible for the widows. Then they resolved, "We will turn this responsibility over to them and will give our attention to prayer and the ministry of the word." They did not neglect their social responsibility, but they *did* have priorities.

Of course the establishing of priorities is a very important thing in any sphere of life. We all have limited time and limited resources, and we display our priorities in the way we use our time, energy and money. But the establishing of priorities is nowhere more vital than in the personal and corporate life of the people of God, for the sake of mission. And the apostolic priorities were prayer and the ministry of the Word.

Here then are the waves of opposition that arose against the missionary expansion of the church of God. This is real spiritual warfare, and we really do need to learn how to wage it. One of the vital questions in all warfare is to find out where the front line is. Do you know where the front line is in spiritual warfare, in the battle for people's souls?

I had an uncle who served in World War 1 and was decorated by the king for his service. And I remember his telling us that one of the greatest problems they had when they were sent out to the front line in France was finding it. They would go around asking, "Where's the front line? Do you know where it is?"

I believe that in the church of God in our generation we have forgotten where the front line is. That's why there are so many unreached people in the world today. We have lost the front line. The testimony of the apostolic church is that the front line is in the place of prayer and the ministry of the Word and the purity of the church. If God should in his great goodness lay his hand on you this week and call you into his service, you may be sure the devil will do all in his power to make you believe that the front line is somewhere else. He will want to divert you from the great apostolic priorities. That is why it is vital for you to have your priorities fixed and settled now and by God's grace not to be moved from them.

6

Mission and the People God Calls: Acts 6—7

Eric J. Alexander

In Acts 6 and 7 we find that, despite opposition, the church in Jerusalem has now grown from the small beginnings of chapter 1 to three thousand in chapter 2:41, five thousand in chapter 4:4, multitudes both of men and women in chapter 5:1, and in chapter 6:7 we read that "the number of the disciples multiplied greatly."

At this stage, God is preparing to expand the frontiers of the church's mission into Judea and Samaria. The events of chapters 6:8—8:3 describe how these frontiers were expanded. The whole passage is dominated by the life, ministry and death of Stephen, the first Christian leader to lay down his life for the gospel. Stephen's death significantly introduces us to the figure who is going to dominate the rest of the book of Acts: "The witnesses laid down their clothes at the feet of a young man named Saul" (7:58). That Saul was of course Saul of Tarsus, who was to become Paul, the apostle to the Gentiles.

There are three things we are told about Stephen in this passage. First, we find Stephen is a man who reflected the beauty of God. Second, he is a messenger who preached the whole counsel of God. And third, he is a martyr who laid down his life for the glory of God.

A Man Who Reflected the Beauty of God

In Acts 6:15 we are told that "all who were sitting in the Sanhedrin looked intently at Stephen, and they saw that his face was like the face of an angel." There was a glory and beauty that belonged to another world about this man. Professor Howard Marshall comments in his commentary on Acts that "the description is of a person who is close to God, and reflects some of His glory as a result of being in His presence."

The one word which Luke repeatedly uses to describe Stephen is the word *fullness*. In Acts 6:5 he is described as "full of faith and of the Holy Spirit." In 6:8 he is "full of God's grace and power." And in Acts 7:55 he is said to be "full of the Holy Spirit." Now what this adds up to is that Stephen is a man who is just full of God and of the glory and beauty of the character of God revealed in Jesus Christ. So if you want to know what the fullness of the Holy Spirit primarily does for someone, here is one of the most striking examples in the New Testament.

The evidences of the fullness of the Holy Spirit are very significant. There are at least five evidences which can be identified, and they can all be seen in Stephen's character. There we find the kind of moral glory God longs to produce in future missionaries.   Let me just emphasize that there is a vital principle embedded here. The point is that in the work of God you can never divide what a person *does* from what he or she *is*. You can do that in many other spheres of life. For example, I went to school with a fellow who was a man of great ability. He took several degrees and went to London and became a senior partner in a firm. He led what appeared to be a very normal family life and rose to the top of his profession. But one day it was discovered that the man was leading a double life. In private he had created utter moral confusion and negatively affected the lives of many others. He kept his professional life and his private life quite separate. His effectiveness in the business world was apparently not altered by his moral duplicity.

But you can never make that division when you are serving God. The person you are determines the work you do. This whole issue of moral character is of the very essence of what God means to do in our lives. I know of no place where this is put more clearly than in a book which I read every year because it is such a blessing to me. It is the book *Power through Prayer* by E. M. Bounds. In it he says:

> We are constantly on a stretch, if not on a strain, to devise new methods, new plans, new organizations to advance the church and secure enlargement and efficiency for the gospel. This trend of the day has a tendency to lose sight of the man, or sink the man in the plan or organization. God's plan is to make much of the man, far more of him than of anything else, because men are God's method. The church is looking for better methods; God is looking for better men. The Holy Ghost does not flow through methods but through men; he does not come on machinery but on men; he does not anoint plans but men. It is not great talents nor great learning that God needs but men great in holiness, great in faith, great in love, great in fidelity, great for God. These men can mold a generation for God.

So the fullness of the Holy Spirit produces a *vigorous godliness*. There is a remarkable strength of character in Stephen. His angelic countenance must not be interpreted as weakness. He towers above his detractors: "They could not stand up against his wisdom or the Spirit by which he spoke" (6:10). His manliness and godliness leave them silent and send them scurrying to find false witnesses (v. 11). That same strong godliness is seen in his steadfastness in all kinds of circumstances, so that when they were gnashing their teeth and plotting his death, he was more aware of the glory of God than the wrath of men (7:54-55). There is nothing weak or affected about true godliness.

The fullness of the Holy Spirit also produces *godly wisdom*. In Acts 6:10 we learn that Stephen's opponents could not stand up to his wisdom. This wisdom is not the weight of his own learning. It is not an academic thing at all really. It is instead the fulfillment of the promise of Jesus in Luke 21:15 "For I will give you words and wisdom that none of your adversaries will be able to resist or contradict."

This wisdom is not something which comes from the university or the seminary. It comes with a growth in grace and godliness with a life filled with the fullness of God. As a result it is possible for a person to be greatly gifted intellectually and to have a string of degrees, and yet to be a baby in wisdom. And it is possible for another person to have very little intellectual training, and yet in godly wisdom be a giant.

In the fellowship in which I was brought up after I became a Christian, I can still vividly remember a young man who had scarcely had any education at all. He worked as an assistant to a laborer in a foundry, and his job was to do those things that no one else would do. He was not even thought intelligent enough to be a regular worker. But God laid hold of him one day and transformed his life by his grace, and he was born again by the Spirit of God.

I watched that man grow. In prayer meeting the first several times he came people could not even understand him because he couldn't articulate the Queen's English. But slowly this man grew in wisdom so that today there are hundreds of students in Glasgow who have sought his counsel and sat at his feet while he has opened God's Word to them. And I heard him pray one day, "O thou that makest wise the simple ..." This is the wisdom of God.

What is the secret of this wisdom? I think it is to be found in such places as Psalm 119:

Oh, how I love your law!
   I meditate on it all day long.
Your commands make me wiser than my enemies,
   for they are ever with me.
I have more insight than all my teachers,
   for I meditate on your statutes.
I have more understanding than the elders,
   for I obey your precepts. (vv. 97-100)

How much that godly wisdom is needed in the missionary world today!

The fullness of the Holy Spirit produces vigorous godliness, godly wisdom and *a love of Scripture.* The most obvious conclusion one can draw from the seventh chapter of Acts is that a person who is filled with the Holy Spirit will also be filled with holy Scripture. It just poured out of Stephen. He did not have time to go and prepare a discourse for the Sanhedrin, but his love for the Word of God was such that his heart and mind were steeped in it. He was a student of the Scripture, not a scholar. But his love for God's Word was a vital index to his character.

We are not called to be scholars but we are called to be students, to the limit of our God-given abilities. We must give ourselves to the Word of God and to the earnest, systematic and serious study of it. Put yourself under a regular ministry of the Word of God. Pray for a new love for Scripture and a deeper appetite for it, so that, in Spurgeon's words, your very blood may become "bibline." This is the way God's fruitful servants are made.

The fullness of the Holy Spirit also produces *a willingness to serve in ordinary and insignificant ways and places.* Remember the background from which Stephen emerged (Acts 6:1-4). He was one of the seven chosen to serve tables, to do an ordinary, everyday task in order to free the apostles for the ministry of the Word and for prayer. We do not know how long he did this, but that was his first ministry, and God proved him there, and he proved God there.

I want to suggest that this is a lesson we greatly need to learn. I think Christian character is most tested, and servants of God are frequently trained and proved in situations of obscurity and in tasks which may appear menial and even distasteful.

Coming from a large, urban church, I often have people asking for a task. "What can we do?" they say. "We see the need for ministry here. How can we help?"

They want to help in some dramatic way. But what we really need is for them to wash the dishes after a church supper or sweep the floor.

I remember a young woman who became a most fruitful

servant of God overseas. The first evidence of her godliness which struck people was that she was faithful in the ordinary and insignificant and distasteful tasks.

We should not spend our time looking for some exalted place of service. We must first be willing to wash our fellow disciple's feet.

Finally, the fullness of the Holy Spirit produces a *selflessness which is ready to sacrifice everything for the glory of God*. In this, as in so much else, Stephen is like Jesus. His life is marked not by self-indulgence and self-interest, but by self-denial and self-sacrifice. And it is all because the vision at the end of his life (7:55) had been his vision throughout his life, namely, the glory of God.

These then are the great hallmarks of Stephen as a man of God. This kind of godly character is immensely important in fruitful Christian service. The resources which will really be tested and sifted wherever you serve God will not be your physical resources or your intellectual resources or even your emotional resources. The resources which will really be tested will be your spiritual resources.

One of the most important resources we have is fellowship. Let us look at the church fellowship out of which such a character as Stephen's was produced. It was a church which, according to Acts 6:4, had settled its priorities. "We . . . will give ourselves to prayer and the ministry of the word."

Such a church, with such priorities, produces people, like Stephen, of remarkable Christian character. And that is the vital importance of this matter of priorities within the life and ministry of the church of God. It is not just that we are interested in maintaining patterns. It is that we are interested in producing men and women of God whose lives will witness to God in a needy world. And such men and women are born out of a church situation where the priorities are apostolic.

For the days in which Stephen lived, they needed men of such caliber. And I believe that for the days in which we live, we need men and women with a vigorous godliness, a godly wisdom, a love of Scripture, a willingness to serve in insignificant ways and places, and a selflessness which is ready to sacrifice everything for the glory of God. But there are more things we can learn from Stephen's character.

The Messenger Who Preached the Whole Counsel of God

Stephen's ministry is so important that a whole chapter, the longest in the book of Acts, is devoted to it. The vital thing about the whole of Stephen's public ministry seems to have been that the power of God was upon him and in him—not the power of natural gift or of personality, but the power of God. And that power was such that people felt it. There was about his speech a persuasive wisdom which had nothing to do with man but with God's reaching down into the conscience and bringing conviction. This is what the servant of God ought to covet more than anything else. It has nothing to do with noise or oratory. It is God, and it is ministered through the Scriptures, as it was here with Stephen, for his address is just a prolonged exposition of Scripture.

Stephen draws a straight line from Abraham (7:2) to Christ (7:52) and sees the whole of biblical history as the unfolding purpose of God's preparing the way for Jesus.

There are two main themes running through Stephen's exposition, and they correspond to the two charges made against him in 6:11 and 13. These charges are that he had spoken against Moses and the Law, and that he had said Jesus would destroy the temple.

It is the Jews who through their history had rejected Moses (v. 25 and 35ff.) and the prophets (v. 52) and through them, God himself (v. 51). So their rejection of Jesus (v. 52) was not a new or isolated act, it was the culmination of their consistent rejection of God and resistance of the Spirit (v. 51). Stephen turns the charge against him around and uses it against the Jews. They are the ones who reject Moses.

Again Stephen argues from chapter 7:2 onward that the Most High does not live in houses made by men. He dwelt with Abraham in Mesopotamia (v. 2) and Haran (v. 4) and Palestine (v. 4), with Joseph in Egypt (v. 9), with Moses in Midian (v. 29), and with the Israelites in the wilderness (v. 44). Although they had a tabernacle or tent (v. 44) and later the temple (v. 47), yet God has declared that he is not tied to these places (vv. 48-49). If therefore Stephen had been saying that God is not tied to a place, he was simply consistent with Scripture.

You see Stephen's address fulfills two functions. It answers with biblical force the false charge brought against him, and it opens up the biblical basis for the expansion of the church's mission to the Gentiles. The Jews, to whom the gospel was first preached, have rejected it, and the way is now clear for the church to turn from Jerusalem and the temple to evangelize the Gentiles. So this is a biblical address on mission, taking in the whole scope of Old Testament revelation.

A Martyr for God

Finally, we see that Stephen's death was for the glory of God. There are clearly two sides to the *fact* of Stephen's death. There is on the one hand the wrath and hatred of the people as they gnashed their teeth against him (v. 54). That, I suppose, is the basic physical fact behind Stephen's death, and it issued in their casting him out of the city and stoning him. But there is another side to it. In verse 55 we read that, full of the Holy Spirit, Stephen gazed into heaven and saw the glory of God and Jesus standing as his advocate at the right hand of God. And he said, "Look, . . . I see heaven open and the Son of Man standing at the right hand of God" (v. 56).

I really think that was what settled Stephen's death. I do not think it would have been possible for him to come back and live in this world after that. I think that experience spoiled him for earth for ever. He had seen the glory of God. How could a man or woman be satisfied with life here on earth after that? So Stephen cried, "Lord Jesus, receive my spirit" (v. 59). That is the fact of his death.

But we cannot leave Stephen without also noticing the fruit of his death. Again this is twofold. It is first the occasion for the outbreak of persecution, led by Saul of Tarsus, whose conscience was clearly inflamed (8:3). And that persecution drove the church out of Jerusalem, to Judea and Samaria. Stephen's sermon drove them out theologically and biblically, and Saul's persecution drove them out physically and geographically. So the martyrdom of Stephen was, in the gracious purpose of God, bearing fruit already. It is not without significance that the very word for witness in the New Testament is the word from which we get our English word *martyr*. There is a profound cost involved in true and effective witness.

But the other fruit of Stephen's death is even more significant. In Acts 7:58 we read, "the witnesses laid their clothes at the feet of a young man named Saul," and in 8:1, "and Saul was there, giving approval to his death." I would reckon that Saul was there under that spirit-anointed ministry as Stephen preached, and the Word of God stabbed his conscience. However, conviction is a fire which, depending on the wind it catches, may burn into repentance and faith or into resentment and bitterness.

Even as Saul was on his way to wreak destruction on the church at Damascus, the heavens opened again, and the Lord Jesus called, "Saul, Saul!" And Saul fell down blinded by the same glory that had drawn Stephen's soul to heaven. And he cried out, "Who are you, Lord?" And he was a fruit of Stephen's death. Think of that. Some might have said of Stephen that his life was prematurely cut off and untimely thrown away. But indeed his life was the fuel which set alight a flame for the glory of God in the ancient world that has no parallel.

I think I would lay down my life if it would produce a life of the caliber of Paul's. Wouldn't you?

You see, whatever the hour and whoever the servant, there is a death to be died if we are going to be fruitful for God and for his glory—a death to self, to pride, to our own plans and our own ambitions. "Except a corn of wheat fall into the ground and die," says Jesus, "it abideth alone: but if it die, it bringeth forth much fruit" (Jn 12:24 KJV). What a glorious thing to be the kind of life that is fuel to be set alight for the glory of God in our generation, as Stephen's life was fuel to set such a fire in the life of Saul as had never been seen. We need to be thinking not only of the impact we are having on our own generations; we need to be thinking about the generation we are producing to come after us. We need a sense of history and destiny. What kind of people are we producing?

Do you know these words of Jim Elliot, who gave his life
bringing the gospel to the Auca Indians? He prayed: "Saturate
me with the oil of the Spirit that I may be aflame. But flame is
transient, often short-lived. Canst thou bear this, my soul, short
life? In me there dwells the Spirit of the Great Short-Lived,
whose zeal for God's house consumed him. 'Make me thy fuel,
flame of God.' "

It is for people such as this that God is looking in these days.
May he find us to be his fuel for the flame of his glory.

7
Mission and Vision: Acts 13:1-4; Acts 16

Eric J. Alexander

There are two passages in the second half of Acts which deal particularly with the missionary expansion of the early church: Acts 13:1-4 and Acts 16. These two passages are particularly significant as the beginning of the fulfillment of Jesus' command and promise that they would witness to him not only in Jerusalem, Judea and Samaria but to the ends of the earth.

In these two passages there are three principles for all missionary endeavor which I want to draw out and expound. They could be put in this form:

—the motive for missionary endeavor lies in the glory of God
—the responsibility for missionary endeavor lies with the church of God
—the strategy for missionary endeavor lies with the Spirit of God

First then we will look at Acts 13:1-3. Howard Marshall in his commentary on Acts says this about this passage:

The importance of the present narrative is that it describes the first piece of "planned" overseas mission carried out by representatives of a particular church, rather than by solitary individuals, and begun by a deliberate decision, inspired by the Spirit, rather than somewhat more casually as a result of persecution. Luke thus describes in solemn detail how the missionaries were appointed at a church meeting, under the guidance of the Spirit. He is well aware that he is describing a crucial event in the history of the church.

That crucial event is also something of a pattern for every missionary venture in any church which seeks to be biblical in its thinking and practice. There are two principles embedded in these verses to which I want to give attention.

The Motive for Missionary Endeavor: the Glory of God

I think it is of central importance that when this missionary endeavor was first conceived, the church at Antioch was worshiping with such a seriousness of purpose that they abstained from food (v. 2). It was while they were worshiping that the Holy Spirit said, "Set apart for me Barnabas and Saul for the work to which I have called them."

Worship and mission are so bound together in the economy of God that you really cannot have one without the other. The reason for this is that true worship is rendering to God the glory which is due his holy name. And this is the great end and purpose for which all things exist. God created the world as a theater in which to display his glory. He created man and woman in order that they might reflect the image of his glory. He sent Jesus in order that the glory of God might be seen in the face of Jesus Christ. He redeemed sinners in order that they may be changed into the image of his glory. There is nothing beyond this for us: it is the terminus of everything in the universe. And that is why worship is the highest employment of our faculties: it focuses on the glory of God.

But when we come to know God, we discover that he is jealous for his glory. He will not give it to another, nor his praise to graven images. He desires his glory to be declared among the heathen (Ps 96:3). Do you see the logical corollary which must be drawn from these premises? No Christian man or woman worshiping God and desiring his glory can be unmoved by the fact that there are areas of the world and nations where God is being robbed of his glory. That is why true worship and true mission always go together, and it is why the glory of God is the only ultimate missionary motive. There are, of course, others: compassion for the lost, obedience to the Great Commission and so on. But these are not the ultimate motive. The ultimate motive is the glory of God.

Do you remember how Paul displayed this in Acts 17 when he arrived in Athens? He saw the city full of idols, and the New International Version says, rather weakly, that "he was greatly distressed." The Revised Standard Version says, "He was provoked in spirit." But the verb here is the one from which our word *paroxysm* is derived. There are suggestions that it was the medical term for a heart attack. It is certainly the word used in the Septuagint of God's being provoked by Israel's idolatry. And what provoked God in Israel and Paul in Athens was the same thing: idols were being worshiped in place of God and, the Lord was being robbed of his glory.

Do you know the story of Henry Martyn? He was a Cambridge scholar of the last century who turned his back on academic glory to go to India, and he died as a young man. In India he watched people prostrating themselves before pagan images and heard someone tell of a vision of Jesus bowing before Mahomet. Martyn wrote: "I was cut to the soul at this blasphemy. I could not endure existence if Jesus was not glorified: it would be hell to me if He were also thus dishonoured."

I think you and I would have to confess that in the contemporary church and in our own hearts there is little of this deeply felt emotion of a burning jealousy for the glory of God. And I am persuaded that this is the root cause of our relative indifference to the missionary task.

It is in this sense that we need to remind ourselves again and again that worship without a missionary burden is humbug. "While they were worshipping the Lord and fasting, the Holy Spirit said, "Set apart for me Barnabas and Saul for the work to which I have called them." It is of great significance that the

China Inland Mission, now the Overseas Missionary Fellowship, was born out of the agony in the heart of Hudson Taylor over worship divorced from mission. In June 1865 he was in Brighton one Sunday morning worshiping in church when he suddenly seized his hat and left. He explained in his diary what happened to him: "Unable to bear the sight of a congregation of a thousand or more Christian people rejoicing in their own security while millions were perishing for lack of knowledge, I wandered out on the sands alone in great spiritual agony." And there on Brighton beach he prayed for twenty-four willing, skillful laborers, and the China Inland Mission was born.

Mission and worship belong together. No one can truly worship God and at the same time have an apparently total indifference to whether anyone else is worshiping him or not. This is the answer to the foolish questions which so often arise: "Is your church interested in missions?" or "Are you a missionary-minded church?" or "Are you a missionary-minded Christian?" The question is preposterous. You might as well ask: "Are you interested in the glory of God? Do you care about that?" When a church has no deep care for mission, its true sickness is that it has no vision of the glory of God and no jealous regard for the honor of his name.

The motive for missionary endeavor is the glory of God. When I was ordained, one of the questions I had to answer in our Presbyterian ordination service, was: "Are not zeal for the glory of God and a desire for the salvation of men, so far as you know your own heart, your great motives and chief inducements in seeking this ministry?" That question puts things in the right order, as zeal for God's glory is the great motive of all missionary endeavor. It is the one thing that will keep you on track.

Give yourself today to learning something of the glory of God and of his jealousy for the honor of his name.

The Responsibility for Missionary Endeavor: the Church of God
In Acts 13:2-4, the Holy Spirit is clearly laying the responsibility for missions on the church at Antioch and particularly on its leaders. There are four stages in this commissioning of Barnabas and Saul to a missionary task.

First, the Holy Spirit calls them (v. 2). That may have been an individual and private experience for Barnabas and Saul. They may have had God open their eyes to the needs of the gentile world. And they may have heard him say, as he did to Isaiah, "Whom shall I send and who will go?" They responded, "Lord send me." What they were saying to God was this: because we care for your glory in the gentile world, because we desire the honor of your name, we have given ourselves to your lordship and want to hearken to your voice. Send us that the Gentiles might glorify God.

But then the next stage was that the Holy Spirit came to a worshiping church and said, "I have called these men; now you set them apart for this work to which I have called them." From the tense that is used here, we might assume that the Holy Spirit called these men personally prior to his approach to the church. In verse 3 the church responded with great seriousness to this responsibility. They fasted and prayed before they did anything. And I do not think they were just praying for the blessing of God for their two brothers. I think they were also doing something else. They were saying, "Lord, we are really serious in bringing this issue before you, and as a token of that we abstain from food. We want you to confirm to us their call."

Let me spell out how I think this should affect practically a church's response. When men or women believe they are called by the Holy Spirit to a particular service, their next question should be, "Does the church where I worship, and of which I am part, recognize that call so that the Holy Spirit is placing upon them the responsibility of setting me apart for this work?" And the church should take that person's sense of call with the utmost seriousness, and the leaders of the church should come together and wait upon God to know his will, so that they may take the responsibility for this missionary and be able to say, "It seemed good to the Holy Spirit and to us to set apart this person for this ministry."

Let me illustrate this from our own experience in Glasgow. We have a large number of students and other young people in the church in which I have the privilege of serving. Many of them have been called to the Lord's service at home and overseas. But in our board of elders we became troubled by the fact that many of these young people would go off to Bible college and then come back after three years and tell us of a call to a certain missionary situation and suggest that we have a commissioning service. But we had no confirmation of that call. We had not been involved in the student's life for three years. So we realized that the church ought to have been involved from the early stages of the receipt of that call.

The result is that now, before people go to Bible college, a group of our elders spend most of an evening interviewing them and praying with them. A report is then made out and discussed by the board. If necessary the person is seen again and only after that are we able to say, "At this stage we can discern the Lord's call and will be ready to take responsibility for this servant of his." This seems to be an important biblical pattern.

The third stage in this commissioning is that, once the Holy Spirit has called and communicated that call to the church, the church now lays hands upon Barnabas and Saul. The laying on of hands is not an ordination or a conferring of gifts, since Paul and Barnabas have already exercised a ministry and displayed the necessary gifts. Instead F. F. Bruce explains its significance in this way in his commentary on Acts:

> It is evident that the laying on of hands in this instance imparted no qualification to Barnabas and Saul which they did not already possess. But by this means, the church at Antioch through its leaders expressed its fellowship with Barnabas and Saul and recognized them as its delegates.

In other words, the church was accepting responsibility for them. That was the responsibility to support the missionary spiritually by prayer, emotionally and psychologically by continuing encouragement, and financially by money.

The fourth stage in the commissioning is when the church sends Saul and Barnabas out (v. 3). You will notice that it is really the Holy Spirit who does this: "sent on their way by the Holy Spirit" (v. 4). But he commits the execution of the task to the church.

So the biblical pattern is that the church is the sending agent, not the missionary society. The missionary society is the *servicing* agent of the churches at home and overseas, but the *sending* should be done by the church. And in Acts 14:26-27 it was to the church that they reported back when they returned from this first missionary journey.

Before we leave this issue of the church's responsibility, let us look briefly at who these men were that were sent. William Arnott, a Scottish scholar of last century, pointed out in his *Studies in Acts:*

> *The men chosen for foreign work in accordance with the mind of the Spirit, were the mightiest men. They did not send out some persons who had turned out useless at home. The foreign field always needs, and in the age actually obtained, the ablest labourers. I suspect the chief obstacle to the success of modern missions lies here. The church at Antioch sent the cream of the ministry abroad; if they had sent the grounds, their success would have corresponded to their effort. Here and there in our own time, when the Spirit has descended in power, some men mighty in word and deed have taken the field, and the result has been a gain corresponding to the outlay; but it is the grief and the weakness of the church at the present day that her chiefs are for the most part occupied at home.*

So we need a new vision of the church.

The Strategy for Missionary Endeavor: The Spirit of God

Finally, we also need a new vision of the strategy of God. The strategy for missionary endeavor lies with the Spirit of God. I want to draw some conclusions about the divine strategy in missionary endeavor as illustrated in the ministry of the apostles, especially Paul.

The strategy that matters is God's strategy. The thing that should really matter to us is God's strategy, God's will, God's perfect purpose. When I was in a little rural church in Ayrshire for fifteen years, I would often get letters when things were difficult and discouraging, and these letters would make things worse. I was in an area which was spiritually barren —a place desperately in need of ministry. But people would write and say, "Isn't it time for you to be in a more strategic place? Isn't it time that you got a strategic situation somewhere else?" I used to respond, sometimes with a fair bit of vigor, "Whose strategy is it that we are talking about? God has set me down in this place, and he has given me no liberty to leave it."

So whose strategy are you talking about? Whose strategy are you interested in? The vital thing is that your life as God has created it and redeemed and prepared it should be in God's will. That's what matters.

From the apostolic teachings and experience we can draw out three things about missionary evangelism: the preparation, the pattern and the goal. The preparation for missionary evangelism as seen in the book of Acts relates God's providence to our evangelism. The apostles clearly had a deep sense of being caught up in the purpose of God, and they found themselves being sent not so much to people they had chosen as to people God had chosen for them. Again and again their own inclinations were corrected either by direct, divine intervention or by providential circumstances. Persecution was a providential circumstance used by God to reveal the divine strategy for evangelism. Phillip went to Samaria, and Paul and his band frequently left one city to go to another because they were harried out. But in these persecutions the purposes of God were working out the strategy for evangelism.

At other times there was direct, divine intervention, such as that which took Philip from a situation of great blessing in Samaria to meet a solitary Ethiopian statesman in the Gaza road. Or again there was God's action in removing Peter's prejudices against bringing the gospel to the Roman centurion Cornelius. Time and again in Acts we see the right hand of God bringing the person with the need and with the prepared heart, such as the Ethiopian. The left hand of God brings the person with the answer, and they are brought together for salvation. That is the glorious providence of God working in the strategy of God in evangelism.

Perhaps nowhere is this principle illustrated more vividly than in Acts 16 where Paul and his company experience frustration after frustration. In Acts 16:6-8 you will remember that doors were closing against them rather than opening, frustrating their plans rather than fulfilling them. And I think in those days they were learning how vital it was to be wedded to the divine strategy and to be ready to die to their own plans for the sake of God's plans, to die to their own timetable in order to live by God's. Have you learned that lesson? Are you wedded not only to the will of God but to the timetable of God?

In Acts 16 as God was opening up an amazing new door into Europe the apostles were experiencing frustration. This of course does not mean that God does not want us to have plans, but that there is a deeply rooted biblical principle imbedded in apostolic history. It is the principle of divine initiative. Do you remember how Joshua had to learn this? In chapter 5 Joshua is preparing to assault Jericho, and he sees a man in mighty armor standing before him. He did not realize that it was the living God coming to meet with him. But he says to the figure, Now whose side are you on? Are you for us or for our enemies? (v. 13).

And the man gives him a very unhelpful reply. He says, "Neither."

And Joshua wonders, "Now which side is this man on? Can we really trust him?"

And the man says, "As commander of the army of the LORD I have now come."

So the vital question the Lord asks is not, "Am I on your side or on their side?" but "Are you on mine?" That is the question.

When you have this vision of God's sovereignty in evangelism, what a transformation it brings. What a glorious thing to think of the living God not only preparing you for a people, but preparing a people for you. This was a vital element in Paul's ministry. The sovereignty of God is what made evangelism possible even when Paul was discouraged. When Paul had just about given up on the Corinthians (Acts 18), God came to him and said, "Go on, Paul. Carry on preaching the gospel here because I have many people in this city." But those people hadn't yet been converted! But God said, "I have many people here. You go on preaching." And Paul stayed there eighteen months and many came to Christ. God prepares hearts for the message of the gospel.

But God also offers a pattern for missionary evangelism. This pattern has two features. One: it was a presentation of truth. That is the inescapable conclusion one must draw from Paul's ministry. For example in Acts 17 in Thessalonica even the vocabulary Paul uses proclaims this. In Acts 17:2 he reasons with them out of the Scriptures, explaining and proving that the Christ is to suffer and rise from the dead. And the result was that some of the Jews were *persuaded*. This is the characteristic of apostolic evangelism: truth presented to the mind. But the apostles were not aiming merely at an intellectual conquest. They were aiming for a conquest of the will and the emotions, too, but through the mind.

It is a very significant thing that the old order of sin makes its appeal through the senses—a tree desired for food, arousing the appetite, and so on. By the new order of grace in Christ, however, the appeal is made to the mind—be transformed by the renewing of your mind. So the apostles spoke the truth to the mind.

The second element in the pattern was the proclamation of the whole counsel of God from Scripture. This is how Paul describes his ministry as he meets the Ephesian elders at Miletus. In Acts 20:27 he says, "I have not hesitated to proclaim to you the whole will of God." When you listen to apostolic preaching—whether of Peter or Paul or Stephen—the amazing thing is how much there is in it, a wide presentation of doctrine covering God, man and Christ.

People often ask these days about the irreducible minimum of the truth that we must present: What is the irreducible minimum of the gospel that we need to present to people? I find that the apostles were not concerned with the irreducible minimum but with the indispensable maximum. That is what they were interested in. And we must go to a needy world with a *whole* Christ from the *whole* Bible for the *whole* person in these days.

Finally, the goal of missionary evangelism in Acts is the goal that Paul speaks of in his letter to the Colossians (1:28-29). He says that our concern is to present every man mature in Christ.

This is why Paul stayed eighteen months in Corinth and two years in Ephesus. This is why he retraced his steps on his first missionary journey back through Lystra and Iconium and Antioch (Acts 14), strengthening and encouraging the disciples. This is why he planted churches and ordained elders to care for the flock. It was so that he could bring people to maturity in Christ. That is our great concern.

This brings us back to the whole issue of the goal of missionary evangelism, which is the glory of God. The great concern of God is for his own glory—and this is what salvation is all about. God wants to take people who are broken and shattered by sin, in whom the image of God is defaced, and he wants to mold and form and recreate them so that on the day of glory we can go into God's presence and say, "Here I am, and the children you have given me, changed from glory into glory." That is the goal of missionary evangelism. And God knows how much we need to learn it in our day.

So, to conclude, we need to ask God to give us a vision: a new vision of the glory of God as the motive of mission; a new vision of the church of God as the agent of mission; and a new vision of the strategy of God the Holy Spirit as the sovereign Director of mission. May God make us fruitful for his great glory.

Part III

The Church and the Witness

8
The Sending Church

Gordon MacDonald

I would like to direct your attention to a passage of Scripture which is found deep in the New Testament—the little letter that Peter wrote to a group of scattered Christians and congregations.

Peter writes, "But you are a chosen race, a royal priesthood, a holy nation, God's own people, that you may declare the wonderful deeds of him who called you out of darkness into his marvelous light. Once you were no people but now you are God's people; once you had not received mercy but now you have received mercy" (1 Pet 2:9-10 RSV).

In the stories that General Eisenhower used to tell about his associates in the military and the government, there was one favorite in which he indulged at the expense of one of his chief aides, a man named George Allen. George Allen had the dubious distinction of having played in a record-setting football game in Cumberland College, during the early 1930s, I believe. The team lost 222 to 0. Allen played halfback on the losing side. After about three quarters of the game, when the score had begun to mount and the team was dramatically demoralized, there came a moment, in one of the few plays in which they had the ball, when the ball was snapped back to the quarterback who immediately dropped it. The opposing linemen came charging in, the ball was trickling around the backfield and the quarterback screamed out to Allen, "Pick it up. Pick it up." Allen took one look at the charging linemen and said, "You pick it up! You dropped it!"

Pick Up the Church

There are many people today who would say those same sorts of words about the church—people who, having come out of a background in which the church has been a bad experience or in which the church has seemed irrelevant, apathetic or simply stifling in its institutionalism, have said, "Let somebody else pick up the church and run with it. I want nothing to do with it."

This is particularly true of the younger generation. And I say that with some degree of authority because it was true of me when, in my early twenties, I first made a commitment to Christian ministry. Entering seminary, I found myself saying on many occasions, "I will do anything but be a pastor. I will involve myself in anything but a church because that is not where the action is."

Now I start on that somewhat pessimistic note because I need to make contact with many of you who think the same thing and who have not yet understood that Christian mission, over the long haul, cannot happen until we understand the nature and the place that God has made for the church throughout all of history. The reason I quoted Peter's words is that they constitute one of the noblest yet briefest statements about the nature of the church. Not only is there something special about the content of those two verses, but there is something we need to notice about the man who wrote those words. For I have a suspicion that in some very general way he shared at one point in his life that rather dim view of the church.

1 Peter 2:9-10 easily breaks into two halves. The first is a state-

ment of affirmation or esteem that Peter has as he writes to a group of people who have been dispersed apparently through persecution across the large reaches. Remember, these people are quite poor. They are perhaps a bit demoralized from having been wrenched from their homelands. They have few resources apart from their own energy. So when Peter talks in words like these, he is trying to lift them and build their understanding of their role in God's history. Look at the phrases he uses. "You are a special race. You are a holy nation. You are priests of royalty. And most important, you are God's own people."

Peter seems to be struggling to find the most magnificent, the highest of phrases, to elevate the self-esteem of these poor congregations. Peter recognizes, as we often do in the twentieth century, that people don't think that they have much to give until they see the specialness that God by grace has created in them. I hear Peter saying to these congregations, "You are really important in God's view of things. You are special; you are royal; you are holy; you are chosen; you are part of his family." In a whole different sense I recall the words of Paul to the elders in the Ephesian congregation: "Take heed for the church which God has purchased with his own blood." I worry about being critical of the church when I see it described like that.

In the second half of 1 Peter 2:9-10 Peter moves from the privileges of the people—their great position—to what they are privileged to do. You are going to proclaim the praises of the one who called you out of darkness into light. Remember who you were not; now see who you are. But the thrust of the challenge is this: you must act on the privileges that have come to you through the grace of God.

Where did Peter get such theology? such insight? Did he always think that sort of thing? Emphatically not. Remember Peter's encounter with Jesus a few years earlier, one of the first encounters that he had with Jesus. Remember the miracle of Christ's power in filling the nets with an overload of fish. Peter's mind was boggled. He was awestruck over the power of Christ. And he fell before Jesus and said, "Depart from me for I am a sinful man. I am not worthy of being a part of this equation. I can't live up to this stuff. Pick somebody else."

Jesus made a beautiful reply: "Peter, don't be afraid, because from this point forward, you are going to be catching men." From that moment a countdown began and Peter moved from wanting to run from Christ to the place where, years later, he could with his pen articulate magnificently what it means to be special people doing a special thing.

The Sending Church

My question lies between those two extremes: What were the insights, what were the inputs, that led Peter to think the sort of things which called him to issue his challenge to the church? For Peter is attempting to take ordinary churches and create out of them "the sending church."

There are many churches; there are relatively few sending churches. Let me define a sending church. We can do it by way of a historical model, the church in Acts 13 in which the Holy Spirit was free to speak because he would be heard. That church called Saul and Barnabas and sent them out to the uttermost parts of the earth. That was a sending church. It was a church marked by intercession for world evangelization, marked with caring for the needs of hurting people, marked with a hunger for the teaching of the Word of God. It was a church marked with leaders who really believed the mandate of Acts 1:8. Into that sort of atmosphere the Holy Spirit can quickly move. So when he said, "Set apart these men," the church laid hands on them and "sent them off" (v. 3). It was a sending church.

By way of contrast let us look at what is not a sending church. We are not a part of a sending church when we are just part of a Bible study. Don't assume that you have built yourself into the sending congregation if you are just part of a Bible study. Bible studies are great, but they are not the church in its fullest form. We are also not a part of the sending church if we attend church on Sunday morning via television.

Moreover, the sending church is not just the American church. Let's not set in motion our North American chauvinism and equate the sending church with the church in the West. The sending church can be found any place where certain conditions are met. A sending church can be found anywhere: in

North America, Africa, the South Pacific, Asia and so on. Likewise, nonsending churches can be anywhere.

So why is Peter so concerned in this epistle with the sending mentality of that congregation? First, he has learned something about the nature of a sending God. Second, he has learned something about the mission of Jesus Christ. Third, he has learned something about the enablement of the Holy Spirit.

The Sending God

Luke 24:27 has a magnificent description of a postresurrection encounter between Jesus and the disciples: he opened up the Scriptures to them and taught them from Moses and the prophets what his mission was all about. What could be a more marvelous experience than having the Savior open up the Word of God, as he must have done that day? And there was Peter sitting in the front row hearing Christ unfold God's work in history. I suggest that one topic Christ had to talk about was the movement of the heavenly Father in history. I can hear Peter thinking about things like this. Sending first happened when God created the heaven and earth. The first thing that we know about the God of the Bible is that he created, that he sent something out of himself. By his word he created things out of nothing. We know also from Psalm 19 that when God created the heaven and earth, he created everything with a purpose—to proclaim his glory. Everything in all of creation that is untouched by the power of sin has one basic function—to raise and elevate the consciousness of all who see the glory of God. "The heavens are telling the glory of God; and the firmament proclaims his handiwork" (Ps 19:1 RSV). There is no language where their voice is not heard. All of creation was built to send, to confront people with God's creation glory.

In all of creation today the only objects that are not *sending* by instinct are human beings controlled by sin and those things in our environment which we have touched and exploited. When we exploit our environment, we diminish its capacity to send out messages about the glory of God. What God creates, he creates to send, and what does not send is a corruption of God's creation.

As Christ unfolded the Old Testament to Peter, as he developed that sending mentality for the old apostle, I can hear Jesus talking also about God encountering men like Abraham and Moses. What is the first thing God says to Abraham when he encounters him in his homeland? "Go forth to a land that I will show you. You are going to be a blessing wherever you go, and nations are going to be touched wherever you are willing to touch them." The whole encounter with Abraham is built upon sending.

So it is with Moses, too. "Moses, I send you to Pharaoh. I send you to create the conditions in which the liberation of our people will happen. I have been among the people. I have heard their cry. I know their affliction. I've walked among them. Now Moses, you go. I'm sending you."

When Moses and his people gather (in Exodus 19 and 20) at the foot of Mt. Sinai to hear the first orders which God is giving them as he forms those people, he uses the words which Peter quotes in 1 Peter 2. "You will remember," he says to Israel, "how I've drawn you out of Egypt on eagles' wings, and I have made you a special people, a kingdom of priests. Now I am going to send you into the world." Israel's great mandate was to be a sent people—a mandate that they never fully understood or put into action.

In Isaiah 43 when Isaiah is bemoaning the role of God's people in the world, he talks about a people who were formed at God's pleasure to be his very own. And for what purpose? To declare the praises of the God who created them. Wherever God touches people and things, *sentness* is the issue.

The great Methodist missionary, E. Stanley Jones, tells this delightful story:

> I arrived one day at Pahlevi, Persia, which is now Iran, on the shores of the Caspian Sea, to go up to the capital, Teheran. The man in charge of transportation had difficulty because there were more passengers than cars. I saw a car with only one passenger in it, so I said to the man in charge, "Why can't I go in that car. There's only one man in it."
>
> He said, "I'll see." But he came back crestfallen and he said, "I'm sorry, sir, but the man said you couldn't ride with

him because he is a French diplomat and you are only a mis-sionary."

I suppose I should have felt squelched but inwardly I straightened up and I said to myself, "If he is a French diplo-mat, then he represents a shaky French kingdom, which has had twenty-six governments in thirty years. I am an ambassa-dor of the unshakable kingdom of God which has had one government since the foundation of the world and will have one government to the end."

Later on the way across the Caspian Sea by boat the diplo-mat got caught by a treacherous lock in the men's room and couldn't get out. He waved at me frantically and he said, "Please sir, extricate me!" So the ambassador of the kingdom of God extricated the diplomat from France.

Is that what the ambassadors of the kingdom must do–ex-tricate the diplomats of the world who have boxed themselves up in impossible ways of life? If they only knew it, they are crying out in more ways than one, "Please sirs, extricate us." And we must humbly but assuredly say, "Brother, this is the way. Walk in it, the way of God's kingdom."

The role of God the Father in sending from the foundation of the world: that is the sort of thing that must have swelled the soul of Peter, as he sat at the foot of Christ.

The Mission of Christ

The mission of Jesus Christ was also part of the sending mindset of Peter. Recall again that incredible moment when Peter resists the notion that Jesus can use him. Christ said to him in what must have been gentle but firm words, "Peter, don't be afraid, don't be afraid." And when Jesus invades that group of men in those first days they have anything but a sending mindset. These men are relatively poor, probably uneducated, come out of the simpler professions and have a provincial view of the world and of history. Moreover, they are poorly organized.

There is a "Peanut's" story in which Linus, Lucy's younger brother, is watching television. Lucy walks into the living room, looks at Linus's choice of program and says, "Change the chan-nel!"

Linus looks up and he replies, "What makes you think that you can walk into this room and just say like that, 'Change the channel'?"

She says, "You see this hand? Individually these five fingers don't amount to much, but rolled together tightly into a ball-like fist they become a weapon formidable to behold."

Linus changes the channel. And after Lucy is comfortably ensconced and watching her own program, Linus looks at his own hand and fingers, and says, "Why can't you guys get organized like that?"

This could have been said to the disciples to challenge them to become apostles. I wouldn't have picked one of those men. Indeed, there must have been moments when in his humanity Jesus must have said, "Why can't you guys get organized?"

The answer is simple. In the earliest stages of their walk with Jesus they were like many of us. They loved him; they were following him, but they did not yet think of "sending." They had to grasp the notion at the very beginning that God the Father so loved the world that he gave his only Son; he sent him into the world that the world through him might be saved. Until they knew that Jesus was the sent One from the Father, and that they in turn were to be sent by him, they could never mature and get organized as they were to be.

Jesus was drilling this deeply into their spirits month by month in experiences of discipleship, failure and success, slowly unfolding to them this enormous concept that we are trying to grasp that every person is sent.

John 4 tells the story of the woman at the well in which Jesus talks to the woman and her life is scoured and changed. The disciples come back thinking that Jesus would be hungry for food. But he said, "Look, food is good, but that's not the important priority today. My food is to do the will of the one who sent me. Look out upon the fields and see these people coming. They are the most important thing." That's the way we think when we are sent.

Slowly, rhythmically, like a sledgehammer pounding at the resistance of their innermost spirits, Peter and the men around him are taught what it means to be sent.

In John 17:18 Jesus says, "Even as you, Father, have sent me into the world, so I have sent them." And in John 20:21 he says, "As the Father has sent me, even so send I you." Over and over, each time he is in the presence of Jesus, this great consuming theme touches Peter's life. He begins to see it as the important issue.

The Enablement of the Holy Spirit

When Peter in the next phase of his life comes under the promised power of the Holy Spirit, the polishing work of understanding the nature of sending becomes complete. Jesus returns to heaven. Peter and the others go to a room in Jerusalem. For days they tarried and they prayed. And they are not yet sure what this is all about.

One of my favorite stories is about an old New England prep school headmaster in the 1930s. One day he booted from school a boy who was the son of a wealthy Boston businessman. When the message reached Boston that the boy had been booted, the father caught the first train to Concord, New Hampshire, and rushed unannounced into the office of the headmaster. Now the words I'm about to say no preacher would ever say of his own volition. But he walked into that office, looked at the headmaster and said, "You damn well act as if you're running this school all by yourself."

The headmaster stood to his feet and looped his thumbs into his vest so his Phi Beta Kappa key could easily be observed. And he said in a good New Hampshire accent, "Your language is coarse, your grammar is despicable, but you have grasped the idea!"

That is exactly what happened when the Holy Spirit came upon those one hundred or more men and women in that room. When he filled them to the uttermost, they grasped the idea that men and women following Jesus Christ, filled with his Spirit and walking in the nature of the God of the Bible are sent people.

This is demonstrated in Acts 2. God the Spirit proved before scores of people from different lands, cultures and languages that world evangelization is a possibility for all who want to be filled with the Spirit. Throughout the book of Acts the story is the same. Wherever the Holy Spirit is in charge, wherever the Holy Spirit controls the lives of women and men, there is a compelling sense of sentness. We see it in the story of Peter's going to the house of Cornelius and later on justifying his act to the church at Jerusalem. We see it finally in one of its peak moments in Acts 13 when Paul and Barnabas were set apart by the Spirit.

As we read 1 Peter 2:9-10, we say, "Peter, how far you've come from Luke 5! How far you've come in understanding this concept that men and women are special only as they are sent to reveal the glory of God like all of creation and proclaim that glory in the salvation from Jesus Christ. And now, Peter, you are trying to convince churches that they should see the same vision."

No burden weighs heavier on my heart as a Christian pastor in these days than to help convince as many of you as possible of the importance of being part of a congregation that sends. And I suspect this message is relevant to every Christian. If we are part of a sending congregation, we will be called either to go out to plant sending congregations or to stay behind to release others to go. If we want to walk in the way of Christ, there are no other alternatives. No human organization is more beautifully fit to send than a congregation in which there is a cross section of generations, a broad mix of resources, abilities, talents, gifts, a steadiness in the disciplines of worship and spiritual growth, and a caring fellowship which spans young and old alike. Out of this environment, when the church is doing its job, come the sent ones who invade the world around that congregation and go all the way to the uttermost parts of the earth.

George Allen, the football player in a losing game, looked at a fumbled football and didn't want to pick it up. Likewise, some of you may not want to pick up what you think to be a "fumbled" church because it seems to have mistreated you, or it seems too stodgy, stale or irrelevant for your taste. Don't fall for that perspective! If the congregation which you are a part is not a sending congregation, then assume with the words of Peter that you are sent to light a fire in that congregation until it discovers its grand possibilities of sending. If you are part of those who are to go, then draw from the sending congregations in your world, the strength, the sustenance, the intercessory power that they must give you. So as you consider your role in the sending church, I challenge you as a new generation to pick up its possibilities and potential. Pick the church up. Please, in your generation, pick it up!

9
The "Receiving" Church
Isabelo F. Magalit

We heard about the sending church from a North American pastor. Tonight a pastor from Asia has been asked to speak about the receiving church. The implication is obvious: the North American church is the sending church, and the church in Asia is the receiving church.

That conclusion is false, and I will attempt to show why. Nevertheless, Urbana 81 is a North American missions convention, so I will seek to speak from the perspective of the churches that receive missionaries from North America.

My thesis is quite simple and can be summed up in three short sentences: (1) the church is international; (2) our task is immense; and (3) partnership is imperative.

The Church Is International
Today the Christian church exists in every corner of the globe from Afghanistan to Zimbabwe, from Thailand to Turkey. In

every nation there are at least a handful of believers who are disciples of Jesus Christ. Some of them meet as bustling congregations in Rio de Janeiro or Seoul, Korea, while others can only meet as underground groups as they do in Albania or North Korea. But the church is everywhere.

Thus, the church is everywhere so long as we include as church (or *ekklesia*) any two or three who are gathered in Christ's name (Mt 18:20). The word *ekklesia* also describes believers gathered in a house (Col 4:15). This church in the house of Nympha seems also to have been part of a larger grouping, the local congregation in the city of Colossae (Col 1:2). Ephesians 3:10 also refers to *ekklesia*, and I understand this fourth meaning as including all true believers in all the world—past, present and future.

Perhaps we cannot completely understand all the relationships among these four meanings of the word *church (ekklesia)*. It is a difficult problem, which the philosophers call that of the one and the many. No matter. We know that the church is indeed international, so that it is possible to speak of the American church or of the Filipino church.

These two churches—the American church and the Filipino church—are not the same. One is wealthy materially; the other is relatively poor. One has at least a hundred fifty years of experience in sending out foreign missionaries; the other has just barely begun. There are other cultural differences. But the two churches belong to the one body, for there is only one body (Eph 4:4). Ephesians 4 also tells us that Christian unity is not uniformity. It is a unity in diversity with a multiplicity of gifts. It is not dull uniformity, but multicolored diversity that is characteristic of God's church. Thus, the American church ought to be thoroughly American, and the Filipino church ought to be truly Filipino. Each is to be indigenous, native to its own soil.

There is a beautiful verse in Revelation 21 (v. 24) that speaks of the New Jerusalem and how the kings of the earth shall bring their glory into it. I believe that means that there is a distinctive American contribution to the holy city as well as a Filipino contribution.

Each church is distinctive, and both churches are equal in

standing before God. While Jerusalem, the older congregation which counted the apostles among its members, sent Barnabas to investigate the church in Antioch (Acts 11:22), Jerusalem never lorded it over Antioch, never regarded Antioch as a ward, responsible to her. There was rich fellowship between them, but it was a fellowship of equals. The apostle Peter, writing to early believers, could speak of their faith as of equal standing with his (2 Peter 2:1).

Each was distinctive, but both were equal and called to a common task. Our priority today remains the same as that of the early church: worldwide witness (Acts 1:8). We are to be witnesses to Christ, not only in Jerusalem, but also in all Judea, and in Samaria, and to the end of the earth.

But now that the church is worldwide, where is the end of the earth? From Jerusalem, the end of the earth, or halfway round the globe, is Honolulu. From Manila, the end of the earth is Buenos Aires. From Chicago, the end of the earth is Dacca, Bangladesh. The point is: your end of the earth may be my Samaria; my end of the earth may be your Jerusalem.

The mandate for worldwide witness is given to the whole church, and therefore to every church. That is why it is wrong to think of the North American church as the sending church, and of the Asian church as the receiving church. Every church is a sending church. Likewise, every church must be willing to receive.

The theme of the 1967 Urbana Convention expressed this truth well: God's Men from All Nations to All Nations.

Our Task Is Immense

We need to understand that it is the whole church that is sent, not just an elite—a corps of apostles, or a specially called group called "missionaries"—but the whole church. Acts 1:8 was fulfilled, not only by the eleven, but by others as well—Stephen and Philip, evangelists from Cyprus and Cyrene, Paul and Barnabas.

I do not agree with the thesis that all are called and only some are sent. I prefer to say that all are sent—but we are not all sent in the same way, or to the same places. I see in Acts 8, 11 and 13 three kinds of missionaries. M1 missionaries are the stay-put missionaries, like the apostles who stayed put in Jerusalem in spite of the persecution (Acts 8:1). M2 missionaries are the share-as-you-go missionaries, like Philip and the nameless evangelists from Cyprus and Cyrene who preached the gospel wherever they went (Acts 8:4; 11:19; 20). M3 missionaries are the set-apart missionaries, like Barnabas and Saul, whom the Holy Spirit and the church in Antioch set apart for the evangelization of Cyprus. We need all three types of missionaries, for our task is immense.

There are nearly 4.5 billion people on planet Earth, distributed among 220 territories, which include 160 sovereign states or nations. Only about a billion of this large number are Christians—including all those who are Christians only by name. While the church is growing in many places, there is tough resistance to the gospel and severe difficulties for the church in others. In our day, the church not only has to face the restrictions of communist governments, it also has to contend with militant Islam.

China, India and the world of Islam already account for more than two billion people. The overwhelming majority of Chinese and Indians are not Christians. Mission experts also tell us that about forty nations, representing nearly two billion people, do not admit foreign missionaries at all, nor give much freedom to native preachers of the gospel.

Even if the church grew to keep pace with the rate of world population expansion, the absolute number of non-Christians would continue to grow. *There are more non-Christians in the world today than at any other time in the entire history of the world.*

Our task is immense.

Partnership Is Imperative

We must work together. Unless we learn to work together, we will not get the job done. The marketplace may be good for keeping the quality of consumer products high and their prices low, but the spirit of competition is not for the church of Jesus Christ. Cooperation is what we need, not competition.

We must learn to be partners. Otherwise, we will not only discover that we are getting in each other's way and wasting precious resources, we may actually end up working at cross-purposes. We also deny the unity of the one body. Thus partnership is not only required because the task is immense but also because the international church is only one body.

I think we all agree that partnership is imperative. But we have different ideas about the meaning of partnership. One form of partnership, or working together, is represented by the multinational corporation. Let us take McDonald's as an example.

McDonald's was started on April 15, 1955, by an astute Chicago businessman named Ray Kroc. Since he opened his first restaurant in Illinois, Mr. Kroc has sold 40 billion McDonald's hamburgers! In 1966, the clown Ronald McDonald was created and appointed spokesman, and he now speaks eleven languages. In 1970, McDonald's set up an international division, and what started as a uniquely American concern became truly global, with branches in twenty-five countries.

On October 17, 1981, McDonald's opened with fanfare in Manila, right in the middle of university belt, only five hundred meters from my office. It is now doing well under a Filipino manager, who was first trained in McDonald's Hong Kong but went on to finish a bachelor of hamburgerology and a minor in French fries at Hamburger University in Oak Brook, Illinois. We are told that Hamburger University is the ultimate in training and development for aspiring managers of McDonald's stores.

That's the McDonald's story. Now I am not critical of McDonald's; in fact, I like their hamburger. But not for my *daily* diet. It is not indigenous; not Filipino. I prefer rice. Still, McDonald's is welcome in Manila. It is indeed a form of partnership.

Unfortunately, however, this pattern of internationalizing is often the same pattern adopted by Western mission agencies. I find this hard to accept. What I can welcome in the matter of marketing hamburgers, I find unacceptable in relation to the gospel. For I believe that the gospel must become truly indigenous everywhere it is preached.

For the gospel to really grow deep and spread widely among native populations, it must become native to the soil. When Western mission agencies appoint a Filipino head who acts simply as their "agent in Manila"—to spread their westernized form of the gospel, their theological distinctives, their particular methods—they do not help the gospel really take root. The Filipino head may be called national director, or even president, but he is only "a little brown American."

My brothers and sisters in Christ, please do not make us into little brown Americans. We want to be, we need to be, genuinely Filipino Christians for the sake of the gospel.

Models of Partnership

There are patterns of partnership to commend. Let me tell you about the Presbyterian Church of Brazil and the Presbyterian Church of the USA.

The Presbyterian Church of the USA has had missionary work in Brazil for more than one hundred years. As the Presbyterian Church of Brazil has taken root and grown over the years, the relationship between the two churches has also evolved.

In 1973, the two churches drafted the First Agreement of Cooperation to govern the relationship between them. In October 1980, representatives from both churches met to work out an amended agreement which was submitted for approval to the two General Assemblies in 1981.

What is the goal of the agreement? The preamble says: "the goal [is] full integration of our missionaries in the midst of the church in which they will be working . . . recognizing that the work will be more fruitful for the glory of God when undertaken together and with the same objectives."

The central principle in the agreement is the principle of mutuality. Mutuality is recognizing each other as full partners in the common task of proclaiming the Lord Jesus Christ. This mutuality is worked out under the different headings: The Missionary, Finances, Freedom of Action, Properties, and so forth.

The common task is "proclaiming the Lord Jesus Christ in word and deed in Brazil, within the United States of America, and, God willing, in other countries as well." Notice it says "in other countries as well," not just to help each other evangelize their own peoples, but to work together to bring the gospel to third parties. Or, in the precise words of Article 26: "Both partner churches shall seek opportunities for cooperation in mission to peoples in countries other than their own."

Of course, the agreement raises questions. For example, this church to church relationship does not cover all situations. What about parachurch agencies and missionary societies? The agreement also implies a theology of the church and of church polity that is not acceptable to everyone.

Nevertheless, the agreement presents us with a specific model of cooperation that spells out what a partnership of equals means. If we are committed to the principle of mutuality, whatever our situation, we can work out the applications to ourselves.

There are other models of partnership. There is the Lausanne Committee for World Evangelization. Fifty men and women from across the world, representing many nations and the whole spectrum of evangelical faith, have pledged to work together for the speediest evangelization of the world. The chairman of the committee is evangelist Leighton Ford from North America, and the executive secretary is Gottfried Osei-Mensah from Ghana, Africa.

Then there is the International Fellowship of Evangelical Students. Let me commend the IFES to you. The IFES is composed of 75 national evangelical student movements working together to reach students for Christ. Some of the member movements are large and established, like Inter-Varsity Christian Fellowship, USA. Others are small and struggling, like the Thai Christian Students. But each national movement is autonomous, and the international fellowship is a fellowship of equals.

Every four years, each national movement sends representatives—a maximum of three—to a General Committee which meets also as an international conference to survey the worldwide student scene, to learn together from God's Word, to pray and consult together, and to make plans for further outreach. The General Committee also chooses eleven men and women— from all six continents—to serve on an Executive Committee which meets every year and carries out the mandate of the General Committee, as well as guides the decisions of the General Secretary. The General Secretary and all the staff of IFES are appointed by the Executive Committee, subject to confirmation by the General Committee.

The IFES General Secretary is Mr. Chua Wee Hian from Singapore, and the present chairman of the Executive Committee is Dr. John W. Alexander of the USA.

There is no uniformity of methods or structures, or even name among the 75 members of the IFES. The Brazilian movement is called Alianca Biblica Universitaria do Brasil (ABUB); the German movement is called Studentenmission in Deutschland (SMD); the fellowship in Nigeria is simply known as the Nigerian Fellowship of Evangelical Students (NIFES); the Japanese movement is called Kirisutosha Gakusei Kai (KGK); the Indian movement is called Union of Evangelical Students of India (UESI); the affiliate in Papua New Guinea is called Tertiary Students Christian Fellowship of PNG (TSCF). Each one is distinctive and indigenous, but all are committed to the one Lord and to his one gospel. The IFES—the international fellowship of equals.

Many of you will not have much to do with the IFES or the Lausanne Committee or the Presbyterian Church of the USA. That does not matter. We want you to come and help us. But please come as real partners. Let us work together to make the Lord Jesus known to the ends of the earth.

There is another lovely passage in Revelation:

After this I looked, and behold, a great multitude which no man could number, from every nation, from all tribes and peoples and tongues, standing before the throne and before the Lamb, clothed in white robes, with palm branches in their hands, and crying out with a loud voice, "Salvation belongs to our God who sits upon the throne, and to the Lamb!" (Rev 7:9-10 RSV)

What a magnificent scene! What is even more remarkable is this: God has called you and me to be his coworkers in bringing that scene to reality. As we work together as real partners in bringing the good news to the end of the earth, we ensure that there will be people from every tribe and people and tongue and nation who will sing glory to God and to the Lamb.

"Amen! Blessing and glory and wisdom and thanksgiving and honor and power and might be to our God for ever and ever! Amen" (Rev 7:12 RSV).

Part IV

The Preparation of the Witness

10
Mission Impossible: Your Commitment to Christ

Billy Graham

I well remember 1939 when Hitler invaded Poland, going after the city of Gdańsk, which was then called Danzig. Hitler wanted a corridor to the sea, and his action marked the beginning of World War 2. Poland sits astride Europe and has been the battlefield in many of the great wars that have raged across Europe. And many of those wars have started because of Poland.

One year ago I was in Poland, lecturing and speaking at the University of Warsaw. They were kind enough to give me an honorary doctor's degree. I talked to the foreign minister and many other people whose names are in the news right now, including the late Cardinal Wyszynski. I went to Rome and saw the Pope. We talked about Poland and other events in the world, and I could see his concern over world peace. My own conviction was strengthened that Christians cannot be silent and allow the world to approach the point of the genocide of the human race without at least speaking and praying for peace. That has

not always been my position, but it is my position now because
a new factor has dawned upon my thinking—the terrible technology that science has created that can destroy the human
race. Of course, we know from the Bible that ultimately God
will intervene, and the Prince of peace will come back, and the
kingdoms of this world will be his. But till then, whatever hour
that may be—a thousand years from now or a hundred years
from now or tomorrow—we ought to be about the business of
witnessing to the whole world of the gospel of the Lord Jesus
Christ.

"But you will receive power when the Holy Spirit comes on
you; and you will be my witnesses in Jerusalem, and in all Judea
and Samaria, and to the ends of the earth" (Acts 1:8). Jesus gave
the apostles an impossible task, a mission impossible. It was
geographically impossible because a great part of the world
had not been discovered at that time; physically impossible
because they had no airplanes or radios or television or printed
pages, and they had no way to get about except by donkey or
camel. Numerically, there were too few of them: one hundred
twenty at most at Pentecost. Financially, Josephus says the
wealth of the church at that time was equal to about 50,000
American dollars. Legally, in most parts of the known world
it was against the law to talk about Jesus Christ. Logistically,
Jesus Christ had told them to go to Jerusalem, Judea and Samaria
all at once. They couldn't do it. By every rule of the book, it
was impossible to do what they were told to do except for one
thing—the power of God that came in the Holy Spirit on the day
of Pentecost. People were there on that day from every part
of the world, and they heard the gospel and carried it back to
the then known world. And I pray that God will do it again in
the decade of the '80s.

Two and a half weeks ago I was a guest at the house of Vice
President Bush and his wife who have been friends of ours for
many years. He gave a little dinner party on the evening of December 12th, and in the middle of the party he was called to
the telephone. Jim Baker was on the phone from the White
House and said that he must come to the situation room immediately. He had to leave and explained to us that things had

broken loose in Poland. I waited up for him that night and, when he came back about 11:00 or 12:00, he had a grim look on his face. I'll never forget when I went to bed that night. I said, "Oh, God, you have told us to pray for those in authority." I did not pray just for President Reagan that night and Vice President Bush. I prayed for Mr. Brezshnev; I prayed for the leaders of Poland, that God would give them wisdom because God puts up one and takes down another. God is sovereign. Three times Jeremiah talked about a pagan king by the name of Nebuchadnezzar and quoting God he said, "Nebuchadnezzar my servant." We are to pray for our enemies.

Though we are different geographically, denominationally, racially, all those who know Christ as Savior are members of one body. The apostle Paul said, "The body is one and has many members" (1 Cor 12:12 RSV). It is to this "one body" that the Lord Jesus Christ committed his work. Christian service is not limited to a certain group within the church. We all have one or more gifts that the Holy Spirit has given, and it is the privilege and duty of each member to serve him.

Proclamation and Service

Throughout the world today there is a lot of discussion within the Christian church about the mission of the church. And it is our purpose to consider missions and the mission of the church.

I believe that mission can be summed up in two words: proclamation and service. These represent the essence of the Christian mission at home and abroad, and they are inseparably linked together as the key both to evangelism and to penetrating culture with the gospel of the Lord Jesus Christ. During this century many evangelicals have removed themselves from culture instead of penetrated it. Many of them have accommodated themselves to the culture. We have all but lost our sense of what it means to be "in the world but not of the world." Our Lord, I believe, taught that we are to be spiritually insulated but not isolated. We have created two separate realms, sacred and secular. The sacred is the realm of the church and things that pertain to our faith, and we place a high value on it. The secular realm is that of the world and culture, and we place a lower value on that. Richard Halverson, who is chaplain of the U. S. Senate, said recently that we have created a "destructive polarization between secular and sacred, a distinction you do not find in the New Testament. It is this thinking that makes the assumption that teaching public school is secular; teaching in Bible school or a Sunday-school class is sacred. Running the business of a corporation is secular; running the business of a church is sacred." When lay persons have little time to do these so-called spiritual things, they experience a low-grade frustration, and they don't think of themselves as ministers of Christ.

If we are ever going to do the job of getting the gospel to all the world, we are going to have to break out of our false distinction between the sacred and the secular. We as citizens of Christ's kingdom are living in occupied territory—as sheep among wolves. It is terribly dangerous, but what a challenge! It is even exciting! I would not want to live in any other period of history if I had a choice. We are in this world with a mission—the mission of proclaiming the good news of Jesus Christ and serving a needy world. If we try to separate proclamation from service we are doomed to failure. Jesus never separated the two and we dare not either. Jesus said, "I am among you as one who serves" (Lk 22:26-27 RSV).

A friend of mine who is writing a book on coping with culture asked a youth worker whether he thought a young person today could be a committed Christian and still be an accepted, respected member of his or her school scene. And the youth worker replied that he could think of only one person he knew who was successfully doing that—a young lady who had made it her aim not to be popular at the university, not to be accepted, but just to serve others. In her school she had assumed the role of a servant and gone about doing the jobs that no one else would do. She was willing to work hard and let someone else have the credit, to do things behind the scenes in order to make others look good. Later she became very open in her proclamation of her Christian faith, and people listened and respected her for it. She had won the right to be heard through her service.

The Scriptures teach, "Have this mind among yourselves, which is yours in Christ Jesus, who, though he was in the form of God, did not count equality with God a thing to be grasped, but emptied himself, taking the form of a servant" (Phil 2:5-7 RSV).

As members of the body of Christ, we are called to the same radical commitment exemplified by our Lord. Put bluntly, it is a "self-emptying" ministry. He said we are to deny ourselves and take up a cross.

I have had the privilege of being with Mother Teresa in Calcutta. When she received the Nobel Prize a few years ago, the world was brought face to face with just such a "self-emptying" ministry. But I have met others and heard of others around the world who are carrying on similar ministries of sharing and giving that we will never hear about until we get to heaven. They haven't gotten a Nobel Prize. Their names have never been in the paper, but there are thousands of them out there. I have met them in the four corners of the world as it has been my privilege to travel and preach. That calling to be a servant of Christ and others begins right now. It is a calling given to everyone who names the name of Christ. It's a calling that begins right where you are—in your home, on your campus, in your dormitory, your apartment, your sorority, your fraternity. It begins in the school where you teach, the hospital where you are a nurse, the office where you work.

I believe that God wants many of you to go overseas in the future as his ambassadors around the world. I am also convinced that he wants all of us to live lives of total commitment and service to him right now where we are. Christ has loved us totally. He demonstrated this by his death on the cross and the shedding of his blood, and only total commitment to him and to his command to go into all the world is a reasonable response to that kind of love.

But first you must face the question of your own personal relationship with Jesus Christ.

Many of you could not stand up and say, "I know my sins are forgiven. I know that I have received Christ as my Savior. I know that I am going to heaven if I should die. I know that I have eternal life." You couldn't say that because you are not sure.

Perhaps you're like I was—reared in the church, vice president of the young people's society. But I didn't know Christ. I thought I did. I had churchianity and Christianity, but I didn't have Christ. Many of you are like that. Has there come a time in your life when you repented of your sin and received Christ as Savior? Are you sure of it? Are you certain of it? He died on the cross; he rose again for you, but he also demands a response and that response must be repentance of sin. That means that you acknowledge that you are a sinner and you are willing to change, to turn from sin and live a new life in Christ. It means that by faith you receive him. This is not something you can understand fully intellectually, because your mind has been affected by sin.

Perhaps you don't know why you chose to attend Urbana, but God is speaking to you about your relationship with him. You know that you need to repent of your sin, to quit living for yourself and give your life to Jesus Christ in response to his love. That is the first step in becoming a part of God's mission in the world. The Bible says, "Let the redeemed of the Lord say so." Only those who have been redeemed and have experienced salvation through faith in Christ have a message to proclaim. As God's own people, he gives us the privilege of declaring his gospel to the world.

Last year we held six crusades in Japan, and in every place we saw thousands of students. In some places, like Osaka, the police had to lock the big stadium gates. The governor of one city, a man who governs thirty million people, gave a reception for us with about a thousand guests. These included both Christians and non-Christians. When the governor stood up to speak, he suggested that there must be a reason why Christianity has not grown much in Japan since the seventeenth century. He said, "I think I know the reason. The gospel has not been made clear to us here. And I'm challenging you, when you travel throughout Japan, to make the gospel clear." That's all I needed. I had the authority from God and the authority from the government to make it clear!

Worship and Vocation

As you consider your role in world mission and how God is leading you to express your proclamation and service for him, I would like to focus on several aspects. First, *worship*. Worship is not a passive activity, but an active one. It's not something we watch, but something we do. Worship, both corporate and private, is necessary to sustain us in the midst of the sometimes hostile culture in which we live now or will be called to penetrate in the future for Christ. The apostle says in Philippians 2 that we should be "blameless and pure, children of God without fault in a crooked and depraved generation" (v. 15). Certainly this is a crooked and depraved generation if there ever was one. And you and I are to live in the middle of it as pure children without fault. We are to shine like stars in the universe as we hold out the Word of life. Where do we get that kind of life? Where do we get that kind of power?

Many of you are wondering how you will face the temptations and discouragements of your campus culture. How can you maintain the closeness to God that you found at Urbana? You are going to have to make worship a part of your daily life—worship alone or with a few other Christians. You are going to have to establish a disciplined devotional life. Satan will oppose you at this point more than any other. In addition to systematic Bible study and prayer each day, get a good devotional book to help you, such as Oswald Chambers's *My Utmost for His Highest*. You will be a flop and a failure as a Christian witness unless you have a disciplined, daily devotional life. I have never met anyone around the world that was growing as a Christian that did not have a daily devotional life.

Besides worship, another aspect of your ministry of proclamation and service is *vocation*. Whatever vocation you follow, you are called to Christian service in it. You may be a politician, a business representative, a laborer, a teacher. But you are first and foremost a Christian. Everything you do is to be sacred in the sense that it is done for the Lord. "Whatever you do, work at it with all your heart, as working for the Lord. . . . It is the Lord Christ you are serving," said Paul (Col 3:23-24).

We often lament the fact that certain countries are closed to

Christian activity. But there are opportunities and possibilities we have never even imagined. I heard about a young man and his wife who are right now living in a country that is closed to Christian missionaries. They are there at the invitation of the government to teach the young people of that country to play a certain sport. When they applied for their visa, they openly told the government that they were Christians and that they often spoke of their faith. The government told them, "We don't care what you talk about in private on your own time as long as you teach our young people to play your sport." There is another couple that we got a Christmas card from in code. They cannot talk openly about Christ, but they can bear the fruit of the Spirit. People often come and inquire what makes the difference. Thus, because of their dedicated service, they are able to give a quiet proclamation.

In Galatians 5 the apostle Paul lists the fruit of the spirit: "love, joy, peace, patience, kindness, goodness, faithfulness, gentleness, self-control." Then he goes on to say, "against such there is no law" (vv. 22-23 RSV). There is no law in the world that can keep you from loving somebody. There is no law that can keep us from having the joy that is produced by the Holy Spirit within us. In some countries you may not be able to use the public gifts like preaching or teaching, but you can bear the fruit of the Spirit. And when people see you are different, they will ask your secret. This is what brought Solzhenitsyn to Christ in a concentration camp in Siberia. There were three men from Estonia in the camp. They never talked about God, they just lived it. They were the hardest workers; they went out of the way to befriend everybody. One day when Solzhenitsyn was ready to commit suicide, one of them came up and drew the picture of a cross and walked away. Solzhenitsyn said, "I realized for the first time that that was the only place where real freedom was to be found." These men were the freest people in the world, even though they were in a concentration camp. When you know Christ, you can be free anywhere.

Is Jesus Christ the Lord of your vocation? Are you training so that you can serve him or yourself? Are you willing to let Christ change your motives and be the master of your life's work? If you're willing to go anywhere and do anything for him, you will be amazed at the opportunities he will open up to you. Whatever your vocation, it is part of your ministry of proclamation and service for Christ in the world.

Go

Jesus said to his disciples, "You did not choose me, but I chose you and appointed you that you should go and bear fruit" (Jn 15:16 RSV). The first word of Jesus' Commission to his disciples was go. "Go therefore and make disciples of all nations." The Bible is filled with God's calling and God's sending. God calls us to his son the Lord Jesus Christ and then he sends us into the world. God does not reveal himself to satisfy our curiosity. He summons us that he may send us.

When the thought that God had chosen me for ministry first struck me, I was humbled and astounded. I agonized and said, "No, Lord, no. I'll do anything except be an undertaker or a preacher!" Yet here I was. Having just accepted Christ as my Savior, I had said in a service of dedication, "Lord, I'll be what you want me to be."

And God was saying, "I want you to preach my gospel."

"But Lord, I stutter when I talk. I can't even stand up and talk in a classroom. How can I ever preach? I'm nothing— I have no education; I have no money; I have nothing."

But God said, "I want you; I have chosen you." He didn't say it in so many words. I did not hear an audible voice. There was no lightning, no thunder. There was just that quiet working of the Holy Spirit, and one night on the eighteenth green of a golf course near Tampa, Florida, I knelt down and said, "Lord, I'm yours." And I began to realize the consequence of that—my whole life changed. Even the person I was to marry changed. Everything changed. He took me step by step.

Then after I had prepared four sermons, nobody wanted me to preach. And I said, "Lord, please send somebody to ask me!" Finally, I got a chance to preach in north Florida. I had prepared those four sermons, and I had spent weeks practicing them. I figured each sermon could last forty-five minutes. But that night, in a cold little Baptist church in north Florida, with thirty-six people present, I preached all four sermons in eight minutes!

The biblical emphasis upon God's call is a call to responsibility. It is an appointment to the exacting, rigorous work of God's redemptive mission—calling men and women to faith in Christ. Dietrich Bonhoeffer said, "When Jesus calls a man, he bids him come and die."

In the April 1981 issue of the Wycliffe Bible Translators' publication *In Other Words*, the editor, Hyatt Moore, wrote: "Years ago, a group of missionaries was sailing to a remote place where they hoped to minister the gospel. The captain of the ship warned them that they were risking their lives and tried in every way to dissuade them. Their response was simple: 'We died before we left.' ... On March 7, 1981, Chet Bitterman, in the service of the Lord, died his 'second death.' After being held for 48 days by terrorists, Chet had been executed and his body left in a bus in Bogota, Colombia. He was 28 years old. He left a wife, Brenda, and two young daughters."

Those who have already given their lives to the Lord can say with Paul, "I eagerly expect and hope that I will in no way be ashamed, but will have sufficient courage so that now as always Christ will be exalted in my body, whether by life or by death. For to me, to live is Christ and to die is gain" (Phil 1:20-21). That is the call of God to your life. That is the great question you are going to have to face. Is God calling you to that kind of life and death? God is now appealing to your will. You are going to have to come to the point. I find that many students at many campuses do a lot of talking and debating and arguing, but they never come to the point of commitment. Someone said this is the generation of the uncommitted. I was at Oxford last year and met a sociologist who told me that the '60s was the generation of trouble; the '70s was the "me" generation; the '80s is the generation of survival. Will we survive?

God has chosen; God has commanded. Are you ready to answer? If we do not respond in obedience, we can become castaways, as Paul was afraid of: "But I keep under my body, and bring it into subjection: lest that by any means, when I have preached to others, I myself should be a castaway" (1 Cor 9:27

KJV). What about you? Have you given God every key to your heart and life? Are there spaces and rooms and closets into which he has not fully come because you have not given up all the keys to your spiritual house?

I am asking you to make an affirmative response to the call of God, a response in which you dedicate your life totally to Christ. God is looking for young men and women who will say to him, "I'll go where you want me to go; I'll be what you want me to be, at home or abroad. I'll put my life under the lordship of Christ and recognize that I am in Christian service, even though I may be working at a secular job. And I am willing to go to the ends of the earth if that is your will, Oh Lord."

A missionary is anyone, anywhere, who will obey the command to faithfully witness at home or abroad. You may be doing it under the auspices of a mission board or as a private citizen, working and supporting yourself. But God has given you a spiritual gift, and he calls you to put that to work in his service. But first you must be sure that the Lord Jesus Christ lives in your heart.

In 1913 a young man arrived in Cairo, Egypt. He was twenty-five years old, a graduate of Yale University and Princeton Seminary. He was tall, strong, handsome, intelligent, single and very rich. His name was William Borden from Chicago. He had come to Cairo as a missionary. Eventually, he wanted to go to China. Many people had difficulty understanding why a young millionaire wanted to spend his life like that, but those who knew Borden for any length of time understood it. The consuming desire of his life was to share the love of Jesus Christ with people wherever he was, whatever he was doing. He loved people and longed for them to know the God who loved them and died for them and who had changed his life.

After only a few months in Cairo, Borden contracted cerebral meningitis and died. In the weeks following his death, the world began discovering some remarkable things about him. He never owned an automobile because, he said simply, "I can't afford one." Yet during his three years at Princeton he gave away hundreds of thousands of dollars to Christian work. Borden's biographer said of him: "No reserve, no retreat, no regrets had any place in Borden's consecration to God." Borden had planned to go to China as a missionary. But God had other plans. And the story of his life and early death became a rallying cry for hundreds who went to the mission field because of Borden of Yale.

There are a thousand things you can do with your life, a thousand things for which you can spend it. But how many of them will enable you to say at the end of your life, "No reserve, no retreat, no regrets"?

Sandy Ford was only twenty-one. He was a great athlete, a great student and had just been elected president of the largest Inter-Varsity chapter at any university in the United States—the chapter at the University of North Carolina, Chapel Hill. He was one of the most remarkable young men it was ever my privilege to know.

One day Sandy Ford was running the mile and winning. Then he stumbled and fell. He was having a heart attack. He looked back and saw that he was still ahead, and he got back up and went a few more yards and fell again. He saw that he was still ahead, so he crawled across the finish line. Newspapers carried his picture on the front page.

Last month, on November 27th, after an eight-hour operation to fix it, Sandy's heart stopped beating. It never started again. Many people said, "What a tragedy." The newspapers in our state wrote editorials about his life. Thousands who knew him or didn't know him were shocked—but challenged. Sandy's life had not been cut short at twenty-one. It had been completed.

I was in New York the Sunday before Sandy's operation. Something said to me, "Go by Duke University and see Sandy." He was at the Duke hospital. I went seven hours out of my way to see him. He was my nephew, the son of Leighton Ford, who married my sister. I went into the hospital room, and we had two wonderful hours together. And Sandy said an interesting thing: "You know, Uncle Billy, I believe that my illness here has something to do with those meetings at the University this next fall." Little did he know that by death he would be bringing the first message. God had a plan. Sandy had put himself totally and completely in the hands of Christ: "no reserve, no retreat, no regrets."

You do not know how much time you have. But that is not the important thing. The important thing is that from this moment on, you decide to be God's man or God's woman—without reserve, without retreat, without regrets—wherever he sends you. Jesus hung on the cross publicly with hundreds of people watching him and mocking him. He hung there for you, and if you were the only person in the whole world that needed him he would have died for you. That's how much he loves you. He is ready to come into your heart and forgive you, cleanse you, make you a new person and start you on a new road. Will you dedicate yourself to him without reserve, retreat or regrets?

11
Characteristics of a Witness
Samuel Escobar

Urbana 81 is a clear example of how the mission of God in today's world has become a truly international, interracial and multicultural enterprise. It is far from being anymore "the burden of the white man" or the ideological side of imperialism. I am thankful to God for this new reality, and as a brother in Christ I want to reflect on a basic question: What is a witness? What are the characteristics of a missionary here or elsewhere?

These basic questions have to be asked from time to time. We cannot take for granted that we know the answers. Some time ago as I was arriving in Argentina after one of my missionary trips the revenue department of that country was conducting a survey of travelers to find out their sources of income. A girl came to me with a questionnaire and a pen. "Where are you coming from, sir?"

I answered, "From the United States."

"What is your occupation?"

I answered, in Spanish of course, "Soy un misionero," "I am a missionary."

Now that word *misionero* in Argentina means a person born in the province of Misiones. So she looked at me a little baffled and said, "I am not asking where you were born, but your occupation."

It was not easy for me to explain to her an occupation that was not on her list. And I have not found it easy to write this paper to ask these basic questions because in a way it has been a painful but spiritually rewarding exercise in self-examination.

Several times in the New Testament, the apostolic writers define what a witness is. We will go to these passages as a source to answer our questions. We will not ask the social scientists or the experts in international relations what a missionary should be. They have their own presuppositions and their own expectations, and they like to tell us what we should be. But the Christian way, the evangelical way, of asking basic questions is to ask them of God's Word. God is the Source and the Master of the missionary enterprise. He has revealed his standards and set up his models.

One important feature of the New Testament answer to these questions is that the writers always base their theory of missions on their experience of missions. Their way of stating what a witness or a missionary should be comes out of their daily practice and experience. They are not so much teaching a course for people to get a Ph.D. in missiology, but rather explaining, defending or establishing the revealed basis of their practice, sometimes passionately and polemically. We find in Scripture the joy, the sense of adventure and the surprises of the missionary experiences of Peter, Paul, Barnabas and Philip. We also find passages in which these men *explain* what they do and why they do it. In their words we find constant references to the example and inspiration of Jesus, the missionary par excellence. We find both facts and interpretation in Scripture, and both are the Word of God that should rule our lives.

Let me then outline five characteristics of a witness that I have found in Scripture.

A Living Testimony of God's Initiative

The idea of witness in the Bible presupposes that there is a God who is active in the world as a creating, saving Lord, and that the very existence of some people is a testimony to the fact that God is in action. In Isaiah 43 we have a beautiful passage about the missionary existence of the people of God. The chapter begins with words attributed to "The LORD, your creator . . . He who formed you, O Israel." The writer goes on to say, "Do not fear, for I have redeemed you," and then we read:

"You are My witnesses," declares the LORD,
"And My servant whom I have chosen,
In order that you may know and believe Me,
And understand that I am He.
Before Me there was no God formed,
And there will be none after Me.
"I, even I, am the LORD;
And there is no saviour besides Me.
"It is I who have declared and saved and proclaimed,
And there was no strange god among you;
So you are My witnesses," declares the LORD,
"And I am God.
"Even from eternity I am He;
And there is none who can deliver out of My hand;
I act and who can reverse it?" (Is 43:10-13 NASB)

This passage was key in the self-understanding of Jesus and of the apostolic preachers of the gospel. What it teaches should be key in our own understanding today. Who is the source of our identity and reward? God. And here we are facing the heart of the matter. There is a God, and every missionary is a witness to that fact. That means that the personhood and action of the missionary cannot be explained by any other thing but the mystery of God.

This is important in an age of explanations, in a technological culture that wants to put mystery aside. It is important also in an age of quantification where every conceivable aspect of reality is to be reduced to figures and numbers, scales and graphs. You as a person are a witness to the reality of a God who is there.

Now there are some consequences to this. The witness has to be the reality of God to others in places where God is denied. I have thought of two instances that illustrate this recently.

One student from my church, after the difficult entrance examination, entered the university. The first day she went to class, she came in the evening to the Bible study at the church, and I asked, "How was it?"

She said, "It was interesting."

"What was the class about?"

"It was about dialectical materialism. The teacher said to us, "We are going to teach you here how to think, and once you learn how to think you won't believe in God and in all that religious stuff."

So there are some people that have to say *there is a God* in that kind of environment. There are many professors who believe the way that one did.

I also was reading, just before coming, an article written by a Spanish journalist in El Salvador. He was telling the story of a group of soldiers who came to the little hut of a poor family. The men entered violently, shoving aside the old grandmother. They put the men against the wall, then grabbed the three teenage girls to rape them. The old woman in despair cried out, "Don't you have fear of God?"

The men replied, "God doesn't exist. He is dead. Don't you know that? We are gods now."

That is also a situation in which some people have to be witnesses to the existence of God. This reminds me of all the places in the Bible in which we find people like Job. His wife said, "Forget about it. Forget God."

But he replied, "No. I cannot understand myself nor the world around me apart from God." A witness is first a living testimony that there is a God.

This gives to the witness a tremendous sense of freedom. I think this is necessary in the missionary enterprise. As in any enterprise, the missionary enterprise has its own structures, and sometimes people can become prisoners of structures. So everyone who is working inside a structure needs this conviction: "I exist because there is a God. I am a witness because there is a

God. It is God to whom I have to respond first."

Paul talked about this in the midst of a discussion about his apostolic call. Writing to the Corinthians he says: "To me it is a very small thing that I should be examined by you, or by any human court; in fact, I do not even examine myself. I am conscious of nothing against myself, yet I am not by this acquitted; but the one who examines me is the Lord" (1 Cor 4:3-4 NASB).

This kind of awareness is necessary today. Let us remember that it is not arrogance but humility that was characteristic of Paul. But he was very clear about a deep relationship with God that had a liberating effect in his relationship with people. Can we say honestly, "The one who examines me is the Lord"?

There is a second consequence to the fact that God is the source of our identity. In a practical way witnesses are experts in communication with the God of whom they give testimony. They have attentive ears to the Word of God. They have active lives of communication with God. The missionary enterprise has become so complex and sometimes demands such a degree of specialization that it is easy to forget these basics. I was recently talking with national leaders of a denomination in Perú who could not understand some of the newly arrived missionaries. They were so highly specialized that they were unable to lead a small Bible study in a home or a prayer group. In a country that is so open to the gospel that we have to mobilize every lay person to cope with the task, how can we afford to have missionaries who do not have the basic qualifications of a witness?

Holiness is another aspect of lifestyle. If my existence is explained only by the action of a holy and just God, my existence must be marked by a continuous movement toward holiness and justice in the large and small aspects of life. Every time in the Old Testament that Israel is reminded of her God, two consequences result: *worship* and *obedience,* love of God and love of neighbor.

Latin America is a religious continent. There are signs of religious life everywhere. In architecture, archaeology or art you find expressions of this love and worship of God. But we have not had love for our neighbors expressed in social and political institutions. We have had centuries of injustice, greed and cruelty instead. And now we are faced with a tremendous revolutionary movement that says, "No more talk about God; let us change our societies so that a minimum of love for the neighbor is possible." But revolution without God is as bad as religion without love for neighbor and equally destructive. So we need witnesses of Christ that in their very existence show how the two demands of God go together.

A Steward of the Mysteries of God

"Let a man regard us in this manner, as servants of Christ, and stewards of the mysteries of God. In this case, moreover, it is required of stewards that one be found trustworthy" (1 Cor 4:1-2 NASB). We are stewards because, as Paul argues a bit later, everything we have is something we have been given, so there is nothing to boast about.

A witness is not the originator of truth. Witnesses are not the makers of truth. We are not the clever, superior people that patronizingly pass on our knowledge to the poor pagans. Humility is the first mark of a witness because he or she is only a steward. Grace is a key to the witness's attitude. He or she does not feel superior nor does act as such. We live in a moment of history in which this attitude is needed again. Paul gave this advice to a church that was so proud, so full of gifts, so rich that they needed to be constantly reminded to practice humility. And Paul pointed to his own style which was marked by a sense of grace, of humility, of sharing what he received with the attitude of a debtor paying a debt.

What are the mysteries of which we are stewards? What are the facts to which we are witnesses? In one of the first uses of the word *witness*, to describe himself, Peter mentions the death and the resurrection of Christ: "This Jesus God raised up again, to which we are all witnesses" (Acts 2:32 NASB). Jesus' death and resurrection is the basic core of the gospel, the *kerygma*, the fact to which the apostles were witnesses. This gospel had to be passed on faithfully, without additions or deletions, without heretical twistings. That is what Paul meant by faithful stewardship. But if we leave it at this point witnessing would be an intellectual exercise—the communication of some information according to basic rules of respect for the integrity of the sources. But there is far more to it.

When Peter speaks the words we have just quoted, the death and resurrection of Jesus Christ have already taken place and are historical facts. And the power of God is a living reality in his life and in the things that are happening around him and his hearers. The church is being empowered and shaken; the sick are being healed; sinners are being convicted. So what Peter is saying is that the power they are seeing in action is the power which can only be understood by reference to Jesus Christ, his death, his resurrection and the fulfillment of his promise of the Holy Spirit. Unless we know that power and learn to acknowledge it, we do not have a basis for our task of witnessing. We must be able to say that we have seen that power in action in our own lives. We may need to learn how to see it and recognize its presence. Let us remember that we are not asked to produce this power nor to produce a system of ideas about this power. First and foremost, we are called on to tell the world that this power is in action because we have seen it in action.

Paul refers many times to the power of God active in his own life: correcting, changing, sanctifying, empowering, guiding. Before you start to witness through words you have to become aware of this power of God in your own life, in your church, in your student chapter, in big and small things, in the ordinary and in the extraordinary. You mortify your flesh and fight for purity because you are crucified with Christ (Gal 2:20; 5:24). You overcome temptation and advance because the power of the resurrection is active in you (Rom 8:11-18).

John Stott in *Basic Christianity* offers a clear understanding of the kerygma, its content and consequences. John White's *The Fight* will help you experience God's power in daily life. *Rich Christians in an Age of Hunger* by Ronald Sider will give you a grasp of the wider consequences of being taken over by God's power in the complex world in which we live. But unless you bend your knees and say *yes* to the Spirit that wants to take you over, all that theory will not help.

A Servant and a Prophet

Another instance in which Peter used the word *witness* to describe himself is when he spoke at the home of Cornelius. A new shade of meaning is brought up by this context. You know that Cornelius was a centurion and a devout man who feared God. In other words, Peter is witnessing to a military chief, a pragmatic man used to giving and receiving orders and making things happen, a representative of the established empire of those days.

Two things call my attention to what Peter says on this occasion. He mentions the kerygma of Christ but he adds new elements. First, he refers to the life of Jesus: "You know of Jesus of Nazareth, how God anointed Him with the Holy Spirit and with power, and how He went about doing good, and healing all who were oppressed by the devil; for God was with Him. And we are witnesses of all the things He did both in the land of the Jews and in Jerusalem" (Acts 10:38-39 NASB).

Then Peter goes on to mention the death and resurrection of Jesus Christ, adding: "And He ordered us to preach to the people, and solemnly to testify that this is the One who has been appointed by God as Judge of the living and the dead" (Acts 10:42 NASB).

Is it significant that in talking to a military chief Peter stresses these two facts? I can witness, says Peter, that Jesus spent hours and days and weeks and years doing good to people in need, that he was the anointed servant who preached the manifesto of Luke 4. In other words, he was not an unoffensive and unobtrusive "guru" teaching transcendental meditation surrounded by flowers and incense and soft cushions. Because he served the people, especially the poor, who needed him more, he entered in constant conflict with the governing elite. "We are witnesses of all these things," says Peter. And if we think of the pattern that we described in the previous section, we might imagine Peter adding, "and because we are witnesses of Jesus, who did all this, we are also servants. We organize our communities so as to fight poverty. By attacking idolatry we go to the root of some structural economic evils. We teach masters and slaves how to find a totally new ground for fellowship and coexistence. This is what the power of God does in our midst, making us servants as Jesus was in his day."

Another fact that Peter adds about Jesus is that he has been appointed judge; that over and above any human or superhuman principality and power there is a Judge that will have the final word in history. Powerful, military men need to hear this. And so we come to the point at which the witness becomes a prophet, witnessing to people in power that there is a power over and above them—that they will be judged.

What I have found in both the history of missions and the modern missionary situation is that many witnesses start by being servants and are pushed by the logic of their call into being prophets, even if they do it reluctantly. Missionaries who take God's Word to the ends of the earth can tell you the kind of world we live in. If you as a witness cross the boundaries of social class, race and status in order to take God's Word, even to an American city, you will be, as the Lausanne Covenant says, "shocked by the poverty of millions and disturbed by the injustices which cause it." The experience and conviction of thousands of witnesses in the evangelical ranks has been very aptly summarized in these words of that Covenant:

We affirm that God is both the Creator and the Judge of all men. We therefore should share his concern for justice and reconciliation throughout human society and for the liberation of men from every kind of oppression. Because mankind is made in the image of God, every person, regardless of race, religion, color, culture, class, sex or age, has an intrinsic dignity because of which he should be respected and served, not exploited. Here too we express penitence both for our neglect and for our having sometimes regarded evangelism and social concern as mutually exclusive. Although reconciliation with man is not reconciliation with God, nor is social action evangelism, nor is political liberation salvation, nevertheless we affirm that evangelism and sociopolitical involvement are both part of our Christian duty. For both are necessary expressions of our doctrines of God and man, our love for our neighbor and our obedience to Jesus Christ. The message of salvation implies also a message of judgment

upon every form of alienation, oppression and discrimina-
tion, and we should not be afraid to denounce evil and in-
justice wherever they exist. When people receive Christ they
are born again into his kingdom and must seek not only to
exhibit but also to spread its righteousness in the midst of
an unrighteous world. The salvation we claim should be
transforming us in the totality of our personal and social
responsibilities. Faith without works is dead.

A witness is a servant and a prophet.

A Ready Sufferer

How many times have we been surprised by suffering? "If we
are doing God's will in the world, how can this happen to us?"
we have been tempted to think more than once. But suffering
is closely linked in the Bible to witnessing. A turning point in
the usage of words according to some Bible scholars comes in
Acts 22 when Paul describes himself and Stephen as witnesses.
He is speaking to a hostile Jewish mob. He refers to the call
spoken to him by Ananias at the time of the restoration of his
eyesight and his baptism: "You will be a witness to all men of
what you have seen and heard" (Acts 22:15 RSV). And then Paul
goes on to give us a bit of an autobiographical sketch, opening
the veil of his intimate life and experience. He tells us of an
ecstatic trance in which he confessed in prayer to God: "When
the blood of Thy witness Stephen was being shed, I also was
standing by approving, and watching out for the cloaks of those
who were slaying him (Acts 22:20 NASB).

Stephen was a witness and a martyr, and we can say that at
the point of his death he was recruiting another witness, because
it was at that point that God's Spirit started to melt the cold
fanaticism of the Pharisee from Tarsus. The fire of Stephen's
commitment was used by God to start a chain of events that
would turn Saul into Paul. This courageous witness could write
to the Corinthians, "God has exhibited us apostles last of all,
as men condemned to death; because we have become a spec-
tacle to the world, both to angels and to men" (1 Cor 4:9 NASB).

Readiness to suffer is part of the commitment to Jesus Christ
that makes a witness. We cannot produce this in ourselves.

It is a gift from God that operates in us through God's Holy Spirit.

The task of witnessing around the world is surrounded by risks of all kinds. Modern missionary enterprises have developed many ways of helping missionaries to avoid unnecessary suffering. Public relations expertise, insurance of various kinds, selective choice of areas for action, even technological gadgets can help. Maybe that accounts for how costly missions have become in some places, and maybe we seem to know better than Paul in these things.

But in spite of all this, witnessing in our daily lives on our campuses, within our churches or abroad in professional or nonprofessional missionary service will involve some degree of suffering for the Lord. And we should not be surprised, because that may be the only way our witness will get across to some people.

I find it very significant that when our modern world wants to give a prize to people who have served humanity, to call attention to actions that deserve imitation, for instance in the Nobel Peace Prize, it is often Christian witnesses who are nominated as candidates. These are people who have not only tried to serve their neighbors in the name of Christ but who have been ready to suffer for it like their Master. Three such people come to my mind as outstanding for the actions they performed: Martin Luther King, Jr., for his nonviolent fight for justice; Mother Teresa for sacrificial service among the poorest of the poor; and Alfredo Perez Esquivel for prophetic and courageous endurance in the face of terror and torture.

On March 7, 1981, Chester Bitterman, a young linguist working in Columbia, was assassinated by terrorists. In this age of suspicion in which we live, his supporting activity in the translation of Scripture was interpreted by extremists as politically dangerous. Witnesses today in some parts of the world where the gospel is desperately needed cannot avoid becoming first suspects and then maybe scapegoats in the clash of empires that is tearing the world apart. But they know well who they serve, and they know that the hands of their Master have scars from nails. May his Spirit help us to be witnesses under trial.

A Hopeful Sense of History

Witnesses need a hope-filled sense of history. A sense of history means that we understand our life today in relation to events of yesterday and in the light of our hopes for tomorrow. Christians believe that history is important because we see God in action in history. In the same way in which the simple thirty-three years of a life in Palestine divided history so that we date events from that life, we cannot understand our world or our own life apart from Jesus Christ.

We have quoted God's call through Isaiah to the Israelites to be his witnesses in the world. That task was taken up by Jesus as the perfect Servant of the Lord and passed on by him to his disciples. At a crucial moment in his missionary life, Paul became aware that he had been commanded by God through Isaiah to become a witness, and thus he said when he left the synagogue at Pisidian Antioch: "Thus the Lord has commanded us, 'I have placed You as a light for the Gentiles, that You should bring salvation to the end of the earth" (Acts 13:47 NASB). The words keep resounding through the nations and generations: Isaiah, Israel, Jesus, Paul, you and me. That is the way we enter into history, because the Lord who spoke through Isaiah keeps speaking to us.

A very practical consequence of this is that you and I are part of a large company of people that spreads through history in time and space. We are not individualist free-lancers or snipers. To begin with, your local church is the visible expression of that "cloud of witnesses surrounding us" to which the author of Hebrews refers. Maybe it is there that you first heard the gospel, and certainly there you are called to live out your experience of conversion and growth. I was a bit shocked at Urbana 79 by the large number of students who came to talk with me about missions and who could not give me a clear answer about their church situation. If you have not experienced fellowship in a local church here, what do you expect to be the outcome of your missionary work abroad?

It is very important for us to grasp that the new life brought by the gospel is experienced in a community, a fellowship. This is clearly presented by Paul in 2 Corinthians 5 where he speaks of the new creation of our lives by Christ and deals with the consequences—a different way of looking at others and relating to them. Commenting on this, John Howard Yoder says:

> *Because Christ has taken the place of all, now all persons can be seen in the image of Christ. Instead of seeing people as what they were, what their past had made them, I see them (Paul says) as what they became in the reconciliation worked by Christ. . . . So what Paul says is not centered in the changes that take place within the constitution of the individual person, but on the changed way in which the believer is to look at the world, and especially on overcoming the "carnal standards" in which he used to perceive men in pigeonholes and categories and classes.*

Some of you may feel a bit unhappy by the people that are not "like us" in the local church. You may prefer the homogenous unit of your Inter-Varsity chapter. But that is a passing, temporary situation. Jews and Gentiles had to learn how to cope with the problems of their heterogenous churches, and in the process they discovered the great new thing that the gospel had brought into history as Paul explained in Ephesians. Classes and races met and experienced the melting power of God's Spirit. What a tremendous testimony that same power is to a world torn apart by hate and terror!

Hope is the other element of your sense of history. Witnesses must be possessed by a sense of hope. They are moving with Jesus in the direction of that glorious freedom of the children of God toward which creation and the church move. Jesus the Lord and Judge will come. And the final defeat of all evil will be manifested. In Paul's life and teaching we see at least two consequences of this hope.

First, in his speech before Felix, Paul describes himself as a witness and says, "having a hope in God . . . that there shall certainly be a resurrection of both the righteous and the wicked. In view of this, I also do my best to maintain always a blameless conscience both before God and before men" (Acts 24:15-16 NASB).

The hope of the witness reflects itself in the moral quality of his life, in the tested quality of his behavior. In view of our hope you and I are also called to be witnesses that do our best to maintain blameless consciences.

Another consequence is spelled out by Paul in the passage of 1 Corinthians that we quoted before: "Therefore do not go on passing judgment before the time, but wait until the Lord comes who will both bring to light the things hidden in the darkness and disclose the motives of men's hearts; and then each man's praise will come to him from God" (1 Cor 4:5 NASB).

Your task as a witness, your missionary effort is to be evaluated by none other than God himself. What a challenge! What a liberation! Missiologists are devising scientific and technical ways of evaluation. Some of them may help you to see your witness in perspective. But there are many hidden things and undisclosed motives that only God knows. Learn to trust him for the true evaluation of your life and your mission.

12
Knowing God's Will

Robert B. Munger

A fine young Christian man came to see me in my study one afternoon. "I'm faced with some important decisions. I've been reading my Bible and praying but don't seem to be getting any guidance from God. Time is running out on me. I don't know what to do!"

"Tell me about it," I replied.

"Well, I graduate in June in engineering. I have two good job offers—one in Utah, the other in California. I'm also considering giving two years as a short-term missionary overseas before making a long-term career commitment here in the United States. Most importantly, I've been going with a wonderful young Christian woman. She doesn't graduate for another year. I need to know if this relationship is of God before I graduate and leave."

There they were—three crucial, life-determining decisions cascading down on him at once. Vocation . . . location . . . marriage. You may not be in such a crisis of decision but most students are making major choices. How do we make right decisions? What process will assure us that we are doing God's will?

Carrier pigeons have an amazing sense of direction. Across hundreds of miles of strange territory they make their way home with unerring instinct. When released from their box, they circle high in the air until they have gained their sense of direction then fly straight toward their home nest. To gain a sense of God's direction for our lives, we must heed Jesus' words in his call to his disciples. First note some of the basic principles of guidance found in his call. Then consider some practical procedures to help us get a good start toward fulfilling his intention for our lives.

Look at Mark 1:14-18. "After John was put in prison, Jesus went into Galilee, proclaiming the good news of God. 'The time has come,' he said. 'The kingdom of God is near. Repent and believe the good news!' As Jesus walked beside the Sea of Galilee, he saw Simon and his brother Andrew casting a net into the lake, for they were fishermen. 'Come, follow me,' Jesus said, 'and I will make you fishers of men.' At once they left their nets and followed him."

We Are Called to Follow a Person

"Come, follow me!" Jesus said. He did not say where he was going to lead them nor did he designate the place of their ministry. The call of Jesus is first and always to himself, to walk with him and be at his side. His first call is not to a particular mission or movement. He does not hand us a plan telling us exactly where we are to be or what we are to do at any particular moment. Rather, he offers himself, saying, "Follow me."

Martin Luther confessed, "I do not know the way that I take but well do I know my guide." I would rather have an experienced guide than a detailed map. More than once, with a map in front of me, I have sat in my car completely confused and utterly lost. A good guide may not inform me how he is going to arrive at the desired destination but he will give me directions when needed so that I may make the right decision at the proper time. Best of all a guide is a companion on the journey. "This God is our God for ever and ever; he will be our guide even to the end" (Ps 48:14). To gain a clear sense of direction and move out in today's world according to the will of God, we therefore begin with a wholehearted, irrevocable decision to follow Jesus Christ, to live for him, to be his. We offer ourselves up to him, to serve and please him supremely.

We Are Called to a High Purpose

We are to be fishers of men and women. That is big business. I am convinced that the greatest work in all the world is to make Jesus Christ known as Savior and Lord. Give yourself so that you will have a part in fulfilling his reign in the world and establishing to the glory of God the power of the Holy Spirit. Nothing is bigger than that. Nothing will give you more significance or meaning or dignity or value than the commitment to be Christ's and to be in his will and to do his work in the world. No matter how insignificant an individual may feel, to be linked to that purpose gives supreme significance. "He that does the will of God abides forever" and you are in a forever work that will last.

What are you going to do ten years from now? How much is it going to count a thousand years from now? If you have not thought through your intended vocation, you have some business to do with God. Are your plans what God wants you to do so that you may fulfill his supreme purpose? You may not go to a foreign mission field, you may not be in a church vocation, but be sure that you are led by God so that his purpose will be forwarded. Let us be fishers of men and women, in business for God.

A man sitting next to me during a plane flight said he was on his way to conclude the purchase of property for his company which was opening a new branch in a West Coast city. He was very excited about the company he represented. Then he turned and asked, "What is your business?"

It was a very natural opening and I replied, "I am engaged in the greatest business in the world."

"Really?"

"Yes, the greatest business, with the most important product, the greatest future and, above all, the greatest president and manager."

It was during the summer following my graduation from the university that I started seriously to follow Christ. Since that time I often have reflected, Suppose I had chosen to live as many of my colleagues and pursue my own plans? I had been accepted at the University of California School of Dentistry. The Depression had knocked out prospects of a future in business, convincing me to switch my major from economics to pre-medicine. Dentistry appealed to me more than medicine because there were no house calls and one could play golf Wednesday afternoons and Saturdays. A good living and a good time were my primary goals. Suppose I had settled for an easy life and personal happiness. What would be my satisfactions now, fifty years later? I am overwhelmed at the incredible grace of God that has given me a part in that work which abides forever, the mission that fills an ordinary life with extraordinary meaning and brings immeasurable rewards in both time and eternity. To follow Christ is to be in the biggest and most rewarding business in all the world.

We Are Called to a Powerful Partnership
"I will make you fishers of men," Jesus said. In his Gospel Luke relates that the call of the disciples followed the miraculous draft of fishes (5:1-11). All night the men had fished and had caught nothing. Reluctantly, at the command of Christ, they pushed out into the deep and let down their nets. They were amazed to pull in a catch of fish beyond the capacity of their boats to contain. They knew it was not a stroke of fisherman's luck. Certainly it was not due to their wisdom or skill. All they had done was to shove out at the word of Jesus and let down their nets. It was Christ's doing. The Master Fisherman had been at work. In that moment, still overcome by the awesome display of Jesus' authority and power in the miraculous draft of fishes, they heard him say, "From now on you will catch men." Could he mean it? Ordinary people? Sinful people? With hang-ups and failures and average abilities? "Yes," we can hear the Lord reply. "As I worked through you to catch fish, so I will work

through you to catch people and do the greater work of God I come to fulfill. Simply follow me. Trust me and do what I say." They forsook all and followed him. As we now know, he did for them what he said he would do. He will do the same for you and me.

Jesus Christ is the Master Fisherman enabling those who follow him to do the work of God. He is also the good shepherd who guides and cares for those who trust themselves to him. Blaine Smith puts this well in *Knowing God's Will*. The shepherd in New Testament times was an autocrat over the sheep. He took charge of their care, saw to their needs, led them out to pasture and brought them back again to the fold. Sheep were utterly dependent on him. Sheep are stupid animals, not knowing how to find their food or make their way back to the sheepfold. But the halting, helpless, even wayward sheep are not left to their own way. They are shepherded. The shepherd's call and encouragement, his rod and staff, make sure every sheep in the flock will safely arrive.

We too need not be anxious about getting the right directions from God or be concerned about whether we will have the courage to follow his direction. Instead, we are simply to put ourselves in the shepherd's care. If we want to do his will, he will see to it that we have the necessary information and put within us the desire and the energy to move out with him. He is able even to overrule past mistakes and in the process mature us in Christian life and service. The words of the apostle Paul encourage us: "God is always at work in you to make you willing and able to obey his own purpose" (Phil 2:13 TEV).

If you are not certain that you have launched in response to the word of the Master Fisherman, or question whether you have placed yourself in the care of the good shepherd, consider this suggestion. Prayerfully and carefully draft a statement of ownership recognizing the lordship of Jesus Christ over all you are and have, authorizing him to take whatever steps necessary to accomplish his will in and through us. Sign it. Settle it once and for all. Then continually remind him of his responsibility to keep that which has been entrusted to him (2 Tim 1:12).

We Are Called to a Close Companionship

"He appointed twelve... that they might be with him," to be his companions (Mk 3:14). He wanted them to be alongside in his saving mission, as personal friends. He was teaching and training them to one day carry his mission to the world. More than that, he loved them and desired their companionship just as he loves you and me and wants us to be close to him forever. He calls us not only for what he may bring to us and through us to the world but also for what we may bring to him. Incomprehensible but true. He loves us. He wants me to be with him because he loves me. Do we ourselves not want to be with a close friend just for companionship? Do not loving spouses have joy simply in each other? So God, the source of all true love and the one who is love, delights in the companionship of our heart with his.

In following Christ, my first struggle was to be willing to give up my personal plans, to leave family and friends and boats and nets, to follow him wherever he might lead. My second great struggle, which to me was more difficult than the first, was to give up my ambition to be a successful servant of Christ and humbly be and do whatever he wanted. I was eager to achieve great things for God, to preach to crowds of people with numbers of converts and applauding saints. Lovingly he brought me to a deeper level of commitment, a desire to be and do whatever pleased him. One thing, however, I know I must have—the light of his face and the smile of his favor. Here is the key to guidance: we must be willing to do God's will before we know what it is. To trust ourselves to him. To be taught, shaped and led as he shall choose.

Roll Up Your Sleeves

Now let us turn from principles to a few practical procedures. First, *offer yourself daily to God.* "Present your bodies as a living sacrifice" (Rom 12:1 RSV). As a recruit presents himself for military service, confident that he will be developed into an effective soldier, we are to present ourselves to our commander, Jesus Christ. Each morning report to him, saying, "Here I am. Do in me whatever needs to be done. I give you full authority to take whatever steps are necessary that I may be all that you want me to be and do all that you want me to do." He is completely reliable. He will take us at our word. "In all your ways acknowledge him, and he will make your paths straight" (Prov 3:6).

Second, *pray for guidance and grace.* Ask him to make his way plain to you and put his desires within you. (See Lk 11:9-13.) In *Affirming the Will of God,* Paul Little recalls his own experience:

At the Urbana Convention in 1948, Dr. Norton Sterrett asked, "How many of you who are concerned about the will of God spend five minutes a day asking him to show you his will?" It was as if somebody had grabbed me by the throat. At that time I was an undergraduate, concerned about what I should do when I graduated from the university. I was running around campus—going to this meeting, reading that book, trying to find somebody's little formula—1, 2, 3, 4 and a bell rings—and I was frustrated out of my mind trying to figure out the will of God. I was doing everything but getting into the presence of God and asking him to show me.

May I ask you the same question: Do you spend even five minutes a day specifically asking God to show you? (pp. 17-18)

Third, *inform the mind.* We are guided by the truth of God's Word. The Scriptures are a primary source of our knowledge of God. Here we learn about him and his will for our lives. Here Christ's Word addresses us and his grace promises to support us. John White states it well in *The Fight:*

God does not desire to guide us magically. He wants us to know his mind. He wants us to grasp his very heart. We need minds so soaked with the content of Scripture, so imbued with biblical outlooks and principles, so sensitive to the Holy Spirit's prompting that we will know instinctively the upright step to take in any circumstance, small or great.... Through the study of Scripture you may become acquainted with the ways and thoughts of God.

We are guided also by the facts of God's world. People are guided by what they know, not by what they do not know. How does it happen that so many earnest, committed Christians labor in the familiar field here at home and so few choose to place their lives for Christ in the areas of the world where the "laborers are so few"? May it not simply be that they have never been adequately exposed to the facts of the world today or allowed them to penetrate deeply into consciousness?

William Carey placed a map of the world above his cobbler's bench and marked on the map the number of people in those areas unreached by the gospel in order that he might pray for them. The modern missionary movement began. The five students of Williams College who took shelter under a haystack to pray and offer themselves to God for world evangelization had been studying geography. The hard facts of lost millions pressed them to pray and to act. Robert E. Speer, who for fifty years was a powerful force for world missions, placed a map of the world above his desk while an undergraduate at Princeton University and made it his business to learn more about the world. He prayed that God would help him know where and how he might best serve Jesus Christ. They all fed the fire of their faith with the fuel of facts.

We keep in touch with God's world through persons, periodicals, programs, missionary conferences and workshops. Do not neglect to read missionary biographies. Personal conversation with those in cross-cultural missions is also helpful. Best of all is a short-term or summer mission assignment, many of which are now available through various agencies and mission boards. Continually we are to lift up our eyes to see the fields of God's world, ripe, ready to harvest (Jn 4:35).

A fourth thing we should do to find God's will is to *join with other world Christians.* Jesus called his disciples to a committed company. We must not presume to be solitary followers of Jesus Christ. Seek the counsel of trusted believers. To move out step by step alongside our Lord with bright faith and a warm heart, we need one another as fellow followers—praying for one another, supporting one another, seeking to love one another even as he has loved us (Jn 31:34-35).

If, as a result of Urbana, there were to be one thousand teams of three to five students committed to Christ and to one another praying, learning and serving so that God may be glorified and his will done on earth, I believe the impact would be immeasurably greater than if we were to scatter across the land as earnest Christian individuals. We all need a support team around us, caring enough to hold us accountable and providing encouragement and love.

Finally, you should get going! Start now right where you are! Sam Shoemaker had a hard-hitting formula for Christian living: "Get right with God! Get together! Get going!" We are called in Christ. We have been given his message. We are now in his service, entrusted with the everlasting gospel. Wherever our lives are touching people there is a God-given ministry with opportunity to listen, to love, to lift, to share and to serve. Flight across an ocean into another country or culture does not somehow change us. The statement is true, "Wherever you go, you are there!" When Jesus called his first followers he said, "Follow me" and started walking. If they were to follow him they had to move. They left everything and followed. World opportunities are before us. Jesus is striding to enter them in love and power. He is calling us to follow.

It was the summer following my graduation from the university that I ventured to follow Christ. It was quite a change for me. I was about to enter a dental school. It seemed to offer a good living without the demands of the medical profession— no house calls, with Wednesday afternoons and Saturdays for golf. In a Bible class, however, the truth of 1 Corinthians 15:58 struck me: "Always give yourselves fully to the work of the Lord, because you know that your labor in the Lord is not in vain." To give myself fully to the work of the crucified and risen Lord was both right and reasonable. At that time none of my friends were preparing for Christian service. None of my close companions were professing Christians. I moved in one step from a fraternity house to a Bible school. The "culture shock" was profound.

When my train was pulling out of Berkeley for school in Chicago, I saw from my train window the lights of San Francisco disappear around a bend of the track. I was profoundly lonely and anxious. Where in the world was I going? In what part of that world would he choose to place me? And what did he want me to do? Would I really find joy and fulfillment serving Christ or would I regret the step I had taken? Suddenly I sensed Another's presence. It seemed as though Jesus himself came alongside, saying, "I am with you and will be with you always. I'm in charge here. Relax. Trust me! Enjoy the journey!" So the years have gone by. I have no regrets at all, only profound gratitude. He has kept faith.

13
Christ the Lord of My Life
Simon Ibrahim

When in 1969 I became a student at Ahmadu Bello University of Zaria, in northern Nigeria, I resolved that I would never again work with the church—yet here I am the general secretary of an indigenous church in Nigeria. What happened? I based my resolution on common sense. After serving the church for seven years, I left to go to the university. I figured out that when the church consulate heard that instead of going to seminary I was going to Ahmadu Bello University, which is a strongly Islamic university, they cancelled a plan to license me as a pastor. I was happy about that. I also figured out that Jesus served only three years and I had been serving for seven. It was time now for me to have it easy. Of course I attended the Wednesday prayer meetings and Sunday worship. That was the best I could do the first year at the university.

Then came a bombshell. The district church council sent an elderly church pastor to me to ask if I would like to pastor one of the churches part-time while I was a student at the university. I could not understand this. I replied, "You wait until tomorrow and I'll let you know my decision." That night three things happened.

A Bittersweet Day

First, I could not sleep at all because I remembered how I became a Christian and how the Lord took control of my life. I reflected on how I was born and reared in a Christian home after a missionary came into our tribe. As a little child, I was taught to memorize verses and sing songs. But I was also taught not to retaliate when attacked, and that made a coward of me. When one of my schoolmates slapped me and I did not slap him, he would slap me again. Those years were an eternity for me. Worse yet, in those days Christianity denied me a long list of good food including meat offered to idols, which was the cheapest means of getting a meal in those days. Here I was, a little Christian kid, and I could not partake of such food, except of course in secret when my grandfather took me. Therefore I resented Christianity. If I hated anything as a little boy, it was Christianity.

Then one day I heard that my father was excommunicated from the church. For me it was a great day of rejoicing. To celebrate the occasion, I started to sing non-Christian songs. Immediately my father rebuked me. The following day he summoned the whole family and said, "I am born again; my name is in the book of life. My sins are forgiven. If I died today I am going to heaven. You are not going to heaven unless you repent of your sins and believe in Jesus." A new conflict began in my life.

One day I was so miserable, so unhappy, I climbed a big rock on our compound and viewed the whole village. I wondered, "What is the purpose of life? Why am I so miserable? Where can I find peace?" Someone seemed to tap me on the shoulder and say, "Repent of your sins and believe in the Lord Jesus Christ. You will have peace with God." I turned around and looked, but I was alone.

I descended from the big rock, went inside our house and took out the Tangale New Testament. I read verses on forgiveness, knelt down, prayed for forgiveness and believed in Jesus. When I opened my eyes, I could not believe the flood of joy that came into my life. I could not understand it. Immediately I ran back to the rock. I climbed it and surveyed the whole village. To my amazement it was beautiful.

That night during my university years, I remembered the flood of joy. I said, "How dare I refuse to be a part-time pastor in the church of Jesus Christ?" I began to weep.

God's Guiding Hand

Second, that night the Lord showed me my path of decisions after I became a Christian. I remembered that at the age of nineteen I was admitted into teachers' college, but I was shown a need for someone like me to teach little children in primary school. My teacher used a song to help me understand the need for me to be involved in Christian service right now instead of rushing to college to get a certificate. The song was:

> *I have nothing to do with tomorrow,*
> *Its sunshine I never may see.*
> *Though today with a plough in the furrow*
> *In the vineyard I faithful would be.*
>
> *I have nothing to do with tomorrow,*
> *My Savior will make that his care.*
> *Its grace and its strength I cannot borrow,*
> *So why should I borrow its care.*

I quit teachers' college and became involved in the vineyard. The Lord showed me something else; I was readmitted to college the next year but the Lord made it plain that he wanted me to go to Bible school instead. I told my father and he approved of it. But all of my uncles became very angry. One of them said, "Son, you are lost if you are going to Bible school."

So my father called me and said, "If I die two verses are going to be your inheritance." One was, "But seek first his kingdom and his righteousness, and all these things shall be yours as well" (Mt 6:33 RSV). The second was, "Take heed, and beware of all covetousness; for a man's life does not consist in the abundance of his possessions" (Lk 12:15 RSV). "Your uncles don't understand. All those things they are worrying about if you go to Bible school—God is going to add all of them to you. So you have my blessing. Go to Bible school."

It was also brought to my memory that at the age of twenty-two I resolved never to write to any girl. In those days the "in" thing was to write to as many girls as you could; the more letters you got from girls, the better off you were. But the letters contained many lies. You write to Naomi and call her "darling." You write to Ruth and call her "darling." I could not understand it as a young Christian and felt something was wrong. I told my friends that I was not going to write to any girl. If any girl wanted me to marry her, she could write to me. (In those days it was impossible, because the girls were cleverer than the boys.) For more than a year I never responded to any girl. I was seeking the Lord's will for marriage. To cut a long story short, he guided me to the right woman. And today we have three children.

That night the Lord said to me, "If you could trust me then, why don't you serve me now?"

At age twenty-eight I gave up a well-paying job to work with the Sudan Interior Mission. The Lord blessed me richly. But at the age of thirty-four I was saying that I did not want to serve the Lord anymore. I wept bitterly that night.

Who's Responsible?

After weeping, then a third thing happened. It dawned on me that although I was in the university on a government scholarship, God was the one who arranged for me to get the scholarship and be there at the university at his service.

When I woke up in the morning, I contacted the elderly pastor and said, "Go and tell the district church council that I will pastor the church part-time." Immediately I took over the pastorate while going to school.

During this time the Lord urged me to open the door of the church for university students to meet there. We decided to invite the leaders of the church and the leaders of Sudan Interior Mission to come for what we called "a meeting with the firing squad" in which we asked them various questions. Of course, as university students we saw lots of problems in our nation, our society and our community.

Nigeria had just come through a civil war, and we students debated the issues which had threatened to divide our nation. The main thing which concerned us, however, was the future of the church. Anti-Christian feelings were gaining strength, especially in the university. We realized that many of our local churches could not meet the needs of the youth. Also, what was the future of the missions which had brought our parents and us the gospel? Nigeria would never be the same after the civil war, and missions had to realize that they had to adapt to a new day if they were to continue.

Becoming Available to the Church

Several of us Christian undergraduates met after that meeting with the church and mission leaders. We decided that we must be available to the church. In those days government and industry were competing to get university graduates, but we thought the ministry of the gospel was more important than all the prestige and the positions we would get in industry and in government. Today, of those who met at the church, some of us are in full-time church work, some are in government, some are in industry, but we all serve together with God. We realize that all education, including university education, is to be used to the glory of God if Christ is the Lord.

Recently a state government sent a team here to North America to recruit four hundred teachers, but to date only three have been recruited. What an opportunity for three hundred ninety-seven Christians! I thank God that he worked in my life so that I was willing to say, "Jesus, you are still Lord of my life." I would have missed out on the drama that is taking place today in the church. There is great excitement because of what God is doing through the church. The church with which I am involved is so committed to cross-cultural evangelism that we see hundreds coming to Christ every year. We have over a hundred couples who are preaching full time in the rural villages. During early 1982 by the will of God we will place a hundred couples in other villages.

Northern Nigeria is culturally different from the rest of the country. There are those who have been raised in Islam. There are two groups who have run away from Islam. The word for one group means "we are running away from Islam." The second means "better dead than become a Muslim." Both groups are coming to Christ in the hundreds. We cannot meet the demands for evangelists for them. The couples we are sending are going mostly to these two groups.

I would have missed this drama if I had refused to obey my Lord. I thank God that his Spirit compelled me to obedience.

14
The Spirit's Enablement
Helen Roseveare

The seventh chapter of the book of Judges tells us how God called Gideon to lead the people of Israel against the hosts of Midian, who were encamped around them. This chapter tells the wonderful and dramatic story of how God chose and equipped and led his chosen crack regiment against the forces of evil:

The LORD said to Gideon, "The people with you are too many for me to give the Midianites into their hand, lest Israel vaunt themselves against me, saying, 'My own hand has delivered me.' Now therefore proclaim in the ears of the people, saying, 'Whoever is fearful and trembling, let him return home.' " And Gideon tested them; twenty-two thousand returned, and ten thousand remained.

And the LORD said to Gideon, "The people are still too many; take them down to the water and I will test them for you there; and he of whom I say to you, 'This man shall go

with you,' shall go with you; and any of whom I say to you,
'This man shall not go with you,' shall not go." So he brought
the people down to the water; and the LORD *said to Gideon,*
"Every one that laps the water with his tongue, as a dog laps,
you shall set by himself; likewise every one that kneels down
to drink." And the number of those that lapped, putting their
hands to their mouths, was three hundred men; but all the
rest of the people knelt down to drink water. And the Lord
said to Gideon, "With the three hundred men that lapped I
will deliver you, and give the Midianites into your hand; and
let all the others go every man to his home." So he took the
jars of the people from their hands, and their trumpets; and
he sent all the rest of Israel every man to his tent, but retained
the three hundred men. (7:2-8 RSV)

Then we read that Gideon went down and overheard a conversation in the camp of the enemy. He also heard an interpretation
of a dream. God greatly encouraged him and he went back and
said to the 300 men, "OK, we're ready to set forth." He divided
them into three companies and put trumpets—not swords, not
guns, but trumpets—into their hands. He also gave them empty
jars. One each, each hand. No room for anything else. He put
trumpets into their hands and empty jars with torches inside
the jars. Then he said, "Watch me; do as I do. When I come to
the outskirts of the camp, do as I do. When I blow the trumpet,
all that are with me blow the trumpets on every side of the camp
and shout." They did it. They went forward and did exactly
as he did. When he blew the trumpets, they blew the trumpets.
When he smashed the jars, they smashed the jars. When he
threw the torches in the air, they all did it. And God wrought
a fantastic victory which threw the enemy into disarray. They
cried out and fled.

God's Grammar
The *subject* of this convention is you, the potential missionary
in your service for the Master. I shall now briefly describe how
God chooses, equips and leads his soldiers, just as he did in
Gideon's day. The *object* is undoubtedly the world, the 3,125
billion in total ignorance of the Savior. There must be a *verb* that

brings subject and object together to make a practical concept. That verb, the indwelling Holy Spirit, speaks through every aspect of our convention, making it possible for the subject (you and me, the missionary) to relate to the object (the Christless world).

A *subject* should be a clean vessel filled with the Holy Spirit of Christ. In 2 Corinthians 4:7 we read, "We have this treasure in jars of clay to show that this all-surpassing power is from God and not from us." We are jars of clay filled with precious treasure, the indwelling Christ. But how can that treasure satisfy the thirsty souls in a dying world if the containing jar stays at home? Or remains sealed up? Or is concerned only with its own state of cleanliness and fullness, and not with the needs of those around?

We have seen in the seventh chapter of Judges how God chose his crack regiment; and then how he equipped them (vv. 15-16) with a trumpet in the right hand to sound out the battle cry, a torch in the left hand ready to scare the terrified enemy, and an empty jar containing and covering the torch through the preparatory period. This jar had one purpose only—to be shattered to smithereens to reveal the light. Christ says, "I am the light of the world" (Jn 8:12). He is not the Light only of the United Kingdom or the United States, but of the world. This Light indwells us, as in clay jars. As we allow God to "smash" us his Light can stream forth to enlighten the world in its darkness (Phil 2:15; 2 Cor 4:4-7). The woman in Mark 14:3 smashed the alabaster jar of very expensive perfume in order to pour it on the head of Christ, her only means of showing her deep love. I ask myself, "What is my alabaster jar? What can I give to Christ to show my love? Is it my rights? My very self? My hopes and ambitions, desire for success or popularity, for public image, or for security? For privacy or the right to lead my own life, care for my own family? God so loved me that he gave ... himself ... as my ransom. What have I ever sought to give him to tell him that I love him?

The two-sided biblical concept of love shows God as always giving and us as obeying. God only gave Adam and Eve one way in which to show that they loved him—obedience. In John 14, three times the Savior says, "Whoever has my commands and obeys them, he is the one who loves me." Will I give God my unquestioning obedience? The Lord Jesus Christ, the first great missionary, left his Father's glory in heaven to come to this earth, to die and shed his blood at Calvary to redeem us. He made himself of no reputation, gave up his rights to equality with the Father, became a slave and was obedient—even unto death. Surely he was a clay jar in his Father's hands, willing to be smashed that I might be saved (Phil 2:5-8).

What is my response? Am I willing to be made like-minded, to give up my rights, to let go of my reputation, to become wholly submissive to God's will, to allow him to make my choices? Am I prepared to give unquestioning obedience to his authority, willing actually to die if by that means the gospel might reach dying men and women?

The Holy Spirit can work that desire into me, making me "like-minded to Christ" if I want him to. If only I will ask the Holy Spirit to make me like Jesus, conforming me to his image, he will move heaven and earth to do it. He will use every means —people, circumstances, events—to mold me to the Image, that others may see Jesus only. But it will—it *must*—involve the death of self, which I often call the "crossed-out I" life of Galatians 2:20 where Paul states that he is crucified with Christ though he lives. Yet, he says, "I no longer live, but Christ lives in me." This is the New Testament explanation of the Old Testament imagery of the smashed jar: the contents, the treasure, may stream out to others. As the "I" indwelling my body accepts the smashing sentence of death, then Christ is enabled to take up his residence in place of me. Then and only then can he love through me, act through me, reach out to others through me. I am God's subject.

The *object* of Urbana 81 is mission. The need of mission is the fact that two-thirds of the world—3.125 billion unreached hidden people who are beyond all present missionary endeavor —have never yet heard the name of Jesus. Not even once. Not even by radio. What shatters me most is not the awful plight of those billions, terrible though that is, but the complete apathy of the Western church—of you and me—to do something about it. Conferences are fuller than ever. Everywhere crowds flock to hear, to talk, to debate in student Christian fellowships throughout the English-speaking world; in Bible schools, colleges and seminaries; in world gatherings such as at Lausanne, in Thailand, at Edinburgh. But then what?

Every missionary society is crying out for dedicated full-timers, those willing to throw away everything else as "worth less than nothing" that they may serve Christ (Phil 3:7-10). Where are they? "Fields white unto harvest" is the cry all round the world. Today should be a day of reaping in South America, Central Africa, India and Indonesia, but the reapers are so few. Talk will not do the task. We need men and women, sold out to God in an abandonment of love, "to serve him without counting the cost of seeking for any reward save that of knowing that we do his will."

We hear of the need for workers in every part of our world:
- *teen-agers and students in your own country, brainwashed by humanism and marxism*
- *Communist lands, with doors open to go in though not always open to come out*
- *one-sixth of the world given over to the militant fatalism of Islam*
- *two-thirds of the world in hunger and abject poverty who need our loving compassion*
- *one-half of the world, which we call "youth," in all their eager enthusiasm, needing guidance*
- *the two-fifths in vast cities, the work force and the unemployed, students and outcasts, politicians and prisoners, the virtually unreached masses of every continent.*

Why haven't we got the needed laborers? Why are you hesitating? Is it because our God has become too small, and you're scared that you couldn't go through with it?

Finally, if the *subject* is the missionary and the *object* is mission, the *verb* of this conference must be the missioner: the gracious indwelling Enabler, the Holy Spirit of God, the third Person of the Trinity. Only the indwelling Holy Spirit can enable the clay jar to be willing to go, and to carry and ultimately to be smashed in order to release the treasure to those in need.

"But how," your heart cries, "How can I know for myself for sure that he wants me, and that he'll enable me? Who can give me the assurance I need to step out into mission?" Have you ever stopped to think of the twelve disciples that evening after Jesus had been teaching the crowd all day? The Lord took the lad's picnic of rolls and sardines, blessed it, broke it . . . and then? He gave it to his skeptical disciples and said, "Right about turn, forward march! Your mission is to feed this waiting crowd!" And they did not know the end of the story. What a fantastic step of faith as they faced the cynical crowd with their backs to the Savior and only half a loaf in their hands! Did God fail them? Of course not! He never can. His grace is always sufficient for every need, theirs and yours. It worked: 5000 were fully fed, twelve baskets full of leftovers were gathered up.

This same God wonderfully kept us at peace during the five months of rebel captivity in the Congo in the sixties. But if I described it someone would say to me, "I couldn't possibly go through what you went through!" and I'd realize that you had missed my whole point. My God is your God. He kept me; he'll keep you. He who saves us and calls us and sends us out into his service is the great almighty omnipotent Creator God. His grace is sufficient for you and for me. Reading Foxe's *Book of Martyrs* in 1964, I found it made me sick and I couldn't sleep at night. I knew I wasn't made of the stuff of martyrs. "If God should ever ask me to be burned at the stake," I thought, "I would do it, but I sure won't be singing." Six months later, I was among seven who stood before a firing squad of rebel soldiers. They were told to take their sights and we were singing the praises of Jesus. How? What had happened to change me in those six months? Nothing. But when the moment of need came there was his grace available, always in present-tense sufficiency, made real to us by the indwelling Holy Spirit.

"But," you cry, "that's OK for the twelve disciples—they were actually with Jesus. Maybe it's OK for you—you're a missionary. But for me? Could he do it for me, just an ordinary nobody? Can I be sure?"

Yes. Unreservedly yes. By the presence and power of that same indwelling Holy Spirit he cannot fail you. The same Holy

Spirit who enabled the disciples, who enabled me and countless others, he will enable you. Everything you understand about God the Father—his almighty power as Creator, sustainer and upholder of the universe—and all you know of the Lord Jesus Christ and the redemption he bought for you by his death at Calvary—is only made real to you by the quiet ministry of the Holy Spirit, the third person of the Trinity. You already have the Spirit indwelling you if you know and love the Savior, because you couldn't do even that without him.

He is indeed the verb that relates the subject to the object. It is he who will enable you, the subject, to tackle and fulfill mission, the object. The Spirit alone can make mission urgent in the heart of a missionary so that it is real, vital, essential and all-important. He alone can stream out to meet the need of the millions as the clay jar is willing to go and be smashed.

How It All Starts

Initially, the Holy Spirit puts the urge in our hearts which starts us wondering whether God wants us to be missionaries. It is the same Spirit that causes that urge to grow into an earnest desire that becomes in turn a burning necessity to go and tell others. The Holy Spirit takes over our inadequacy and shrinking fear, and reveals to us God's adequacy and power for the task to which he calls us. Steadily, the Holy Spirit will make plain to us God's plan so that every occurrence becomes God related. We realize that nothing ever occurs in a Christian's life by accident; all is under God's infinite control (Rom 8:28).

In Africa, I soon realized that as I asked for his guidance the Spirit could breathe into my conscious thinking solutions to complex and difficult church situations. He could direct my hands in an unknown surgical procedure, as the team prayed his enabling. I am sure that it was the same Spirit who showed us what to do when the truck broke down, hundreds of miles from the nearest garage, as soon as we acknowledged our own ignorance. He also was the one who gave me cultural insights and divine sense as I sought to teach national students without giving offense or appearing proud.

As we grow in our trust of the Holy Spirit, we find that we

can accept situations without always demanding explanations, always recognizing God's sovereignty. Eventually he enables us to thank God for trusting us with each and every circumstance, even if he never tells us the "why." I do not believe that Scripture teaches us nor that God asks us to thank him *for* each and every circumstance, but rather to thank him for trusting us *with* the circumstance (even if he does not always tell us the reason for it).

This particular ministry of the Holy Spirit in our hearts, making us into the "more than conquerors" of Romans 8:28-39, appears to me to be the most important aspect of his enablement in the life of a missionary. He enables me to accept every detail of my life from the hands of a loving Father with thanksgiving (Phil 4:6). But is this really possible in the face of suffering? Frustrations? Misunderstandings? Even death? I certainly believe so.

A young missionary couple during their first term of service in a foreign land were expecting their first baby. The mother wrote to me from the hospital a few days after the birth to tell me that her baby had died. Can you thank God for *that?* She went on in her letter, "Local women whom I've been trying to reach with the gospel visited me yesterday, and their loving sympathy was very touching. Then one of them said to me, 'Now you are the same as us. Now we will listen to what you tell us.' I find my heart rising above my sorrow and that I can actually thank God for the unbelievable privilege of being trusted with this deep sadness, if it means that I can identify with the local community in their daily sufferings and so be able to share Christ with them." Surely only the Holy Spirit in her gave her such triumphant peace.

When I first arrived in Congo/Zaire to serve as the only doctor for a half million people, I longed to give them the best possible medical service. But there was no hospital. So we started by learning to make bricks, burn bricks, put bricks on bricks to build one. Called one day from the brick kiln to the temporary mud and thatch hospital, I needed to do an immediate emergency operation to save a woman's life. My torn and bleeding hands (from working at the kiln) smarted under nail brush and antiseptic alcohol. I was puzzled as to why God allowed such an apparently intolerable situation. Later that week, local church elders helped me understand God's ways by explaining that when I was "being a doctor" they were scared of me as the "white witch-doctor." They were not listening to the message that I sought to bring them. "But," they continued, "when you are at the brick kiln, your hands as sore as ours, using our tribal language and making jokes that we all laugh at, *that* is when we have come to love you and trust you, and so to listen as you tell us of God." On that occasion, within one week, God showed me the "why" of the situation in which he had placed me and enabled me to thank him for the privilege of being a builder. At that stage I had needed to see the "why" before I praised him. He will surely lead us beyond that.

During the Simba uprising of the sixties, I was the first white woman in our area to be savagely attacked and beaten and taken away to guerilla headquarters. Through that unforgettable night God asked me to thank him for trusting me with that suffering. The Holy Spirit enabled me to do so, though at the time I could see no possible reason for it. The next morning, members of the local Greek trading community found me and ministered to my physical needs. Nine weeks later, I was taken by armed soldiers from the cell where I was imprisoned to a house where over fifty of these same Greek traders, with their wives and children, had been rounded up the day before after several hours of vicious brutality.

While apparently tending the sick among them, by using several languages (some known, some unknown to the soldiers) I was enabled by God to preach the gospel to them in their hopeless despair. As I prayed with them, before being taken back to my prison cell, they were very responsive and I believe that some may have opened their hearts to the Savior. For twelve years I had preached to the same people with no response. Why did they listen to me now? If, nine weeks before, I had not suffered the same cruelties that they had just endured, I do not believe that they would have listened. "What does she know about it?" would have been their bitter cry. But I *had* suffered —and they knew it—so they listened. Was God showing me part of the "why"?

Ten years later, while I spoke to a university Christian group, God prompted me to allude to the suffering of rape, the most dreaded experience of any girl or woman. I told them how during that very night God had given me the wonderful enabling of his Spirit to thank him for trusting me with this experience, even if he never explained "why." I continued my message. At the close, when all but two students had left the hall, one of them came to me and asked if I would speak to her teen-age sister. "She was raped five weeks ago," the older girl explained, "and none of us can reach through to help her. For five weeks she has not spoken to anyone." I turned and looked at the younger girl, who slowly started toward me.

She quickened her pace, ran to me, threw her arms around my neck, and we cried together. Then she poured out her story in the next two hours. She ended with a sob: "Nobody ever told me I could thank God for trusting me with such an experience, even if he never chooses to tell me why!" I looked back ten years and knew a little more of the "why."

This is the wonderful enabling of the Holy Spirit, working in our hearts and lives, that makes it possible for us to receive everything—good and bad—from the hands of a loving heavenly Father. We can do this in humble submission and unquestioning loyalty, knowing "that in all things God works for the good of those who love him, who have been called according to his purpose" (Rom 8:28). This is the ministry of the Holy Spirit, the missioner: working in the heart and life of the missionary to conform us to Christ's image; making real to us the enormous and overwhelmingly urgent task of mission; enabling us to step out into God's purposes, trusting in his promises.

15
Jesus Christ
Is Lord

David M. Howard

Thirty-five years ago 575 students were gathered in the Great
Hall of the University of Toronto for the New Year's Eve service
of the first Inter-Varsity Student Missionary Convention. That
convention would later become known as *Urbana* when it
moved in 1948 to the University of Illinois campus. But that
night students were tired at the end of five heavy days of ses-
sions. They had walked for miles through ice and snow across
campus. Yet there was a spirit of expectancy as we came to the
end of that historic gathering. I say "we" because as a college
sophomore, I was privileged to be one of those present.

It is strange how some little aspect of an event will stick in
one's mind. My major recollection of that meeting was how
fast the speaker talked in trying to keep within his time limit.
He got going faster and faster, every couple of minutes whip-
ping out his wrist watch to see the time. Then he would speed
up noticeably, raise his voice and rush on—until he looked

at his watch again, whereupon the process would be repeated.

I don't know what you will remember of tonight, but my guess is that you may remember having caught a cold or cough in the freezing weather of Urbana. You will have a vague recollection of dragging your exhausted body and foggy mind to this final session, dreading the long bus trip, drive or flight home, and hoping desperately that the speaker would be short but fatalistically expecting that he will be long and dry. Whether the speaker disappoints or confirms those fears will be yours to judge.

The convention planners requested that I not give a theological treatise but that I share from my heart what it has meant in my life for Jesus Christ to become Lord. Therefore, if I seem to talk about myself, I am doing so at the direction of the convention planners. Addison Leitch, my late brother-in-law, was fond of saying that he never apologized for sharing personal experience, "because it's the only kind I've ever had."

I still have the small, faded, brown evangelism decision card which I signed in 1946, indicating that I would pray for missions and seek God's will for my own life. The card used to be green. I can tell that by the small green circle where a thumb tack held this card above my desk throughout my college days. It stared me right in the face every day and served as a daily prayer reminder that I had committed myself to serve God overseas unless he clearly directed otherwise. The fact that my wife and I had fifteen years of wonderfully satisfying service in Latin America is attributable in large measure to prayer—much of which was stimulated by that little card.

Upon returning to college after the Toronto convention, students began to meet regularly to pray for missions. This was the post-World War 2 era. Many of the students were veterans who had returned from overseas and were now in college. These men were older than the average college student. They had been in the military for two, three or four years. They had faced life and death. They had seen men die at their sides. They had seen the world in a way that no previous student generation had seen. There was a seriousness about them not usually found in college freshmen. Many of them came to college with a firm desire to return to those lands which had been so devastated by the war in order to share the love of Christ and demonstrate his healing power to their former enemies. Various mission boards were founded by these men and women. Some who had been trained in the air corps decided that they could put their pilot and mechanical skills to use in flying missionaries. Thus Mission Aviation Fellowship came into being. Far Eastern Gospel Crusade was founded by men who had fought in the Pacific and now wanted to return and take the gospel there. Greater Europe Mission was founded at least partly by veterans of the European theater of war who also wanted to return and help restore those nations to health and spiritual life. Jesus Christ was Lord in their lives, and they wanted to obey him completely.

Praying for Missions

My closest friend in college was Jim Elliot. Jim was only to live for a few years beyond college, but in that short life he would leave a mark for eternity on my life and the lives of hundreds of others. Exactly ten years to the very week when the Toronto convention ended, Jim and his four companions were speared to death by Auca Indians on the Curaray River in Ecuador. In his death he would speak to thousands. But we did not know any of that in our college days. Jim encouraged a small group of us to meet every day at 6:30 A.M. to pray for ourselves and our fellow students on behalf of missions. This became a regular part of our college life together.

Jim Elliot also organized a round-the-clock prayer cycle on our campus of Wheaton College, asking students to sign up for a fifteen-minute slot each day to pray for missions and for mission recruitment on our campus. The entire twenty-four hours were filled in this way. Thus, every fifteen minutes throughout the day and night at least one student on that campus was interceding for missions. Alarm clocks were set for unearthly hours such as 3:00 A.M. A student would rise to pray for fifteen minutes, knowing that as he or she returned to bed, another alarm was going off somewhere on campus as another student rose to pray.

Art Wiens was a war veteran who had served in Italy and

planned to return as a missionary. He took it upon himself to pray systematically through the college directory, praying for ten students by name every day. Art followed this faithfully through his college years, and then shortly after graduation returned to Italy, where he still serves as a missionary. I did not see Art for twenty-five years after that until we met in 1974 at the Lausanne Congress on World Evangelization in Switzerland. As Art and I renewed fellowship and reminisced about old times, he said, "Dave, do you remember those prayer meetings we used to have at Wheaton?"

"I never forget them," I replied.

Then Art said, "You know, Dave, I am still praying for five hundred of our college contemporaries who are now on the mission field."

"How do you know that many are overseas?" I asked.

"I kept in touch with the alumni office and found out who was going out as a missionary, and I still pray for them."

Astounded, I asked Art if I could see his prayer list. The next day he brought it to me—a battered old notebook he had started in college days with the names of hundreds of our classmates and fellow students. I found my own name there and my wife's name.

For Art Wiens, Jesus Christ as Lord in his life meant a commitment to faithful prayer. Art is a quiet man who will never see his name in marquee lights as a great orator or leader. But when the crowns of faithfulness are given out in heaven, Art Wiens will surely wear a bright one.

Jim Elliot not only prayed for other students; he went to work on them. He took several campus leaders as special prayer projects, and then he went after them aggressively on behalf of missions. Our senior class president was Ed McCully, star football player and track man, holder of the college record in the 220-yard dash and winner of the national oratory championship. Ed planned to be a lawyer, and he would have made a good one. He was smooth, articulate, brilliant and spell-binding in his oratory.

One day shortly after Ed had won the national oratory championship, he came into the locker room following an athletic workout. Jim was there getting ready to shower, and I recall seeing him go up to Ed, grab him by the back of the neck, and say, "Well, McCully, so you won the national oratory championship, did you? Great stuff, McCully. You have a lot of talent. But who gave you that talent? God did, and you know it. So what are you going to do with it? Spend it on yourself all your life? You have no business doing that, McCully. You owe it to God to give it back to him. You should be a missionary, and I'm praying that God will make you one." Then turning to me Jim said, "Howard, we've got to pray for this guy."

McCully stood there looking rather sheepish and not saying much. Following graduation he entered law school for a year, but was uncomfortable. God was at work in his heart. At the end of that year he dropped out of law school and joined Jim Elliot in going to Ecuador, where he also died under the Auca spears in 1956.

It was this combination of prayer and action, faith and works, that acknowledged Jesus Christ as Lord in the lives of students.

My Call to Missions

Students and others often ask me to describe my missionary call, rather expecting that I will tell of some monumental or dramatic experience when the heavens were opened with a flash of lightning and a voice called to me. Such an experience has never been mine, although I do not doubt that on rare occasions God may choose to work that way. Normally, the Lord seems to direct us in very practical, day-to-day steps.

Two major factors came into focus during my college years. First, it became clear that if Jesus Christ is truly Lord of my life, then I must obey his commands. His last command given on earth—the one that was repeated at least three times during the forty days following his resurrection—was that we are responsible to tell the gospel to the whole world so that every tongue may confess that Jesus Christ is Lord. No Christian can get away from that command.

Second, I saw that most of those who have never heard that Jesus is Lord are not located in the area of the world where I live. There are about 4.5 billion people in the world today. We are told that there are nearly 3 billion who have never had an adequate opportunity to hear of Jesus Christ or respond to his claims. The population of North America (Canada and the U.S.) is only about 250 million, and the large majority of these have at least had an opportunity to hear of Jesus Christ. The gospel is available to them on radio, TV, through the Bible that can be bought in any bookstore, through churches in every city and almost every town. But the large majority of the 3 billion unreached people have no opportunity in their own culture to hear of Jesus Christ. If every tongue is to confess that Jesus Christ is Lord, someone must cross those cultural and geographic frontiers to carry the message to them.

As I pondered these two great facts—the final command of Jesus Christ and the fact that most of the unreached are outside of North America—it occurred to me that I could not go wrong by at least trying to get to some area of the world where the gospel had not been given. If God didn't want me there, he could close the door. But if he wanted me there, and I was sitting with folded arms at home waiting for a spectacular call, it would be far more difficult to acknowledge Jesus Christ as Lord in my life. To accept his lordship meant obeying his commands. Jim Elliot used to say to us with fire in his eye and some agitation in his voice, "We don't need a call; we need a kick in the pants!"

Therefore, I came to the conclusion that if Jesus Christ was to be Lord of my life, I should plan to obey his commands by heading for some area of the world where the gospel had not yet been freely given, trusting God that if he did not want me there, he would close the door.

Recognizing Jesus Christ as Lord also meant a commitment to action on behalf of missions. During my junior year, in prayer with others, I felt that God would have me spend the summer traveling throughout the midwest to present the challenge of missions to other young people. This would be done peer to peer, one young person to another. So I did something which makes me shudder today when I think of it. Having now planned such massive events as several Urbana conventions and a world congress on evangelization, and knowing how many things can go wrong in such a venture, I am not sure I would have the sim-

plicity and audacity of faith that I had as a twenty-year-old student. But somehow God responded to my step of faith. Having
worked out a detailed itinerary, I wrote to churches and Bible
conferences across the midwest from Michigan to Montana,
announcing that on a certain date a group of four young missionary volunteers would be in their area and would be glad
to share the challenge of missions.

The only problem was that I was the only one prepared to go.
I had no other student committed with me and no car. As late
as April I still had no idea how we would fulfill this commitment. However, God answered prayer and that summer four
of us—Jim Elliot, Rodger Lewis (who has now served in Indonesia for thirty years), Verd Holsteen and I set out with no
money, a small car, an itinerary and a great expectation of what
God would do. God graciously responded to our somewhat naive
faith, and we have good reason to believe that God touched many
hearts through this tour. At least I know he touched the hearts
of the four of us in deep ways. If you have read Jim Elliot's
diaries in *Shadow of the Almighty,* you have read some of the
lessons he learned on that trip.

Prayer and action continued to be a key in understanding
that Jesus is Lord in my life. Following college, I joined Inter-
Varsity staff for one year as missions staff member. I don't think
I made much contribution to Inter-Varsity during that year, but
they made a great contribution to me. I learned much from God
in traveling across the U.S. and Canada visiting something
over a hundred campuses to stimulate missions interest.

The next year I was married and entered graduate school for
theological preparation for missionary service. My wife had
already settled the question of missions in her own life before
we were engaged. While in graduate school we decided to investigate several different mission boards to give God a chance to
lead us in whatever direction he wished. I had set my sights
tentatively on Central Asia, but had no clear sense of calling
from God in terms of geography. It seemed more important that
we find a group with which we could work happily than it was
to worry about geography. I was impressed that God's call in
the Scriptures was primarily to himself and not to a place. Geog-

raphy was always secondary. God called Abraham not to go to Canaan, but primarily to obey him, and then to go to Canaan. God called Paul and Barnabas and others to himself and then secondarily to a geographical location. So, it seemed it was far more important to be obedient to the Lord in moving ahead, finding a group with which we could work happily, and letting geography take care of itself.

Thus we wrote to boards in the Far East, India, Africa and Latin America. By the process of elimination we reduced our options down to two boards. They seemed to be fairly equal. Finally, in a step of faith we said, "Lord, we will apply to the Latin America Mission. If you don't want us in Latin America, you can close that door, and we will look elsewhere." When we were accepted by the Latin America Mission, in response to Jesus Christ as Lord in our lives, we went first to Costa Rica and later to Colombia. I can remember thinking as we got off the plane in Costa Rica for the first time, "Now I can say with certainty that God has called me to Latin America, because this is where I am by his grace as I have taken one step at a time. He has opened the door so I take this as his place."

One Step at a Time
Years later in Colombia I experienced something which helped me understand how Jesus Christ as Lord usually leads us just one step at a time. Psalm 119:105 says, "Thy word is a lamp to my feet, and a light for my path." The only kind of light the psalmist could have known about was a little clay lamp which could cast a small circle of light. Searchlights and flashlights with strong beams were unknown to him. In the backwoods of Colombia the people make small lamps out of tin cans. They fashion a simple handle, put a wick through the top, fill it with kerosene, and light it. This little lamp is functionally identical to the clay lamp that the psalmist must have known.

This type of lamp casts just enough light for one step. There have been times when I have walked along a jungle trail at night in Colombia with one of these tin lamps in my hand. Sometimes I have wished for a powerful flashlight which would illuminate the trail far ahead. It would be a bit more reassuring to be able to see what lies beyond that next log! But when one is limited to a small kerosene lamp which casts a circle of light, the only way to get more light is to step into the circle. As soon as we do that, the lamp casts enough light for one more step. Thus we can go on in faith one step at a time.

Jesus Christ as Lord does not usually lead by means of a searchlight. Rather, he normally takes us one step at a time. As we obey the light we have and take that one step, he provides enough light for one more. This is a lesson I am still learning. But I am convinced it is biblical and practical.

If Jesus Christ is Lord of our life, we must be prepared for that circle of light to lead us in ways we had not expected. God has said, " 'For my thoughts are not your thoughts, neither are your ways my ways,' declares the LORD. 'As the heavens are higher than the earth, so are my ways higher than your ways' " (Is 55:8-9). As Lord he has the right to lead in his ways, not ours.

Having been led of God to Latin America, my wife and I fully expected to spend the rest of our lives there. We loved the people and the land. Three of our four children were born there, and it was home to us. I still get a massive dose of homesickness and nostalgia when I hear Spanish spoken or when I have an opportunity to return to Latin America.

But Jesus Christ as Lord chose to lead us in ways we had not expected. Fourteen years ago tonight, at the end of the final service of Urbana 67, I was standing on the floor of the assembly hall when Dr. John Alexander approached me. He indicated that Inter-Varsity was seeking a new missions director and director of the Urbana conventions. And he wondered if I would be available for the position. While the challenge of the student world was great, I told him that there was no way I could accept such an invitation. We were due back in Latin America within three weeks, and I felt I could not change plans. Dr. Alexander asked if I would pray about this for a while anyway. I agreed to do so. Over the next six months I prayed earnestly about this invitation and turned him down three times. Each time he came back patiently asking me to pray some more. Finally, the fourth time around, Jesus Christ as Lord of my life made it clear that he was trying to get a message through to me and that I was to return to the U.S. to work among students on behalf of missions. This was a total surprise to me, but I accepted it as God's leading. Thus we returned to North America and have enjoyed happy and stimulating years of ministry among students through IVCF. In recent years God has led again in new directions, but each time it has been a matter of taking one step at a time as he opens the way and gives the light.

Let us get one thing straight right here. Do not assume that God will lead you in the same way he led me. The lordship of Jesus Christ is foundational to all of our lives, but you must respond to that lordship as he speaks to you. I have tried to share something of what it has meant for Jesus Christ to be Lord in my life. I cannot claim to have been obedient at all times. Quite the contrary, I could stand here for hours and relate to you the failures on my part and the times I have disobeyed my Lord. Yet he has remained faithful. He is Lord, and my job is to let him control my life.

As we come to the close of Urbana 81 we quite appropriately conclude with a focus on Jesus Christ as Lord of our lives. He wants to speak to all of us. He wants to speak to us corporately as representatives of the body of Christ. But more than that, he wants to speak to us individually.

I am awed by the tremendous potential of what it would mean if each of us truly made Jesus Christ Lord in our lives. I think of the great host of messengers that God could raise up and send out to declare to all the world that Jesus is Lord.

Two and a half years ago I sat in this assembly hall at the University of Illinois in entirely different circumstances. I was part of the crowd looking down on what was going on. There was no speaker's platform. Rather there were wrestling mats laid out on the floor. The Illinois state high-school wrestling championships were being held. The competitors were boys who had won district and sectional tournaments and had qualified to compete for the state championship. Among them was one of my sons. Although I was just part of a faceless mass of thousands of people, he was the center of my attention. Every time I heard the loud speaker announce, "Now, wrestling on mat number two, in the 132-pound class, representing Wheaton North High

School—Michael Howard," I lost interest in every other person in the huge crowd. My attention was riveted on one person—my son, who is my own flesh and blood and whom I love with all my heart. I agonized with him through every move he made in every match. When he won (and he won six or seven matches in the tournament), I was elated. When he finally lost by one point, I still loved him with all my heart. When he stood on the victory stand to receive a medal, even though it wasn't the gold medal, my heart burst with pride. He had done well. But even if he hadn't, my love for him would not have changed. I looked down with the proud and loving eyes of a father.

Although the wrestling match is over now, there may still be struggles going on as you wrestle with what it means to make Jesus Christ Lord in your life. You can be assured that the eyes of a loving Father look down on you right now. He knows every move you are making, and he agonizes with you in your struggle to obey his Son as Lord. You can also be assured that you have an enemy. You may lose some matches along the way, but your Father still loves you; he is still faithful to you; he still calls you to follow him in total obedience. He wants to be Lord of your life.

Our Hope Is in the Lord

As 1981 passes into history, I want to leave you with a great word of hope. We go out, not in our own strength, but in the strength of the loving Father who is looking down on us. And we go out with a great hope.

Let me tell you a story that has helped me and has given me great hope as I look at the scene of world evangelization. In the last three or four years it has been my privilege to travel a number of times, perhaps six times, around the world. And in the course of my travels I have gone to many different countries. I have gone twice into East Berlin. Each time my heart has been deeply moved and I have been terribly depressed as I have sensed the awful oppression in which the Christians behind that terrible wall must live. I have traveled over the top of the wall in an elevated train that takes passengers into East Berlin from the freedom of the West. Until that train ride I was not aware that there are two walls in Berlin. One is right smack next to the West Berlin side. It is a very high wall, and there's no barbed wire on it because that looks bad to the tourists who come from the West to take pictures of this infamous wall. They have a very ingenius system. Instead, at the top there are large, stainless steel drums set on ball bearings. You can imagine what it would be like to try to climb over a wall and get a handhold on a drum that spins on ball bearings. After that wall then there is an open space of a hundred or a hundred fifty yards filled with tangled barbed wire, all kinds of obstacles and fierce police dogs. And then there is a second wall, also very high. This one is topped with barbed wire and heavily armed guards with machine guns and searchlights flashing all night long over that area.

One passes over that wall and sees how impossible it would be to get through there. As I passed through immigration to get into East Berlin, I also had a depressing experience. I have been in and out of many countries where I don't speak the language. But I know that a smile is a universal language, and so I smiled. I thought, "I'll see if I can get anyone in immigration to smile." And as I went through about seven different checkpoints, I smiled and thanked them as they handed my passport back to me, but all I got were stony stares and frowns until I got through and my dear friend, a German pastor, rushed up, gave me a big bear hug, welcomed me to East Berlin and smiled from ear to ear.

On one occasion my friend took me to spend the day with a group of German pastors that I came to know and love. They told me what it was like to be a Christian in a land like East Germany. My heart was heavy for them when I saw what they had to live under.

Before I had gone there, I had said to someone who knew the Eastern countries, "How can I minister to these people?" And my friend said, "Dave, don't try to minister to them. You'll find that they'll minister to you. Just tell them what is going on in the world because they don't have a chance to hear what is happening elsewhere in the world. You do that for them, and they'll minister to you." So I tried to share with them what I had seen

God doing around the world. Then they began to minister to me. They asked me about myself, my family and my work, and they prayed for me and laid hands on me.

As it came time for me to go back, one pastor who used to live in the West but had voluntarily gone to the East because of the need there, said to me, "I'll never get out again. I came here because of the need of my people in the East, and I know that I'll never live again in the West, but I'm here by the will of God." He drove me back to the train station and he said, "I'm taking you now to the tear palace. They call it the tear palace because people from the West are allowed to come in to visit their families, but as they leave their families are not allowed to go with them. The families go to the train stations to say good-bye, and you see a lot of tears in that train station. I'm taking you now to the tear palace." As we got to the tear palace he reached over and laid his hands on me. I was the one who was leaving, who was going to the freedom of the West, but he laid his hands on me and said, "I give to you the peace of Jesus Christ." My heart was deeply moved.

As he drove me back to the station I recalled what I had seen on the day I arrived. When I came out of immigration with my friend, he pointed something out to me which I had seen at a distance from West Berlin but now saw more clearly. There's a great high television tower that sits right in the center of East Berlin. They tell me it's the highest structure in all of Germany, and it was built by the Marxist government to show symbolically the superiority of Marxism over all other forms of government. This great tower looms up and there is a big ball on top with reflectors on it for television reception. The architects did not realize when they made this great structure what would happen when the sun shone on it. When the sun shines it forms the perfect sign of a cross. When the government officials saw this they ordered the architects to change it. They tried and were unsuccessful. It's still there.

As we walked out of the immigration building, my friend pointed, "See that? See the cross?" And he told me the story about the tower. He said, "You know, we live under terrible oppression, but we also live under the sign of the cross. Every morning when the sun rises in East Berlin, it shines on that tower and goes right around the sky and the cross follows it around. Every day we live under the sign of the cross."

I say to you that today we go out to serve Jesus Christ as Lord in our lives. And we go under the sign of the cross.

Part V

The Task
of the Witness

16
Being a Witness
Rebecca Manley Pippert

One day on a plane I happened to sit next to a rather intellectual-looking professor. We leapt into a stimulating conversation, and I intended to tell him about my faith—at the appropriate time. But abruptly he asked me what I did for a living. I said, "Well, I'm in Christian work." (It's one thing to be a Christian; another thing to do it for a living.)

A look of amazement spread across his face. He was clearly thinking, "funny—she *looked* so normal!" Immediately his demeanor changed, and he was clearly trying to find the appropriate words to use for a "Christian type." He asked, with the slightest condescension, "Well, what's the name of your little organization?"

"Inter-Varsity Christian Fellowship," I replied. He looked bewildered. I asked, "Is something wrong?"

He said, "Oh, nothing really. It's just ... well ... you don't *look* like a Christian athlete."

"Thinking at that point that he was joking, I said, "Well, yes, I play basketball for Jesus. It's a living."

Without a hesitation he said, "Oh, I'm sure it must be very rewarding."

It was a great temptation to play along with his feigned religious behavior and say, "Yes, well, it's such a little blessing. You know, we never lose a game." However, with uncommon restraint, I told him, "No, actually that was a joke. We make jokes sometimes. However, you asked if my work is rewarding. I would prefer to say it is terribly intriguing."

And almost in spite of himself he asked, "Intriguing? Well, why is that?"

I answered, "Because I work with students. And we constantly face the question of: 'How do we know anything is true? How do we know that we aren't taking our own little world and labeling it reality? Is there any basis for our faith or is it mere wish fulfillment?' "

He answered, "You may not believe this, but those questions were going through my mind as well. OK. What kind of evidences do you have?" And so we talked about the evidence for Christian faith. Then he said, "You know, besides the evidence I think what impresses me most in this conversation is that you seem to be a person of hope and not despair. Why is that?"

Then I was able to share for the last five minutes of our descent that the reason is Jesus Christ.

How do we get to the point of discussing Jesus Christ with our friends? As communicators we have been made "agents of reconciliation." The word is reconciliation not confrontation. We are summoned to be fishers of people, not hunters. When we listen carefully to where they are, when we pique their curiosity, when we discern what their defenses are against Christianity and cite them before they do, when we do these things we reveal that we care.

My experience with the professor was vastly different from my embarrassment in first sharing the gospel. At that time I was in Spain as an undergraduate student. I knew God had called me to be a witness, but for the first several months I allowed the fears and insecurities of sharing Christ, as well as my discomfort with being a witness in a different culture and a different language, to intimidate me.

For example, one day I was reading the Bible for my devotions when a cynical friend entered my room unexpectedly and said, "¿Que estas leyendo?" (What are you reading?)

I was sure she would think I was a religious fanatic—not only reading my Bible but on a weekday! So I quickly slipped my Bible under other books and tried to look as cool as possible. "Oh nothing, really."

"Yes you were, what were you reading?"

"Oh, not much," I answered.

"Becky, what were you *reading*?" she demanded.

"All right! It's the Bible!" I confessed. And I behaved this way so she would not think I was strange!

I slowly began to realize that we are called to expose our faith, not impose it nor hide it. As I read the Gospels and saw how beautifully Jesus dealt with people, it began to free me up.

It's a long story, but God gave me an antidote for my fears and timidity about sharing my faith. By the time I left Spain, and through the great encouragement of my roommate, Ruth Siemens, God used a Bible study that I was leading to win five people (including avowed atheists and one Marxist) to Christ. Until that time I had not seen one person ever become a Christian. Today I am the godmother of one of the former atheist's children. If you had asked me at the time if any of those five students seemed open to God, I would have laughed out loud. But I could not see their hearts, nor the power of God's Spirit to penetrate their hearts. So remember: *All* of you are potential "Godparents"!

Compassion: Called to Love

I learned that even more than our words, God uses the way we love others with Christ's love to build his kingdom.

Jesus constantly taught that if we are followers of his, our lives will bear the stamp of profound love: to God, to our neighbors and to ourselves. If we are to be effective witnesses, our lives must be dominated by his love, not merely religious activity. Our sociology ought to reflect our theology. How we treat others will be the clearest signal of what we think God is like. The first Bible most people read will be our lives, long before they ever read the book.

And as I travel to campuses, I see students moving away from the "us-and-them" mentality that isolates Christians from the world except for an occasional evangelistic meeting. There is less of the "holy huddle" syndrome; less of the local "God squad" mentality; less manipulation and fewer gimmicks and more genuine involvement with the people we want to win. There is more real sharing of our lives—the strengths and the warts—than mere preaching and leaving.

John Stott says we must not be "rabbit-hole Christians." The rabbit-hole Christian is the one who leaves his Christian roommate, runs to class, and looks around the whole room to find a Christian to sit next to—which is a rather odd way of approaching a mission field. He or she then goes to the cafeteria for lunch and sits with all the other Christians. "Praise God," these Christians say, "all sixty of us here eating together. What a witness to all those people out there eating alone!" And then the rabbit-hole Christian goes to a prayer meeting and prays for all the unsaved on his floor. To me that is the most insidious reversal of what it means to be salt and light. How can you be the salt of the earth if you have never gotten out of the saltshaker? We are called to love with the love of Christ. I think we all really know that. So why do we still struggle?

First, we are too *complacent*. My pastor, Edward Bauman, tells the story of three devils. They were discussing what strategy to use to keep Christians from being effective. One of them said, "I know! Let's tell them that there is no hell—no punishment. Then they won't feel any fear."

The other said, "No, I've got it! Let's tell them that there is no heaven, no prospect of reward."

But the third devil said, "Wait, I know! Let's just tell them that there is no *hurry*. It's all true, but there's no need to rush, no urgency. They have plenty of time."

But there is a pressing urgency. The kingdom of God is at hand.

Gabriel Fackre says that to be an effective witness you must get the story *straight* and get the story *out*. So, second, we must get the story straight. That means that you must understand what you believe; you must know what you speak of and not simply recite Bible verses. Be involved during the school year in Bible studies. Use your summers to deepen your knowledge of the faith by attending Inter-Varsity conferences or institutes like New College in Berkeley or C. S. Lewis Institute in Washington, D.C. Go where you can deepen your understanding so you have got the story straight.

Third, we must also get the story out. We need to work on building our communication skills. Analyze your style of communicating, find your areas of strength and work on your areas of weakness. For example, are you timid and shy? Do you feel intimidated because you never can think of how to begin a conversation, much less how to get the conversation around to God? Do you frequently miss detecting people's needs? Ask God to make you more perceptive and sensitive.

But remember that the key to communication is the ability to love as Christ loved. Jesus put a child in the midst of the disciples and said, "Whoever receives one such as this receives me." That is powerful. It means any person we touch, Jesus is touching too. It also means that when I touch a person, I am touching something of Jesus himself, no matter how distressing the disguise. How are you treating Jesus as you see him everyday?

C. S. Lewis understood this well. In "The Weight of Glory," he said:

It is a serious thing to live in a society of possible gods and goddesses, to remember that the dullest and most uninteresting person you talk to may one day be a creature which, if you saw it now, you would be strongly tempted to worship, or else a horror and a corruption such as you now meet, if at all, only in a nightmare. . . . There are no ordinary people. You have never talked to a mere mortal. . . . It is immortals whom we joke with, work with, marry, snub, and exploit—immortal horrors or everlasting splendours. . . . Next to the Blessed Sacrament itself, your neighbour is the holiest object presented to your senses.

A witness has compassion.

Cost: Called to Do Justice

A witness also must know the cost of the faith. One year and one day ago, I sat in a courtroom and heard a judge sentence to prison someone who means a great deal to me. This person remains behind bars even now. It was a shattering experience for me. Very few incidents in my life have reminded me so grimly of the reality of evil and the inevitable consequences of sin. But I discovered something else. Suddenly verses about caring for the prisoner—verses which had always seemed so distant and remote—now seemed pointed at me. And I marveled anew at the depth of Jesus' identification with the poor and the oppressed. The Son of God, the Prince of peace was also a convict.

It was a personal crisis that sensitized me to Jesus' words that we are to be concerned about prisoners as well, of course, as other persons in need. I told God I wanted to do something to follow his words. One week after the sentencing, I received an unexpected call from Charles Colson's Prison Ministries asking me if I would consider teaching inside the prisons, as well as speak to a seminar for eleven women convicts who had recently been converted to Christ. That seminar experience became one of the highlights of my teaching ministry this year. When I arrived I was scared. But they put me at ease by their warmth and humor and by their understanding of my own wounds, still fresh from my friend's court sentencing. I received far more from them than I gave. They demonstrated the miracle of people who had come from great darkness into Jesus' glorious light. They are praying for me now from their cells.

This is a small example of how each of us must respond in individual ways to Jesus' mandate to visit the prisoner, to feed the hungry, to clothe the naked, to set at liberty those who are oppressed. But this, in a way is easy because it is one-dimensional. Those women still must return to the wretched living conditions in prison which too often does more to improve criminal skills than to reform criminals. And when they get out, will they be able to find jobs? How will they handle a society that is suspicious of them? What about the strained personal relationships to which they return?

In other words, it is not enough that we love the prisoners

simply by befriending and witnessing to them. We are equally duty-bound by Scripture to attack the problem of crime and punishment at the structural level. Institutions as well as individuals are sinful. And God calls us to redeem the structures of our society as well as the persons in it. There are so many issues we must confront with biblical light: nuclear arms, poverty, racism, sexism, terrorism. I can't tell you where to start, only that you must—both at the individual and societal levels.

We live in a country where our poodles eat better than many people on this earth; where the evident goal of our culture, and sadly of many believers, is to obtain the comfortable life and self-fulfillment. I sat in my family's church on Christmas eve, the First Presbyterian in Champaign, Illinois, and the words that the pastor, Rev. Malcolm Nygren, read leaped out at me: "I the Lord love justice."

We must love in intimate ways those persons God puts close to us. But others, whom we may never meet, God also calls us to love through the pursuit of justice, mercy and fairness.

You may ask, but what does God's call to justice have to do with evangelism? Make no mistake, the way you seek mercy and justice for others authenticates the message that you speak. If your witness to God reflects nothing of God's concern for the oppressed and needy and suffering people of this world, there will be little authenticity to your testimony.

Jesus tells us that we do not need to worry about what we wear and eat, because God cares for us and will provide. How do you reconcile that with the grotesque pictures of starving children with bloated bellies and two tiny pegs for legs? Was Jesus wrong? Or is it that God only provides for the converted —that is, if the children were not Hindu or Buddhist they wouldn't be starving? No. We are God's agents in caring for a hurting world. We are *his* agents of mercy and healing. Could it be that we must share part of the guilt for the world's misery because instead of giving from our abundance, we have thought mostly of our own needs and luxuries?

I wonder if among all the people attending Urbana there are even five persons who have been truly anxious over the necessities of life. In my entire life, I have never been anxious over *anything* that was not a luxury. Do you realize what percentage of the world population that puts me in? and the awesome responsibility? Scripture is persistent on its emphasis of special concern for the poor, the lonely, the oppressed. Make no mistake —we cannot separate the call to justice from true discipleship.

"Ah, but I'm in college now. I need to study and worry about these world concerns later," you say. Yes. You are at the university to study. But how you approach these years will largely determine the shape of your future life. The Son of man came not to be served but to serve, and to give his life as a ransom for many (Mk 10:45).

What makes the call of simple justice hard to hear is that it requires us to look at our own culture, to give and serve the poor instead of living fully out of our glut. And that is difficult because our culture is the very lens through which we see the world. We do not see our culture's excesses because we see *with* our culture. However, if we are not discipling ourselves to be honest about the strengths and flaws of our culture as we try to wean biblical faith from Americanism, we will make colossal mistakes on the mission field. It is painful to hear the stories of North American missionaries living on a far grander scale than those to whom they minister.

Consider these two Christian conferences where I spoke in order to get some understanding of our own culture. One was in South America, the other in North America. In Bogota, Colombia, a group of Christian student leaders (whom we would call the "exec") decided to create their own summer conference. Their summer consisted of going to the poorest of the poor in Colombia where they lived in a small pueblo with intolerable heat. They taught skills, talked about salvation, and simply loved the people by enjoying them. I saw them the first night they returned, glowing from their service. They gave not merely out of their abundance. They gave until it hurt. But they felt only joy.

Contrast that with the conference for young people I spoke to in North America. I was told that night that when the announcer said, "Here's Becky!" I was to walk through a revolving door under an arch of flashing lights. I was to be followed by a lion who was to come through the revolving door with flashing lights and roar on cue. It and I were followed by a Christian rock band and a born-again, stand-up comedian. It was like Johnny Carson's "Tonight Show." When I asked why they had all of this raucous entertainment the answer was, "It really catches their attention." I couldn't decide which was more depressing: that they felt they needed flashing lights, lions, stand-up comedians and rock bands to get students interested in Christ, or that they felt I would be the ideal speaker for such an occasion.

Christians used to be fed to lions because we were so radical that the world wanted to silence us. Now we bring lions on to entertain us! I was told that some—not all—of the entertainers demanded exorbitant fees and a specified number of bottles of Perrier in their dressing rooms. One young singer sat at dinner with earphones on. She removed them only for grace. She seemed to be saying, "The only one that has access to me is this object—not people."

As I sat there I wondered what my Colombian brothers and sisters would think if they saw the hundreds of dollars the American students spent on records and tapes and little puppets that were selling for $2.50 saying, "Smile, Jesus loves you." I am afraid they would think the Marxists were right, that American Christianity all too often is just materialism with a little God sprinkled in. There is no cost, no sacrifice, no sensitivity to the massive needs of the poor, but we sure know how to have fun.

It is not that we are to be ungrateful for what God has given us. It is not that we must insist on drabness. This is God's world, and we celebrate and enjoy his beauty and his gifts. But that celebration must be tempered by a sensitivity to the poor, those who are starving to death, the crying needs of the city streets. I have been mugged at gunpoint and had my home burglarized twice this year. The perpetrators are, at least partly, products of our age.

So what can you do? Begin by desiring to be a good steward of your resources. Take seriously Jesus' words on money. You do not have to dress without style or taste. But make it a game. Try to beat the system. Find great sales, and what you save you can give in constructive ways to the poor. Remember that your money is a gift from God to be used for his purposes. Think about how you will use your career and salary and professional position to be a servant of Jesus Christ, not a consumer American.

There is no other area where I tremble more before God in my life than this one. If your life does not reflect the justice of God's call, if it does not reflect service, your evangelism will be void of power and authority. A witness is one who knows the cost of the faith.

Character: Called to Holiness

Lastly, we are called to holiness. Jesus tells the disciples in the Sermon on the Mount "do not be like them." He tells us we are to be different from the world.

I have a beautiful friend who, in the midst of personal crisis, posed for *Playboy* magazine. Sometime later she made a commitment of faith to Christ. Recently *Playboy* asked her to do another series and offered her enough money to have bought three or four Rolls Royces. She asked me with great sincerity, "Can I pose for *Playboy* now that I'm a Christian?" I do not condemn her. She was genuinely trying to understand what it means to be a Christian. I grieve instead for us—the body of Christ—that our model of godliness has been so shabby and weak that she would have to ask that question. Why shouldn't she ask it, when there are people who claim to be born again and continue to "live it up" (for example, the star who claims to be born again and demonstrates her witness by singing in someone's arms, "with you I'm born again." Senate Chaplain Richard Halverson says, "Evangelicals seem to have been more influenced by the world than they have influenced the world."

We need desperately to re-examine what it means to be a holy people. Too often we have settled for a narrow understanding of holiness. It used to be that "I don't smoke, drink, dance or chew, or go with girls who do" characterized our understanding of holiness. Today we pat ourselves on the back, feeling smug because we are not as legalistic as our parents used to be. Yet in our attempt to identify with the world (which we *must*

do) have we ignored the call to be different? To identify is not
to be identical. We are to walk alongside our neighbor empa-
thetically, but without compromising our difference.

So what is holiness? To understand holiness we must look
at the cross and the resurrection. Therefore, the first essential
is we are called to die. Paul says in Romans 6 that when Jesus
died you died too. He means by this that everything that has
ever kept God at a distance from you—all the junk, the brokenness—must no longer control you. To be a Christian we need
to look at the cross. To be a witness is to be holy, and if Jesus
is our model, then the first thing we must be willing to do is
to die.

John 12:24 says, "Unless a grain of wheat falls into the earth
and dies, it remains alone; but if it dies, it bears much fruit"
(RSV). The way to experience the power of our risen life with
Christ, the way to be an effective witness, is to die to sin. You are
to die to that which is destructive and keeps God at a distance.
It would be a lot simpler if I said, "When you see things in your
life that shouldn't be there, try to tiptoe around them or ignore
them." But the Bible says we must die. It's an absolute in an age
that loves angles. We have angles for everything: how to catch
a lover, how to have fresh breath, how to witness. But God says
our only angle is our bankruptcy, when we finally admit that
we are broken, that we cannot control our lives. We live in a
narcissistic "me" decade. With its emphasis on my rights, my
desires, my needs, our culture says that to be strong is to be in
control. The Bible says that our strength comes from the realization of our brokenness.

When we finally admit that for all the manipulating we attempt, when it comes down to it, we cannot control the things
that matter most to us—life, health, our spouses or friends, our
children's destiny. Even our money is subject to the vicissitudes
of the president's economic policies. God is pleased to receive
our resignation as self-appointed managers of the universe.
Sometimes we must be wounded before we learn relinquish-
ment. We cannot die unless we are willing to be honest about
our sin. We have become too nice to believe in sin.

Bishop Fulton Sheen addressed the National Prayer Breakfast

shortly before he died. "Good morning, fellow sinners," he told this politically and spiritually elite audience. They squirmed. He said, "I'm a Catholic. For years our Doctrine of the Immaculate Conception set us apart. I never thought I would live to see it, but that doctrine now seems to be universally embraced, for today everyone appears to believe that they were immaculately conceived." One of the difficulties in being willing to die is that there are so few Christians who are honest about their sinfulness.

I know an elderly Christian woman. She is a saintly inspiration to me in the genuineness of her faith. But she is not a helpful model in other ways for she feels it is unspiritual to admit she struggles. Consequently, she seems very naive about the lurking presence of evil. She once told me, "Well, I'll be honest and tell you what I really need to confess. I don't write enough letters." Is that why Jesus died? He should have saved his blood if that is all that was wrong!

I heard a television evangelist not long ago say, "People ask me if I ever struggle with sin. I said, 'Maybe I do, but you'll never hear about it. I just go to God.' Others say, 'But don't you ever struggle with being godly, or loving your wife, or being a good father?' I answered, 'Maybe I do, and maybe I don't, but you'll never hear about it. I just go to God.' " Then he read a poem entitled "Be a Man." It went something like this: "Feel tired or discouraged? Don't let it get you down—be a man! Feel like throwing in the towel? Feel tired of struggling? Be a man!" I counted the refrain "be a man" twenty times. Now I have a little trouble identifying with a poem that is called "Be a Man." But more importantly, why did this man exhibit such resistance to even admitting that he struggles? I can understand his not wanting to bare himself in front of millions of viewers, but does he bring glory to Christ by refusing to say that he is even tempted?

Paul claimed boldly, "I am the chief of sinners." When Paul told of his struggle with his "thorn in the flesh," I do not recall Jesus' answer being, "For heaven's sake, Paul, would you buck up and be a man!" No. Our Lord's answer was, "Paul, I won't take away this trial of yours because I am glorified in your weakness." I fear our TV evangelist has done what we often do. He has taken a popular image from our culture and spiritualized it. It's the myth of the Lone Ranger. All I need is God and my horse. I don't need anybody else. If I make mistakes, I'll just tell my horse and ride off into the sunset.

Beware of the cults within Christianity, including the cowboy cult.

A holy person must know how to die. So do not be naive about the evil around you. Be honest about what tempts you. Know what things by God's grace you must put to death. See clearly what motivates you. Augustine claimed that what caused most sins were pride (self-aggrandizement) and sensuality (self-indulgence). Luther felt that the cause of most sins was unbelief. Many modern theologians believe sin is manifested by anxiety, insecurity and alienation. Know yourself well enough so that you confess the root problem, not just the symptom. I wish I could tell you that you will find lots of Christian models who will encourage you in this. But you may not.

Whether you find positive models or not, Christians have no excuse for naiveté about evil. Heaven will not be filled with innocence! The Christian in heaven will say, "By God's grace and my sweat I am home at last!" Never be ashamed of struggling with sin. To struggle may mean you are alive to God! You have seen what God desires and how far you are from being the person he wants, and you are willing to enter the fight, to allow God to change you and make you into his own likeness.

You will have to be gritty. You must learn what to say no to. If you sin now you may hurt only yourself or another. But part of the complexity of growing older is that your life becomes intimately interwoven with several others. If you aren't willing to say no now, when future temptations come, and I assure you they will, your inability to say no will hurt not only you, but those whom you love the most.

We have to be tough-minded. That is difficult because we are pampered. We know very little of the cost of our faith. While a vast number of believers around the world have had to sacrifice their families and jobs and food, we respond like martyrs if God calls us to live with any unfulfilled desires at all. While others live in gratitude that they can read a Bible without going to prison, we complain that as singles we are required to be sexually celibate.

Most of you wonder if you will marry. The vast majority of you will. But that is not the real issue. To spend so much time anticipating marriage is needless. If God wants you to be married, he will take care of it. He has done it for centuries. The real issue is, once married, what kind of a spouse will you be? Will you be faithful? Will you be a servant? Will you live in harmony? Raise godly children? Your answers will be determined by the character you are building now, whether you are mastering godly patterns for living. If you think this has little to do with evangelism you are mistaken. Wes and I have heard of a distressing number of divorces among evangelicals and often the same statement is made by the spouses: "It's a shame, but God forgives." It is certainly true that God forgives. But why don't we also hear that God cares about promises? Whatever became of duty? What kind of witness to the power of God to reconcile and heal is the evangelical divorce rate?

We are to be a holy people. That includes being willing to admit we have sinned. That is the negative. But there is also a positive side to being holy. The positive is seen in the resurrection. We find in Romans 6 that when Jesus rose we rose too. The moment has come that we can be new people. The old has passed away. We can forsake the old and identify with the new. We can say, "My past has been covered by the blood of Jesus Christ and I have become new. The resurrection becomes my mindset. It shapes everything that I am and that I do." If we want to be holy we must know how to die and how to live. We must die to sin so we can live to Christ.

Satan will do anything in his power to convince you that you are not new. For example, suppose you are shy and you go to a party. Someone starts talking to you, but you cannot think of anything to say back. There are some long silences. What is your thought during those silences? Do you think, "I am such a turkey. I can never think of anything to say. This is so embarrassing. I don't know why I come to these gatherings"? Or do you think, "My shyness is showing again, that's true. But isn't it thoughtful of the hostess to invite royalty to a party"? (Because

I am a child of the King.)

Or suppose there is someone that you want to witness to, but you are afraid. One day she corners you and says, "Boy, am I glad to see you. I've been meaning to ask you a question about your faith for a long time." So the question is asked and you can't answer it. What do you think? Do you think, "I knew it! I should have run when I saw her coming. Why do I get myself into these things"? Or do you think, "Isn't this exciting! I don't know the answer to this question. God says that he is glorified in weakness, and look how much he has to work with"?

Or let's say you leave Urbana, and during the first week on campus you blow it—you sin. Is your thought, "There, you see, you want to know what I'm really like, look at what I just did. That's me—one step forward, ten steps back"? Or do you think, "Isn't it amazing that I did that! Isn't it amazing that I would do something so contrary to my new nature"?

There is a world of difference displayed in these various responses. One reflects the kingdom; the other reflects the world.

You are called to be a witness. And you will be a witness when you have compassion; you will be a witness when you realize the cost that reflects the justice of the gospel; you will be a witness when your character is so shaped by God that you know how to live and you know how to die. And what has appealed forever to the world but is ours alone is *joy*. "Joy," G. K. Chesterton says, "which was the small publicity of the pagan is the gigantic secret of the Christian."

17
Pioneering for Jesus Christ
Marilyn Laszlo

It is quite a shock to come to Urbana. There are more people here than trees in the jungle where I live. For the past thirteen years I have been working in Hauna, a little village which is 500 miles up the Sepik River in the heart of the jungle and swamps in Papua New Guinea, an island just north of Australia.

On the island of Papua New Guinea there are over 700 distinct languages, most of which are unwritten. There are thousands of dialects. Can you imagine never having seen your language written? Try to picture English as only a spoken language. How different our lives would be. It would mean, for one thing, that you would not have one single verse of the Bible. Actually, there are over 3,000 language groups in the world that have no written language. They do not even have an alphabet, much less any books.

That is the way it was in Hauna Village, home of the Sepik Iwam people. They had no idea that the words that came out of their mouths could be written down. What an exciting and rewarding adventure it has been for me to learn their language, develop an alphabet and to teach the people to read and write their own language—their own talk. Come with me now to Hauna Village and I'll show you how it all began.

We first entered the village in a sixty-foot dugout canoe powered by an outboard motor. What a trip! We had 100 miles to travel, heading upriver from the little grass airstrip where we had landed in our single-engine plane. It is beautiful country with hundreds of varieties of orchids and other flowers, hundreds of varieties of birds—it's a bird paradise of the world. There are also hundreds of varieties of snakes and crocodiles. Quite an exciting trip indeed.

I hope that you know more about motors than I do, because the only thing I know to do is to change the spark plugs. If that doesn't work we just drift back downriver, and it takes four and a half days to drift back to the airstrip. But all went well on that first trip, so after about ten hours we arrived at the village. The village people were not at all sure about my translation partner and me. They wondered whether we were male or female. They decided we were neither. We were "its," spirits that had fallen from the clouds. They were very afraid of us, which was quite handy because we were afraid of them too. But there were 410 of them and only two of us.

Carving with a Thorn

We were given training in linguistics and we began learning the language one word at a time by pointing to objects and by acting out concepts. One day I was walking through the village, my paper and pencil in hand, pointing to objects and trying to "gather" words. I discovered the word for *pencil* is *nimid* meaning "thorn," the word for *paper* is *yokwo* meaning "banana leaf," and the word for *writing* is *wini* meaning "carving." Ever since I have been going through this village with my thorn and my banana leaf, carving the language of these people, learning one word at a time so that someday I could teach them to read and write their own language. Eventually I will have translated the entire New Testament into their own "talk."

I remember one day I was trying to get the word for tree. I was pointing to a tree and trying to get them to say the word for tree. Finally they said, *"ana,"* so I wrote that word in my notebook. Now I had my first word in the dictionary. Later, as I went around the village practicing the words that I had collected, I pointed to a tree and said, *"ana."* They all shook their heads and laughed. I had obviously said it wrong. Later I discovered that they hadn't given me the word for tree. They had given me the word for finger!

Next I tried to get some verbs. I sat down to get the word for sit and stood up to get the word for stand. I ran to get the word for run. One day I was talking with Nokiyan. He is one of many of our older men who wear no clothes—only a vine around their waists. That's it. They're all dressed up. I was trying to get the word for jump so I was jumping up and down in front of him. He kept staring at me and finally said, *"korawhowaenanae."* Try writing that! I thought, That's a long word for jump! But some of their words are long and he kept saying, *"korawhowaenanae,"* so I wrote *korawhowaenanae.* One problem we have when we are first learning a language is not knowing where the word breaks are. I didn't know whether it was a phrase or a sentence or two words or even one word, but I wrote it in my dictionary as I heard it. Several months later I discovered that wasn't the word for jump. He had said, "Why are you acting so stupid?" The worst part is that I had gone around the village practicing to everybody I met. This is no way to win friends and influence people.

"What Are You Doing?"

Even after we were there for several months they still did not fully understand what we were doing. One day Nokiyan said, *"K+ra, +i kw+r +i kw+r om napr+r+ri. Dimu w+ni?"* He said, "Marilyn, every day I see you walking through the village. What are you carving on that banana leaf with that thorn?"

I replied, *"K+rir yaig+n t+ yokwo w+ni*—I'm carving your talk."

"Oh, my talk! You mean the words that are coming out of my mouth you're carving on that banana leaf like the white man's talk?"

"Yes." He had the idea that only the white man's talk could be carved on the banana leaf. That was the white man's power. I said, "No, Nokiyan, your language and your talk can be carved on the banana leaf too. God created your talk."

He became so excited and he came up very close and he looked on the carving and said, "*Dimuwinin?*—What does that say? What have you written?"

"This is what it says, '*Kara omaka nami.*'"

He rubbed his wrinkled fingers over the carving and kept saying, "My talk—that's my talk." He was thrilled. He called a village meeting and he gathered the 410 people together and he said, "Do you know what the two white misses are doing? They're carving our talk on the banana leaf, the talk of our ancestors. But they can't speak the language fluently."

They had no idea what was involved in learning a language. They could not understand why we were not speaking the language fluently. Nokiyan got up in front of all the people in the village and said, "The problem is that their lips and their mouths and their tongues are too thick and heavy. And that's because they're not eating enough of our food."

So they brought us all sorts of delicacies. One day Wongomar, one of the men of the village, came into the house and he had something wrapped up in the banana leaf. After being there for thirteen years I know now that when they bring you something wrapped up in a banana leaf, it is some special delicacy from the jungle that they want to share with you to eat. At this stage I have established a good rapport, so I usually accept it and tell them, "Thank you very much. I'll eat this later." Much later.

But at this stage I was still establishing a friendship so there was no way that I wanted to risk offending Wongomar. So I opened the banana leaf and there they were—still wiggling around, alive. I learned later that they were a certain kind of flying termite. I didn't know what I was supposed to do with them. I didn't know if I should throw them on the floor and stomp on them first and then eat them, or just what I was supposed to do. So I asked Wongomar, "How do I eat these?"

He replied, "Oh, you just take a handful and you pop them into your mouth."

I decided I would take just a little pinch, just a couple of
them. I looked at their beady eyeballs and their scrawny legs
and their wings still flapping around. I tried to get just a couple
of them, but they were all clinging to each other—a whole big
line of them. I tried to shake them loose but they would not
come loose. I said, "Oh, Lord Jesus. Please. I want to give thanks
for this food. Lord, I'm willing to eat these all up if you're willing to keep them all down."

So I popped them into my mouth and they really were not
too bad—once they stopped wiggling. They tasted a lot like
walnuts and I discovered later that they have a lot of protein in
them. You are probably thinking right now, "I could not do
that." I was thinking that too. But the God who helped me to do
it is the same God who would help you to do it. You could do
it. Those are the kinds of things that break through the barriers,
that win the people, that soften their hearts and prepare them
for the gospel.

The Green Belt

Nokiyan, with just a vine around his waist, came to the house
every day to teach us the language. Our house stands on stilts
because we work in the swamps where the river floods. In fact,
the whole jungle area floods six months out of the year so all
the houses are on stilts. He would come up the ladder of our
house, plop himself down in a chair and begin to teach us the
language. He was a good teacher. When he was finished, after
three or four hours of teaching, we would usually give him
some salt, fishing hooks or the like for his pay. One day I said
to my partner, "Judy, instead of giving Nokiyan some salt, why
don't we give him a pair of trousers. Everyday he comes to our
house and, you know, it's kind of . . . distracting."

Among some used clothing we had available we found a pair
of Bermuda shorts that were bright green with a nice green belt.
We gave them to Nokiyan. He was so excited! It was the first
time he had ever held a pair of pants. He went down the ladder
of our house, got into his canoe and paddled across the river,
holding those trousers close to him. Then he walked through
the village telling everyone, "Look what the two white misses

gave to me!" He was impressed.

The next day he came back to the house and he shouted up, "Marilyno, *kara b+di y+t+n. Kara y+uw+s b+di y+i*—Marilyn, I have come. I am ready for work and I'm all dressed up in my new clothes."

I said, *"Wadega Kira*—Good, you come on up." So he climbed up the ladder and walked in the door. He was so proud as he strutted across the room and plopped himself down in the chair. Judy and I stared at each other in disbelief. He had on only the green belt! I told the Lord, "If you want clothes on these people you'll have to do it your way and in your time because I am going to just concentrate on learning the language." That was that. Thirteen years later, Nokiyan is still running around with that green belt.

Isn't it wonderful, though, that eternal life and salvation do not depend on what we wear or what we don't wear? Nokiyan became one of the first Christians in our village, one of the first older people to accept Jesus into his throat. For these villagers, the heart is not the center of emotion but the throat. You ask Jesus into your throat. The throat is the center of life. I had talked to him many times, trying to witness to him in his language, and one day I said, "Nokiyan, I want you to ask Jesus into your throat." He became very excited about that even though he could not read. We had very little of the Word of God translated, and yet in simple childlike faith he asked Jesus into his throat. There is something very powerful about the name of Jesus. There *is* just something about that name.

After figuring out the alphabet and working out the grammar and building our dictionary, we were ready to teach the people to read and write their own language. That was an exciting time. The Lord gave us twenty sharp fellows to teach. In about six months these young men were reading and writing for the very first time in the history of that tribe. From there I got a team of fourteen translation helpers to work with. All are now beautiful Christians and they have been with me from the very beginning.

When we went home to the United States on furlough, we left plenty of stationery and stamps for the villagers to write to us.

We told them, "You know how to read and write now. We want to hear from you." We were home four months and received over 650 letters.

The Word Comes to Hauna

We started translating God's carving, one verse at a time. As we were translating, I became very burdened about the older people in the village. The witch doctors, the leaders, were being left out. There were over fifty witch doctors in our village, one for every house. They are the most powerful men in the village and are always busy because there is always someone sick or someone dying. That is why there are so many of them. The life expectancy in this village is about thirty-five years. The infant mortality rate is ninety per cent.

One day I shared this burden with the translation helpers and said, "Somehow we have to reach out to these leaders— these big men of the village—with the Word, the carving. We must involve them in our work so that someday they too can become believers."

"Well," they said, "What can they do? They don't know how to read or write. They don't know anything about Jesus, or anything about this book or this carving. They know nothing about God. How could they possibly help us?"

I looked at those fellows and said, "When I first came here you knew nothing either. In fact, you didn't even know which end of the pencil to hold. If we aren't going to have some kind of an outreach to these older people in the village and to our enemy tribes that are all around us, then I might as well pack up my bags and go home."

I knew their problem. They were afraid of those older men, those powerful medicine men. But I challenged them to follow our pattern in Bible translation in which we work in teams of two. "I want you to go out as a team and go to each clan in the village and get one older man from each clan to join your team." There are four clans in the village so each team had to get four older men. They were very hesitant but they went.

To their surprise they had little trouble getting these older men to come. My partner and I wanted them to come from 6:30

to 7:30 every night, after we had finished translating, and listen to the verses that we had translated that day. Our goal was to listen and see if they understood it, to see if it was meaningful and whether we were using the right verbs and the right idioms. When these older men came in I met with them and said, "This is great. We are so delighted to have you to help us with this task. You are the leaders of the village. You are the big men. You are the powerful men. You are the men who understand the language best. These young fellows who are working with me, these fourteen translation helpers, they don't really understand the language that well." Of course, they were all nodding their heads in agreement, saying, "Yes, you really do need us." They were excited and later decided to label themselves "the judges."

Every day these old men came to listen to the Word of God. An important concept to know is that our word for *doctor* is *inkam hiiswoki*, which means "the man who spits" or "the spitter." These spitters are busy all the time. If you have malaria they take a sharp bamboo and cut your forehead where it hurts to let out the bad blood. Then they chew on a plant that supposedly contains a very powerful spirit, and they spit and blow into those cuts. This is the power that will help heal you. The same is true for pneumonia—they cut your chest and your ribs and then they spit and blow.

We had translated a portion of the Gospel of Mark, chapter eight. As we were reading these verses we came to verse 23. Those men were listening carefully when we read,

He took the blind man by the hand and led him outside the village. When he had spit on the man's eyes and put his hands on him, Jesus asked, "Do you see anything?"

He looked up and said, "I see people; they look like trees walking around."

Once more Jesus put his hands on the man's eyes. Then his eyes were opened, his sight was restored, and he saw everything clearly.

We have blind people in the village, but never ever has a medicine man been able to heal them. So when we read this verse those big men said, "He did what? He spit! He spit! And the blind man's sight was restored." Those big men rose from their chairs and said, "Wow." Then leaning over and looking at the carving they said, "Why Jesus must be the most powerful spitter in the whole world!" Right then and there those big men identified with Jesus. They went out into the village and went to the men's spirit house and spread the story throughout the village. "Do you know, Jesus spits! And he heals blind men and he heals all kinds of other sicknesses and diseases." What a tremendous breakthrough! From that day those big medicine men started coming to church. They wanted to know more about this man Jesus, the spitter.

I believe that that particular story in Mark was specifically for the Sepik Iwam people. God created them and God created their language. Jesus did not have to spit. That's rather gross. He did not have to spit. I believe that Jesus used that method because he wanted to reach the throats of the Sepik Iwam people. Today over half of the people in Hauna village are believers. Let's never doubt the power of the Word of God in our own language.

The Good News Travels

As we translated and as we taught the people to read and write their own language, we became very burdened for all of the enemy tribes around us. Hundreds of villages were untouched. Hauna was becoming a shining light throughout the swamp and the jungle in that area because people started to hear about the work that was going on in this village. One day a canoe loaded with fifteen people came. It had taken them four days to paddle to our village. They were coming for medical help. They spoke another tribal language and I was trying to communicate with them through the trade language, Pidgin English. It was very difficult because only one of them spoke it, and his proficiency was scant. They came into our house and the smell of their rotten sores, tropical ulcers and other diseases made me sick to my stomach. I have never seen anything so pathetic. I am not a nurse and the last thing on earth I ever wanted to be was a nurse, but I was overwhelmed by their needs. I told them that they must stay in our village at least a week so that I could give them a penicillin series to treat their sores. We fed them and they stayed for a week.

While they were there they watched what was going on. They saw 200 people coming to school learning to read and write their own language. They saw us carving on the banana leaf God's talk in their own language. They went to church and heard the people singing in their own language; and they listened to our own Sepik Iwam pastors giving forth the Word of God in their own language.

When it was time to leave Hauna and return home, the leader with whom I communicated asked, "Do you think you could come to my village and put our talk on the banana leaf so that we might know about God too?"

I had to shake my head and say, "I'm not finished here yet. I have several more years of work in this place. There is no way I can come." I could tell he was very disappointed, but before he left I promised that someday I would at least come and visit his village.

Several weeks later we organized a party to find that village. After traveling all day up a small tributary back into the mountains, we parked our canoe and walked through the jungle for about half an hour. We finally got to the village. The leader of the group that came into our village was thrilled to see us. He called everybody to come and see the two white misses. We were the first white people to come into that village, so we were quite a sight. As we were walking through the village I noticed in the center a new building, very different from their regular houses. I knew it had just been built and I asked, "What is that building there in the center of the village?"

He said, "Oh, that is God's house—that's our church."

"Your church? Do you have a mission here?"

"Oh, no, we have never had a mission here."

"Well, do you have a pastor here—you know, someone that comes to preach God's Word?"

"Oh, no, we've never had a pastor here."

"Well, is there someone here in the village that can read and write Pidgin English who holds services in your church?"

"Oh, no! There is no one here that can read or write. And we have no books."

I looked at him and said, "Then what is that building for?"

He said, "Well, we saw the little church in your village and our people decided to build a church too. Now we're waiting. We're waiting for someone to come and tell us about God in our talk, in our language."

I turned and started crying. I have never seen that kind of faith. Friends, out in the middle of the jungle stands that little church and today they are still waiting—waiting for someone to come and tell them in their own language about Jesus. There are over three thousand groups just like them, waiting to hear the Word of God in their own language. They are waiting for you.

18
Professing Christ in the City

George D. McKinney

A careful perusal of holy history reveals that God has an un-ending love affair with the city. From the biblical record of the building of the first organized population center by Cain, which he named after his son Enoch (Gen 4:17), to the building of planned cities in our own day, God's concern and love for the city is evident. In spite of the historical evidence that cities have been the scene of man's greatest sin of arrogance, pride and re-bellion against God, both the New and Old Testaments affirm that God unhesitatingly seeks to redeem the city and its in-habitants.

Abraham's plea for God's mercy on wicked Sodom would have been granted if only ten righteous persons could have been found. The great city of Nineveh was spared when Jonah delivered God's message to its king who led that city-state in repentance. Nehemiah, under God's anointing, left the high position of cup bearer to King Artaxerxes in Persia to lead a discouraged, defeated and disunified Jewish remnant in Jerusalem to rebuild the walls and to restore the city from ruins.

In the Gospels Jesus made the public announcement of the beginning of his ministry in the synagogue in the city of Nazareth. His first recorded miracle was in the city of Cana of Galilee. While much of the work was done among the rural folk who heard him gladly, he nevertheless was a familiar figure in the temple in Jerusalem and the synagogue and streets of the cities of Caesarea, Jericho, Bethany, Capernaum and Bethesda. It was in the city that Jesus met his greatest opposition. He wept openly because of the injustice, oppression and sins of the city of Jerusalem. After his arrest, he was taken to the city hall and tried before Pilate. His vicarious and redemptive death occurred just outside the city walls. After his resurrection, he returned to the city and appeared to five hundred astounded citizens. Jesus directed the disciples to return to the city of Jerusalem and to wait there for the promised infilling of the Holy Ghost, and instructed them to begin their ministry in the city. And from the city, they were to reach the world.

Cities Today

The reported death of God in the cities of America and the world is a false report. Those of us involved in ministry in the city are here to testify that in the city, as everywhere else, where sin doth abound, there doth grace much more abound: Christ the Wounded Healer is present in the concrete jungles, in the overcrowded, rat- and roach-infested projects, in the halls of justice and in the jails. As our eternal contemporary, he is wherever there is human hurt and suffering. Since God in Christ has never forsaken the city, neither must the church.

A careful examination of the urban scene reveals that the church and all social institutions are both challenged and threatened by conditions which are widespread in twentieth-century America. These conditions, though observable because of physical manifestations, are primarily spiritual and theological. They are basically problems arising because of man's rebellion and alienation from God, and man's pride, selfishness and contempt for others.

In recent years the church too often has been in captivity to the society for economic and social reasons and has not been free to fulfill its prophetic role of pronouncing judgment as well as proclaiming the way of healing and salvation. Rather, the church has too often simply reflected the prevailing social or political attitudes and values and has said by its silence and complicity that the majority and might make right. Consequently, the church in many cities of the United States has lost its effective witness.

In spite of the dismal failures of the past, the Christian church in America must apply to itself its teachings regarding forgiveness, healing and redemption before it can respond constructively to the desperate cry of the larger society for direction and meaning.

The city was the first laboratory for the testing of the power of the gospel to create a fellowship of love and forgiveness that cut across racial, language and socioeconomic barriers. The success of the experiment is well documented in the book of Acts. The Jerusalem church was multiracial, multicultural. Christians from Africa, Asia and Europe sat at the communion table together and had "all things in common." It may be that much of the distress in our cities is due to the abandonment of the cities by the many representatives of Christ, leaving only a remnant of his followers to do what Christ has commissioned all of us to do together.

Social diagnosticians, demographers and prophets of doom have researched the city and its problems, and they have all concluded that the major cities are in a serious state of decay. The diagnosis generally includes information that:

1. Great population shifts have resulted in loss of tax bases and the concentration of poor and ethnic minorities in the city.

2. Modern technology has rendered many old skills and jobs obsolete and created need for new skills and professions.

3. Automation and robots are replacing people, without concern for their future livelihood. Thus many people are rendered obsolete.

4. Environmental pollution is worsening (air itself is poison and hazardous).

5. The criminal justice system has broken down in face of increased crime, violence and lawlessness.

6. The delivery of necessary social services to the poor and the defenseless has broken down.

7. The educational system is breaking down, resulting in schools that do not teach.

8. There are other demographic changes: (a) Growing percentage of older people in the total population, and (b) growing youth population without skills.

In addition to this socioeconomic diagnosis of the urban problems, we must add a spiritual analysis. The city is the scene of spiritual confusion. Cults and nonbiblical religions have proliferated, often led by opportunists and spiritual pimps who capitalize on the ignorance, spiritual hunger and vulnerability of many city dwellers who have never heard the gospel of Jesus Christ. Moreover, there are those in the city, like their counterparts in suburbia, who heard the gospel but rejected its claims on their lives. Consequently, God's wrath is revealed against those who renounced his grace. Paul gives a clear statement of the spiritual conditions in the major cities of America today:

> For the wrath of God is revealed from heaven against all ungodliness and unrighteousness of men, who hold the truth in unrighteousness; Because that which may be known of God is manifest in them; for God hath shewed it unto them. For the invisible things of him from the creation of the world are clearly seen, being understood by the things that are made, even his eternal power and Godhead; so that they are without excuse: Because that, when they knew God, they glorified him not as God, neither were thankful; but became vain in their imaginations, and their foolish heart was darkened. Professing themselves to be wise, they became fools, And changed the glory of the uncorruptible God into an image made like to corruptible man, and to birds, and to four-footed beasts, and creeping things. Wherefore God also gave them up to uncleanness through the lusts of their own hearts, to dishonour their own bodies between themselves: Who changed the truth of God into a lie, and worshipped and

served the creature more than the Creator, who is blessed for ever. Amen. For this cause God gave them up unto vile affections: for even their women did change the natural use into that which is against nature: And likewise also the men, leaving the natural use of the woman, burned in their lust one toward another; men with men working that which is unseemly, and receiving in themselves that recompence of their error which was meet. And even as they did not like to retain God in their knowledge, God gave them over to a reprobate mind, to do those things which are not convenient; Being filled with all unrighteousness, fornication, wickedness, covetousness, maliciousness; full of envy, murder, debate, deceit, malignity; whisperers, Backbiters, haters of God, despiteful, proud, boasters, inventors of evil things, disobedient to parents, Without understanding, covenantbreakers, without natural affection, implacable, unmerciful: Who knowing the judgment of God, that they which commit such things are worthy of death, not only do the same, but have pleasure in them that do them. (Rom 1:18-32 KJV)

The biblical diagnosis of the spiritual climate in the city is also the correct diagnosis of the spiritual climate of suburbia. God is no respecter of persons. When the person with power, wealth and prestige rejects the lordship of Christ and the authority of God's Word, the personal, social and spiritual results are the same as when a person without power, wealth and prestige rejects Christ. The Bible says God sends strong delusions on both the rich and the poor, the minority and the majority. The powerful and the powerless believe lies and are damned.

Please note that the wealthy who reject Christ and the poor who reject Christ believe the same lie—that gain (wealth) is godliness. Greed and covetousness motivate both. When the godless, poor and greedy become godless, middle class and greedy, their behavior is hardly distinguishable from the old-line godless, rich and greedy. Thus, we conclude that the mere shifting of wealth from one segment or group in the society to another is not the solution to the problem of poverty in the city. There must be a spiritual change in the heart and consciousness—a new birth wrought by God that brings a new philosophy of wealth ("The

earth is the Lord's and the fullness thereof") and a new understanding of stewardship ("To whom much is given, from him
much is required"). Those who have wealth and power are
stewards, accountable to God for its responsible use and distribution. Believing this, we issue a Macedonian call to our
brothers and sisters who fled the city, "Come back and help us."

A second lie that many urbanites and suburbanites believe
is that God has forsaken or abandoned the city. The poor, powerless and disinherited demonstrate that they believe God is gone
from the city through their criminal behavior, hopelessness,
ruthlessness, murder, suicide, drug and alcohol addiction,
abandonment of families, trafficking in human flesh, and the
loss of reverence for all life and truth.

While the powerless unbelieving urbanite tends to turn in
upon himself and resort to self-destructive behavior, the powerful unbelieving suburbanite tends to deify himself and give
greater value and worth to his property and power than to the
life of the poor. Thus, he fails to use political and economic
power to deal compassionately and creatively with the problems of the inner city.

The profit motive dictates major political decisions, and financial interest overrules human interest. The poor and powerless in the inner city are often manipulated by various welfare
programs designed to perpetuate dependency and hopelessness. Also, there are so-called urban renewal schemes which
some astute observers have rightly called "poor people removal
programs." Among the latest manipulation scheme by the
powerful suburbanites is the regentrification movement.

Notice the arrogance of the term *regentrification*—the re-
population of the city. The use of the term suggests that those
who never left the city—the poor, the minorities, the powerless
are nonpersons. Are not the present inhabitants of the city
people? Are only those of Anglo-Saxon or European ethnicity
considered people? It is ironic that many who were responsible
for dissecting the city with freeways to facilitate their escape
to suburbia are now returning. The victorious escape from
ethnic minorities and the poor was a pyrrhic victory. In simple
terms, the two to three hours per day in bumper-to-bumper

traffic, burning gas that costs $1.50 per gallon, made the city seem attractive again.

Having discussed the sins of the powerless in the city as self-destruction and the sins of the powerful as self-deification and oppression, I ask, "Is there a word from the Lord for the city?" Yes, in the Sacred Book there is a word for the church in the city.

A word of judgment: In Galatians 6:7 we read, "Be not deceived; God is not mocked: for whatsoever a man soweth, that shall he also reap."

A word of instruction: In 1 Timothy 2:1-4 we read, "I exhort therefore, that, first of all, supplications, prayers, intercessions, and giving of thanks, be made for all men; For kings, and for all that are in authority; that we may lead a quiet and peaceable life in all godliness and honesty. For this is good and acceptable in the sight of God our Saviour; Who will have all men to be saved, and to come unto the knowledge of the truth."

A word of hope: In John 12:32 we read, "And I, if I be lifted up from the earth, will draw all men unto me."

Ministry in the City

Since God has not forsaken the city, the church must maintain a dynamic, compassionate servant ministry there. While the church holds to eternal and unchanging principles and truths, the application of the principles and methods of interpreting truth must have existential meaning. The church must maintain and preserve all that it can of its past liturgy, its worship, its teaching and healing ministry. Yet it must not be afraid of using new forms, methods and categories to proclaim the good news that God was in Christ, that the kingdom of God is come in Jesus of Nazareth, that Christ is Lord and he is Emmanuel —God with us—to show us all a better way.

Now the call to ministry in the city, like every call of God to ministry, is initially a call to prepare for service. The suffering servant must be "thoroughly furnished unto all good works" (2 Tim 3:17). The equipping of the servant must follow a baptism of love which will enable the servant to represent Christ in a multicultural, multiethnic, multireligious environment. This baptism of love will prepare the servant to remain faithful in the face of violence, hostility and other Satanic forces. The inner city provides an excellent opportunity for the Christian soldier to gain wounds for the cause of Christ. Here is an opportunity to get to know Christ in the fellowship of his suffering and to imitate Christ as a wounded healer.

In recent years God has raised up strategic training camps for preparing workers for urban ministries. These training facilities are already making a significant contribution in the spiritual, theoretical, practical preparation of workers. Special note is made of the Rev. John Perkins and the Voice of Calvary in Mendenhall and Jackson, Mississippi; Rev. Tom Skinner and Associates in New York; the late Bishop Jessie Winly of God's Soul Saving Station in Harlem; the Rev. Reuben Conners of Black Evangelism Enterprises in Dallas, Texas; the Rev. Lloyd Lindo in Chicago; the Rev. Don Green of the San Francisco Christian Center. Simpson College in San Francisco under the direction of Dr. Craig Ellison has initiated an academically bold and innovative program that is theologically sound and evangelical and at the same time sensitive to the multicultural, socio-economic condition in the inner city. And there are others.

At St. Stephen Church of God in Christ, where I have been the founding pastor since 1962, we have seen the good hand of God prosper a work in the inner city. The congregation was organized with seven people in southeast San Diego, an area in transition. White flight had resulted in a leadership vacuum. In an area with approximately ninety thousand citizens, there was an influx of minorities, primarily Afro-Americans and Mexican-Americans. These newcomers were generally poor, unemployed or underemployed with limited skills and education. As an eyewitness to the community transition that took place during the decade from 1959-69, I observed that not only did the established churches abandon the area, but so did major grocery chains, some doctors and health-care professionals. Even the Little League program was moved.

St. Stephen responded to the needs of the neighborhood in transition and began to develop a ministry to the whole person. This ministry to the whole person has three broad emphases:

(1) preaching, the proclaiming of the Word of God; (2) teaching, the explaining and application of God's truth to the human situation; and (3) healing, the work of the Holy Spirit in the life of the believer and community of faith. Hereby the Word becomes flesh and "dwells in the ghetto" for healing and reconciliation.

The Central focus of our ministry is to win the lost to Christ, nurture the faithful and make disciples. We have worship in the church on a daily basis and Holy Communion on Fridays. We also conduct worship services "on location" for shut-ins and sponsor small groups for nurture and Bible study.

We have educational programs including a traditional Sunday school plus a Christian school—kindergarten through twelfth grade with college courses added in cooperation with local colleges. Family life must be emphasized in ministry in the city, for there the family breakdown is devastating. So we have couples conferences, singles conferences and other meetings for those who are engaged and those who are married. And we have seminars and workshops and drama activities, family camps, vacation Bible schools and summer camps. All of these tend to strengthen the fabric of the family in the inner city.

Special ministries are necessary for army camps, prisons, high schools, elementary schools, colleges and universities.

There are, as well, street ministries to those who are completely down and out. Every Friday night when the action is going thirty or forty of our young people go out in teams of twos to the red-light district and the bars and places where flesh is being peddled. Just their presence is a significant witness to God's love because many of them have come from that situation and they are going back. It is not uncommon for gangsters to bring in their dope—heroin and the needles and the spoons—on a Sunday or during the week. One Sunday we were frightened to death when somebody brought his sawed-off shotgun to church, turning it in because he didn't need it anymore. The gospel is reaching those in the city.

We also provide emergency food and housing and counseling for families, drug abusers and alcoholics. We have job referral services and we train people for lay ministry in the city. We even have an outreach into Mexico and are helping to plant churches there.

Finally, we run a bookstore and a couple of halfway houses for rehabilitation. Then we also have the usual scouting program and family activities.

Indeed, God is not abandoning the city. In spite of moral and spiritual decay and socioeconomic deterioration, in spite of every prophet of doom's statement that the city's sickness is unto death, the church must believe that even if there is spiritual, moral and economic death in the city, our God is able to speak life into it and call for resurrection in the city.

A spiritually dead and doomed Nineveh received God's word through Jonah, repented and lived. A desolate, disunited, despairing remnant in Jerusalem heard God's word through Nehemiah, and the walls and the gates were restored, and the city of Jerusalem came alive again. The Sanballats, Tobiahs and Geshems couldn't stop the movement of God. In more recent times, England was thrown into a deep spiritual darkness. Sin was rampant and the church to a great extent had become God's frozen people. But God's word came through John and Charles Wesley until millions of hearts were strangely warmed and revival fires were ignited in the cities and hamlets throughout England. The course of history and the life of a nation were changed.

The destiny of our nation is inextricably bound to the destiny of our cities. God hears the cries and the hurting of the hopeless in the city. He sees the growing tensions between the powerful and the powerless, the haves and the have-nots. He knows the sin and the uncleanness that prevails there. The only power to save the city is God's redemptive power. And he uses human instruments.

God has asked a simple question. Who will go for us? Jesus gave a commission. In Acts 1:8 he has directed and commanded that we shall be his witnesses in the city of Jerusalem. We must go to Jerusalem, that great city that is explosive with religious bigotry and sectarianism. Jesus says go back to that city and preach the gospel of reconciliation, restoration, deliverance and healing, the gospel of power and hope. Go to Samaria with its riots and racial strife, with its slums and ghettos, with its immorality and sins, its poverty and filth. Go into the byways and the hedges and tell them that the good master has prepared a supper and has invited the poor and the disinherited and the halt and the blind to come to be blessed.

Finally, Christ has told us to go to the uttermost parts of the earth. Let neither sea nor mountain, language nor custom, color nor race, suffering nor sacrifice hinder you. Don't let "nobody" turn you around. Take the message of Christ everywhere.

19
Acknowledging Christ in a Suffering World
Eva den Hartog

Luke 9:23-24 says, "If anyone would come after me, he must deny himself and take up his cross daily and follow me. For whoever wants to save his life will lose it, but whoever loses his life for me will save it."

A few months ago, I was reading some old newspaper articles, and I found an article in which the students of the University of Texas described and characterized their own times. The words they used to characterize their own times were these: apathy, dullness, deadness and crisis. Nobody spoke of the joy and the hope of living. One of the students said, because he was facing unemployment, that humanity would either annihilate itself or that we could look forward to increasing bureaucracy, dull jobs, a lower standard of living, and what he called "being packed in ever smaller boxes."

One of the professors said, "It takes but one glance at my students to see that they are not happy and that they are in a

sense empty. They have no ideas, no passions, no dreams and
no vision."

We live in a depressing time, and nobody can promise us a
decent job or the end of inflation or better TV programs. We
live in a world of evil passions. We live in a world of racial
hatred, jealousy, selfishness, envy and greed, mistrust and
suspicion, the threat of a third world war. Every hour the world
spends thirty million dollars for weapons and defense.

We live in a world of pain and suffering and desperation. We
live in a world where two-thirds of the world's population has
not enough to eat. This same two-thirds has only fifteen per
cent of the world's income. And we have eighty per cent of the
world's income. Twenty million people, mostly children, die
of starvation each year. Millions of children are mentally disturbed and physically handicapped due to a shortage of nourish-
ing food. For millions of people, life is short; hunger is daily;
disease is rampant, and hope is in short supply.

Meanwhile, one-third of the world's population, and you
and I belong to that third, have too much on our plates.

Being now in a part of the world where I am well looked after,
my thoughts go to the hundreds of refugee camps in the poorest
parts of the world where I have worked. In my opinion we are
living in heaven and many millions who are living in inhuman
conditions are living in hell.

We say today in the United States that there are no more
slaves. But I think the millions of people who are living as rats
in pitiful conditions in refugee camps are like millions of slaves
without names. And it is my personal experience that it is easier
to write about and to speak about a hungry world than to satisfy
a hungry world.

On the Thai-Cambodian Border

Let me tell you a little bit about life in a refugee camp. For many
people, those who live in refugee camps in Thailand, for example, the situation is desperate. There is unbelievable suffering in the refugee camps. I have been utterly horrified by
the conditions in which the people were living. I could not
understand and I cannot understand today how man's inhu-

manity to man could be so extreme. I have seen refugees living in tents, in the streets, in the railway stations, in the open air; people dying of starvation, dying of communicable diseases, dying of great epidemics. I remember refugees begging for food, mothers sitting on the roadside or walking with their skinny children lying limp in their arms. I will forever think of children going blind, covered with sores that will not heal due to a lack of vitamin A and other vitamins. I think of the babies I have delivered under trees, in a railway station in Calcutta, in the refugee camps in tiny pathetic shelters.

I can never forget the lonely children whose fathers and mothers committed suicide. My mind remembers the fathers digging little graves for their children; or Cambodian children shaking with fear after seeing their mothers and fathers killed and tortured before their eyes.

I will never forget the boys and girls who came to our house in Bangladesh during the famine when millions died. They came offering themselves for adoption because they knew that we were adopting babies. They were shivering and shaking; they were fearful of being sent back into the streets to beg for food. There were people standing in line for hours in the pouring rain and in the tropical sunshine for a cup of watery milk or a little bit of rice or some water. Do you know what it is like to have no water at all during the day? Once in Cambodia I saw in one village some pathetic little plastic bottles with a bit of dirty water, and it was all the water that these people had.

In Somalia in the refugee camps there are three to twelve liters of water a day for the people *if* the truck that brings it has not broken down. And that three to twelve liters of water is for everyone's washing, bathing, drinking and cooking. In the United States and Western Europe, in all the rich countries, we have all the water we need—not just enough for drinking, washing, cooking and bathing, but extra. Are you counting your blessings? One by one? They are many. Yet millions of people are dying of starvation, living in tiny little shelters built of rags, old blankets, iron, plastic, anything that can be found. Outside these little shelters there is a stench of cholera, open latrines, dysentery.

Christian mission is not only a spiritual mission, but also a social one. Feed the hungry and not the full. Clothe the naked and not those who are well dressed. Give the shelterless shelter. Heal the sick; protect the oppressed and help them to help themselves. That is our mission.

But how do we do this? Some young people asked me if in my faith I was always on a mountaintop. Let me tell you, there is no missionary in the world who has not been in the valleys. I have been there often. My faith has been on trial many times. I've felt spiritually depressed, lonely, doubtful.

One morning while standing in the pouring rain looking on a mass of suffering people, seeing mothers sitting on the roadside covering with their naked bodies the babies on their laps, protecting the babies from getting cold, not realizing that they would probably soon be dead of starvation or cholera, I said to myself, "Does God really care? Is there a living God? And is there a God of love in the world? Is God in the Far East? Is God in the refugee camps? Is God in the concentration camps? Why is God not doing something about it? Why do we have so much and they have nothing?" I came one day to the conclusion that good was only imaginary. I was in despair, spiritually depressed. There were moments for me when I came to the end of myself and moments when I came to terms with Jesus Christ. I was facing forces that were too difficult for me to understand. And I learned in a tragic way that trusting Christ is not only believing, but also obeying: doing his will in my life. He wanted me to be there.

My question to you who call yourselves born-again Christians is, "Can you see beyond the barrier of your comfortable, luxurious lives? Can you be more aggressive as Christians to meet the tremendous needs in the world?" Many of us are deeply touched by everything we hear, by the tremendous suffering of the people in refugee camps. We are shocked when we hear stories of Christians who have been tortured and killed because they confessed that Jesus Christ is Lord of their lives. But I wonder, are we really deeply touched, have we shed real tears for the suffering in the world? We who live in countries where there is freedom of religion, freedom of expression, political freedom, find it difficult to understand what it means to be unwanted, without a country, without a future, forced to live in refugee camps in inhospitable surroundings.

There are countries where an open profession of Jesus Christ may lead to loss of work, loss of country, loss of liberty, loss of loved ones or even loss of life itself. People today in many parts of the world, not only men and women but even boys and girls, have faced death for a refusal to deny their Lord. In one of the refugee camps on the Thai-Cambodian border where I was with my medical team, there were Communists and non-Communists herded together in pitiful conditions. That caused many problems. Non-Communists were faced with the destroyers of their country, the killers of their loved ones. Many of them felt traumatized by the cyclone of murder that had swept through their country. Many showed signs of despair. The stories they brought were horrifying, and the expressions on their faces and in the eyes were shocking. They showed fear, hate and despair.

And did you hear about the child soldiers in Cambodia? Young boys and girls in their early teens made up part of the army. Once I was stopped by some of these young soldiers when a television man who was interviewing me called me "major." And the children asked, "What are you? Are you a major of an army?" And I had to explain that I was a major of an army of peace and that I was coming to help them. But these boys and girls, who were carrying guns, rifles, antitank explosives and cannons, were interested only in protecting themselves and the people in the camps. If you ask these children why they are killing, do you know what they say? "They have destroyed our country. They have killed our loved ones. If we do not kill them, they will kill us." These few years of fighting have made of these boys and girls young men and women without any warm, loving feelings.

The members of my team in Thailand were all Christians, and in spite of our different backgrounds, we were determined to put God's love in action. We came from the Salvation Army, the Assemblies of God; we were evangelicals, Southern Baptists, Baptists. You could not name a group that was not there.

We never mentioned our church background; we focused on
preaching the gospel, putting God's love in action. It was a wonderful thing to see thousands of people coming to Christ; Communists and non-Communists coming to Christ and forgiving
each other. They did not even think of the possibility of being
killed by other Communists. They started going out to preach
the gospel to other people, and they brought people in and asked
us to pray with them. It was a miracle. It was also a tremendous
experience to see doctors and nurses who were not only giving
medical care, but also spiritual help. I once saw a doctor in my
team kneeling in the mud and dirt near someone who was dying,
praying with the dying person.

When Jesus Christ came into the heart and lives of these Cambodian people all their fear was gone. They knew that with
Christ they could pick up the pieces of their broken lives and
start anew. What an experience to see frightened, broken-
hearted, despairing people changing to exciting, happy, smil-
ing people.

If I was ever unsure of the truth of the living Savior in my
life, I was sure of it as I saw it in the change that came over these
newly reborn refugees. There was a change in their eyes, voices
and faces. I think of Danny. Forty-six of her relatives were bru-
tally killed or died of starvation. When she came into camp she
said, "As long as I live, I will cry." Danny not only found Christ,
but was moved by the Holy Spirit to dedicate her life to preaching the gospel and teaching the Bible. Forgiveness stood central
in her life from the moment she became a Christian.

How was this possible? Because the love of God was over-
flowing in her heart and in her life. Where love is, there is no
place for hate. Love is never going without giving, but love is
never going without forgiving also.

In many parts of the world there are weak Christians. Two
of the biggest reasons for this are the absence of forgiveness
and the lack of unity. Booker T. Washington once said that we
must fear more the weakness of the Christian than the strength
of the Communist. And Ghandi said that if Christians had lived
as the Bible tells us to we would have won India for Christ.

These newly reborn Christians in Thailand realized that

the Bible was not written for sightseers but for those who were
tired of the wrongs in their own lives, in the lives of others and
even in the lives of their enemies. They taught me that where
there is forgiveness for each other, there is love for each other.
They realized that burning love must be the ruling passion of
their lives—not a cold, half-hearted love, but a changeless,
quenchless, burning love, because love, only love, can save the
world from more disasters.

In Zaire

Let me tell you one of my personal experiences in Zaire. One
day I was asked by the authorities to go to an area outside Kin-
shasa. It was during the struggle for independence. I was a little
scared to go, but they gave me papers and said everything would
be all right. Smallpox had broken out in one of the villages where
the Salvation Army had a mission school. I departed early in
the morning, and I promised my leaders I would be back before
dark. It was only thirty miles out of town. In the morning the
soldiers were very nice to me, and I got to the station early. We
worked hard. In the afternoon I started back in my car. I went
through the first four check points and then came to the fifth. A
soldier came and stopped the car and said, "Your papers!" So I
gave him my papers. And he said, "You are arrested. Your papers
are not in order." He was holding the papers upside down. But
I couldn't tell him he was wrong because he might get excited.

On the other side of the road were fifteen soldiers. Their beer
bottles (it was Western beer) were empty. And they had smoked
a kind of opium, and their eyes were popping out. They were
not in their senses. I was sitting in my little car, and the soldier
came with his rifle and said, "Don't move! I shoot!"

I was trembling with fear. I was so scared, and I think I was
in such a panic if I had had a gun in my car. . . . As I was sitting
there a little hate came and then there came more, and I thought,
"Those guys have killed so many people; they've killed so many
fine Africans and missionaries and now they're going to kill
me."

And then I imagined someone saying to me, "Hey, you hate
them. You who are preaching as a Salvation Army officer, as

a missionary, with the S on your collar. You are preaching that
you want to serve God. You love your enemies. But you are not
loving your enemies now."

So I sat in that little car and I prayed, "Oh God, please forgive
my sins. Give me that love, that burning love, and I will go where
you want me to go and I will do what you want me to do. I will
work with these violent people or I will go somewhere else if
you want me to. Lord, if they want to kill me please let them not
rape me. Please let them not torture me. Let them shoot me
straightaway. I am scared." And as I was sitting in that little
car and praying it was as if somebody came to me with peace.
I felt the rifle of the soldier. The fear was not gone, but I was
not alone. I felt that I was in a dark tunnel, but there was someone with me and I could sing, "He walks with me and he talks
with me and he tells me I am his own." Sometimes after a church
service when I speak the people come and shake my hand and
say, "Oh, that was tremendous. God was with us." And I always
wonder, God was with you? Where was he before and where is
he after the service? Is he not there? He lives within us and walks
with us and talks with us along the narrow road. And that day
I felt that I was in a dark tunnel, but at the end of the dark tunnel
was God's glorious light: maybe it was the chance to do a better
job with my life or to be in heaven.

Suddenly I heard a jeep stop behind my car. Two heavily
armed soldiers with stars on their collars came out. I knew they
were officers. One went to the soldiers on the other side of the
road and one came to me. And he said to the soldier, "Go away,
I want to speak to the lady." "What are you doing?" he asked me.

I said, "I'm arrested."

And he said, "Do you know what that means?"

"Yes, sir. That means being tortured, being raped and being
killed and I'm scared."

He said, "Don't panic, don't panic. I will go to the soldiers
on the other side of the road." So he went to the soldiers on the
other side and told them that missionaries are good people. He
came back with my papers in hand and gave me my papers
through the little window of my car. "Here are your papers,
he said. "Nobody will kill you. They will first kill me."

"Sir, thank you. You saved my life." He looked at me, and I said, "God bless you."

A big smile came across his face and he said, "When I was a little boy in the bush of Africa there was a woman missionary who taught me to write and read and speak French (which he did beautifully), and she said to love God, love my neighbor, love my enemies, love everybody. Today I feel that God is putting my faith to the test, and I am putting my love for God in action."

The faith of that soldier influenced his thinking, determined his actions and shaped his whole life. When I think of that soldier I still see in his eyes the light of God; I hear in his voice the authority of God, and I know that in his soul was the passion of God. The faith of that soldier was implanted in him by a woman missionary. (I'm proud that it was a woman. The founder of the Salvation Army said, "My best men are women.")

That man became a man of very strong convictions for which he was prepared to live or die. He was certain that the God in whom he believed would stand by him, whatever the cost may be. God was able to steer the hearts of the rebels.

In these years of ongoing disasters the reputation of Jesus Christ is at stake. His reputation is in your hand and in mine. During my years of Christian service, I have become a woman with very strong convictions. I know that behind me is a power that nobody can take away from me, even the devil. That is the power of the Holy Spirit. We who call ourselves Christians are all being watched, observed. Everything we do is of tremendous importance in the world today. The situation in the world is our concern. Time is desperately short. We must act now or be prepared to witness many greater disasters than ever before in world history.

What to Do

We should ask ourselves, "What can I do to make the world a place in which everyone can live in peace? What can I do?" What can I do to save someone from starvation? We should start in our own communities. I read once in a pastor's manual that there are more barmaids than college girls. There are two liquor stores for every church in America. A hundred thousand young boys and girls enter "white slavery" (prostitution) every year in the United States. We want our boys and girls to have pure, intelligent minds, yet we allow millions of filthy magazines to be published in this part of the world. And our young people read them. They should be reading their Bibles, committing themselves to God's rules and God's love.

Do not make the mistake by feeling guilty because you live in a rich country. Living in a country of abundance is not a sin. God said in his Word that Jesus' ministry was to heal the sick and feed the hungry. But what else does he say? He said he would give his people a land of milk and honey and affluence if they kept his commandments to love him, love their neighbors, love the unloveable. The unloveable are found even in your surroundings. Don't withdraw from them. Love them. Pray for the world situation. Pray for the people who are hungry and for those that have to work in that situation. That's what you can do.

You can also change your lifestyle. Arthur Simon says in *Bread for the World*, "Lay aside your pattern of living. Consume less, waste less, eat less, drink less, drive less, air condition less." And I tell you: fast once a week and give that money to an organization or church that is involved in effective world hunger projects. In short, live more simply so that those who are dying of starvation and disease may receive food, medical care and life itself.

Christians in the Western world still have a chance to show our compassion for our neighbors by significant actions. But unless you and I commit ourselves to concrete plans for increasing giving I feel that God will come one day and find that we have all died from suffocating luxury. Our giving should not be giving out of our surplus, but sacrificial giving.

Ask yourself, How can my life be used to the glory of God? and for the good of humanity? It is not just some of your money that God wants. It is not some of your time, not some of your talents, not some of your possibilities. It is *yourself*. All you have and all you may become. That is sacrificial giving. God does not promise you an easy time or freedom from suffering, disappointments, misunderstandings or danger. But you will never carry your burdens alone. That is my experience. You never face anything alone if you are committed to the Lord.

Our Bible reading was about losing our lives for Jesus' sake. Are you and I ready to do that? If each of us here today has the courage to acknowledge the love of Jesus Christ in this suffering world, we will be able to deal with a suffering, dying world, to pray for a suffering, dying world, to give to a suffering, dying world and to go to a suffering, dying world to preach the gospel of love in words and in deeds because God's love can save the world from disasters. And if men and women will submit to his rule, God will break down the barriers between classes and races, between the rich and the poor. God can stop wars and create racial equality, economic justice, political freedom and, what is most important, a world of peace. And we can be his instruments. May God bless us with his task.

20
Student Witnesses around the World
Chua Wee Hian

I would like to present to all of you several word portraits of the family of the International Fellowship of Evangelical Students (IFES). Our first picture is dated 1946. In the background are the tall spires and towers of Oxford University. And there leaders from nine member movements are pledging together to form a worldwide evangelical fellowship. The following year, 1947, the scene changes. We are looking at the yard at Harvard University where these member movements ratified the formation of an international body.

The dream of our founding fathers was that on every campus, every university on planet Earth, there would be groups of committed students who would live and witness the power of Jesus Christ. And over the last thirty-five years this dream has been fulfilled. There are now seventy-five national movements that are linked to IFES. Together we seek to pool our resources and manpower to pioneer work, to establish Christian fellowships in another thirty countries. Let me take you to this picture gallery of IFES. Artists would need large canvases—photographers, wide-angle lenses—to capture the size and activity of our groups in West Africa.

Africa

Let's look at snapshots of Nigeria. There you find Christian fellowships of over a hundred members each. There are over one hundred student groups in Nigeria, involving nearly 15,000 students. This year when they held their annual conference, four thousand students were there. These students are deeply committed to aggressive evangelism, disciple making and to building up God's people.

Let's go to Ghana. There we see pictures of students sharing their faith fervently with their contemporaries. Last year at the Universities of Legon and Kumasi two evangelistic missions were held. At the end of those intensive days of evangelism five hundred students became Christians. We see reaping in Africa.

Let's go to Zambia. About fifteen months ago the student leaders in Lusaka were praying that God would use them to speak his word to the leaders of their nation. So they prayed; they also wrote to the president of their nation. President Kenneth Kaunda graciously invited them to conduct a Sunday service at the presidential palace.

A team of students was being prepared and trained by the national staff worker Derek Mutungu. And so on Sunday morning this team went to the palace. They sang and gave their testimonies, and the student president preached from the book of Amos. He spoke of righteousness exalting the nation. He called on the president and others to "let justice roll down like waters, and righteousness like an everflowing stream" (Amos 5:24 RSV).

The president himself was deeply moved. There were tears in his eyes, and later he said to this team, "If there are more young people like you in Zambia, we have hope for our nation." And turning to this team he said, "Look, I'd like all of you to come back again. I want you to hold another service in this palace. And next time I'm going to invite all my cabinet ministers, all their permanent secretaries, and you preach the same message to them."

In Africa many of these students who hear the Word of God go back to their villages and towns to proclaim that same Word.

The Middle East

Let's now go to the Middle East, to Israel. In September it was my privilege to speak at the second national conference held on Mount Carmel. At the time of the last Urbana there was no organized student work in Israel. But today, praise God, there is an emerging fellowship. It was a joy to see Arab Christians and Hebrew Christians praying together, reading God's Word together, holding hands together to go forward to proclaim the gospel of Jesus Christ. The gospel has triumphed. We thank God for that.

In neighboring Egypt we have experienced a significant breakthrough. A small team of workers built bridges to various churches in Egypt. Last year one of the bishops of the Coptic Orthodox Church invited one of our coworkers to prepare Bible study materials for all the young people in his diocese, and in addition he was asked to train the youth leaders in Bible study methods. This worker wrote to me in my office in London saying, "Wee Hian, please send all the Bible study materials you have. It's a tremendous privilege to teach Coptic Christians how to read and obey God's Word." This is a very significant move because there are over six million Coptic Christians in Egypt.

Europe

Let's go now to Europe. As we look at this portrait gallery of Europe, we see on one hand large student groups in Britain, in Germany, in Norway, in Finland and in Sweden. And yet in Latin Europe—in countries like France, Portugal and Italy—the groups are still small and struggling, but, thank God, he has now rallied and mobilized Italians, French, Belgians and Spaniards who are committed to him, who want to see their movement grow.

We have also developed work among international students

in France, in Germany and Italy. But I want to bring our minds and hearts to a country which is occupying the headlines of the world press—Poland. This year we have been able to send several Bible expositors to minister to camps and conferences in Poland. This summer we were able to invite several Polish students to our international student conference. In September, at our Oxford Training Conference for staff, we were able to welcome five Polish graduates. Four were studying in theological seminaries; one was a research student. I'll never forget the research scientist from Gdańsk. He said to me and to many others, "Please, please give me paper; give me paper to print Bibles because my people are hungry for God's Word."

We weren't able to fulfill his wish; we didn't know that freedom was going to be limited so soon. But we were able to scrounge around and find him Bibles to take back to his city. We said to him, "Look at our booktable and take all the commentaries you want." And he took quite a few IVP commentaries. And then, just after the clampdown in Poland, we received a cryptic note from him and his friends. It read something like this: "Stand and pray with us in solidarity. We have God's Word. He will never leave us or forsake us. Brethren, pray for us."

How good to know that some of our brothers and sisters in Poland have been trained and equipped in God's Word! They have Bibles and they have Bible study materials which they can now use. We are so glad that we have been able to help our Polish brothers and sisters.

Latin America

Let's move on along our picture gallery. This time we're going to see some rather unpleasant and even gruesome portraits. El Salvador—a land of suffering. There we find pictures of bloodshed. We can almost hear the machine-gun fire in the background. In that country IFES staff worker Mardoqueo Carranza ministers to a growing group of students. Together they have to mend broken lives by the power of God. They seek to provide relief aid; they try to heal the wounds of refugees and comfort those bereaved of loved ones.

In Brazil, the largest nation in South America, there is a thriving student work. Sometimes the ABU, our national movement, receives news of Roman Catholic students in high school suddenly becoming committed Christians and saying to the staff workers, "Come and teach us to read the Bible; help us to grow." There's a great hunger in that land. We have been able to lend to the Brazilian movement Bill McConnell, who used to be a staff member of IVCF-USA. He has been able to help them get a literature program off the ground, and recently many important books and commentaries have been published in Portuguese both for student work and for the churches.

Let's turn our attention now to another country—Bolivia. Now we see only the snow-capped mountain peaks in La Paz. Underneath that picture we see small portraits of four different people. First there is Felicity Houghton, an English missionary whom God used to start student work in Chile. Then there is a picture of Marcello. He was converted while studying in Brazil. He is from Bolivia, but he wants to go back to his home country to pioneer student work. The third picture is of a Norwegian family that is closely linked with IFES and our student ministry. And finally there is a Scottish missionary now training in Lima, Peru. Four of these great friends will join together in the next few months to go to La Paz, to Cochabamba, to start a student movement in that land. Will you please pray that in God's good time a strong student movement will be established in Bolivia?

Asia

Then let me take you to Asia. There are many, many portraits. One of the exciting trends in Asia today is this: God has been calling Asian missionaries to take the gospel to many lands. Many people have caught the missionary vision and received the call through missionary conventions and meetings organized by our member movements. And now they are scattered throughout Africa, Latin America and other parts of Asia. I could tell you thrilling stories of Korean and Japanese and Chinese missionaries working in Africa or in Central Asia. For example, some people in Nepal were very surprised when they encountered Japanese missionaries for the first time. They were surprised because Japanese were known only as the people who sold Hondas, Datsuns, Sanyo and Sony products. But instead they went to Nepal with the treasure of the gospel. They worked hand in hand with missionaries from the West.

After this quick tour of the IFES portrait gallery, let us return to ourselves and to our responsibilities. What should be our response to these real-life portraits? For a start, we could pray for our brothers and sisters in other lands. Let's identify ourselves with our brothers and sisters in Poland. Let's pray for those who struggle and work in difficult lands where Islam and Hinduism hold sway. Let's praise God for the vibrant African groups. Let's thank him for the new spirit of missionary commitment among the Asian movements. We can pray intelligently using the Praise and Prayer letters published monthly by the IFES (10 College Rd., Harrow, Middlesex, HA1 1BE England). We will be pleased to enroll you as a prayer partner. Your IVCF chapter can adopt a staff worker of the IFES or one of our national movements by giving, by caring, by writing to this worker or the movement. Then we also invite you to go and spend three to four years studying and working in a country of spiritual need, and at the same time helping in our student work. Let us work together in prayer, in giving, in going, so that the whole student world will confess Jesus Christ as Lord.

21
Seven Witnesses Tell Their Stories

During the convention, seven people shared briefly how God has heightened their awareness of his calling. Their testimonies are reprinted here in alphabetical order.

Sir Frederick Catherwood is a member of the European Parliament who lives in Cambridge, England, and helps with the work of the International Fellowship of Evangelical Students (IFES). Eldon Claassen is a nurse who has served in India, East Africa and Thailand. He is now a student at Case-Western Reserve University. Elizabeth Cridland, a former missionary to the Sudan, Somalia and Kenya, works with the Bible Alliance Mission which provides the Bible on cassettes.

Jack Kopechek, a medical-school student, spent a summer working in Malta and hopes to be a medical missionary. Naty Lopez, a Filipino nurse, serves as coordinator for Nurses Christian Fellowship International (NCFI) in the Pacific East Asian region.

Dwight Nordstrom is a graduate student at Fletcher School of Law and Diplomacy who is looking forward to "tentmaking" work in mainland China. Dr. Charles Smith, a radiologist who has done short-term missionary work in Israel, the Ivory Coast, Liberia, India and Pakistan, has written a booklet on missions entitled *What If I Don't Go Overseas?*

Sir Frederick Catherwood: IFES

When I was your age, I asked the Lord if he wanted me to be a missionary, and he said, "No. You don't have the qualifications. You don't have the patience, for instance." And I was bound to agree that it was so. Instead, he called me into industry and then into politics. But recently, especially in the last ten years, he has allowed me to get into overseas missionary work, into student work through the back door.

Now I would like to tell you why I see this student work as by far the most important work I do. As a politician I have been around the world. As a businessman I have been jetting around the world since jets were invented, and before that in propeller aircraft. I've been to China, southeast Asia, India, the Arab oil countries, as well as this great continent of North America. I've worked with food aid, development, famine relief and trade, and I am convinced that the future of every country I visit does not depend on political systems or economic systems. It depends instead on the tiny minority of dedicated Christians, especially those in leadership roles such as pastors, teachers and businessmen. They alone understand the truth of our God's creation. They understand about man's relationship to God and man's relationship to other people. These are the leaders of the Christian church, that church which is there to preserve their society. The Christian church is the salt flavoring the world and the light chasing away the darkness.

All these Christian leaders that I meet have come through universities. So in concentrating our evangelistic and pastoral work in the universities, which is what the International Fellowship of Evangelical Students does, we are concentrating on the future leadership in countries around the world. We are focusing on those who will lead their societies—in ways that are right and good and true.

When these students, these future leaders, return to their home countries, they cannot be expelled for cultural imperialism; they are citizens. They do not need expensive furloughs or elaborate fund raising. They earn their own keep. They do not suffer culture shock. They put the Christian view into the idioms of their own native language. You cannot therefore make more of an impact than you do when you give to student work overseas. And, you will be giving your money economically—since we are simply in the pump-priming business.

The IFES does not have an elaborate and permanent mechanism to cover every country to which we go. We instead bring into being new movements and make those new movements self-sufficient and committed to doing the same thing in neighboring countries. And beyond that we are there to help the weaker movements and to keep them in touch with the stronger ones through conferences such as the annual international student conference where my wife, Elizabeth, and I often act as house parents.

In addition, we send staff to some countries who cannot come to us. Chua Wee Hian, our general secretary, went by invitation of the Lutheran church this year behind the Iron Curtain to talk to students in East Germany. Two hundred students attended a prayer meeting there. These students have had to decide whether they will go to the university and compete for the five per cent of the places that are the only ones available to those who are not members of the Communist Party. They do not feel that they can belong to the Party because it is atheistic.

Students from Hong Kong are beginning again to visit students in mainland China, and they have a radio program beamed into mainland China and are getting an increasing flood of correspondence from listeners.

This year we had a large group of Polish students at our international conference, along with Czechs, Hungarians and Yugoslavs. We all pray that the Poles may be able to come again. These students have as much to teach us as we have to teach them. They have had to face trials of faith that we never dream of. One newly trained teacher, dedicated to children, felt compelled in applying for jobs to say that she was a Christian. Predictably, she was turned down for every single job. But then she was sent for by a headmaster who was also a local Communist Party secretary. He told her, "Your application has a rare honesty. I have never seen anyone prepared to put her position on the line in such an honest way. I'm surrounded by people who will tell me anything to get a job. But I value honesty." After two hours of negotiation about what she could actually say and in what circumstances about her Christian faith, he gave her a job and exempted her from all political duties.

This year at the conference we had students from Israel, both Jewish and Arab, and we are trying to set up a Christian student movement in Israel for Jews and Arabs together.

This year also a woman named Toyome came from Japan. Toyome told us that she lived in a Shinto temple. We thought that was rather remarkable. "My father is the chief priest of this temple in Kyoto," she said.

"How did you ever become a Christian?" we asked.

"I was dissatisfied with Shinto, dissatisfied with life, dissatisfied with everything, and I thought there must be something that was true. I sent away for a book about the Christian faith. It told me I was a sinner. I knew that I was a sinner. It told me that Jesus Christ was God and had come to save me from my sins, and that I had to trust him. I told Jesus that I trusted him, and a great peace came over me. I then prayed to him. Since he had sent me the right book, I asked if he would now send me to the right church. So I sent away for literature about churches, and he sent me to the right church."

That sovereign Lord of ours was able to pick out a girl from the middle of a Shinto temple and without any apparent aid to bring her to himself through the power of the Holy Spirit. But someone had financed that literature that she sent for. Someone had founded that church. And the Japanese student movement was there to receive her into its fellowship, which is essential in that country where only two per cent are Christians. As she sang to us, "I'd rather have Jesus than silver or gold," we thought of the beauty of the Shinto temples. She had given up that old culture which is part and parcel of Japan, turned her back on her own culture, and become a Christian. That brought tears to our eyes.

In the U.S. and Britain we have not only Jesus; we also have silver and gold. Let us remember those who do not have Christ and let us share our silver and our gold for them.

Eldon Claassen: Thailand

For each of us, the early years of our lives are very important. My life has been very much influenced and enriched by having lived the better part of my life in India. During my growing years there, my missionary parents taught me very carefully that I needed the saving work and authority of Jesus Christ. I understood that, and I was renewed in salvation early in life. Even as a youngster in India, I remember being impressed by the distinct contrast between life in Jesus and the obligatory and fearful homage paid to idols by Hindus. This contrast was made especially clear on occasions such as when, with my family and Indian Christians, we jolted along in our dusty jeep toward a distant village to visit and fellowship with an isolated Christian family. The welcome was always loving, and Christ's Spirit of peace and joy in these village believers was clearly different from the sadness and fatalism which held their Hindu neighbors.

I am thankful to God for my bicultural experiences. I feel at home in both India and the United States. But having grown up in a country other than my own does not validate nor make automatic my call to be a missionary. Since early in life, God has been creating in me a desire to live and serve people who do not know him. Where and with whom is in God's plan. He may use my love and understanding of India to lead me there.

I am coming to understand more clearly that my motivation to obey the call of Jesus Christ to go into all the world is directly related to my understanding that the Father desires the nature of his Son, Jesus Christ, to be formed and active in me. The command of Jesus to go into all the world and preach the gospel is clear. The issue of my obedience is a question of motivation. As a young person under the lordship of Jesus Christ, and a student, with my whole life before me, my motivation can stem

from either obligation or from a profound love for Jesus Christ
and his kingdom. His purposes must motivate me to obey him,
love and serve him in the same way that he obeyed, loved and
served.

In 1979 and '80 I spent six months on a medical team in Thailand, working in refugee camps for Cambodians. During my time
there something began to nag at me. Why was I not a refugee?
Why have I never had to flee for my life? Why have I never gone
for days without anything to eat? Why was I born into a situation
where I could learn about Jesus' and the Father's love so easily?
These questions made me stop and see how fortunate I am. But
the questions still weren't answered.

During this time I began to see my answer in the example
of Jesus. Philippians 2:3-8 made this clear to me:

Do nothing out of selfish ambition or vain conceit, but in
humility consider others better than yourselves. Each of you
should look not only to your own interests, but also to the
interests of others.

Your attitude should be the same as that of Christ Jesus:
Who, being in very nature God,
did not consider equality with God
something to be grasped,
but made himself nothing,
taking the very nature of a servant,
being made in human likeness.
And being found in appearance as a man,
he humbled himself
and became obedient to death—
even death on a cross!

I began to see a parallel. Although Jesus was fully God, he was
willing to be humbled in taking on the nature of a servant. He
was completely humbled in being obedient to death. This was
done from a heart of obedience and love.

I began to understand not why I was not a homeless, starving
refugee, but that I have been placed in life, like Jesus, to be a
servant. My background, my education, my abilities, my health
have been given to me so that I may minister to those of this
world who have not heard that there is a God whose love wants

to produce in them the greatness of Jesus Christ. I have been richly blessed so that I, in turn, may be used to benefit others. I like what Paul said in Romans 14:7-9:

> For none of us lives to himself alone and none of us dies to himself alone. If we live, we live to the Lord; and if we die, we die to the Lord. So, whether we live or die, we belong to the Lord.
>
> For this very reason, Christ died and returned to life so that he might be the Lord of both the dead and the living.

Jesus is our example. I want to follow him. I hope you do too.

Elizabeth Cridland: Africa

"Has anyone been overlooked in the passing of the bread?" It was the voice of the young Dr. Robert C. McQuilkin, the man who sparked the Student Foreign Missions Fellowship, conducting the Sunday communion service at the first young people's conference I had ever attended. I was a recent high-school graduate who had been a less than regular attendant at church. I was there because two friends had insisted on my going. Neither knew the questions deep down in my heart about the apparent futility of the life my "set" was living—from one good time to the next. At that conference I made major decisions. Like the apostle Paul, I had a personal confrontation with the Lord Jesus Christ—not so spectacular, but just as definite.

From that time the Lord Jesus became the greatest reality in my life. The Bible, which I had never read, became alive, and I learned I could claim promises from it for myself. I still make a practice of doing that daily. Like Paul, I bowed the knee and confessed Jesus Christ as Lord by saying as did he, "What wilt thou have me to do?" The answer came in the form of a promise: "As thou goest step by step I will open up the way before thee." To Paul he said in substance: "You will be my first frontier missionary—to the gentile world." At that conference I was startled by hearing that 500 million people in the world had never once heard the name of Christ. That is the reason for my beginning question, "Has anyone been overlooked in the passing of the bread?" All I could see were masses of hands being raised as I related that figure to masses of people answering, "I have been overlooked." Knowing very little of what it was all about, and with great struggle of heart over my decision to obey the directives of the Lord Jesus, I went away with my thinking about life completely changed.

I had learned for sure that God had a plan and that he would show that plan to anyone who would follow his leading step by step. Thus one of my inward fears was quelled: my life would not be wasted. My life could count for all eternity if day by day I lived with the attitude, "Lord, what will you have me do?" Also I gradually learned from many whom God brought across my path as well as from Paul's writings that I did not have to do everything by myself. When I faced an impossible task alone I saw the pattern of Paul was for me. By having the Lord Jesus live in me I became a new person with God dwelling within. I had a life goal and the power to accomplish it: the power by which God raised the Lord Jesus from the dead was available for me to appropriate.

By following his directives I found myself transported to the heart of the tall grass jungle of the southern Sudan with teammate Mary Beam. As a member of the first party that had opened that area to the gospel seven years before, she was assigned to be with me in the Uduk culture. By now the statistics indicated 700 million people had never heard of Christ, and in the Sudan 243 tribes were without one word of the Bible in their language. Before the missionaries arrived no one had ever tried to learn the language of the Uduks, who possessed an ancient lifestyle. The women wore small loin cloths, beads, grass and a red oily mixture covering their bodies. We were five hundred miles from the nearest grocery store, shut in for six to eight months every year. I was there because I had become available to the Lord to help put the Word of God into that language. And I had unspeakable joy in the constant assurance I was where God wanted me to be. That has been my story for fifty years of Christian service.

German psychologists from Heidelberg confirmed the fact that the Uduks lived in deep-seated fear. They often accused us of causing a death or threatened us for saving twins, whom they buried alive because of fear. Sometimes for days and nights we were confined to a house with Muslims robed in jalabiyyas marching round and round outside day after day. Muslims marched on the station to desecrate the new church or get rid of the missionaries or build a mosque on the compound.

One day Mary, a coworker named Barbara Harper and I saw a great truck arriving at our station. Mary had built a small bridge over the drainage ditch so that we could more easily reach the government road. The schoolchildren had said their "livers shivered" as they watched Mary test the bridge by driving our jeep over it. Now there was a huge trailer truck about to drive over that bridge. When it held I couldn't stop thinking of the strength of the church in the face of attacks by Satan.

It turned out that truck was bringing our New Testaments. Just four days after those New Testaments arrived the three of us were escorted from the mission station by armed police sent from the central government of the Sudan. The arrival of those New Testaments at that time was a miracle.

When the first missionaries had arrived the Uduk people had assigned them a place at the foot of the huge baobob tree, which was the center of all evil-spirit worship in the tribe. The Uduks let the shiny, white-faced people build their houses under that tree so that they would have to fight the evil spirits and would not harm the people.

Later, when the first New Testaments in their language arrived, the young pastor called for a dedication service. He asked all the men to bring their spears. He put an image of Satan, who had blinded their eyes for so many years, on the baobob tree. Then, in front of people from all the surrounding villages, the men threw their bamboo spears at that image on the tree.

Three weeks after the government expelled us (the last of the missionaries were sent away in 1964), that great tree fell down one evening as the believers gathered for their weekly Bible study. "No more spirit worship," shouted the believers. Unbelievers from miles around came to see what had happened because they had felt the earth shake that night. Today the absence of the great baobob tree at the Chali station confirms that God's church stands there in its place.

Altogether we had been able to train a thousand men to read

and had established seventeen church centers. What God did in that isolated area he is able to do among the nearly three billion people who are still without an adequate knowledge of the Lord Jesus Christ.

When we found that we could no longer remain in the field in Africa, we began working with a new tool for spreading the gospel. The Bible Alliance Mission provides the New Testament on cassette for those who cannot read. These cassettes are provided not only in English but in many other languages of the world. The Word of God is recorded on cassette and distributed in some areas with a special cassette player that uses a hand crank instead of electricity. These tools are now being used in nearly seventy countries.

"Has anyone been overlooked in the passing of the bread?" May we remember the uplifted hands of those who have been overlooked because no one has ever told them of the Savior. As we think of this, may we do all we can to carry out God's work in the world so that all may confess Jesus Christ as Lord.

Jack Kopechek: Malta

The Lord has helped me hear his call for workers in the Muslim world a little bit at a time. I would like to share a few of the stepping stones I have encountered along the way.

Three years ago, while I was a sophomore at the University of Cincinnati, I had very little knowledge of God's world outside my own campus. Fortunately, my roommate at that time talked me into subscribing to a popular news magazine. I began to read the world news section each week. At first I was a little lost with all the strange names, but I soon began to piece the news together and get more interested in it.

It was about at this time that I was asked to be missions chairman in our local Inter-Varsity fellowship. I was lacking tremendously in knowledge and experience for this position, yet the Lord used this opportunity to show me his great concern for the world I was beginning to learn about. One afternoon my Inter-Varsity staff person sat down with me and asked me to recount all the passages in the Bible that spoke about missions. I was stumped at first, but soon began to see how the basis for missions extended through the whole Bible. I was particularly struck by the greatest missionary act of all time, our heavenly Father's sending his sinless Son into a sin-filled world for our sakes. This search in the Scriptures convinced me that our God is a missionary God.

Sparked by God's Word, I began to delve into books on missions and missionary biographies. I saw the way that God was calling others to obey him overseas and the great ways he stood with them and blessed their work. I also attended local missions conferences and met a variety of missionaries at these. This brought the stories in my books even closer to home.

The great need for laborers for Christ in the Muslim world first hit me while preparing for Urbana 79. I was reading through Dr. Ralph Winter's article "The Need in World Mission Today." I felt sad as I looked at the diagrams for the Chinese and Hindu peoples and saw how small the Christian circle was in comparison to the circle for non-Christians. But my heart was stung most when I turned the page and saw that there was no Christian circle at all in the Muslim world. Within me, God's Spirit cried out for this to change.

Yet it is difficult to love people you have only met in diagrams, so the Lord led me to apply to Inter-Varsity's Student Training in Missions (STIM) program last year. I was assigned to a team of six to take the good news of Jesus to Muslims on the Mediterranean island of Malta. I met about a hundred Muslims during our summer there and developed friendships with about a dozen of them. We heard their misconceptions concerning Jesus and saw their futile attempts to win a place in God's heaven by their good works. But most importantly, we met some with a hunger for truth. One Palestinian Muslim, on receiving the New Testament in Arabic, stayed up all night and read the whole thing.

We spent our final week of the summer in Tunisia, North Africa. One morning God led me to climb a large hill in the city of Tunis. After a time of private worship and prayer I looked out over the city. Seeing no churches, the Word of God began to echo in my mind: "Everyone who calls on the name of the Lord will be saved. How, then, can they call on the one they have not believed in? And how can they believe in the one of whom they have not heard? And how can they hear without someone preaching to them? And how can they preach unless they are sent? As it is written, 'How beautiful are the feet of those who bring the good news!' "

May God bless you all with beautiful feet.

Naty Lopez: NCFI

It is a privilege to share with you what God is doing among nurses in different parts of the world, especially in the Pacific and East Asia region.

After I became a Christian through an IFES conference, I thought of becoming a missionary. I became a nurse and dreamed of working among tribal people. The Lord had other plans for me, however, and led me to work among students and nurses in the Philippines for eleven years. During this time he gave me the privilege of attending two regional conferences of Nurses Christian Fellowship International in the Pacific and East Asia region. It was during these two conferences that I learned what God was doing in the lives of students and nurses in other countries. It was also here that I came to appreciate the training we had in the Philippines IVCF in inductive Bible studies and the quality of our Bible exposition. The Lord convinced me that we Filipino nurses had something to share with the students and nurses in the region. We were not always to be on the receiving end. We could share something in developing the work among nurses in our region.

We were invited to do a Bible study workshop at the 1974 regional conference. There I learned the excitement of discovering the Bible, studying it through the eyes of a nurse. For example, the leper in Mark 1 did not only have physical problems. Because he was isolated from people, he had emotional and social and spiritual problems as well. We learned how Christ cared for the whole person. We learned how we could model our nursing after him and how he cared for the whole man. The Lord increased any involvement with students and nurses of other countries, and I joined the staff of NCFI to work in the Pacific and East Asia.

This region is made up of twenty-one countries. Eight have growing work among students and nurses. Only four have full-time national staff. We thank the Lord that through our fellowship in NCFI some national groups are learning to apply biblical principles in their practice of nursing. Workshops help nurses give total patient care, paying specific attention to the neglected spiritual dimension. As a result, professional nursing organizations are recognizing that Christian nurses can make a specific contribution to nursing practice. In the Philippines, for example, the NCF is now a recognized interest group called the Philippine Nurses Association and will be consulted on decisions about nursing practice in the country.

In New Zealand, Australia and Korea, nursing instructors have made considerable progress in incorporating the spiritual dimension in the education of students. The NCF in Singapore and Malaysia have sponsored conferences on spiritual care; as a result some of their nurses have started to apply what they have learned.

We are thankful in NCFI for the fellowship of both younger and older fellowships. NCF in Fiji and New Zealand were instrumental in giving birth to the NCF in Tonga. NCF-Philippines has grown in part from the help of NCF-USA. God gave the vision to establish work among students and nurses to Filipino leaders of IVCF. An NCF staff member from the United States came to help train national leaders—I was one of these nationals. What she told me in my first year of staff work greatly encouraged me. She said that there was a nurse in the United States that prayed for me every day. That helped me through the growing pains of staff work.

In approximately a dozen countries in the Pacific and East Asia region Christian nurses are asking for help to train their leaders, to get them organized, or to help them integrate the Christian faith in the practice of nursing. A nurse from one Pacific island asked me to come to help her start Bible studies in their hospital. She knows about Christian nurses in other islands but does not know how to get them organized so that they can study together and encourage one another. During my visit in Taipei a student nurse asked me if I could help her give spiritual care to her patients. She had included an assessment of spiritual needs in the nursing care plan that student nurses are required to do, but her instructor did not accept it. He considered her assessment an intrusion.

NCF-Hong Kong told me last month that they are dying to have a staff worker, so I talked for five hours with one of the advisors about how she could help them. She was a very busy community health nurse who wanted to help meet that need for staff work. The needs are great; national leaders have to be trained. We have regional conferences next year in Europe, Asia, Latin America and the Pacific and East Asia region. The youngest movement of NCFI, Peru, is hosting the regional conference in Latin America. The leaders need wisdom. Pray for us and consider becoming a partner with us to reach students and nurses worldwide.

Dwight Nordstrom: China

I've always desired that people come to know who Jesus is. I have also always desired to be a missionary. The Holy Spirit has taken these desires, guided my intellect and experientially confirmed in my life a love and desire for service to the Chinese people.

During high school I thought the skills of missionaries determined where they would go. I hunted for my missionary skill. At times I wanted to be a missionary pilot or doctor, and at other times I didn't want to do these professions. Upon entering college still not knowing what type of missionary I should be, I chose three "safe" fields of study: religion, psychology and political science. These studies were "safe" because they would keep my options open, particularly if I did not go on to seminary and instead became a nonprofessional, tent-maker missionary where a strong biblical background would be essential.

While a college freshman I started reading missionary histories. Since before 1949, thousands of missionaries had gone to China. I naturally came across their stories. I was humbled by their sincerity and devotion, but troubled by their apparent statistical lack of success. If God truly was Lord of this world, why hadn't the Chinese recognized him? Did this mean Jesus was not their God? This intellectual questioning was guided by the Holy Spirit into a personal challenge—either I would assist the Chinese people to come to hear that Jesus the Christ is God's universal and highest revelation, or else I must disclaim Christianity's universality.

Now instead of searching for my missionary profession, I began to acquire the skills necessary to assist Chinese Christians. I studied Mandarin at Middlebury's summer language program. I devoured all the information I could get on China, particularly the newsletters and commentaries put out by praying Christians. In addition the Holy Spirit encouraged me to befriend internationals. By my junior year many, if not most, of my friends were internationals, and I was becoming adept at crossing cultural barriers. (Incidentally, I knew I was really good at crossing cultures when at a Baptist Student Union's retreat for internationals I became friends with a gorgeous Iranian student who just happened four years later to become my wife and partner in missions.)

By university graduation time I knew I needed a closer look at the Chinese situation before I could confidently choose what schooling and tent-making task would be most effective in helping the Chinese Christians. With the Holy Spirit again directing my thoughts, I went to Hong Kong where I could learn from discerning Christians about China. While there I did further language and cultural studies. Since no short-term mission service allowed for such outside studies, I went as a student funded by a scholarship, by my dad, and by money from my old summer jobs. In Hong Kong I participated actively in a racially mixed, but English-speaking church and led Bible studies at the university. During our study of David Bryant's book In the Gap I saw how the vision of being a world Christian powerfully gripped the more apolitical Hong Kong students. Eventually, when my language proficiency allowed, I translated and taught basketball at a refugee camp.

After six months in the British camp in Hong Kong the Holy Spirit told me that I now must live, eat and sleep Chinese. Since I did not think I was fully equipped to live an effective and vibrant witness within the spiritually isolated environment in China, I went to Taiwan. Having only the phone number of a basketball coach who had seen me play, the address of a friend's parents' house, and the hearsay that I could earn a living teaching English if all else failed, I was glad God was in charge and not me. While in Taiwan I played professional basketball. The opportunities to witness were everywhere. I joined a local Chinese congregation and became a youth sponsor. I helped out at a Christian radio station's youth programs. I observed and learned principles and methods of church growth from the Chinese pastors and laity, the missionaries and visiting Koreans. I can't forget my deep joy when, using only Chinese, I led and discipled to Christ a Chinese basketball player.

After ten months in Taiwan the Holy Spirit had further clarified what roles I could play in China. I needed to come back to America to study international law, business and finance at the Fletcher School of Law and Diplomacy. Now in Boston I've worshiped at a Chinese church, and among my Chinese friends I'm continuing to look for successful patterns of communication with which I can share who Jesus is.

Though time does not permit me to explain fully, in the next few years while my wife pursues her studies in medicine, I want to work in investments so that I will have the contacts to help the Chinese do better business with Americans. After that my wife wants to practice and teach medicine in China while I do consulting and perhaps manage a boarding home for students.

Though these dreams may not be fulfilled, I am utterly convinced that as of this moment God is moving me to reach a specific people group: the Chinese.

Dr. Charles Smith: Giving

I was six years old when I had my first lesson in stewardship. I had just earned my first dollar passing out handbills house to house in the blocks around my home. My father set me down and said, "Now look, son, you need to set aside ten cents of this dollar and give it to the Lord." I'm not sure how correct his theology was in asking a little non-Christian boy to start tithing, but I am happy for his teaching me to recognize the importance of God's ownership in this way.

I must have been twelve or thirteen when I took a fascinating course in stewardship at a Methodist summer camp. There I realized for the first time that I needed to do more than just meet my obligation to God if I was to express to him something of the gratitude that I had in my heart for his having saved me. I resolved then and there to start giving twenty per cent of my income to the Lord's work. God rewarded that childlike but sincere commitment with an abundance of jobs throughout high school, college and medical school.

It was not until I got into Inter-Varsity Christian Fellowship in medical school that I learned to dig into the Scriptures for myself. I learned that the New Testament guideline for giving was proportionate, giving as God had prospered me. I learned that having given twenty per cent did not mean that I was free to spend the other eighty per cent on myself as I pleased; it all belonged to him. I was being asked to be a steward, responsible to him for the way I managed all that he had entrusted to me.

By this time I knew that God had commissioned the church to take the gospel, the good news of redemption from sin through Jesus Christ, to the whole world. But it was several years before I understood that a missionary is a person and not an overseas assignment and that all Christians have been commissioned by God to bear witness to Jesus Christ in that geographical area where God has placed them.

Many years ago David Adeney said something that has helped me: "God's estimate of the value of our service does not depend upon the geographical area in which we serve, but on the extent it is in conformity to his will."

God leads many disciples to a cross-cultural witness. Our love for those who have never heard of Jesus Christ and our desire to see him honored and glorified should certainly cause us to give this option top priority. But God for his own reasons does give some of us home assignments. Missions is a team effort. In football there is an offensive team and a defensive team. Like-wise, God has those who go and those who send. All of us are on God's team. That should encourage those who feel guilty about having a home assignment.

Many of you this week are beginning to become concerned for the world. But let me warn you that Satan is disturbed by this and will do all that he can to cloud that vision. One of his favorite and often successful methods is through the appeal of material things—the good life, the comfortable life. Once you graduate, get out into the world and start earning a little money, you will be amazed how fast things will get a grip on your life, and how hard it will be to keep world evangelization as the central goal of your life.

In 1951 I came to Urbana as a grad student from the University of Minnesota in thirty-below-zero weather. But God spoke to me through Dr. Bob Smith, a young philosophy professor, and I saw the absolute necessity of obligating myself for missions if I was to avoid the pitfall of materialism in my own professional life. As soon as it became apparent that we were not going to find a mission board to send us overseas, we immediately committed ourselves to the full support of a young couple ready to go to Pakistan, a former Inter-Varsity staff member. (Inter-Varsity loses a lot of staff members to the foreign field.)

That was just the beginning of an exciting life of being involved in the support of other missionaries whom God had given overseas assignments. Now I can only be a Christian physician in Kokomo, Indiana, where I practice, but I can be involved in the medical ministry of Dr. Phyllis Irwin in Pakistan, Dr. Steve Dillinger in the Ivory Coast, Dr. Tom Cairns in Zaire, Dr. Steve Foster in Angola. I am not even confined to medical work. I can be involved in the work of the linguists in Papua New Guinea, a student worker in Europe, a jungle pilot in the Cameroon, a pastor in Bangladesh, a seminary professor in the Philippines.

Now I personally can only serve the Lord 10, 12, 14, maybe 16 hours a day, but those that I am supporting are serving the Lord every one of the 24 hours. When I get up in the morning, those that I am supporting in Europe and Africa have already been working for 6-8 hours. And when I go to bed at night I know that those in India, Bangladesh and the Philippines are starting their day and serving the Lord. I think the only way that one can be in full-time 24-hour Christian service is to be involved in a supporting relationship with those who have overseas assignments.

Paul describes this relationship between the missionary and the believer back home as fellowship or partnership in the furtherance of the gospel. The Philippian church was thus involved in Paul's ministry while he was in Rome.

Our family has experienced an extremely close bond between us and those whom we support. Of course, we correspond with them. We pray for them. They pray for us. We have had the privilege of visiting most of them in their fields overseas, and most of them have visited in our home at one time or another. Seven of their children lived in our home while they were in college.

Several years ago when I was just starting in practice and it wasn't going very well, the missionaries we were supporting sent us money for us to live on. That was very humbling. That is missions in reverse. One woman sent two thousand dollars which she had inherited. Another missionary sent seventeen dollars which he said was all he had. It reminds me of Paul's words to the Corinthians: " ... your abundance at the present time should supply their want, so that their abundance may supply your want, that there may be equality" (2 Cor 8:14 RSV). This is the kind of involvement in one another's lives that body life was intended to be.

Being a missionary at home involves a lot more than just giving a portion of our income to support others in their ministries for Christ. The one with the home assignment is just as obligated as the one overseas to be a witness for Jesus Christ in whatever situation God has placed him. We each need to seek creative ways of being salt and light to those for whom God has given us responsibility. God loved the whole world and so should we. All of us should be global Christians regardless of whether we have to cross salt water to get to our assignment.

Convention Speakers

Eric Alexander is pastor of St. George's-Tron Church in Glasgow, Scotland, an urban church with a large student ministry. He is also chairman of the Scottish Council of the Overseas Missionary Fellowship and in 1981 became president of the Universities and Colleges Christian Fellowship (UCCF) in Great Britain. He is the author of the booklet *The Search for God*.

John W. Alexander served as president of Inter-Varsity Christian Fellowship for seventeen years and is now its president emeritus. A former university professor, he is chairman of the executive committee of the International Fellowship of Evangelical Students and the author of several books and booklets, including *Managing Our Work*.

Edgar S. Beach lives in Tectitán, Guatemala, with his family. As a member of the Wycliffe Bible Translators and the Summer Institute of Linguistics, he has worked on preliminary linguistic analysis and alphabet development and has begun translation of the Scriptures into Tectitec.

Chua Wee Hian is general secretary of the International Fellowship of Evangelical Students. In the past he has served as associate general secretary of IFES in East Asia and as a pastor and editor of *The Way*, a quarterly magazine for Asian students. As general secretary he travels widely, speaking to students, counseling staff and seeking new staff members.

Eva den Hartog spent the first five years of her life in a Salvation Army home when her mother could not cope with her three children. At nineteen she joined the Army and began work as a nurse and midwife. She has directed medical relief efforts in Africa, India and Southeast Asia and has been awarded the Knighthood of Orange Nassau by the Queen of the Netherlands.

Samuel Escobar is associate general secretary for Latin America of the International Fellowship of Evangelical Students and the author of several books in Spanish and English. He is also president of the Latin American Theological Fraternity and editor of *Pensamiento Cristiano*, an interdenominational magazine for Christian leaders.

Billy Graham is perhaps the best-known evangelist in the world. Along with his crusades held in the U.S. and around the world, his activities include a newspaper column ("My Answer"), a radio program ("Hour of Decision") and several best-selling books, such as *How to Be Born Again* and *The Holy Spirit*.

David M. Howard is assistant to the president of IVCF. He has directed the Consultation on World Evangelization in Pattaya, Thailand, and two previous Urbana conventions (73 and 76). He served with the Latin America Mission for fifteen years and has written several books, including *The Great Commission for Today*.

Simon Ibrahim is general secretary of the Evangelical Churches of West Africa (ECWA) and former deputy general secretary. The ECWA churches originated with the Sudan Interior Mission. Prior to taking his present position, Ibrahim taught at Kagaro and Billiri Bible Schools.

John E. Kyle is mission director of IVCF-USA and has been the program director for Urbana 79 and 81. He formerly served as director of the missions program for the Presbyterian Church in America and as a missionary with Wycliffe Bible Translators in the Philippines.

Marilyn Laszlo works with the Sepik Iwam people of Papua New Guinea on behalf of the Wycliffe Bible Translators. She has helped to develop an alphabet for the Sepik Iwam people's language and to teach people to read. The goal of her work is to produce a New Testament in the tribal language.

Gordon MacDonald is pastor of Grace Chapel in Lexington, Massachusetts, and an adjunct professor of pastoral ministries at Gordon-Conwell Theological Seminary. He speaks widely to groups of pastors, missionaries and students, and has written two books: *Magnificent Marriage* and *The Effective Father*.

Isabelo Magalit, a medical doctor, is associate general secretary for East Asia for the International Fellowship of Evangelical Students. Living in Manila, Philippines, he is a speaker (at Urbana 76, 79 and 81), writer and director of the first Asian student missions convention.

George D. McKinney is pastor of St. Stephen's Church of God in Christ, a large urban church ministering wholistically to persons in San Diego, California. In addition to his pastoral duties, he carries on a wide counseling ministry for couples, families and children. He has written several books and pamphlets, including *Christian Marriage*.

Robert B. Munger, former professor of evangelism and church strategy at Fuller Theological Seminary, is associate pastor at Menlo Park Presbyterian Church in California. He has had broad experience as a pastor and evangelist and is perhaps best known for his evangelistic booklet entitled *My Heart–Christ's Home*.

Rebecca Manley Pippert, a former IVCF field staff member, is evangelism specialist with IVCF and a popular speaker to a broad variety of groups. She is the author of *Out of the Saltshaker*, a widely read book which tells people in warm and enthusiastic language how to make evangelism an intrinsic part of their lives.

Helen Roseveare served for many years in the Belgian Congo as a missionary doctor with the Worldwide Evangelism Crusade and has since helped set up a new medical center in Zaire which trains African staff. A popular speaker, she has also become widely known through her autobiographical books: *Give Me this Mountain* and *He Gave Us a Valley*.